MEDICAL DICTIONARY AND HEALTH GUIDE

- More than 10,000 words and phrases and their definitions—terms encountered every day when dealing with health-care providers, medical facilities, and the media

- An invaluable consumer health guide that puts useful, important information for you and your family at your fingertips

- A comprehensive listing of alternative medicine and holistic health care procedures and terms

MEDICAL DICTIONARY AND HEALTH GUIDE

EDWARD R. BURNS, M.D.

A Lynn Sonberg Book

HarperTorch
An Imprint of HarperCollins*Publishers*

This book contains advice and information relating to health care. It is not intended to replace medical advice and should be used to supplement rather than replace regular care by your doctor. It is recommended that you seek your physician's advice before embarking on any medical program or treatment. All efforts have been made to assure the accuracy of the information contained in this book as of the date of publication. The publisher, the author, and the book producer disclaim liability for any medical outcomes that may occur as a result of applying the methods suggested in this book.

HARPERTORCH
An Imprint of HarperCollins*Publishers*
10 East 53rd Street
New York, New York 10022-5299

ISBN-13: 978-0-06-072562-4
ISBN-10: 0-06-072562-1

First HarperTorch paperback printing: May 2006

Printed in the United States of America

Visit HarperTorch on the World Wide Web at www.harpercollins.com

10

To my precious wife, Chaya,
the love and inspiration of my life

ACKNOWLEDGMENTS

I am indebted to Deborah Mitchell for her exhaustive research and editorial acumen. Without her skillful contributions and hard work, this book could never have been written.

CONTENTS

INTRODUCTION

Medical care in the twenty-first century is extraordinarily complex and sophisticated. As consumers of health care, it is easy to become confused and frustrated as you try to wade through the jargon and terminology used by your health-care providers and by the media—often in seemingly contradictory ways. But when it comes to your health and that of your family, you want and need accurate, clear information that can help you better understand a symptom or ailment or that can assist you in making informed decisions. The goal of this dictionary and health consumer guide is to help increase your understanding and knowledge of health-related terms, conditions, procedures, and tests so you can, in turn, improve the quality of your life and that of your loved ones.

Here is a handy, comprehensive, and user-friendly resource to help you decipher and make sense of unfamiliar terms and medical-related matters in language that everyone can understand. Our goal has been to use language that is not so simplistic that it loses the essence of the material being presented, but not so technical that it cannot be grasped by a layperson. To that end, the dictionary and tables have been designed to equip you with the vocabulary and knowledge to better understand your health-care providers and to ask them appropriate questions.

This book is divided into two sections. The first and main section is a dictionary. It consists of more than ten thousand words and phrases and their definitions—terms you may encounter every day while dealing with health-care providers or medical facilities, or when interacting with any of the numerous forms of media that report health-related information. For your convenience, pronunciations are provided, as are plural forms of words in some cases. The second section is a consumer health guide, which consists of sixteen tables that cover a wide range of information you can use to maintain or improve your health—from vaccination schedules for infants to how to prepare an advance directive. These two sections complement each other: there are cross-references to the tables throughout the dictionary sections, and terms and phrases you will find in the consumer health guide can be found in the dictionary.

Health care is an ever-evolving and organic field, presenting us with new terms, diseases, procedures, and products and fresh ways to look at familiar ones. Our understanding and appreciation of these things need to change as technology and research bring new information to the forefront. This book can help you and your family navigate through this information and feel more confident when you must confront health-care issues. At the same time, it is critical that you establish a relationship with licensed health-care practitioners for both preventive health maintenance and treatment whenever necessary.

The inclusion of any trademarked and/or patented items in this book does not constitute an endorsement. They are included solely because they enjoy widespread use or are terms you may encounter. Similarly, exclusion of any such items from the book does not mean they are in any way inferior to the ones that are mentioned.

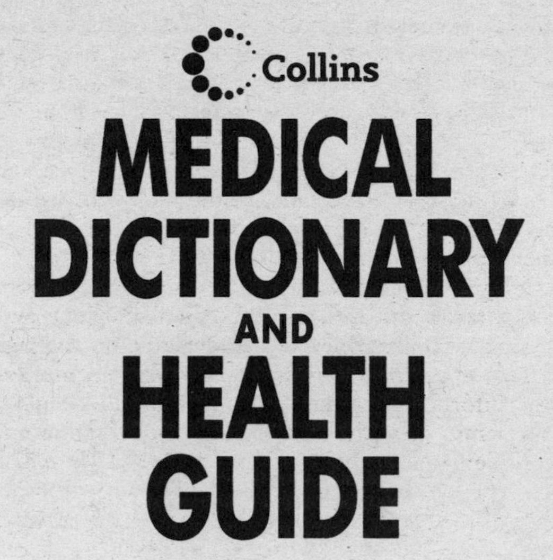
Collins
MEDICAL
DICTIONARY
AND
HEALTH
GUIDE

Part 1

The Dictionary

A

aa. An abbreviation meaning "equal parts of," often seen on prescriptions.

abarticulation *(ăb-ăr-tĭk-ū-lā'-shŭn).* Dislocation of a joint.

abasia *(ă-bā'-zē-ă).* Inability to walk due to impaired coordination.

abate *(ă-bāt').* To become or make less intense or strong (e.g., his headache abated); to cease or make cease.

abdomen *(ab'-dō-mĕn).* Part of the body between the chest and the pelvis, which includes the stomach, large and small intestines, liver, gallbladder, spleen, and pancreas; the belly.

abdominal *(ăb-dŏm'-ĭ-năl).* Pertaining to the abdomen.

abdominal breathing. Type of breathing in which the diaphragm (rather than the ribs) performs most of the effort, pushing the abdominal contents downward and outward when breathing in.

abdominal crisis. Severe pain in the abdominal area; usually refers to the pain experienced during sickle cell anemia crisis.

abdominal delivery. Delivery of an infant through an incision made in the abdomen; *see also* **Cesarean section.**

abdominal pregnancy. Rare occurrence in which the embryo develops in the abdominal cavity rather than in the uterus.

abdomino-. Prefix combining form for words relating to the abdomen (e.g., abdominocardiac refers to the abdomen and the heart).

abdominocentesis *(ăb-dŏm'-ĭ-nō-sĕn-tē'-sĭs).* Surgical puncture of the abdomen, usually done to remove fluid for diagnosis.

abdominohysterectomy *(ăb-dŏm-ĭ-nō-hĭs-tĕ-rĕk'-tĕ-mē).* Surgical removal of the uterus through an abdominal incision.

abdominoplasty *(ăb-dŏm'ĭ-nō-plăs'tē).* Cosmetic surgery on the abdomen in which excess fatty tissue and skin are removed.

abducens nerve *(ăb-dū'-sĕnz).* Sixth cranial nerve. It regulates the muscles that rotate the eyeball outward (to look away from the nose).

abduct *(ăb-dŭkt').* To move a body part away from the median plane of the body (e.g., to move a leg laterally outward).

abduction *(ăb-dŭk'-shŭn).* Moving a body part away from the median plane of the body.

abductor *(ăb-dŭk'-tor).* Muscle that causes a body part to abduct.

aberrant *(ăb-ĕr'-ănt).* Deviating from normal; abnormal (e.g., aberrant behavior); mental disorder.

aberration *(ăb-ĕr-ā'-shŭn).* Deviation from normal.

abetalipoproteinemia *(ā-bā-tă-lĭp-ō-prō-tēn-ē'-mē-ă).* A rare, inherited condition in which fat metabolism is characterized by absence or severe deficiency of betalipoproteins, presence of abnormal red blood cells, and abnormally low cholesterol.

abient *(ăb'-ē-ĕnt).* Tending to move away from a stimulus (e.g., turning away from a flash of light).

abiosis *(ăb-ē-ō'-sĭs).* Absence of life.

abiotic *(ăb-ē-ŏt'-ĭk).* Nonliving; incompatible with life.

ablactation *(ăb-lăc-tā'-shŭn).* Cessation of milk secretion from the breast; weaning.

ablate *(ăb-lāt').* To remove, especially by cutting (incision), suction, or both.

ablation *(ăb-lā'-shŭn).* Removal of a part of the body, especially by surgery, chemical destruction, electrocautery, laser, or radiofrequency.

ablatio placentae *(ab-la'-she-o placen'-ta).* Premature separation of the placenta from the wall of the uterus during the latter stage of pregnancy; *see also* **abruptio placentae.**

ablepharia *(ă-blĕ-făr'-ē-ă).* Congenital condition in which the eyelids are small or absent.

abluent *(ăb'-lū-ĕnt).* A substance with the ability to cleanse (e.g., a detergent).

ablutomania *(ă-blū'-tō-mā'-nē-ă).* An abnormal concern with cleanliness, commonly seen in obsessive-compulsive disorder.

abnormal *(ăb-nor'-măl).* Deviating from normal; aberrant; unusual in development, structure, or condition.

ABO blood group system. The primary system for classifying human blood, used in blood transfusion therapy. It classifies blood into four groups: A, B, AB, and O.

abocclusion *(ăb'-ŏ-kloo'-shŭn).* Condition in which, when biting, the upper teeth and lower teeth do not touch.

aboral *(ă-bō'-răl).* Located opposite to or away from the mouth.

abortifacient *(ă-bor-tĭ-fā'-shĕnt).* A substance used to cause or produce an abortion.

abortion *(ă-bor'-shŭn).* Termination of pregnancy, either spontaneously or through surgical or pharmacologic intervention.

habitual a. Repeated spontaneous expulsion of the embryo or fetus in three or more pregnancies, usually for an unknown reason.

induced a. Deliberate termination of pregnancy, which can be achieved by suction, curettage, hysterotomy, or induction of uterine contractions.

infected a. Abortion accompanied by infection of material retained in the uterus, resulting in fever.

partial birth a. Pregnancy terminated at 20 weeks gestation or later, using an instrument that tears or perforates the skull of the fetus while the fetus is still in the uterus and the subsequent removal of the fetus.

spontaneous a. Abortion that occurs without apparent cause and without intervention; miscarriage.

therapeutic a. Induced abortion performed for medical reasons, usually when the life of the mother is threatened.

threatened a. Appearance of symptoms, including bleeding from the vagina and abdominal pain, that signal an impending loss of the embryo or fetus.

abortionist *(ă-bor'-shŭn-ĭst).* Individual who performs an abortion.

abortive *(ă-bor'-tĭv).* Preventing the completion of something; coming to a premature end.

abrade *(ă-brād').* To rub off, as a layer of skin or mucous membrane, accidentally or deliberately.

abrasion *(ă-brā'-shŭn).* A scrape or scratch of the skin or mucous membrane. It also refers to the wearing away of the substance of a tooth.

abreaction *(ăb'-rē-ăk'-shŭn).* In psychoanalysis, a process by which repressed, painful thoughts and feelings are recalled and relived, with the expectation that the experience will become tolerable because of insight gained during the process.

abruptio placentae *(ă-brŭp'-shē-ō plă-sĕn'-ē).* Premature separation of the placenta from the wall of the uterus during the latter stage of pregnancy.

abscess *(ăb'-sĕs).* A localized collection of pus that results from a breakdown of tissue and which can occur anywhere in or on the body.

acute a. Abscess, with inflammation, that is ready to rupture.

alveolar a. Abscess in the root of a tooth, usually the result of infection caused by a cavity.

brain a. Abscess that develops in the brain or its membranes.

gingival a. Abscess of the gum.

mammary a. Abscess in the female breast, usually seen during lactation or weaning.

abscission *(ăb-sĭ'-zhŭn).* Surgical removal of a body part.

absolute alcohol. Ethyl alcohol with no more than 1% water.

absolute threshold. The lowest level of stimulus that can be sensed (e.g., the lowest intensity of a sound that can be heard).

absorb *(ăb-sorb').* To incorporate within the body; to take in.

absorbance *(ăb-sor'-băns).* The ability of a tissue or material to absorb radiation.

absorbefacient *(ăb-sor'-bĕ-fā'-shĕnt).* A substance that promotes absorption.

abstinence *(ăb'-stĭ-nĕns).* The practice of voluntarily refraining from sex, alcohol, or food.

abstraction *(ăb-străk'-shŭn).* The refinement of a drug from its original form. In psychology, absentmindedness or preoccupation.

abulia *(ă-bū'-lē-ă).* Loss or impairment of the ability to make decisions or to act independently.

abuse *(ă-būs').* Improper or excessive use; misuse.
child a. Emotional, physical, or sexual injury to a child.
drug a. Misuse of drugs; usually refers to narcotics or psychoactive drugs.
elder a. Physical and/or mental mistreatment of the elderly.

abutment *(ă-būt'-mĕnt).* The tooth to which a partial denture is anchored.

ABVD. Acronym for a frequently used combination of chemotherapy drugs: adriamycin, bleomycin, vinblastine, and decarbazine.

a.c. Abbreviation meaning "before meals" used in writing prescriptions.

acacia *(ă-kā'-shē-ă).* A gummy substance used in some medications as a suspending agent.

acalcerosis *(ā-kăl-sĕr-ō'-sĭs).* A deficiency of calcium in the body.

acalculia *(ă-kăl-kū'-lē-ă).* A learning disorder in which one is unable to perform simple arithmetic calculations.

acampsia *(ā-kămp'-sē-ă).* Inability to extend or bend a limb.

acanthesthesia *(ă-kăn'-thĕs-thē'-zē-ă).* "Pins and needles"; *see also* **paresthesia.**

acanthocyte *(ă-kăn'-thō-sīt').* An abnormal red blood cell that has a spiny appearance.

acanthocytosis *(ă-kăn'-thō-sī-tō'-sĭs).* An inherited condition characterized by the presence of many acanthocytes.

acanthoma *(ăk'-ăn-thō'-mă).* A benign tumor of the skin.

acanthosis *(ăk'-ăn-thō'-sĭs).* Skin condition characterized by increased thickness of the prickle-cell layer of skin and warty growths, as in psoriasis.
a. nigricans. An uncommon chronic inflammatory skin disease in adults characterized by pigmented, warty growths on the skin and mouth and sometimes the internal organs.

acapnia *(ă-kăp'-nē-ă).* Condition in which there is less than the normal amount of carbon dioxide in the blood and tissues; also called *hypocapnia.*

acapsular *(ā-kăp'-sĕ-lăr).* Without a capsule; a term often used to describe tumors that have extended out into surrounding tissues.

acardia *(ă-kăr'-dē-ă).* Congenital condition in which the fetus is born without a heart.

acariasis *(ăk'-ă-rī'-ă-sĭs).* Any diseases caused by a mite or acarid (e.g., scabies).

acarid *(ăk'-ă-rĭd).* A mite or tick.

acarophobia *(ăk'-ăr-ō-fō'-bē-ă).* Abnormal fear of small objects such as mites, worms, and pins; also abnormal fear of developing an itch.

acatalepsy *(ă-kăt'-ă-lĕp'-sē).* Deterioration of mental abilities, especially comprehension.

acataphasia *(ă-kăt-ă-fā'-zē-ă).* Inability to speak in coherent sentences, due to a lesion on the brain.

acathexis *(ă-kă-thēk'sĭs).* Lack of emotion toward an idea, memory, or thing that is ordinarily important to the individual.

acaudal *(ā-kaw'-dăl).* Having no tail.

acceptable daily intake (ADL) *(ak-sep'-te-bel).* Estimate of the amount of substances in food or drinking water that you can ingest daily over a lifetime without appreciable health risk.

accessory nerve *(ăk-sĕs'-ă-rē).* One of a pair of motor nerves, the eleventh cranial nerve, that supply muscles involved in swallowing, speech, and certain movements of the head and shoulders.

accident prone *(ăk'-sĭ-dĕnt).* A tendency to have more than the usual number of accidents.

acclimation *(ăk-lĭ-mā'-shŭn).* Process of becoming accustomed to a different condition or environment.

accommodation *(ă-kŏm'-ō-dā'-shŭn).* Adjustment or adaptation of the eye to new surroundings; specifically adjusting for different distances.

accretion *(ă-krē'-shĕn).* Addition of new material to the outside of a structure, resulting in growth.

acebutolol. A beta-blocker used to treat hypertension. Trade name: Sectral.

ACE inhibitors. Acronym for angiotensin converting enzyme inhibitors.

acellular *(ā-sĕl'-yĕ-lăr).* Having no cells or cell structure.

acephalia *(ă-sĕ-fā'-lē-ă).* A disorder in which an embryo develops without a head.

acetabuloplasty *(ăs'-ĕ-tăb'-ū-lō-plăs'-tē).* Surgical repair and reconstruction of the acetabulum.

acetabulum *(ăs-ĭ-tăb'-ū-lŭm).* A cup-shaped hollow in the hip bone into which the femur (head of the thighbone) fits.

acetaminophen. *See* Appendix, Common Prescription and OTC Drugs: By Generic Name.

acetaminophen/codeine. *See* Appendix, Common Prescription and OTC Drugs: By Generic Name.

acetate *(ăs'-ĕ-tāt).* A salt of acetic acid.

acetazolamide *(ăs'-ĕt-ă-zŏl'-ă-mīd).* A drug used to treat glaucoma and high altitude pulmonary edema. Trade name: Diamox.

acetic acid *(ă-sē'-tĭk).* A clear, colorless acid with a pungent odor, it is found in vinegar.

acetone *(ăs'-ĭ-tōn).* A colorless liquid with a sweet, fruity odor found in small amounts in normal urine but in increased amounts in people who have diabetes or other metabolic disorders, and after long-term fasting.

acetonemia *(ăs'-ĭ-tō-nē'-mē-ă).* Presence of abnormally elevated levels of acetone in the blood.

acetonuria *(ăs'-ĕ-tō-nū'-ē-ă).* Presence of excess acetone in the urine, often seen in diabetic acidosis; *see also* **ketonuria.**

acetylcholine *(ăs'-ĕ-tĭl-kō-lēn).* Chemical produced in the body which plays a central role in the transmission of nerve signals to activate muscle contractions.

acetylcholinesterase *(ăs'-ĕ-tĭl-kō-lĭn-ĕs'-tĕr-ās).* Enzyme that stops the action of acetylcholine. It resides in various tissues, including red blood cells, nerve cells, and muscles.

acetylsalicylic acid *(ă-sē'-tĭl-săl'-ĭ-sĭl-ĭk).* Previously used name for aspirin.

achalasia *(ăk'-ă-lā'-zē-ă).* Failure of a muscle, particularly a sphincter (muscular valve or ring) to relax, especially in the gastrointestinal tract.

achievement quotient (AQ). A measurement of a child's education in relation to his age.

Achilles tendon *(ă-kĭl'-ēz).* Tendon that originates from the calf muscles and is attached to the heel.

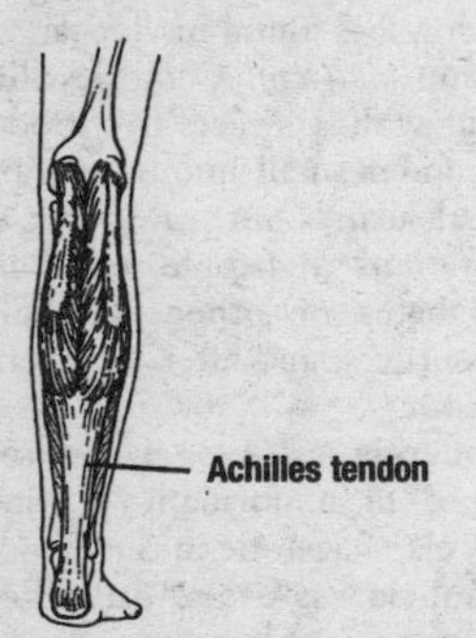

achillobursitis *(ă-kĭl'-ō-bur-sī'-tĭs).* Inflammation of the bursa that lies over the Achilles tendon.

achillodynia *(ă-kĭl'-ō-dĭn-ē'-ă).* Pain caused by inflammation between the Achilles tendon and bursa.

achillotenotomy *(ă-kĭl'-ō-tĕn-ŏt'-ō-mē).* Cutting of the Achilles tendon to correct a contracture or shortening.

achlorhydria *(ă'-klor-hī'-drē-ă).* Absence of hydrochloric acid in the gastric juice, often associated with severe anemia, stomach cancer, and in some normal individuals.

acholia *(ă-kō'-lē-ă).* Condition in which there is little or no bile secreted.

achondroplasia *(ă'-kon-drō-plā'-zhă).* Inherited disorder in which abnormal bone formation results in dwarfs with normal-sized heads but short legs and arms; also called *chondrodystrophy.*

achromasia *(ăk'-rō-mā'-zē-ă).* Absence or deficiency of pigment in the skin; *see also* **albinism.**

achromatism *(ă-krō'-mă-tĭzm').* Colorlessness.

achromatopsia *(ă-krō'-mă-tŏp'-sē-ă).* Complete color blindness.

achromatosis *(ă-krō-mă-tō'-sĭs).* Condition associated with a lack of natural pigment in the skin; *see also* **achromasia.**

achylia *(ă-kī'-lē-ă).* Absence or severe deficiency of hydrochloric acid or other digestive secretions.

acid *(ă'-sĭd).* Any chemical capable of donating a hydrogen proton. Acids react with metal to form salts, and neutralize bases.

amino a. Basic building blocks for all the body's protein production. About 20 amino acids are necessary for human health.

essential fatty a. Polyunsaturated fatty acid required for nutrition and must be obtained from the diet.

fatty a. Large group of organic acids classified as saturated and unsaturated, and further classified as polyunsaturated and monounsaturated.

omega-3 a. Type of monounsaturated fatty acid with a double bond at the end (omega) of the carbon chain.

polyunsaturated fatty a. An unsaturated fatty acid that has two or more double bonds.

saturated fatty a. Fatty acid in which the carbon chain is connected by single bonds and cannot accept more hydrogen.

unsaturated fatty a. Fatty acid in which the carbon chain has at least one double bond and can accept more hydrogen.

acid-base balance. Equilibrium between bases (alkali) and acids in the body which, when normal, maintains a pH of 7.35 to 7.45 (where 7 is neutral and greater than 7 is alkaline).

acidemia *(ăs-ĭ-dē'-mē-ă).* Condition in which there is excess acid in the blood (i.e., blood pH is less than 7.35).

acid-fast. The staining characteristics of bacteria. Acid-fast bacteria retain red dyes when stained, making them easier to identify.

acidity *(ă-sĭd'-ĭ-tē).* Condition of having an acid content or of being an acid.

acidophil *(ă-sĭd'-ŏ-fĭl).* Cell that is capable of being stained by acid stains.

acidophilus *(ăs-ĭ-dŏf'-ĭ-lŭs).* Species of bacteria, *Lactobacillus acidophilus,* that occurs naturally in the body and may protect against some yeast infections and facilitate digestion of milk products by restoring a healthy bacterial environment in the intestines.

acidophilus milk. Milk fermented by *Lactobacillus acidophilus* cultures.

acidosis *(ă-sĭ-dō'-sĭs).* Excessive acidity in the body caused by an accumulation of acids, which may result, for example, from diabetes mellitus, breathing disorders, severe diarrhea, impaired kidney function, or poisoning.

acid phosphatase. Enzyme found in the liver, spleen, bone marrow, plasma, formed blood elements, and prostate gland; determination of serum acid phosphatase activity is an important diagnostic tool.

aciduria *(ă-sĭ-dur'-ē-ă).* Excessive acid in the urine.

aciniform *(ă-sĭn'-ĭ-form).* Resembling grapes, as seen with some tumors.

acne *(ăk'-nē).* An inflammatory condition of the sebaceous glands and hair follicles of the skin, usually on the face and upper body, characterized by papules, pustules, and comedones, and sometimes cysts and nodules.

a. conglobata. Acne vulgaris with abscesses and cysts that leave scars.

a. rosacea. *See* **rosacea.**
a. vulgaris. The most common form of acne, usually seen in individuals from puberty to young adulthood.

aconuresis *(ăk-ō-nur-ē'-sĭs).* Inability to control urination; incontinence.

acoustic *(ă-koos'-tĭk).* Pertaining to sound or to the sense of hearing.

acoustic nerve. The eighth cranial nerve that supplies the ear; *see also* **auditory nerve.**

acquired *(ă-kwīrd').* Something developed as a result of environmental factors rather than inheritance.

acquired immune deficiency syndrome (AIDS). Serious condition, caused by a virus (human immunodeficiency virus, or HIV), in which the immune system does not respond normally to infection. The virus destroys T-cell lymphocytes that are critical in protecting against infections.

acquired immunity. Immunity that develops during life, either through development of antibodies after attack of an infection (e.g., measles), antibodies passed along from the mother, or vaccination.

acroagnosis *(ăk'-rō-ăg-nō'-sĭs).* Absence of sensation of an arm and/or leg.

acroanesthesia *(ăk'-rō-ăn-ĕs-thē'-zēă).* Lack of sensation in one or more of the extremities.

acrocephaly *(ăk'-rō-sĕf'-ă-lē).* Condition in which the forehead is very high, giving the head a pointed look.

acrocyanosis *(ăk'-rō-sī-ă-nō'-sĭs).* Abnormal condition characterized by bluish discoloration and coldness of the hands and sometimes the feet. It is a sign of Raynaud's disease (*see also* **Raynaud's disease**) and poor circulation due to many other conditions.

acrodermatitis *(ăk'-rō-der-mă-tō'-sĭs).* Any skin inflammation that affects the hands or feet.

acrodynia *(ăk'rō-dĭn'-ē-ă).* Now a rare disease, acrodynia (painful extremities) primarily affects young children. Symptoms of irritability, photophobia, pink discoloration of the hands and feet, and nerve inflammation due to chronic exposure to mercury. Also called *pink disease, Feer's disease, Swift's disease.*

acromastitis *(ăk-rō-măs-tī'-tĭs).* Inflammation of the nipple of the breast.

acromegaly *(ăk-rō-mĕg'-ă-lē).* Hormonal disorder caused by excess production of growth hormone in the pituitary gland, resulting in enlargement and elongation of the hands, feet, and face.

acromion *(ă-krō'-mē-ŏn).* The high point of the shoulder blade (scapula) to which important muscles and ligaments are attached. Injuries to this area, including rotator cuff injuries, are common among athletes.

acromphalus *(ă-krŏm'-fă-lĕs).* Abnormal protrusion of the navel, which may indicate the start of an umbilical hernia.

acroparesthesia *(ăk'-rō-par-ĕs-thē'-zē-ă).* Sensation of tingling, numbness, and prickling in the hands and feet.

acrophobia *(ăk-rō-fō'-bē-ă).* Morbid fear of high places.

acrosclerosis *(ăk-rō-sklĕr-ō-sĭs).* Hardening of the skin of the upper extremities, usually following Raynaud's disease.

acrotic pulse *(ă-krŏ'-tĭk).* A very weak pulse.

acrylics *(ă-krĭl'-ĭks).* Plastics used to make dentures and false teeth.

ACTH. Acronym for adrenocorticotropic hormone.

actin *(ăk'-tĭn).* One of two main proteins present in muscles; the other is myosin.

acting out. Expressing feelings and emotions through actions rather than speech.

actinic *(ăk-tĭn'-ĭk).* Pertaining to radiant energy from sunlight, x-rays, and ultraviolet light.

actinic burns. Burns caused by sunlight or ultraviolet rays.

actinodermatitis *(ăk'-tĭn-ō-dĕr-mă-tī'-tĭs).* Skin irritation and inflammation caused by exposure to radiation.

actinomycosis *(ăk-tĭn-ō-mī-kō'-sĭs).* Disease primarily of cattle and pigs but also occurring in humans, caused by a bacteria and characterized by lumpy abscesses, usually around the face and neck; also known as "lumpy jaw."

actinoneuritis *(ăk-tĭn-ō-nū-rī'-tĭs).* Inflammation of a nerve or nerves due to exposure to x-rays or radioactive substances.

actinotherapy *(ăk'-tĭn-ō-thĕr'-ă-pē).* Treatment of disease using sunlight, ultraviolet rays, x-rays, or radium.

action *(ăk'-shŭn).* Performance of a function or process, be it physical, mental, or chemical.

action potential. Local, temporary change in the electrical charge of a muscle or nerve cell membrane after it has been stimulated.

activated charcoal. Purified powder form of charcoal used to treat some forms of poisoning.

activator *(ăk'-tĭ-vā-tŏr).* A substance that causes another substance to become active.

active immunity. Immunity that the body acquires through production of its own antibodies, which can occur naturally after infection or artificially after vaccination.

active transport. Carrying of a substance (e.g., drug, nutrient) across a cell membrane against some resistance, which requires energy to accomplish.

activities of daily living. The self-care, movement, and communication skills people need to live independently; often seen as the acronym ADLs.

acuity *(ă-kū'-ĭ-tē).* Clearness, sharpness.

acupoint *(ak'-u-point).* Any one of the specific sites where needles are inserted in acupuncture.

Stimulation of these points is the basis for acupuncture and acupressure. Also called an *acupuncture point.*

acupressure *(ăk'-yĕ-prĕsh-ŭr).* Ancient Chinese healing technique in which pressure is applied to various points on the body, intended to stimulate nerves, release blocked energy, and restore balance to the body.

acupuncture *(ăk'-yĕ-pŭnk-chŭr).* Ancient Chinese healing technique to relieve pain or treat certain conditions by inserting ultrafine needles into specific sites on the body along channels, called meridians.

acute *(ă-kūt').* Having a sudden and rapid onset, severe symptoms, and a short course; not chronic.

acute abdomen. Any serious acute abdominal condition characterized by pain, tenderness, and rigid abdominal muscles, and for which emergency surgery must be considered.

acute care. Health care for patients who experience acute illness or trauma; generally done in a hospital or emergency room and usually of short duration.

acute lymphocytic leukemia (ALL). Progressive, malignant blood disease in which too many immature white blood cells are made, which prevents production of normal red and white cells and platelets. Mostly occurs in children younger than 10 years.

acute myelocytic leukemia (AML). Progressive, malignant blood disease in which a type of white blood cell (granular leukocyte or neutrophil) reproduces uncontrollably. Usually seen in adolescents and young adults.

acute organic brain syndrome. Sudden appearance of disorientation and confusion in an otherwise mentally healthy individual, usually resulting from drug use, head injury, infection, or chemical imbalance.

acute renal failure. Sudden and often temporary loss of kidney function.

acute respiratory distress syndrome. Often fatal lung disease characterized by difficult and rapid breathing, rapid heartbeat, changes in level of consciousness, and extreme sweating, it most often occurs in young adults who have had a severe infection, illness, or trauma.

acyclovir *(ā-sī'-klō-vĭr).* An antiviral drug used to treat herpes genitalis and shingles. Trade name: Zovirax.

ad. To (Latin); used in prescriptions to indicate that a substance should be added to the formulation up a specified volume.

adactylia *(ă'-dăk-tĭl-ē'-ă).* Birth defect in which fingers or toes are absent. Also called *adactylism, adactyly.*

adamantine *(ăd-ă-măn'-tĭn).* Very hard, referring to tooth enamel.

adamantinoma *(ăd'-ă-măn-tĭ-nō'-ă).* Tumor of the jaw (usually the lower jaw) that may be benign or of low-grade malignancy.

Adam's apple. Normal bulge at the front of the neck formed by the larynx (voice box), more prominent in men.

Adams-Stokes syndrome. Condition characterized by sudden episodes of fainting caused by transient decrease in blood flow to the brain due to disturbances in the electrical activity between the atria and the ventricles of the heart.

adaptation *(ăd-ăp-tā'-shŭn).* Ability of an organism or body part to adjust to changes in its environment (e.g., adjustment of the eye to changing light in a room).

adaptive device. Any tool or device specifically designed to help people with disabilities perform life tasks independently.

addict *(ă'-dĭkt).* Individual who is physically and/or psychologically dependent on a substance (especially alcohol or drugs) or on a behavior (including sex, exercise, or gambling).

addiction *(ă-dĭk'-shĕn).* Condition of being physically or psychologically dependent on a substance or a behavior such that suddenly being deprived of it causes withdrawal symptoms.

Addison's disease. Disease caused by failure of function of the adrenal glands resulting in a deficiency of adrenocortical hormones.

additive *(ăd'-ĭ-tĭv).* Something added to food to improve or enhance flavor, color, shelf life, or other qualities.

adduct *(ă-dŭkt').* To move a part of the body toward the midline.

adenectomy *(ăd'-ĕn-ĕk'-tō-mē).* Surgical removal of a gland.

adenitis *(ăd-ĕn-ī'-tĭs).* Inflammation of a lymph node or gland.

adenocarcinoma *(ăd'-ĕ-nō-kar-sĭn-ō'-mă).* Common type of cancer in which a malignant tumor originates in a gland.

adenocystoma *(ăd'-ĕ-nō-sĭs-tō'-mă).* Nonmalignant tumor of a gland that contains cysts; a cystic adenoma.

adenofibroma *(ăd'-ĕ-nō-fī-brō'-mă).* Benign tumor, usually seen in the uterus or breast, that contains gland tissue and fibrous tissue.

adenoid *(ăd'-ĕ-noyd).* Lymphatic tissue that forms a bulge in the throat behind the nose.

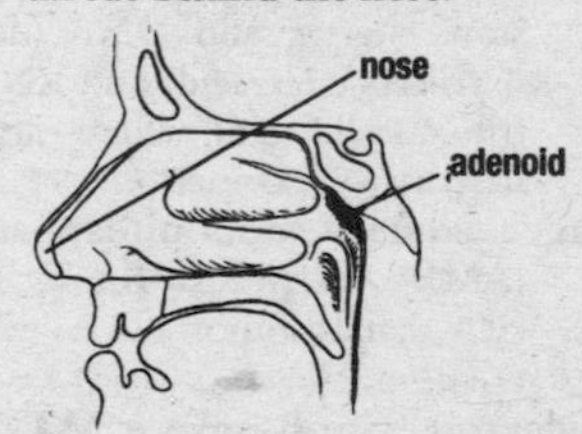

adenoidectomy *(ăd'-ĕ-noyd-ĕk'-tō-mē).* Surgical removal of the adenoids.

adenoma *(ăd'-ĕ-nō'-mă).* Benign tumor either composed of glandular tissue or having a glandular structure. An adenoma may cause the affected gland to become overactive.

adenomatosis *(ăd'-ĕ-nō-mă-tō'-sĭs).* Condition in which many glandlike growths develop.

adenomyoma *(ăd'-ĕ-nō-mī-ō'-mă).* Benign tumor that contains both gland and muscle tissue, usually seen in the uterus.

adenomyosis *(ăd'-ĕ-nō-mī-ō'-sĭs).* Benign condition in which the endometrium grows into the muscular layer of the uterus.

adenopathy *(ăd-ĕ-nŏp'-ă-thē).* General term for gland enlargement and disease, especially of the lymph glands.

adenosarcoma *(ăd'-ĕ-nō-sar-kō'-mă).* Malignant tumor that contains glandular and connective tissue.

adenosine *(ă-dĕn'-ō-sēn).* Compound that is a major component of many important molecules in the body, including adenosine triphosphate (ATP) and adenosine monophosphate (AMP), which help store energy; and DNA (deoxyribonucleic acid) and RNA (ribonucleic acid), which carry hereditary information.

adenosine monophosphate (AMP). Chemical found in cells that is important in metabolism.

adenosine triphosphate (ATP). Compound found in all cells, especially muscle cells, that is involved in the storage and transfer of energy.

adenotonsillectomy *(ăd'-ĕ-nō-tŏn-sĭl-lĕk'-tō-mē).* Surgical removal of the tonsils and adenoids.

adenovirus *(ăd-ĕ-nō-vī'-rŭs).* A group of viruses that can cause infections of the upper respiratory tract, including the common cold.

ADH. Acronym for antidiuretic hormone; *see* **antidiuretic hormone.**

adhesion *(ăd-hē'-zhĕn).* Band of fibrous tissue that holds together two parts or surfaces that are normally separate.

adiaphoresis *(ă-dī'-ă-fō-rē'-sĭs).* Absence or deficiency of sweat.

adipose tissue *(ăd'-ĭ-pōs).* Tissue composed of fat cells; an insulating layer that is under the skin.

adiposis *(ăd-ĭ-pō'-sĭs).* Condition in which there is an abnormal amount of fat in the body because of overeating, a glandular disorder, or a metabolic condition.

adiposis dolorosa. Disease in which scattered areas of painful fat are beneath the skin, usually seen in menopausal women. Also called *Dercum's disease.*

adipsia *(ă-dĭp'-sē-ă).* Absence of thirst or the desire to drink fluids.

adjunct *(ăd'-jŭnkt).* 1. Something that is added. 2. A junior member of a hospital staff.

adjustment disorder. A debilitating reaction, usually lasting less than six months, to a stressful event or situation.

adjuvant *(ăd'-jū-vănt).* Something that helps or enhances a process, such as a drug added to a prescription to increase the action of the main drug.

ad lib. Abbreviation for *ad libitum,* Latin for "as much as wanted," seen on some prescriptions.

adnexa *(ăd-nĕk'-să).* Accessory parts of an organ or structure, usually referring to ovaries and oviducts in relation to the uterus.

adolescence *(ăd-ō-lĕs'-ĕns).* Period between the onset of puberty and maturity.

adolescent nodule. Small, tender lump that appears beneath the nipple in adolescent girls and boys.

adrenal cortex. Outer portion of the adrenal gland, producing various hormones, specifically cortisol, androgens, and aldosterone.

adrenalectomy *(ăd-rē-năl-ĕk'-tō-mē).* Surgical removal of the adrenal glands.

adrenal gland *(ăd-rē'-năl).* One of a pair of triangular-shaped glands, located on top of the kidneys, which manufacture and secrete various hormones.

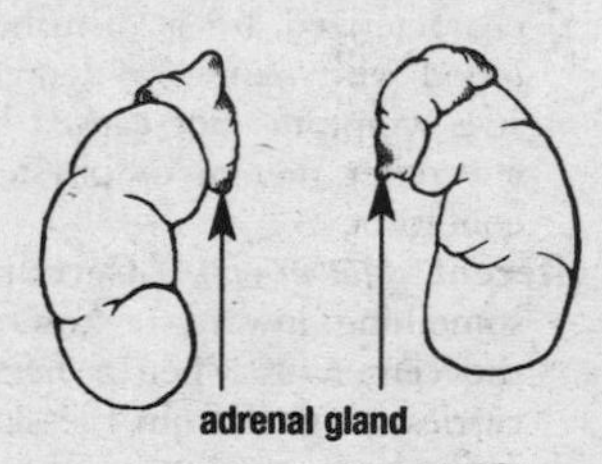

adrenaline *(ă-drĕn'-ă-līn).* Hormone made in the adrenal medulla; also known as epinephrine.

adrenal insufficiency. Abnormal reduced activity of the adrenal gland; also called *hypoadrenalism.*

adrenal medulla. Central portion of the adrenal gland, producing adrenaline.

adrenergic *(ăd-rĕn-ĕr'-ĭk).* Refers to sympathetic nerve fibers that, when stimulated, release adrenaline (epinephrine) or norepinephrine (as opposed to cholinergic, dopaminergic, etc).

adrenergic blocking agent. Substance that blocks transmission of sympathetic nerve signals. *See* **sympathetic nerve system.**

adrenocorticotropic hormone (ACTH) *(ă-drē-nō-kor-tĭ-kō-trōp'-ĭk).* Hormone secreted by the pituitary gland essential for the development and function of the adrenal cortex.

adsorption *(ăd-sorp'-shŭn).* Adhesion of a gas, liquid, or other dissolved substance to the surface of a solid (e.g., the activated charcoal in a gas mask may adsorb gases).

adulteration *(ă-dŭl-tĕr-ā'-shŭn).* Addition or substitution of an impure or weaker (and usually less expensive) substance to a product or formula, which reduces the quality of the product or formula.

adult respiratory distress syndrome (ARDS). *See* **acute respiratory distress syndrome.**

advance directive *(ăd-văns' dī-rĕk'tĭv).* Instructions given by a mentally competent individual that indicate what interventions he or she would accept or refuse after losing the capacity to make decisions. *See* Appendix, Advance Directives: What Everyone Should Know.

adventitia *(ăd-vĕn-tĭsh'-ē-ă).* Outer covering of an organ or other structure (e.g., a blood vessel).

adventitious *(ăd-vĕn-tĭsh'-ĕs).* Accidental; not inherited.

adverse reactions. Undesired side effects or negative responses to a drug or procedure; also called adverse effects and adverse events.

Aedes *(ă-ē'-dēs).* A genus of mosquito including several species that transmit yellow fever, encephalitis, and dengue.

aeration *(ĕr-ā'-shŭn).* Exposure to air. Blood aeration occurs in the lungs, where carbon dioxide is exchanged for oxygen.

aerobe *(ĕr'-ōb).* Microorganism that needs oxygen to live and grow.

aerobic *(ĕr-ō'-bĭk).* Living only in the presence of oxygen.

aerobic exercise. Sustained exercise, using large muscle groups, during which the energy needed is supplied by the oxygen you take in. Such exercise increases heart rate and breathing rate, and improves functioning of the cardiovascular system, thus improving exercise tolerance.

aerodontalgia *(ĕr-ō-dŏnt-ăl'-jē-ă).* Pain in the teeth resulting from a change in air pressure, as when flying or climbing a mountain.

aeroembolism *(ĕr-ō-ĕm'-bō-lĭzm).* *See* **air embolism.**

aerogenic, aerogenous *(ĕr-ō-jĕn'-ĭk, ĕr-ŏj'-ĕn-ŭs).* Gas forming.

aeromedicine *(ĕr-ō-mĕd'-ĭ-sĭn).* Branch of medicine concerned with medical conditions related to aviation.

aero-otitis *(ĕr-ō-ō-tī'-tĭs).* Inflammation of the middle ear caused by changes in altitude during airplane flights.

aerophagia *(ĕr-ō-fā'-jē-ă).* Swallowing of air, which can result in flatulence, belching, and stomach discomfort.

aerophobia *(ĕr-ō-fō'-bē-ă).* Fear of fresh air or drafts.

aerosol *(ĕr'-ō-sŏl).* Compressed gas containing particles of a substance used to treat patients, often those with pulmonary conditions such as asthma or bronchitis.

afebrile *(ă-fĕb'-rĭl).* Without fever, usually refers to an individual whose temperature is normal.

affect *(ăf'-fĕkt).* The observable emotional state; the external expression of mood.

affective disorders. In common usage a group of disorders characterized by a disturbed mood with manic or depressive symptoms not caused by any other mental or physical condition.

afferent *(ăf'-ĕr-ĕnt).* Carrying something inward or toward the center, as when a nerve carries a signal from the skin to the brain.

affinity *(ă-fĭn'-ĭ-tē).* An attraction to or relationship between organisms or substances (e.g., the affinity between an antibiotic and binding sites on the bacteria it destroys).

aflatoxin *(ăf-lă-tŏk'-sĭn).* Carcinogenic toxin produced by some strains of the *Aspergillus* mold that may contaminate some foods, such as peanuts and some grains.

afterbirth. Placenta and membranes that are expelled from the uterus after the birth of a child.

aftercare. Care of a chronically ill, handicapped, or postoperative individual after formal acute medical care has ended, which is needed to ensure the person can function in his/her environment.

aftereffect. A delayed response to a drug or stimulant.

afterhearing. Continuing to hear a sound after it has actually stopped.

afterimage. Continuing to see an image after it is no longer actually visible.

afterpains. Pains from uterine contractions felt by a woman after her infant is born.

aftertaste. Continuing to taste something after the actual substance is gone.

agalactia *(ă-gă-lăk'-shē-ă)* or **agalorrhea** *(ă-găl-ō-rē'-ă).* Condition in which little or no milk is secreted from the breast.

agammaglobulinemia *(ă-găm-ĕ-glŏb-ū-lĭn-ē'-mē-ă).* Condition characterized by a severe reduction in the amount of antibodies circulating in the blood, and associated with an increased risk of infection.

agar *(ă'-găr).* 1. A gelatinlike substance used widely as a medium for growing bacteria in laboratories. 2. Used as a laxative because of its ability to increase in bulk as it absorbs water.

age. The time from birth to the present or to the time of an event cited.

anatomical a. An estimate of age as determined by the stage of development or deterioration of the body as compared with persons of a known age.

developmental a. Age determined by the degree of mental, emotional, and physiologic maturity.

gestational a. Age of embryo or fetus as determined from the date of onset of the last menstrual period.

mental a. Age level at which a person functions intellectually, determined by a series of mental tests.

agent *(ā'-jĕnt).* A substance that causes an effect or reaction.

Agent Orange. A chemical mixture, used during the Vietnam War, that contains the toxic substance dioxin and which may lead to development of cancer.

agglomeration *(ă-glŏm-ĕr-ā'-shŭn).* A grouping together of cells so that they form a mass.

agglutination *(ă-gloo-tĭ-nā'-shŭn).* The joining together of separate particles; e.g., red blood cells.

agglutination test. A blood test used to identify unknown antigens, it is used to help determine susceptibility to certain diseases, in blood typing, and in tissue matching.

agglutinin *(ă-gloo'-tĭ-nĭn).* An antibody that causes agglutination.

aggression *(ă-grĕsh'ŭn).* A hostile action done against others or oneself.

agnathia *(ăg-nā'-thē-ă).* A birth deformity in which the lower jaw is not developed.

agnosia *(ăg-nō'-zē-ă).* Loss of the ability to recognize people or things through one or more of the senses (e.g., auditory agnosia is an inability to interpret sounds, while optic agnosia is an inability to interpret images seen).

agonist *(ăg'-ŏn-ĭst).* 1. The main facilitator, as in one of a pair of muscles that contracts while the other one (the antagonist) relaxes to perform a task. 2. Drug or other substance that assists another drug or augments a body function.

agoraphobia *(ăg-ō-ră-fō'-bē-ă).* Fear of open spaces, often leading to a panic attack, characterized by rapid heartbeat, chest pain, breathing difficulty, sweating, or fear of going crazy.

agraphia *(ă-grăf'-ē-ă).* Loss of the ability to express thoughts in writing, due to a brain disorder or trauma.

agromania *(ăg-rō-mā'-nē-ă).* Irrational desire to live in solitude or in the open.

AHF. Acronym for antihemophilic factor.

AID. Acronym for artificial insemination with sperm from a donor as opposed to acronym **AIH,** which refers to artificial insemination with the husband's sperm; *see also* **artificial insemination.**

AIDS. Acronym for acquired immune deficiency syndrome or disease.

AIDS antibodies. Antibodies found in the blood of individuals with AIDS. A blood test looking for these antibodies helps determine whether someone is infected with the AIDS virus.

ailurophobia *(ă-lū-rō-fō'-bē-ă).* Morbid fear of cats.

air bed. Bed with a special mattress containing air so patients float on a cushion of air; it is used for people who have severe burns or skin ulcers.

air blast injury. Bleeding or rupture of organs caused by being close to an explosion.

air embolism (also **aeroembolism**). Blockage of a blood vessel by a gas bubble in the bloodstream. It can occur accidentally during surgery or an intravenous injection, from a puncture wound, or as a complication of a rapid decrease in atmospheric pressure, as in scuba diving.

air sickness. Motion sickness symptoms of nausea, vomiting, and headache caused by the motion of an airplane.

air swallowing. A condition in which an individual swallows large quantities of air, often associated with anxiety; *see also* **aerophagia.**

airway. Natural passageway for

air to and from the lungs, or a device used to maintain a clear passageway for air in and out of the lungs.

airway obstruction. Any blockage of the respiratory tract that hinders breathing.

akinesia *(ā-kĭ-nē'-zhă).* Complete or partial loss of muscle movement.

akinetic. The characteristic of impaired muscle contraction, often referring to cardiac muscle suffering a lack of oxygen, or damaged by a heart attack.

alalia *(ă-lā'-lē-ă).* Inability to speak (e.g., when the vocal cords are paralyzed or due to brain injury); *see also* **aphasia.**

alanine *(ăl'-ă-nēn).* One of the nonessential amino acids; *see also* **acid.**

Al-Anon. Support group that provides services to friends and relatives of alcoholics.

albinism *(ăl'-bĕ-nĭz-ĕm).* Congenital condition characterized by a lack of normal pigment in the skin, hair, and eyes.

albumin *(ăl-bū'-mĭn).* One of a group of water-soluble proteins found in many plant and animal tissues.

egg a. The white of an egg.

serum a. The main protein found in the plasma.

urinary a. Albumin found in urine, usually an abnormal finding.

albumin-globulin ratio. The ratio of albumin to globulin in the blood or urine; used to help diagnose conditions like liver failure, kidney disease, and hypothyroidism. Also can help assess nutritional status.

albuminuria *(ăl-bū-mĭ-nor'-ē-ă).* Presence of albumin in the urine, found on urine analysis; it may indicate kidney disease but is also seen in some people who exercise vigorously.

albuterol *(ăl-bū'-tĕ-rŏl). See* Appendix, Common Prescription and OTC Drugs: By Generic Name.

alcohol *(ăl'kō-hŏl).* 1. A class of organic compounds composed of carbon (C), hydrogen (H), and oxygen (O) and including a hydroxyl group (OH); includes ethanol, methyl and isopropyl alcohol. 2. General term for liquids that are made from hydrocarbons by distillation.

absolute a. Contains at least 99 percent ethyl alcohol and not more than 1 percent water by weight.

denatured a. Alcohol to which toxic ingredients are added to make it into a solvent and therefore not suitable for consumption.

ethyl a. (or ethanol). Alcohol that is obtained from grains through fermentation of sugars; used in alcoholic beverages.

methyl a. (or methanol). A poisonous alcohol obtained from wood or fossil fuels and used as a solvent, for fuel, and in antifreeze.

alcoholic psychosis. Severe alcohol-related mental disorder (*see* **psychosis**) with predominant hallucinations occurring in many alcohol-related conditions, and often indication of chronic alcoholism.

Alcoholics Anonymous (AA). An international voluntary organization that offers support to people who want to end their addiction to alcohol.

alcoholism *(ăl'-kō-hŏl-ĭzm).* A progressive and potentially fatal disease characterized by chronic excessive alcohol use impairing social and occupational functioning.

alcohol withdrawal syndrome. Signs and symptoms that occur when someone abruptly stops drinking alcohol; typically they include rapid heartbeat and tremors, followed by seizures, and then delirium tremens—hallucinations and possibly violent behavior.

aldosterone *(ăl-dŏs'-tĕr-ōn).* A hormone produced by the cortex of the adrenal gland that regulates salt (sodium and potassium) and water balance in the body.

aldosteronism. Condition characterized by an overproduction of aldosterone from the adrenal cortex; also called *hyperaldosteronism.*

alethia *(ă-lē'-thē-ă).* 1. Inability to forget; 2. Dwelling on the past.

Alexander technique. Program of coordinated simple movements that help people develop more control of their activities and thus improve balance, coordination, and physical functioning.

alexia *(ă-lĕx'-ē-ă).* Loss of the ability to read or understand written words; also called *word blindness.*

ALG. Acronym for antilymphocytic globulin.

algolagnia *(ăl-gō-lăg'-nē-ă).* A disorder characterized by sexual satisfaction derived from inflicting pain on others or by experiencing pain; *see also* **masochism** and **sadism.**

alimentary canal *(ăl-ĭ-mĕn'-tăr-ē).* Tube system through which food passes and is digested and absorbed. It extends from the mouth to the anus. Also called the *gastrointestinal tract.*

alimentation *(ăl-ĭ-mĕn-tā'-shŭn).* The act or process of providing or receiving nourishment.

artificial a. Feeding an individual who is unable or refuses to eat normally; usually accomplished intravenously or by passing a tube through the nose into the stomach.

forced a. Feeding an individual artificially because he/she refuses to eat normally and voluntarily.

parenteral a. Feeding an individual intravenously.

rectal a. Feeding an individual via an enema.

total parenteral a. [*see* **total parenteral nutrition (TPN)**]. Feeding an individual, for an indefinite time, all the nutrients required for life using a catheter placed into a large central vein.

aliquot *(ăl'-ĭ-kwŏt).* Portion of the total amount of a solid, liquid, or gaseous substance; in the plural usually but not always refers to equal fractions of the whole.

alkalemia *(ăl-kă-lē'-mē-ă).* Excessive alkalinity in the blood, in which the pH of the blood is higher than the normal range (7.35–7.45).

alkali *(ăl'-kă-lī).* Any compound that has the properties of a base including pH>7 (vs. an acid, with pH<7). Alkalies combine with acids to form salts and can neutralize acids.

alkaline *(ăl'-kă-līn).* Exhibiting the properties of a base with pH>7.0.

alkaline phosphatase. A class of enzymes measured in the blood to diagnose and monitor various liver and bone diseases; it is usually ordered as part of a "chemistry panel" or "liver profile."

alkali poisoning. Poisoning that results from ingesting an alkaline substance (e.g., lye, ammonia).

alkaloid *(ăl'-kă-loyd).* A natural nitrogen-containing base found in plants.

alkalosis *(ăl-kă-lō'-sĭs).* A condition in which the blood and body tissues are more alkaline than normal.

alkylating agents *(ăl'-kĭ-lā-tĭng).* A class of drugs used in chemotherapy to destroy cancer cells.

allantoin *(ă-lăn'-tō-ĭn).* Substance present in the amniotic fluid that surrounds the embryo.

allele *(ă-lēl').* One of two or more different variants of a gene that can potentially occupy a specific position on a specific chromosome. Human cells generally contain two alleles of each gene, which may be identical (homozygous) or different (heterozygous).

allergen *(ăl'-ĕr-jĕn).* Substance, such as pollen or mold, that causes an allergic reaction or condition.

allergic rhinitis *(ă-lĕr'-jĭk rī-nī'-tĭs).* Inflammation of the membranes in the nose, caused by an allergy.

allergy *(ăl'-ĕr-jē).* Abnormally high sensitivity to certain substances, such as pollen, dust, or foods, that can cause sneezing, rash, and itching.

allograft *(ăl'-ō-grăft).* Transplant tissue obtained from and applied to members of the same species (e.g., from one human to another).

allopathy *(ăl-lŏp'-ă-thē).* A system of medical practice using remedies that produce effects on the body different from those produced by the disease itself, as opposed to homeopathy.

alloplasty *(ăl'-ō-plăs-tē).* Plastic surgery in which foreign material, such as metal or plastic, is used.

allopurinol *(ăl-ō-pū'-rĭ-nŏl).* A drug used to treat gout; trade name: Zyloprim.

allotransplantation *(ăl-lō-trăns-plăn-tā'-shŭn).* Taking tissue or an organ from one individual and grafting or transplanting it into another of the same species.

aloe vera *(ăl'-ō vĕr'-ă).* A species of the *Aloe* plant; the gel or juice from its succulent leaves is used for treating burns and as a laxative, among other uses.

alopecia *(ăl-ō-pē'-shē-ă).* Loss of hair, either partial or complete, temporary or permanent, usually on the head, that can result from certain diseases, hormonal imbalances, use of certain drugs, hereditary factors, or normal aging.

a. androgenic. Also known as male pattern baldness, it is a typical hair loss pattern of males in which the loss begins in the front and proceeds until the only hair left is in the back and temples.

a. areata. Loss of hair in clearly defined patches, usually on the scalp or beard.

alpha blocker *(ăl'-fă).* A drug that blocks the receptors in the arteries and smooth muscle and causes a reduction in blood pressure. It can also improve urinary flow in men with an enlarged prostate (prostatic hypertrophy).

alpha fetoprotein. Protein that is normally synthesized by a fetus, and present at low levels in the blood of children and adults; doctors can check for elevated levels in the amniotic fluid of pregnant women to detect the presence of fetal neural tube defects, and in children and adults to detect and monitor some cancers.

alpha wave. One of four brain wave patterns present in the brain, it is characteristic of people who are awake but relaxed.

alprazolam *(ăl-prā'-zĕ-lăm). See* Appendix, Common Prescription and OTC Drugs: By Generic Name.

ALS. 1. Acronym for amyotrophic lateral sclerosis (also called Lou Gehrig's disease); *see also* **amyotrophic lateral sclerosis.** 2. Acronym for antilymphocytic serum, also known as antilymphocytic globulin.

Alternaria *(ăl-tĕr-năr'-ē-ă).* A genus of fungus that can cause pneumonitis, allergic reactions, and skin infections.

alternative medicine *(ăl-tĕr'-nă-tĭv).* Various preventive or therapeutic health-care practices, such as herbal medicine or aromatherapy, that do not follow generally accepted medical methods and may not have scientific evidence for their effectiveness.

altitude sickness *(ăl'-tĭ-tūd).* Syndrome caused by low oxygen concentration in the air at high altitudes, as can occur during mountain climbing or flying in an unpressurized airplane.

aluminum *(ă-lū'-mĭ-nŭm).* Common metallic element, it is widely used in astringents, deodorants, antiseptics, and antacids.

alveolitis *(ăl-vē-ŏ-lī'-tĭs).* 1. Inflammation of the small saclike structures (alveoli) in the lungs. 2. Inflammation of the socket of a tooth; *see also* **dry socket.**

alveolus *(ăl-vē'-ŏ-lŭs).* 1. A tiny saclike structure in the lungs where oxygen and carbon dioxide are exchanged. 2. Socket in the jaw where the root of a tooth sits.

Alzheimer's disease *(älts'-hī-mĕrz).* Chronic, progressive disease in which individuals lose memory and other mental functions, accompanied by personality and emotional changes.

AMA. Abbreviation for the American Medical Association.

amalgam *(ă-măl'-gĕm).* An alloy of mercury with other metals used to restore teeth.

amastia *(ă-măs'-tē-ă).* Absence of one or both breasts, usually caused by a developmental defect.

amaurosis *(ăm-ăw-rō'-sĭs).* Blindness, usually associated with a brain disorder rather than any obvious eye lesions.

amaurotic familial idiocy. *See* **Tay-Sach's disease.**

ambidexterity *(ăm-bĭ-dĕk-stĕr'-ĭ-tē).* An ability to use both hands equally well.

ambient *(ăm'-bē-ĕnt).* Referring to the surrounding or environmental quality (e.g., ambient air).

ambisexual *(ăm-bĭ-sĕk'-shū-ĕl).* Having a sexual orientation to people of either sex; bisexual.

amblyopia *(ăm-blē-ō'-pē-ă).* Dimness of vision, especially when it occurs in one eye without any apparent physical reason. Also called *lazy eye.*

ambulation *(ăm-bū-lā'-shŭn).* Act of walking or moving.

ambulatory cardiac monitoring. Method of recording heart activity continuously via a monitoring device that the patient attaches to the body.

ambulatory surgery. Surgery performed either in a doctor's office or in a special ambulatory care facility in a hospital.

ameba *(ă-mē'-bă).* A one-celled organism that can constantly change shape. Some can cause disease in humans, such as dysentery.

amebiasis *(ăm-ē-bī'-ĕ-sĭs).* Infection with ameba, especially *Entamoeba histolytica,* the cause of amebic dysentery. It is usually acquired through feces-contaminated food or water.

amenorrhea *(ā-mĕn-ō-rē'-ă).* Abnormal stoppage or absence of menstruation.

primary a. Failure to begin menstruation by the age of 16.

secondary a. Stoppage of menstruation in a woman who previously menstruated.

Americans with Disabilities Act (ADA). Federal statute enacted July 26, 1990, which, among other things, prohibits discrimination on the basis of disability in employment, programs, and services provided by state and local governments, goods and services provided by private firms, and commercial facilities. (ADA is also an acronym for American Diabetes Association.)

amine *(ă-mēn').* A particular class of organic compounds that contains nitrogen.

amino acids. Basic building blocks for the body's proteins and the products of protein digestion. About twenty amino acids are necessary for human health and are divided into two groups:

essential aa. Amino acids that cannot be made by the human body so they are essential to the diet; they include histidine, isoleucine, leucine, lysine, methionine, phenylalanine, threonine, tryptophan, and valine.

nonessential aa. Amino acids that can be made by the human body and so are not essential to the diet; they include alanine, arginine, asparagine, aspartic acid, cysteine, glutamic acid, glutamine, glycine, proline, serine, and tyrosine.

aminophylline *(ăm-ĭ-nŏf'-ĭ-lĭn).* Stimulant used to treat asthma, bronchitis, and emphysema.

amitriptyline *(ăm-ĭ-trĭp'-tĭ-lēn). See* Appendix, Common Prescription and OTC Drugs: By Generic Name.

ammonia *(ă-mō'-nē-a).* Substance normally present in small amounts in the blood with elevated levels causing mental state changes. Blood levels are monitored to evaluate liver toxicity and adverse drug effects. In the salt form combined with other chemicals it has medicinal uses.

ammon(i)emia *(ă-mō-nĭ-ē'-mē-ă).* Excessive ammonia in the blood, which may be caused by impaired liver function.

ammoniuria *(ă-mō-nē-ūr'-ē-ă).* Excessive ammonia in the urine.

amnesia *(ăm-nē'-zē-ă).* Loss of memory caused by injury to the brain, drugs and toxins, degenerative disease, or severe emotional trauma.

amniocentesis *(ăm-nē-ō-sĕn-tē'-sĭs).* Extraction of amniotic fluid from the womb (uterus) of a pregnant woman to analyze it for indications of various fetal abnormalities; it can also predict the sex of the fetus.

amnion *(ăm'-nē-ŏn).* The membranous sac that surrounds the embryo in the womb (uterus). It fuses with another membrane (chorion) by the end of the third month of pregnancy.

amniotic fluid *(ăm-nē-ŏt'-ĭk).* The fluid that surrounds and protects the embryo in the mother's womb (uterus).

amniotomy *(ăm-nē-ŏt'-ō-mē).* Surgical rupture of the fetal membranes to induce labor.

amoxicillin *(ă-mŏks'-ĭ-sĭl-ĭn). See* Appendix, Common Prescription and OTC Drugs: By Generic Name.

amphetamines *(ăm-fĕt'-ă-mēnz).* Class of central nervous system drugs used as stimulants and to reduce appetite, nasal congestion, and depression.

amphoric breath sound *(ăm-fŏr'-ĭk).* Abnormal hollow sound heard through a stethoscope that indicates a cavity in the lung or a collapsed lung.

amphoteric *(ăm-fō-tĕr'-ĭk).* Pertaining to a substance that can react as an acid or as a base (alkaline).

amphotericin B *(ăm-fō-tĕr'-ĭ-sĭn).* Antibiotic used to treat certain fungal infections.

ampule *(ăm'-pūl).* Small glass container that can be sealed and its contents sterilized, usually intended for injection.

ampulla *(ăm-pūl'-lă).* Saclike dilatation of a duct or canal, as in a milk duct or the semicircular canal in the ear.

amputation *(ăm-pū-tā'-shŭn).* Surgical removal of a limb, part of a limb, or other body part.

amputee *(ăm-pū-tē').* Individual who has had a limb removed surgically, as a result of trauma, or due to a birth defect.

amygdala *(ă-mĭg'-dă-lă).* One of two paired almond-shaped structures in the brain located in the temporal lobes and involved in emotion processing.

amylase *(ăm'-ĭ-lās).* Enzymes produced by the pancreas and salivary glands that help digest starch.

amyl nitrite *(ăm'-ĭl).* Drug that, when inhaled, relaxes muscle spasms and can relieve angina pectoris (spasm of the coronary artery).

amyloid *(ăm'-ĭ-loyd).* Protein substance with starchlike characteristics that can cause disease when deposited in various tissues in the body (e.g., the brain in Alzheimer's disease).

amyloidosis *(ăm-ĭ-lōy-dō'sĭs).* A group of diseases caused by excessive deposition of amyloid in body.

amyotonia *(ă-mī-ō-tō'-nē-ă).* Lack of muscle tone.

amyotrophic lateral sclerosis (ALS) *(ă-mī-ō-tŏk'-ĭk).* Degenerative neurologic disease characterized by progressive muscle deterioration and spastic limbs. Cause is unknown and there is no cure. Also known as Lou Gehrig's disease.

anabolic steroid *(ăn-ă-bŏl'-ĭk).* Any one of several compounds derived from testosterone or prepared synthetically, which help in building proteins, fats, and other cell components. Can be used to treat some anemias, promote weight gain, and strengthen bones, but are sometimes used illegally by athletes.

anabolism *(ă-năb'-ō-lĭzm).* Constructive phase of metabolism in cells whereby simple, nonliving substances are converted into complex, living ones.

anaerobe *(ăn'-ĕr-ōb).* Microorganisms that can live and grow without oxygen.

anal *(ā'-năl).* Pertaining to the anus.

analeptic drug *(ăn-ă-lĕp'-tĭk).* Drug used to stimulate the central nervous system or help invigorate the body (e.g., caffeine).

analgesia *(ăn-ăl-jē'-zē-ă).* Absence of sensitivity to pain while remaining conscious, caused by, for example, analgesics (painkilling drugs), acupuncture, or hypnosis.

analgesic drugs. Medications that relieve pain, such as aspirin or acetaminophen.

analogue *(ăn'-ă-lŏg)*. 1. Part or organ with a function similar to that of another part or organ of a different species. 2. Chemical compound similar in structure to another but not similar in composition.

anal personality. In Freudian psychology, a personality in which individuals are characterized as being excessively orderly, stingy, and obstinant.

analysis *(ă-năl'-ĭ-sĭs)*. Examination of the components and nature of a substance; e.g., blood analysis, stool analysis.

analyst. Individual who practices psychoanalysis.

anamnesis *(ăn-ăm-nē'-sĭs)*. 1. Act of remembering. 2. Detailed data collected about a patient used to analyze the case.

anaphoresis *(ăn-ă-fō-rē'-sĭs)*. Condition in which the sweat glands don't function.

anaphrodisia *(ăn-ăf-rō-dĭz'-ē-ă)*. Lack or loss of sexual feelings.

anaphylactic shock *(ăn-ă-fĭ-lăk'-tĭk)*. Severe hypersensitivity (allergic) reaction to the ingestion or injection of a substance to which an organism is sensitive. Symptoms can include shortness of breath, anxiety, weakness, swelling of the throat, and shock.

anaphylaxis *(ăn-ă-fĭ-lăk'-sĭs)*. State of shock or hypersensitivity.

anaplasia *(ăn-ă-plā'-zē-ă)*. Regression of fully developed cells into a more primitive form, as often occurs in cancer cells.

anastomosis *(ă-năs-tō-mō'-sĭs)*. 1. Connection between two blood vessels, lymph vessels, or nerves. 2. Joining together two or more hollow organs or structures (e.g., suturing together two blood vessels), usually done surgically.

anatomy *(ă-năt'-ō-mē)*. Study and science of the structure of an organism and its parts.

ancillary *(ăn-sĭl-lăr'-ē)*. Secondary or auxillary; something that assists another action, but is not necessary.

anconal *(ăn'-kō-năl)*. Pertaining to the elbow.

Ancylostoma *(ăn'-sĭl-ōs'-tō-ă)*. A genus of intestinal parasites, one of which is the hookworm.

androgen *(ăn'-drō-jĕn)*. General term for substances that produce secondary masculine characteristics, such as the male sex hormones testosterone and androsterone.

androgyne *(ăn'-drō-jīn)*. Individual who has both ovaries and testes and undeveloped external male and female sex organs.

androgynous *(ăn-drŏj'-ĭ-nŭs)*. Having both male and female traits and behaviors.

androsterone *(ăn-drō-stĕr'-ōn)*. Male sex hormone, found in both men and women.

anemia *(ă-nē'-mē-ă)*. Condition in which there is an abnormally low level of hemoglobin in the blood, the number or quality of red blood cells is poor, or the number of packed red blood cells is low.

aplastic a. Anemia caused by bone marrow defects or by exposure to radiation, certain drugs, or toxic elements.

idiopathic a. Type of anemia in which the organs that form blood do not function properly.

macrocytic a. Anemia in which the red blood cells are abnormally large.

microcytic a. Kind of anemia in which red blood cells are abnormally small.

pernicious a. Chronic and serious form of anemia associated with a lack of hydrochloric acid in the stomach.

secondary a. Type of anemia caused by a disease, such as cancer or tuberculosis, or by loss of blood.

sickle cell a. Hereditary anemia in which red blood cells have a sickle shape. It occurs mostly in blacks and dark-skinned peoples.

anergic *(ăn-ĕr'-jĭk).* Inactive, sluggish, lacking energy.

anesthesia *(ăn-ĕs-thē'-zē-ă).* Partial or complete loss of sensation, usually induced intentionally to allow a surgical procedure to take place. Can also refer to sensation loss caused by injury or disease.

caudal a. Anesthesia produced by injecting an anesthetic substance into the lower spinal (caudal) canal.

dissociative a. Anesthesia that causes the individual to be partially conscious and feel dissociated from his/her surroundings.

epidural a. Loss of sensation caused by injecting an anesthetic into a space just outside of the spinal column.

general a. Any anesthesia that affects the entire body and causes complete loss of consciousness.

hypotensive a. Anesthesia given when blood pressure has been lowered intentionally to reduce blood loss during surgery.

hypothermic a. Anesthesia given when the body temperature has been lowered intentionally.

hysterical a. Anesthesia that occurs to individuals who experience hysteria.

inhalation a. Anesthesia produced via the inhalation of gases or vapors.

intravenous a. Loss of sensation produced through injection of an anesthetic substance into a vein.

local a. Anesthesia that affects a local area only.

saddle block a. Loss of sensation in the genital region, thighs, and buttocks, caused by an injection into the spine.

spinal a. Anesthesia produced by injecting an anesthetic substance directly into the spinal fluid.

topical a. Local anesthesia produced by applying an anesthetic directly to the body surface to be treated.

anesthesiologist *(ăn-ĕs-thē-zē-ŏl'-ō-jĭst)*. Physician who specializes in administering anesthesia.

anesthetize *(ă-nĕs'-thĕ-tīz)*. To induce anesthesia.

aneurysm *(ăn'-ū-rĭz-m)*. A widening or "ballooning out" of a blood vessel, usually an artery, caused by a congenital defect or weakness in the vessel.

aortic a. An aneurysm that affects any part of the aorta.

dissecting a. An aneurysm in which the arterial wall splits, allowing blood to get between the layers of the vessel wall and separate them.

ruptured a. An aneurysm that has bled through the blood vessel wall into the surrounding tissue.

aneurysmectomy *(ăn-ū-rĭz-mĕk'-tō-mē)*. Surgical removal of an aneurysm of an artery.

angel dust. Street or slang name for phencyclidine (PCP).

angina pectoris *(ăn-jī'-ă pĕk-tor'-ŭs)*. Severe pain in the chest that sometimes radiates from the left shoulder down to the arm, and less often to the back or jaw, caused by an insufficient supply of blood to the heart.

angioblastoma *(ăn-jē-ō-blăs-tō'-mă)*. Malignant tumor of the brain composed of blood vessel tissue.

angiocardiography *(ăn-jē-ō-kăr-dē-ŏg'-ră-fē)*. X-ray of the heart and great vessels after an intravenous injection of a radiopaque solution.

angiofibroma *(ăn-jē-ō-fī-brō'-mă)*. Tumor composed of fibrous tissue.

angiogenesis *(ăn-jē-ō-jĕn'-ĕ-sĭs)*. Development of blood vessels.

angiography *(ăn-jē-ŏg'-ră-fē)*. X-ray of blood vessels after an intravenous injection of a radiopaque solution.

angiology *(ăn-jē-ŏl'-ō-jē)*. Medical specialty that deals with diseases of the blood and lymph vessels.

angioma *(ăn-jē-ō'-mă)*. Tumor, usually benign, that consists of blood vessels and lymph vessels.

angioplasty *(ăn'-jē-ō-plăs-tē)*. Modification of the structure of a blood vessel by surgical procedure or by dilating the vessel with a balloon; *see also* **percutaneous transluminal angioplasty**.

angioradiography *(ăn-jē-ō-ră-dē-ŏg'-ră-fē)*. X-rays taken to study blood vessels and the heart after an injection of an opaque substance.

angiorrhaphy *(ăn-jē-or'-ă-fē)*. Surgical repair of a blood vessel.

angiosarcoma *(ăn-jē-ō-săr-kō'-mă)*. Malignant tumor that originates from a blood vessel.

angiospasm *(ăn'-jē-ō-spăzm)*. Abnormal contraction of a blood vessel, which results in decreased blood flow.

angiotelectasia *(ăn-jē-ō-tĕl-ĕk-tā'-sē-ă)*. Condition characterized by dilatation and enlargement of arterioles.

angiotensin *(ăn-jē-ō-tĕn'-sĭn)*. Chemical in the body that causes blood vessel to narrow. There are three forms: angiotensin I is converted by an enzyme in the lung to angiotensin II (causes blood vessels to narrow and also stimulates water and salt retention), which is then metabolized to angiotensin III (which stimulates aldosterone secretion).

angiotensin converting enzyme inhibitors (ACE inhibitors). Class of drugs that inhibit the angiotensin converting enzyme that transforms angiotensin I to angiotensin II. By blocking this conversion, the blood vessels widen and blood flow improves.

angitis *(ăn-jē-ī'-tĭs).* Inflammation of a blood or lymph vessel.

anhidrosis *(ăn-hī-drō'-sĭs).* Failure of the sweat glands to function.

anhydrase *(ăn-hī'-drās).* An enzyme that promotes the removal of water from a chemical compound.

anhydrous *(ăn-hī'-drŭs).* Lack of water.

aniline *(ăn'-ĭ-līn).* Coal tar derivative used in the manufacture of dyes for medical purposes.

animus *(ăn'-ĭ-mŭs).* 1. Hostility or a grudge; 2. In Jungian psychology, the masculine inner personality in women.

anion *(ăn'-ī-ŏn).* Ion that has a negative charge.

anisocytosis *(ăn-ī-sō-sī-tō'-sĭs).* Condition in which there is excessive and abnormal variation in the size of red blood cells.

anisomastia *(ăn-ī-sō-măs'-tē-ă).* Condition in which the breasts are noticeably unequal in size.

ankle *(ăng'-kl).* The joint that connects the leg and the foot.

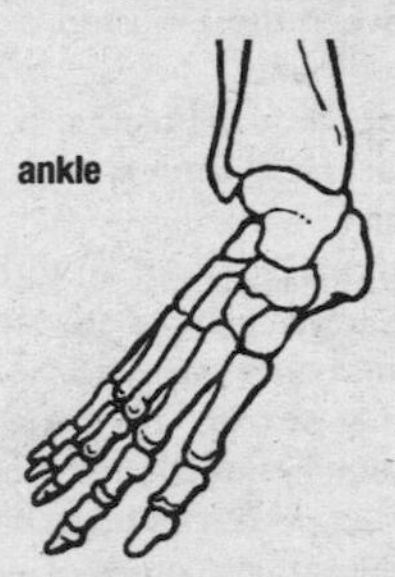

ankle jerk. Result of a brisk tap to the Achilles tendon: contraction of the calf muscles that causes the foot to extend.

ankyloglossia *(ăng-kĭ-lō-glōs'-sē-ă).* Condition in which the band of tissue (frenum) that connects the lower surface of the tongue to the floor of the mouth is abnormally short; also called tongue-tied.

ankylosing spondylitis *(ăng-kĭ-lō'-sĭng spŏn-dl-ī'-tĭs).* Chronic disease, usually seen in males younger than 30, characterized by inflammation and stiffening of the joints and spine, sometimes leading to fusion of the involved joints.

ankylosis *(ăng-kĭ-lō'-sĭs).* Inability of a joint to move.

ankylostoma *(ăng-kĭ-lŏs-tō'-mă).* Spasm of the jaw muscles; lockjaw.

anlage *(ŏn'-lŏ-jhă).* Cells in an embryo that constitute the beginning of an eventual organ or other body part.

annular *(ăn'-ū-lăr).* Circular; ring-shaped.

anococcygeal *(ă-nō-kŏk-sĭ'-jē-ăl).* Referring to the region between the anus and the coccyx (tailbone).

anode *(ăn'-ōd).* Positive pole of an electrical source, such as a battery.

anodyne *(ăn'-ō-dīn).* Any drug that relieves pain; analgesic.

anomaly *(ă-nŏm'-ă-lē).* Something that deviates from the normal, such as a birth deformity.

anomia *(ă-nō'-mē-ă).* An inability to remember names of objects or people.

anomie *(ăn'-ŏ-mē).* Condition in which people feel isolated, anxious, or disoriented.

anoperineal *(ă-nō-pĕr-ĭ-nē'-ăl).* Relating to both the anus and perineum.

Anopheles *(ă-nŏf'-ĕ-lēz).* A genus of mosquitoes that transmits malaria, yellow fever, dengue fever, and other diseases.

anoplasty *(ā'-nō-plăs-tē).* Plastic surgery of the anus.

anorchism *(ăn-or'-kĭzm).* Congenital absence of one testis or both testes.

anorectal *(ā-nō-rĕk'-tăl).* Relating to the anus and rectum.

anorexia *(ăn-ō-rĕk'-sē-ă).* Loss of appetite or the desire to eat.
a. nervosa. An emotional disorder, most commonly seen in adolescent females, characterized by abnormal body image, intense fear of getting fat, and prolonged refusal to eat.

anorgasmy *(ăn-or-găz'-mē).* Failure or inability to achieve orgasm during sexual intercourse or masturbation.

anoscope *(ā'-nō-skōp).* An instrument used to examine the anus and lower rectum.

anosmia *(ăn-ŏz'-mē-ă).* Absence of the sense of smell, which may be temporary (as from a cold) or permanent (resulting from damage to the olfactory nerve or nasal tissue).

anovular menstruation *(ăn-ŏv'-ūlăr).* Menstruation that is not preceded by ovulation.

anovulatory *(ăn-ŏv'-ū-lă-tō-rē).* Not associated or accompanied by ovulation, the development and release of a mature ovum (egg) from the ovary.

anovulatory drug. Drug that inhibits ovulation, as a birth control pill.

anoxia *(ăn-ŏk'-sē-ă).* Absence or abnormally low amount of oxygen in the body, which can occur with anemia or heart failure, at high altitudes, and other situations.

Antabuse *(ăn'-tă-būs).* Trade name for disulfiram, which is used to treat alcoholism. Use of Antabuse causes violent nausea and vomiting when taken with alcohol.

antacid *(ănt-ăs'-ĭd).* Substance that neutralizes excess acid, especially in the digestive tract.

antagonist *(ăn-tăg'-ō-nĭst).* Anything that counteracts the action of something else, as in a drug that neutralizes the effects of another.

antecubital *(ăn-tē-kū'-bĭ-tăl).* Referring to the region of the arm in front of the elbow.

anteflexion *(ăn-tē-flĕk'-shŭn).* Abnormal bending forward of an organ or part of an organ.

antemortem *(ăn-tē-mor'-tĕm).* Before death.

antenatal *(ăn-tē-nā'-tăl).* Occurring before birth; prenatal.

antepartum *(ăn-tē-păr'-tŭm).* Occurring during pregnancy, before childbirth.

anterior *(ăn-tĕr'-ē-ĕr).* 1. At or toward the front of a part, organ, or structure. 2. In people, toward the belly surface.

anterior pituitary gland. Front portion of the pituitary gland, which is located at the base of the brain. It produces important hormones, such as growth hormone, ACTH, follicle-stimulating hormone, prolactin, and others.

anterior resection operation. Surgical procedure performed on colorectal cancer patients in which the lower sigmoid colon and the upper rectum are removed and the remaining sections are joined. It avoids the need for a colostomy.

anterograde *(ăn'-tĕ-rō-grād).* Moving in a forward or normal physiological direction, as when nerve signals are transmitted in their proper direction.

anterograde amnesia. *See* **amnesia.**

anteroposterior *(ăn-tĕr-ō-pŏs-tĕr'-ē-ĕr).* Passing from the front to the rear of the body.

anteverted *(ăn-tē-vĕrt'-ĕd).* Tipped forward, as an anteverted organ.

anthelmintic *(ănt-hĕl-mĭn'-tĭk).* Drug that kills intestinal worms.

anthracosis *(ăn-thră-kō'-sĭs).* Chronic inflammatory disease of the lung that occurs in people who have been exposed to coal dust and soot.

anthrax *(ăn'-thrăks).* Bacterial (*Bacillus anthracis*) disease usually seen in farm animals but which can be transmitted to humans from infected animals and animal products. People have also been infected intentionally through acts of bioterrorism.

anthropoid *(ăn'-thrō-poyd).* Resembling man.

antiadrenergic *(ăn-tē-ă-drĕn-ĕr'-jĭk).* Referring to anything that blocks or counteracts the effects of impulses that are sent by adrenergic fibers in the sympathetic nervous system.

antiarrhythmics *(ăn-tē-ă-rĭth'-mĭk).* Drugs used to control or prevent cardiac arrhythmias.

antiarthritics *(ăn-tē-ărth-rĭ'-tĭks).* Drugs used to relieve the symptoms of arthritis.

antibacterial *(ăn-tē-băk-tĕr'-ē-ăl).* Capable of destroying or stopping the growth of bacteria.

antibiotics *(ăn-tē-bī-ŏ'-tĭks).* Drugs or other substances used to inhibit the growth of microorganisms or to destroy them.

antibiotic sensitivity test. Method to determine how sensitive certain bacteria are to certain antibiotics.

antibody *(ăn'-tē-bŏd-ē).* Protein substance that develops in response to the presence of an antigen (e.g., bacteria or other infecting organisms).

anticancer drug. Any one of many chemicals used to treat cancer.

anticarcinogen *(ăn-tē-kăr-sĭn'-ō-jĕn).* A substance used to block the development of cancer.

anticholinergic *(ăn-tē-kō-lĭn-ĕr'-jĭk).* Referring to blockage of acetylcholine receptors, which results in the inhibition of nerve impulse transmission in the parasympathetic nervous system.

anticoagulant *(ăn-tē-kō-ăg'-ū-lănt).* Any substance that delays or prevents blood clotting.

anticonvulsant *(ăn-tē-kŏn-vŭl'-sănt).* Any medication that helps prevent convulsions or seizures.

antidepressant *(ăn-tē-dē-prĕs'-sănt).* Any substance or type of therapy that helps to prevent, relieve, or cure mental depression.

antidiabetic *(ăn-tē-dī-ă-bĕt'-ĭk).* Any substance used to prevent or treat diabetes.

antidiuretic *(ăn-tē-dī-ū-rĕt'-ĭk).* Any substance that inhibits the formation of urine.

antidiuretic hormone (ADH). A hormone that inhibits the production of urine; also called *vasopressin.*

antidote *(ăn'-tĭ-dōt).* A substance that neutralizes poisons or their effects.

antiemetic *(ăn-tē-ĭ-mĕt'-ĭk).* Any substance that prevents or alleviates vomiting and nausea.

antiestrogen *(ăn-tē-ĕs'-trō-jĕn).* Any substance that blocks or modifies the action of estrogen.

antifungal *(ăn-tē-fŭng'-gĕl).* Agent that destroys or prevents the growth of fungi.

antigen *(ăn'-tĭ-jĕn).* Substance (e.g., a toxin) or an organism that, once it enters the body, causes the production of an antibody.

antigen-antibody reaction. Process by which the immune system recognizes an antigen and causes the production of antibodies specific against that antigen.

antihemophilic factor *(ăn-tī-hē-mō-fĕl'-ĭk).* Substances in the blood that are critical for the clotting process.

antihistamine *(ăn-tē-hĭs'-tă-mēn).* Any substance that works to reduce the effects of histamine and is used to treat allergies, hypersensitivity reactions, and colds.

antihypertensive *(ăn-tē-hī-pĕr-tĕn'-sĭv).* Any agent that reduces high blood pressure (hypertension).

anti-inflammatory *(ăn-tē-ĭn-flăm'-ă-tŏ-rē).* A substance that helps reduce or counteract inflammation.

antiketogenic *(ăn-tē-kē-tō-jĕn'-tĭk).* A substance that reduces acidosis, such as insulin, used by people with diabetes.

antimalarial *(ăn-tē-mă-lăr'-ē-ĕl).* Any substance used to prevent, suppress, or treat malaria.

antimicrobial *(ăn-tē-mī-krō'-bē-ăl).* Any agent that destroys or prevents the development of microorganisms.

antimycin A *(ăn-tē-mī'-sĭn).* Antibiotic used to kill fungi, insects, and mites.

antimycotic *(ăn-tē-mī-kŏt'-ĭk).* Substance that destroys or limits the growth of fungus.

antineoplastic *(ăn-tē-nē-ō-plăs'-tĭk).* Any drug that controls or kills cancer cells.

antioncogene *(ăn-tē-ŏng'-kō-jĕn).* A gene that suppresses growth of cancer cells.

antioxidant *(ăn-tē-ŏk'-sĭ-dănt).* Any substance that absorbs, prevents, or blocks the formation of free radicals.

antiplatelet *(ăn-tē-plāt'-lĕt).* Any agent that destroys platelets, which decreases the body's blood-clotting efficiency.

antiprothrombin *(ăn-tē-prō-thrŏm'-bĭn).* Substance that suppresses blood clotting.

antipruritic *(ăn-tē-prū-rĭt'-ĭk).* Substance that relieves or prevents itching.

antipsychotic *(ăn-tē-sī-kŏ'-tĭk).* Drug used to treat a psychosis.

antipyretic *(ăn-tē-pī-rĕt'-ĭk).* Any substances used to lower body temperature.

antiseptic *(ăn-tĭ-sĕp'-tĭk).* Substance that slows or stops the continuing growth of microorganisms but may not destroy them.

antiserum *(ăn'-tē-sē-rŭm).* Serum that contains antibodies for a specific antigen.

antispasmodic *(ăn-tē-spăz-mŏd'-ĭk).* Substance that relieves or prevents spasm.

antitoxin *(ăn-tē-tŏk'-sĭn).* Any agent that prevents or limits the effect of poison.

antitussive *(ăn-tē-tŭs'-ĭv).* Substance that relieves coughing.

antivenin *(ăn-tē-vĕn'-ĭn).* Antitoxin that counteracts the effects of venom from an animal bite.

Antivert. Trade name for meclizine hydrochloride, an antihistamine used to prevent and treat motion sickness.

antiviral *(ăn-tē-vī'-răl).* A substance, such as the drug interferon, that destroys viruses.

antivivisection *(ăn-tē-vĭv'-ĭ-sĕk-shŭn).* Opposition to the use of live animals in experimentation.

antrum *(ăn'-trĕm).* A cavity or chamber, especially in bone.

maxillary a. The maxillary sinus, located in the cheek.

anuria *(ăn-ū'-rē-ă).* Inability to urinate, which can be caused by kidney or bladder disease or serious decline in blood pressure.

anus *(ā'nŭs).* Opening of the rectum, at which the passage of feces is controlled by a specialized muscle (sphincter ani).

anxiety *(ăng-zī'-ĕ-tē).* Feeling of worry, dread, apprehension, or uneasiness, especially of the future.

anxiety attack. Acute episode of intense anxiety and panic, accompanied by symptoms such as shortness of breath, sweating, palpitations, and stomach complaints.

anxiolytic *(ăng-zī-ō-lĭt'-ĭk).* Any drug or therapy that relieves anxiety.

aorta *(ā-or'-tă).* The large artery that originates in the left ventricle of the heart and carries blood to all parts of the body.

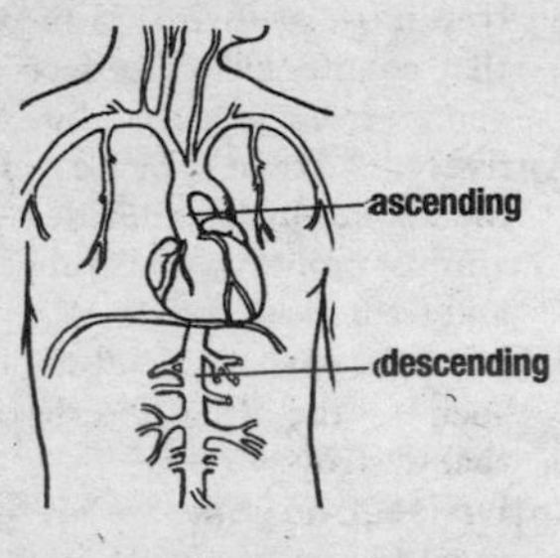

aorta

abdominal a. The final section of the aorta, it starts at the diaphragm and extends to the point where it splits into two to form the common iliac arteries. It supplies oxygenated blood to the abdominal and pelvic organs and to the legs.
ascending a. The first section of the aorta, it starts from the left ventricle of the heart and extends to the bend of the aorta.
descending a. the part of the aorta that starts after the bend of the aorta and ends by splitting into two great arteries that extend to the legs. The descending aorta is subdivided into the thoracic aorta and the abdominal aorta.
thoracic a. The section of the aorta that starts after the bend of the aorta and extends to the diaphragm. The thoracic aorta gives off several branches that supply oxygenated blood to the chest area and organs within the chest.

aortic aneurysm *(ā-ŏr'-tĭk an'-ū-rĭz-m). See* **aneurysm.**

aortic insufficiency. A condition characterized by an aortic heart valve that does not open and close efficiently.

aortic stenosis. Narrowing or constriction of the aortic valve due to congenital defect or a disease, such as rheumatic fever.

aortic valve. Valve in the heart between the left ventricle and the aorta that prevents blood from flowing from the aorta back into the heart.

aortocoronary bypass *(ā-ŏr-tō-kŏr'-ō-nā-rē).* Surgical procedure in which grafts from the saphenous vein in the leg are used to bypass an obstruction in the coronary artery. Also called a *coronary bypass.*

aortofemoral bypass *(ā-ŏr-tō-fĕm'-ŏr-ăl).* Surgical procedure in which a graft is inserted between the abdominal aorta and the femoral arteries to improve blood circulation to the legs.

aortogram *(ā-or'-tō-grăm).* X-ray of the aorta after injection of a contrast fluid.

aortography *(ā-or-tŏg'-ră-fē).* Technique that allows the outline of the aorta to be seen on X-ray.

apathy *(ăp'-ă-thē).* Indifference; lack of interest or emotion.

aperature *(ăp'-ĕr-chĕr).* Opening or hole; entrance to a body cavity.

aperient *(ă-pĕr'-ē-ĕnt).* A very mild laxative.

apex *(ā'-pĕks).* Top or pointed end of a structure.

Apgar score. Evaluation of a

newborn's physical condition—heart rate, muscle tone, respiratory effort, color, and reflexes—made one minute after birth and then repeated five minutes after birth.

aphagia *(ă-fā'-jē-ă).* An inability to swallow due to pain or paralysis.

aphasia *(ă-fā'-zē-ă).* An inability to speak coherently, usually the result of a stroke.

aphonia *(ă-fō'-nē-ă).* Loss of the ability to speak because of a problem with the larynx, such as disease of the vocal cords or chronic laryngitis, or a psychiatric cause.

aphrodisiac *(ăf-rō-dē'-zē-ăk).* Substance that may stimulate sexual desire.

aphthous stomatitis *(ăf'-thūs stō-mě-tī'-tĭs).* Condition seen primarily in infants and young children characterized by a sore throat and sore mouth with many small, white blisters.

apical abscess *(ăp'-ĭ-kăl).* Infection at the root of a tooth.

apis *(ā'-pĭs).* Homeopathic remedy derived from the honeybee that is used to treat stinging pain, swelling, and redness.

apitherapy *(ă-pē-thěr'-ă-pē).* The medicinal use of products from honeybees, including honey, royal jelly, and venom, to treat various ailments, such as inflammation (venom), skin conditions (royal jelly), and wounds and burns (honey).

aplasia *(ă-plā'-zē-ă).* Failure of an organ or tissue to develop normally.

aplastic anemia. *See* **anemia.**

apnea *(ăp'-nē-ă).* Temporary stoppage of breathing that occurs in some people during sleep and in some newborn babies, usually caused by excess accumulation of oxygen or a deficiency of carbon dioxide in the brain.

apocrine glands *(ăp'-ō-krĭn).* Sweat glands, located in the axillary (armpit), genital, anal, and mammary regions, which secrete sweat with a characteristic odor.

apoenzyme *(ăp-ō-ěn'-zīm).* The protein component of an enzyme.

apomorphine *(ăp-ō-mor'-fēn).* Morphine derivative; the hydrochloride form is sometimes used to induce vomiting.

aponeurosis *(ăp-ō-nū-rō'-sĭs).* Type of connective tissue that attaches muscle to bone or other tissues.

apoplexy *(ăp'-ō-plěk-sē).* *See* **cerebrovascular accident.**

apoptosis *(ă-pŏp-tō'-sĭs).* Natural, genetically controlled death of cells.

apparatus *(ăp-ă-ră'-tēs).* 1. Devices designed to help patients mechanically, such as a brace or cane. 2. In anatomy, a group of parts that work together to perform a given function (e.g., parts of the ear work together for hearing).

appendage *(ă-pěn'-dĭj).* Anatomically, something that is attached, such as the nails, limb, external ear, and so on.

appendectomy *(ă-pěn-děk'-tē-mē).*

Surgical removal of the appendix.

appendicitis *(ă-pĕn-dĭ-sī'-tĭs)*. Inflammation of the appendix.

appendix *(ă-pĕn'-dĭks)*. Three- to four-inch fingerlike appendage attached to the first part of the large intestine (cecum) in the lower right abdomen.

apperception *(ăp-ĕr-sĕp'-shŭn)*. Conscious perception and interpretation of stimuli you take in through all your senses.

appestat *(ăp'-ĕ-stăt)*. Area of the brain that controls appetite.

appetite *(ăp'-ĕ-tīt)*. Normal desire, especially for food, but for other needs as well, such as sex.

approach *(ă-prōch')*. Surgical term that refers to the method used to expose an organ or tissue.

apraxia *(ă-prăk'-sē-ă)*. Loss of or an inability to make purposeful movements in the absence of paralysis or any other type of sensory or motor impairment.

akinetic a. Loss of the ability to perform spontaneous movement.

amnestic a. Loss of ability to perform movements on command because of an inability to remember the command.

sensory a. Loss of ability to properly use an object because of a lack of perception of its proper purpose; also called *ideational apraxia.*

aqueduct *(ăk'-wĭ-dŭkt)*. Pathways or channels in the body through which various body fluids are transported.

aqueous *(ā'-kwē-ĕs)*. 1. Watery. 2. Made with water.

aqueous humor. Clear, watery fluid that circulates in the anterior and posterior chambers of the eye.

arachidonic acid *(ă-răk-ĭ-dŏn'-ĭk)*. An unsaturated fatty acid essential for human health.

arachnids *(ă-răk'-nĭds)*. Members of the class of Arthropoda, including spiders, ticks, mites, and scorpions.

arachnoiditis *(ă-răk-noyd-ī'-tĭs)*. Inflammation of the arachnoid membrane.

arachnoid membrane *(ă-răk'-noyd)*. Thin membrane that is the middle of the three membranes (meninges) that enclose the brain and spinal cord.

arachnophobia *(ă-răk-nō-fō'-bē-ă)*. Morbid fear of spiders.

arbovirus *(ăr-bō-vī'-rŭs)*. Large group of viruses found in mammals and insects, such as mosquitos and ticks. Causes yellow fever and viral encephalitis.

arch. Bowlike structure or part of the body; e.g., arch of the foot, although the body actually has many other arches.

archetype *(ăr'-kĕ-tīp)*. In Jungian psychology, an inherited pattern of thought or symbolic imagery derived from past collective experience and present in the individual unconscious.

arch support. Artificial support for the arch of the foot which is usually inserted into shoes.

ARDS. Acronym for adult respiratory distress syndrome. *See*

adult respiratory distress syndrome.

areata *(ă-rē-ā'-tă).* Developing or occurring as patches (e.g., alopecia areata).

areflexia *(ă-rĕ-flĕx'-sē-ă).* Absence or loss of reflexes, a sign of possible nerve damage.

areola *(ă-rē'-ō-lă).* 1. A small cavity or space in tissue. 2. A circular area that has different pigmentation than the area beyond it; e.g., the pigmented ring around the nipple.

arginine *(ăr'-jĭ-nēn).* Amino acid produced by the body that is essential for human life.

arm. An upper limb, especially from the shoulder to the hand and including the upper arm (humerus), forearm (ulna and radius), and wrist (carpus).

arnica *(ăr'-nĭ-kă).* Homeopathic and herbal remedy (*Arnica montana*) used primarily to treat bruising and various types of pain, including dental pain, gout, arthritis, sports injuries, and insect stings.

aromatase inhibitor *(ă-rō'-mă-tās).* An antiestrogen drug that inhibits aromatase (an enzyme), which leads to a reduction in the level of estradiol.

aromatherapy *(ă-rō-mă-ther'-ă-pē).* Complementary medicine practice in which essential oils are used for healing purposes; *see also* **essential oils.**

arrectores pilorum *(ă-rĕk-tō'-rēz pĭl-ō'-rŭm).* Small muscles of hair follicles that, when they contract, cause goose bumps to appear.

arrest. To stop the progress of a disease or of a part in motion (e.g., cardiac arrest is stoppage of the heartbeat).

arrhythmia *(ă-rĭth'-mē-ă).* Abnormal heartbeat rhythm, which can be caused by drugs, disease, or other factors.

juvenile a. Type of arrhythmia that occurs in children.

nonphasic a. Type of sinus arrhythmia in which the irregularity is not associated to phases of respiration.

phasic a. Type of sinus arrhythmia associated with respiration, in which heart rate increases with inhalation and decreases with exhalation.

sinus a. Variation in heart rate related to vagal impulses to the sinoatrial node, which may or may not be associated with respiration. It is common in children and is not abnormal.

arsenic *(ăr'-sĕ-nĭk).* Metallic, highly poisonous element used in the manufacture of medicines, pesticides, and dyes.

arsenic poisoning. Sometimes fatal type of poisoning caused by ingesting or inhaling arsenic. Small amounts absorbed over a long time may cause headache, nausea, and change in skin color; large amounts can cause burning pain in the gastrointestinal tract, swelling, renal failure, shock, and death.

arterial bleeding *(ăr-tē'-rē-ăl).* Bleeding from an artery, in which the blood is bright red and usually comes in spurts.

arterial blood gases. Laboratory

test for levels of pH and the oxygen, and carbon dioxide concentrations in arterial blood.

arterialization *(ăr-tē-rē-ăl-ĭ-zā'-shŭn).* Surgical procedure in which a vein is altered so it will function as an artery.

arterial pressure. Stress exerted by circulating blood on the arteries; *see also* **blood pressure.**

arteriectasis *(ăr-tĕ-rē-ĕk'-tă-sĭs).* Abnormal distension of an artery.

arteriectomy *(ăr-tĕ-rē-ĕk'-tō-mē).* Surgical removal of part of an artery.

arteriography *(ăr-tē-rē-ŏg'-ră-fē).* X-ray of arteries after injection with a contrast substance.

arteriole *(ăr-tē'-rē-ōl).* Smallest branch of an artery that leads to a capillary.

arterioplasty *(ăr-tē-rē-ō-plăs'-tē).* Surgical procedure to repair or reconstruct an artery.

arteriosclerosis *(ăr-tēr-ē-ō-sklĕ-rō'-sĭs).* Disorder of the arteries characterized by hardening of the walls of the arteries, resulting in reduced blood flow.

arteriospasm *(ăr-tē'-ē-ō-spăzm).* Spasm of an artery.

arteriovenous *(ăr-tē-rē-ō-vē'-nŭs).* Referring to both the arteries and the veins.

arteriovenous fistula. An abnormal passage between an artery and vein, which disrupts normal blood flow.

arterioventricular block *(ăr-tē-rē-ō-vĕn-trĭk'-ū-lăr).* Partial or complete interference of the electrical signal that goes from the upper chamber of the heart to the lower chamber. *See also* **heart block.**

arteritis *(ăr-tĕ-rī'-tĭs).* Inflammation of an artery or arteries.

artery *(ăr'-tĕr-ē).* One of the vessels that carries blood from the heart to the tissues throughout the body. Most are named for the body part they cross or reach (e.g., femoral artery runs along the femur).

arthralgia *(ăr-thrăl'-jē-ă).* Pain in a joint.

arthritis *(ăr-thrī'-tĭs).* Inflammation of a joint.

acute a. Joint inflammation that occurs quickly and is accompanied by pain, swelling, and redness.

allergic a. Joint inflammation that occurs from a substance to which an individual is allergic.

chronic a. Joint inflammation that persists for months or years.

degenerative a. Type of arthritis characterized by a loss of cartilage, joint deformity, and stiffness. *See also* **osteoarthritis.**

gouty a. Joint inflammation caused by abnormal uric acid metabolism. It usually affects one joint at a time.

juvenile rheumatoid a. Form of rheumatoid arthritis that affects people younger than age 16.

menopausal a. Joint pain that occurs in women during menopause.

rheumatoid a. Common type of arthritis in which several

joints are usually involved at the same time and characterized by pain and limited motion.

arthrocentesis *(ăr-thrō-sĕn-tē'-sĭs).* Injection of a needle into a joint to remove fluid from it.

arthrodesis *(ăr-thrō-dē'-sĭs).* Surgical procedure that fuses the bones that make up a joint.

arthrography *(ăr-thrŏg'-ră-fē).* X-ray of a joint.

arthroplasty *(ăr'-thrō-plăs-tē).* Surgical procedure to reshape or replace a diseased joint.

arthroscopy *(ăr-thrŏs'-kō-pē).* Procedure whereby an orthopedist uses an endoscope, a specially designed lighted instrument, to view the interior of a joint.

arthrosis *(ăr-thrō'-sĭs).* Degeneration of a joint, which leads to partial or complete loss of use.

articulation *(ăr-tĭk-ū-lā'-shŭn).* 1. Any point of junction between two different parts or objects. 2. Enunciation of words.

artifact *(ăr'-tĭ-făkt).* Synthetic object found in the body, especially something seen on an x-ray or a microscopic slide, that does not normally belong there.

artificial blood. Fluid that carries a large amount of oxygen and is used as a temporary substitute for blood.

artificial heart. Device designed to replace the heart and to pump blood throughout the body.

artificial insemination. Process whereby sperm from a donor (either the woman's partner or an anonymous donor) is placed by syringe into the birth canal to increase the likelihood of conception.

artificial joints. Joints, made of metal, used to replace certain joints in the body, e.g., the hips, elbows, and knees.

artificial respiration. Mechanical or manual way to maintain a flow of air through the pulmonary system in a person who has stopped breathing.

artificial skin. Synthetic covering used to treat burn victims.

asana *(ă-să'-nă).* A physical posture adopted in yoga. There are hundreds of asanas one can learn in the various types of yoga.

asbestos *(ăs-bĕs'-tĕs).* Fiberlike, fire-resistant mineral used as an insulator and roofing material, which has been associated with lung disease and cancer.

asbestosis *(ăs-bĕs-tō'-sĭs).* Chronic, progressive lung disease that results from breathing in asbestos. Symptoms include shortness of breath and cough, with lung cancer as a frequent result.

ascariasis *(ăs-kă-rī'-ă-sĭs).* Infection with *Ascaris lumbricoides,* a roundworm, found in the small intestine and which causes diarrhea and pain.

Ascaris. Genus of worms, including roundworms and nematodes, that can inhabit the intestines of humans.

Aschoff's bodies. Certain cells seen in the hearts of people who have heart disease caused by rheumatic fever.

ascites *(ă-sī'-tēz).* Abnormal accumulation of protein and fluid in the abdominal cavity, often a complication of cirrhosis.

ascorbic acid *(ăs-kor'-bĭk).* Vitamin C, a water-soluble vitamin essential for development of bone, skin, and connective tissue, fighting infections, and prevention of scurvy.

asemia *(ă-sē'-mē-ă).* Loss of the ability to understand speech or writing.

asepsis *(ā-sĕp'-sĭs).* Sterile; having no infection.

aseptic necrosis *(ā-sĕp'-tĭk).* Death of tissue caused by a lack of blood supply.

asexual *(ā-sĕk'-shū-ăl).* Lack of sexual involvement; lack of distinction between female and male.

ashwaganda *(ăsh-wă-găhn'-dă).* Ayurvedic herb (*Withania somnifera*) used as an aphrodisiac and to treat sexual problems, as well as rheumatism, indigestion, and heart disease.

Asiatic flu. A respiratory viral infection, first isolated in 1957, that causes runny nose, body aches and pains, fever, sore throat, and cough.

asocial *(ā-sō'-shŭl).* Not interested in people or the activities in one's life.

asparaginase *(ăs-păr'-ă-jĭn-āz).* An anticancer agent, derived from the bacterium *Escherichia coli,* used to treat certain types of leukemia.

asparagine *(ăs-spăr'-ă-jĭn).* A nonessential amino acid.

aspartame *(ăs'-pĕr-tām).* Artificial sweetener synthesized from the amino acids aspartic acid and phenylalanine. It is about 200 times sweeter than sugar. Trade names: Equal, Nutrasweet.

Asperger's syndrome *(ăs'-pĕr-gĕrz).* Neurobiological disorder, usually appearing in childhood, characterized by repetitive behaviors, limited interests, problems with motor skills, poor communication skills, and impaired social interaction.

aspergillosis *(ăs-pĕr-jĕ-lō'-sĭs).* An uncommon, serious infection caused by the fungus *Aspergillus,* most often seen in individuals who are weakened by disease or who have a compromised immune system.

Aspergillus *(ăs-pĕr-jĭl'-ŭs).* Genus of *Ascomycetes* that includes several species of molds and fungi.

asphyxia *(ăs-fĭk'-sē-ă).* Condition in which too little or no oxygen reaches the tissues; suffocation.

aspiration *(ăs-pĭ-rā'-shŭn).* 1. The action of breathing in, especially inhaling a foreign or unwanted substance. 2. Use of suction to take liquids or gases from an area, such as the lungs.

aspirin *(ăs'-pĕr-ĭn).* Acetylsalicylic acid, a derivative of salicylic acid, that is commonly used to treat pain, fever, and inflammation. It is also used to help prevent blood clots, stroke, and cataracts.

aspirin poisoning. Accidental

overdosage of aspirin, usually seen in children. Acute cases may be characterized by drowsiness, sweating, dehydration, rapid deep breathing, convulsions, and coma; chronic poisoning may cause rash, ringing of the ears, weight loss, and bleeding tendency.

assay *(ăs'-ā).* Test used to determine a drug's potency or the amount of the constituents in a mixture.

assertiveness training *(ă-sĕr'-tĭv-nĕs).* Technique used in psychotherapy in which people are taught how to express positive and negative feelings in a direct manner.

assimilation *(ă-sĭm-ĭ-lā'-shŭn).* Part of the digestive process where the products of food breakdown are absorbed into the body. This process is also known as anabolic metabolism.

association *(ă-sō-sē-ā'-shŭn).* Mental connection of two or more things or events.

astereognosis *(ă-stĕr-ē-ŏg-nō'-sĭs).* An inability to recognize objects by feeling them.

asthenia *(ăs-thē'-nē-ă).* Lack or loss of physical strength.

asthma *(ăz'-mă).* Respiratory disorder characterized by episodes of wheezing, cough, production of thick mucus, and difficulty exhaling.

astigmatism *(ă-stĭg'-mă-tĭz-ĕm).* Defect in vision in which light rays that enter the eye cannot focus properly on the retina because of an abnormal curvature of the lens or cornea.

Aston-Patterning *(ă'-stŏn păt'-ĕrn-ĭng).* A system of movement education, fitness training, bodywork, and ergonomics used to enhance rehabilitation, improve athletic performance, and provide preventive therapy.

astragalus *(ă-străg'-ŭ-lŭs).* Herb (*Astragalus membranceas*) commonly used in Chinese medicine; valued as a tonic, diuretic, antibacterial, and immune system booster.

astringent *(ă-strĭn'-jĕnt).* Agent that causes tissues to constrict or contract.

astroblastoma *(ăs-trō-blăs-tō'-mă).* A type of astrocytoma—a malignant brain tumor—that is of moderate malignancy.

astrocytes *(ăs'-trō-cīts).* Star-shaped neuroglial cells.

astrocytoma *(ăs-trō-sī-tō'-mă).* Type of brain tumor that is composed of astrocytes.

asymmetry *(ă-sĭm'-ĭ-trē).* Lacking equal proportions or size with a mirror-image corresponding part.

asymptomatic *(ā-sĭmp-tō-măt'-ĭk).* Without symptoms, usually refers to someone who previously had symptoms.

asynclitism *(ă-sĭn'-klĭ-tĭz-m).* During childbirth, presentation of the infant's head at an abnormal angle.

asynergy *(ă-sĭn'-ĕr-jē).* Lack of coordination of body parts or organs that usually work together in harmony.

asystole *(ă-sĭs'-tō-lē).* Absence of a heartbeat or contractions of the heart; cardiac arrest.

Atabrine *(ăt'-ă-brĭn).* Trade name for quinacrine hydrochloride, an antimalarial drug.

Atarax *(ăt'-ă-răx).* Trade name for hydroxyzine hydrochloride, a minor tranquilizer.

ataraxia *(ăt-ă-răk'-sĭ-ă).* State of mental calm, serenity.

atavism *(ăt'-ă-vĭzm).* Appearance of a characteristic or disease known to have occurred in a previous ancestor rather than a parent.

ataxia *(ă-tăk'-sē-ă).* Lack of muscle coordination, especially when trying to make voluntary muscular movements.

alcoholic a. Ataxia due to chronic alcoholism.

autonomic a. Lack of coordination between the sympathetic and parasympathetic nervous systems.

Friedreich's a. An inherited degenerative disease that affects the spinal cord and is characterized by ataxia, speech impairment, and paralysis.

motor a. Inability to perform coordinated muscle movements.

sensory a. Ataxia due to interference with sensory signals, especially those coming from the muscles.

telangiectasia a. A hereditary progressive disease in which the ataxia is due to disease of the cerebellum (area of the brain that controls voluntary muscle activity).

atelectasis *(ăt-ē-lĕk'-tă-sĭs).* 1. Incomplete expansion of all or part of the lung in a newborn. 2. Collapse of lung tissue in an adult.

atenolol *(ă-tĕn'-ō-lŏl).* See Appendix, Common Prescription and OTC Drugs: By Generic Name.

atheroma *(ăth-ĕr-ō'-mă).* Fatty deposit in the walls of arteries that occurs during progression of hardening of the arteries (arteriosclerosis).

atherosclerosis *(ăth-ĕr-ō-sklĕ-rō'-sĭs).* Form of arteriosclerosis in which plaques, consisting mainly of cholesterol and lipids, form on the inner lining of the arteries; also called *coronary artery disease.*

athetosis *(ăth-ĕ-tō'-sĭs).* Condition in which the upper extremities, especially the hands and fingers, perform involuntary movements, typically slow and twisting. These symptoms can be caused by various diseases, such as chorea and encephalitis.

athlete's foot. Fungal infection of the foot, caused by various dermatophytes (fungal parasites); also known as *tinea pedis.*

Ativan. Trade name for lorazepam, an antianxiety drug.

Atkins diet *(ăt'-kĭns).* A high-protein, high-fat, low-carbohydrate diet based on the theory that carbohydrates stimulate the production of insulin, which then leads to hunger and weight gain. Other theories say people on the

Atkins diet have a reduced appetite and that their bodies use stored fat for energy rather than burning glucose from ingested carbohydrates.

atlas *(ăt'-lăs).* First vertebra in the neck, which hinges with the skull.

atom *(ăt'-ŏm).* Smallest part of an element, consisting of a nucleus (with neutrons and protons) and surrounded by electrons.

atonic *(ă-tŏ'-nĭk).* Without normal tone or tension, especially in muscles.

atopic *(ă-tŏp'-ĭk).* 1. Pertaining to an allergy. 2. Displaced.

atopy *(ăt'-ō-pē).* An allergy for which one has a genetic predisposition. Only the tendency to develop a certain allergy is inherited, not the allergy itself.

atorvastatin. *See* Appendix, Common Prescription and OTC Drugs: By Generic Name.

atresia *(ă-trē'-zē-ă).* Absence or closure of a normal body opening, such as can occur in the bile ducts, intestines, esophagus, and urethra, for example.

atrial fibrillation *(ā'-trē-ăl fĭ-brĭl-ā'-shŭn). See* **fibrillation.**

atrial flutter. *See* **flutter.**

atrial septal defect. Common congenital heart defect characterized by an abnormal opening in the septum, which allows blood to pass from the left to the right atrium.

atrioseptoplasty *(ā-trē-ō-sĕp'-tō-plăs-tē).* Surgical procedure to repair a defect in the atrial septum of the heart.

atrioventricular *(ā-trē-ō-vĕn-trĭk'-ū-lăr).* Referring to both the atrium and ventricle in the heart.

atrioventricular valve. Either of two valves in the heart through which blood flows from the atria to the ventricles. The left valve is called the mitral valve; the right, the tricuspid valve.

atrium *(ā'-trē-ŭm).* 1. One of two upper chambers of the heart (*pl.* atria). 2. Main section of the middle ear.

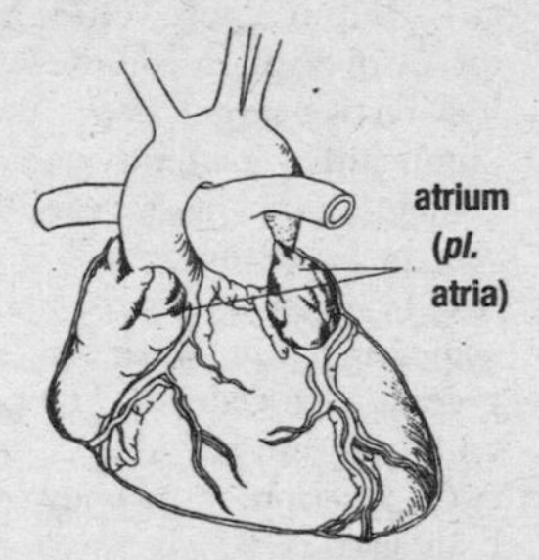

atrophy *(ăt'-rĕ-fē).* Decrease in the size of an organ or other body part, the result of wasting away of tissue due to disease or lack of use.

atropine *(ăt'-rō-pēn).* Antispasmodic drug usually derived from belladonna, it has many uses; for example, treatment of certain heart rhythm abnormalities, dilation of the pupil of the eye, relaxing bowel spasms, treatment of asthma.

Atrovent. Trade name for the inhaled bronchodilator ipratropium bromide, used to treat asthma.

attending physician. 1. The chief of a medical, surgical, or specialty service at a hospital. 2. Any doctor who attends a patient.

attention deficit disorder. *See* **attention deficit hyperactivity disorder.**

attention deficit hyperactivity disorder (AD/HD). Syndrome characterized by inappropriate inattention, hyperactivity, and impulsive behavior, seen in about 5 percent of children (more common in boys than in girls) and in adults. There are three subtypes: predominantly hyperactive-impulsive, predominantly inattentive, and combined, the most prevalent subtype being the latter.

attenuate *(ă-tĕn'-ū-āt).* To make something thin, weak, or less potent, as in reducing the potency of bacteria.

attitude. Position of the body or a body part.

atypical *(ā-tĭp'-ĭ-kăl).* Deviating from the norm; not typical.

audiogram *(ăw'-dē-ō-grăm).* Record of an individual's hearing sensitivity.

audiometer *(ăw'-dē-ō-mē-tĕr).* Electronic instrument that measures an individual's hearing.

auditory *(ăw'-dĭ-tŏr-ē).* Referring to the sense of hearing.

auditory canal. Tubelike structure that runs from the outside of the ear to the tympanic membrane.

auditory nerve. One of a pair of sensory nerves that carries signals from the inner ear to the brain, where they are interpreted as hearing, balance, and sense of position.

Augmentum. Trade name for an antibiotic consisting of penicillin and other ingredients.

aura *(ăw'-ră).* Auditory, olfactory, and/or visual sensations that precede an epileptic seizure and sometimes a migraine.

auricle *(ăw'-rĭ-kl).* Visible portion of the ear.

auscultation *(ăws-kŭl-tā'-shŭn).* Use of a stethoscope to detect certain sounds, such as heartbeat and breath sounds.

autism *(ăw'-tĭz-m).* Pervasive developmental disorder characterized by impaired self-absorption, lack of awareness of the existence of feelings of others, impaired verbal and nonverbal communication, and often highly repetitive behaviors. It usually first appears before three years of age.

autoagglutinin *(ăw-tō-ă-gloo'-tĭ-nĭn).* Substance that causes an individual's cellular elements (e.g., platelets, red blood cells) to agglutinate (clump together).

autoantibody *(ăw-tō-ăn'-tē-bŏd-ē).* Antibody that is produced in and reacts with an antigen.

autoantigen *(ăw-tō-ăn'-tĭ-jĕn).* Antigen that stimulates the production of autoantibodies.

autoclave *(ăw'-tō-klăv).* Container that uses pressured steam to sterilize surgical instruments.

autodigestion *(ăw-tō-dī-jĕs'-chŭn).* Erosion of part of the stomach wall or duodenum by gastric juices.

autoerotism *(ăw-tō-ĕ-rŏt'-ĭzm).* Sexual gratification using one's own body; e.g., masturbation.

autogenic training *(ăw-tō-jĕn'-ĭk).* A form of self-hypnosis in which individuals get themselves into a relaxed state of mind and then complete a series of suggestions and exercises designed to instruct them how to deal with their responses to stress.

autogenous *(ăw-tŏj'-ĕ-mŭs).* Self-producing; originating in the body.

autograft *(ăw'-ō-grăft).* Living tissue transplanted from one site to another in the body of the same individual.

autohemolysin *(ăw-tō-hē-mŏl'-ĭ-sĭn).* Antibody that acts upon the red blood cells of an individual in whose blood it is formed.

autohemolysis *(ăw-tō-hē-mŏl'-ĭ-sĭs).* Destruction of a person's blood cells due to hemolytic substances in the blood.

autoimmune diseases *(ăw-tō-ĭ-mūn').* Group of diseases in which an individual produces antibodies that attack the body's own tissues.

autoimmunity *(ăw-tō-ĭ-mūn'-ĭ-tē).* Condition in which the body produces antibodies against its own tissues.

autointoxication *(ăw-tō-ĭn-tŏk-sĭ-kā'-shŭn).* Condition that results when the body absorbs waste products or toxins it has produced.

autologous *(ăw-tŏl'-ŏ-gŭs).* Pertaining to something that has its origin within an individual; e.g., autologous graft.

autolysis *(ăw-tŏl'-ĭ-sĭs).* Self-digestion of tissue; *see also* **autodigestion.**

autonomic *(ăw-tō-nŏm'-ĭk).* 1. Self-controlling; to function independently. 2. Relating to the autonomic nervous system.

autonomic nervous system. Part of the nervous system concerned with control of involuntary bodily functions (e.g., sweat glands, heart, gastric secretions).

autopsy *(ăw'-tŏp-sē).* Examination of a dead body, usually to determine the cause of death.

autosomal dominant *(ăw-tō-sō'-măl).* Inherited disorder caused by a dominant abnormal gene on an autosome (*see* **autosome**).

autosomal recessive disorder. Inherited condition caused by a recessive abnormal gene on an autosome (*see* **autosome**).

autosome *(ăw'-tō-sōm).* Any one of the 22 pairs of nonsex chromosomes.

autotransfusion *(ăw-tō-trăns-fū'-zhŭn).* Transfusion of an individual's own blood.

auxiliary *(ăwg-zĭl'-ē-ŭr-ē).* Providing additional aid.

avascular necrosis *(ā-văs'-kū-lăr).* Death of tissue due to insufficient or lack of blood supply.

AV shunt. Artificial connection

made between an artery (A) and a vein (V), often done in the forearm to facilitate dialysis.

axilla *(ăk-sĭl'ă).* Armpit.

axillary dissection. Removal of the lymph nodes in the armpit, sometimes done in cases of breast cancer.

axillofemoral bypass *(ăk-sĭl-ō-fĕm'-ŏr-ăl).* Insertion of a graft to extend the axillary artery in the armpit to a portion of the femoral artery in the thigh, usually done because of arteriosclerosis.

axis *(ăk'-sĭs).* Imaginary line that passes through the center of the body or a body part.

axon *(ăk'-sŏn).* Extension of a nerve cell.

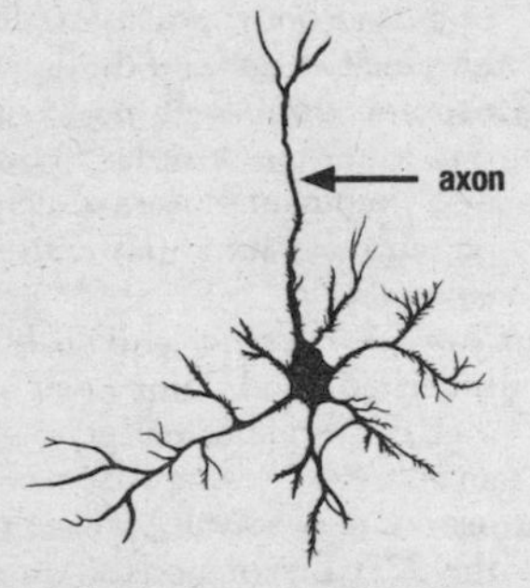

azidothymidine (AZT) *(ăz-ĭ-dō-thī'-mĭ-dēn).* Antiviral medication given to AIDS patients to help prolong their lives.

azoospermia *(ă-zū-ō-spĕr'-mē-ă).* Absence of spermatozoa in the semen.

azotemia *(ăz-ō-tē'-mē-ă).* Excess of urea or other nitrogenous substances in the blood; also called *uremia.*

B

Babinski sign *(bă-bĭn'-skē).* Response of the large toe when the bottom of the foot is stimulated. In this response, the big toe extends instead of turning down when the bottom of the foot is scratched with a pin.

Bach flower remedy *(bŏk).* Alternative/complementary medicine approach in which the essence (the inherent energy) of certain flowers is captured in water and used to help restore emotional balance and physical well-being. This concept is similar to that of *homeopathy.*

bacillary dysentery *(băs'-ĭl-ăr-ē dĭs'-ĭn-tăr-ē).* A diarrheal disease caused by food or water infected with a specific rod-shaped bacillus.

bacilluria *(băs-ĭ-lō'-rē-ă).* Common form of urinary infection in which bacilli are present in the urine.

Bacillus *(bă-sĭl'-ŭs).* Genus of bacteria in which all species are shaped like short rods and are aerobic.

bacillus (*pl.* bacilli). Any rod-shaped microorganism.

bacitracin *(băs-ĭ-trā'-sĭn).* An antibiotic that is effective against certain staphylococci, streptococci, and similar organisms.

backbone *(băk'-bōn).* The spinal column; the vertebral column.

bacteremia *(băc-tĕr-ē'-mē-ă).* Presence of bacteria in the blood; also called *blood poisoning.*

bacterial endocarditis *(băk-tē'-rē-ăl ĕn-dō-kăr-dī'-tĭs). See* **endocarditis.**

bacterial resistance. Development of resistance to a drug, such as an antibiotic, by an organism previously susceptible to it.

bactericidal *(băk-tĕr-ĭ-sī'-dăl).* Capable of destroying bacteria, such as an antiseptic.

bacteriogenic *(băk-tē-rē-ō-jĕn'-ĭk).* Caused by bacteria.

bacteriologist *(băk-tĕr-ē-ŏl'-ō-jĭst).* Individual who devotes him/herself to the study of bacteria.

bacteriology *(băk-tĕr-ē-ŏl'-ō-jē).* The study of bacteria.

bacteriophage *(băk-tē'-rē-ō-fāj).* A virus that infects and kills bacteria.

bacteriostasis *(băk-tē-rē-ōs'-tă-sĭs).* A condition in which the growth and reproduction of bacteria is stopped, either temporarily or permanently. Many antibiotics are capable of producing bacteriostasis.

bacteriuria *(băk-tē-rē-ū'-rē-ă).* Presence of bacteria in the urine.

Bacteroides fragilis *(băk-tĕr-oyd'-ēz fră-jĭ'-lĭs).* Common gram-negative anaerobic (does not require oxygen to survive) bacteria capable of causing many different infections.

Bactrim *(băk'-trĭm).* A drug often successful in treating bronchial, ear, and urinary tract infections.

bad breath. Offensive odor of the breath, also known as *halitosis.*

bagassosis *(băg-ă-sō'-sĭs).* A lung disease that occurs among individuals who work with sugar cane, which emits a moldy, fibrous waste called bagasse dust when harvested.

bag of waters. The liquid that surrounds the fetus in the womb.

Baker's cyst. A sac, accompanied by inflammation, that forms on the back of the knee. Also called *popliteal bursitis.*

balance *(băl'-ăns).* A state of equilibrium, in which the intake and output of substances such as water and nutrients are approximately equal. *See also* **balance, sense of; equilibrium.**

acid-base b. Equilibrium between acid and base production and excretion in the body.

electrolyte b. Condition in which electrolytes, such as potassium and sodium, maintain concentrations needed for normal metabolic and cellular processes.

fluid b. State of equilibrium between the amount of fluids you take in and excrete, especially water.

balanced anesthesia. A method in which an individual is given several different anesthetics, such as muscle-relaxing drugs, intravenous sedatives, and anesthetic gas such as nitrous oxide.

balance, sense of. Physical equilibrium that is maintained through the interaction of different parts of the nervous system: the inner ears, eyes, skin pressure receptors, muscle and joint sensory receptors, and the central nervous system (brain and spinal cord), the latter of which processes information from the other four areas.

balanitis *(băl-ă-nī'-tĭs).* Inflammation of the tip of the penis.

balanus *(băl-'ă-nŭs).* The head (glans) of the penis.

balloon. An inflatable device used to stretch or deflate a constricted artery, as in *balloon angioplasty.*

balloon angioplasty. Coronary procedure in which a balloon-tipped catheter is inserted through an artery in the groin or arm to enlarge a narrowing coronary artery.

balloon tamponade. A procedure in which a balloon is inflated within the esophagus or stomach to apply pressure and stop bleeding.

ballottement *(băl-ŏt-mŏn').* A French term, it describes a technique in which a physician detects or examines a floating object in the body, such as an organ, tumor, or fetus, by pushing against it with his/her finger or hand.

balneotherapy *(băl-nē-ō-thĕr'-ă-pē).* The use of baths, like those offered by spas, to treat disease.

band. A cord or tapelike fibrous tissue that connects or holds structures together.

bandage. A piece of material (e.g., gauze, linen, flannel) used on the body to apply pressure, immobilize a body part, provide support to an injured area, or to hold dressings in place.

Ace b. Trade name for a woven elastic bandage that allows joints to move without loosening the bandage.

adhesive b. A bandage that has a sticky substance on one side.

butterfly b. Adhesive bandage used instead of sutures to hold wound edges together.

compression b. Material applied to help control bleeding.

many-tailed b. Bandage with split ends used to hold a body part tightly; often used on the abdomen following surgery.

triangular b. A sling bandage.

Velpeau's b. Bandage used to hold the upper arm and forearm against the body, often used in shoulder injuries.

Banthine. Trade name for methantheline bromide, a drug that reduces the acid secreted by the stomach and thus treats ulcers and excess acidity.

Banti's disease (or syndrome) *(băn'-tēz).* A syndrome characterized by anemia, an enlarged spleen, hemorrhages, and eventually cirrhosis.

barber's itch. An infection of the hair follicles of the face and neck caused by staphylococci.

barbiturates *(băr-bĭt'-ū-răts).* A group of drugs that suppresses the central nervous system, they are effective in inducing sleep and calming nerves.

bariatrician *(băr-ē-ă-trĭ'-shĭn).* A physician that practices bariatrics.

bariatrics *(băr-ē-ă'-trĭks).* Field of

medicine concerned with weight loss.

barium *(bă'-rē-ŭm).* An opaque substance, administered before undergoing certain x-rays, which allows health-care professionals to better visualize the intestinal tract.

b. enema. The giving of a barium mixture through the rectum before x-raying the large intestine.

b. meal. Swallowing a barium mixture before x-raying the intestinal tract.

b. swallow. X-ray of the esophagus during and after swallowing a barium mixture to view varices in the esophagus.

barosinusitis *(băr-ō-sī-nū-sī'-tĭs).* Inflammation and pain of the sinuses caused by differences in atmospheric pressure inside the sinuses and the air outside.

barotitis *(băr-ō-tī'-tĭs).* Inflammation and pain of the ear caused by a sudden change in atmospheric pressure, as when ascending or descending in a plane.

Barr bodies. Chromatin bodies seen in the nuclei of certain cells, indicating that the individual is a female.

barren. Incapable of bearing children; sterile, infertile.

Bartholin glands. Two small glands, one on each side of the entrance to the vagina, that often become infected, especially in women who have gonorrhea.

Bartholin's cyst *(băr'-tō-lĭnz).* A cyst that develops when the duct of a Bartholin gland becomes obstructed.

basal *(bā'-săl).* Referring to the base; of primary importance.

basal cell carcinoma. A common type of skin cancer that often appears as a one-half-inch or less lesion on the side of the nose or between the eyes. It can be scraped and burned or surgically removed.

basal cells. Small, round cells found in the lower part (base) of the outer layer of skin.

basal metabolic rate (BMR). The amount of energy a person uses when at rest in a comfortable environment.

base. 1. The main substance in a mixture. 2. The opposite of acid. 3. The supporting part of a structure.

baseline. A known or initial value with which subsequent values for different measurements can be compared. For example, a baseline blood pressure compared with a level taken at a later date.

basilar *(băs'-ĭ-lăr).* Referring to the base of a structure, especially the base of the skull.

basilica vein *(bă-sĭl'-ĭk).* The large vein on the inner side of the arm just above the elbow, often selected for blood withdrawal or intravenous injection.

basophil *(bā'-sō-fĭl).* A type of white blood cell that accepts a base stain, used during examination of cells and tissue.

basophilia *(bă-sō-fĭl'-ē-ă).* A condition characterized by a high number of basophils in the blood, seen in some cases of malaria, lead poisoning, leukemia, and other blood conditions.

Bassini's operation *(bă-sē'-nēz).* A type of surgery used to repair an inguinal hernia.

bath. A means and method of cleansing the body or any part of it, or treating it therapeutically with water, air, light, or vapor.

bed b. Bath for a person who is confined to bed.

sitz b. Bath in which only the hips and buttocks are soaked in warm (sometimes cold) water or saline; for patients who have had surgery in the rectal area or to ease pain in rectal/pelvic areas.

battered spouse syndrome. Physical and/or psychological injury caused by a husband or wife.

B cells. Types of white blood cells produced in the bone marrow and found in blood, lymph, and connective tissue.

b.d. Abbreviation meaning "twice daily"; used in writing prescriptions; *see also* **b.i.d.**

bearing down. The effort by a pregnant woman to expel her fetus.

bedsore. Ulceration of the skin, often on the lower back and other bony areas, caused by prolonged pressure in bedridden individuals.

bedwetting. Involuntary urination during sleep; also called *enuresis.*

bee pollen. Pollen plus other unidentified ingredients added by the bees, reported to have healing powers. Highly nutritious, it contains 22 amino acids, 27 minerals, and all the vitamins.

behaviorism *(bē-hāv'-yŏr-ĭz-m).* A psychological theory based on observation of individual behavior and behavior patterns.

Behcet syndrome *(bĕ'-chĕts sĭn'-drŏm).* Condition characterized by recurrent ulceration of the mouth and genitals, inflammation of the eyes, pus-forming skin lesions, joint pain, and central nervous system involvement.

belladonna *(bĕl-ă-dŏ'-nă).* A highly poisonous plant (*Atropa belladonna*), also known as deadly nightshade, used in various forms to relax spasms and stop excess secretions of glands; also widely used as a homeopathic remedy.

Bell's palsy. Paralysis of one side of the face due to swelling of a facial nerve.

Benadryl *(bĕn'-ă-drĭl).* Trade name for diphenhydramine hydrochloride, an antihistamine.

Bence Jones protein test. Urine test in which positive results indicate the possible presence of myeloma (a type of bone tumor).

Bendectin *(bĕn-dĕk'-tĭn).* Trade name for drug used to treat nausea during pregnancy.

bends. Condition characterized by pain in the limbs and abdomen caused by bubbles of nitrogen in the blood and tissues, which result when an individual experiences a rapid reduction in air pressure, as can occur when scuba diving or descending rapidly in an unpressurized aircraft.

Benedict test. Test for presence of sugar in urine.

benign *(bē-nīn').* Not cancerous; not malignant.

benign intracranial hypertension. Increased pressure within the brain in the absence of a tumor; most common in women ages 20 to 50.

benign paroxysmal positional vertigo (BPPV). Balance disorder characterized by sudden dizziness, spinning, or vertigo when moving the head.

benign prostatic hyperplasia. A noncancerous prostate condition in which the normal elements of the gland increase in number and size, resulting in impaired urine flow and related complications.

benzodiazepines *(bĕn-zō-dī-ăz'-ĕ-pēnz).* Class of drugs that act as tranquilizers; they are often used to treat anxiety and to induce sleep.

bergamot oil *(bĕr'-gĕ-mŏt).* An aromatic extract of bergamot orange rind. When ingested in very large quantities, it can block the intestinal absorption of potassium. It is used to flavor Earl Grey teas.

beriberi *(bĕr'-ē-bĕr'-ē).* Disease caused by a deficiency of thiamine in the diet.

berylliosis *(bĕr-ĭl-ē-ō'-sĭs).* Chronic lung condition caused by prolonged inhalation of beryllium (a metal) dust.

beta-agonist *(bā'-tă ă'-gŏn-ĭst).* Type of medication that opens the airways be relaxing the muscles around them; administered for asthma and chronic obstructive pulmonary disease.

beta-blockers. Group of drugs given to reduce blood pressure and to slow heart rate and the force of heart contractions.

beta carotene. Vitamin that is converted to vitamin A by the body; it is the major precursor of vitamin A in humans.

beta-lactam antibiotics. Group of antibiotics (penicillin is one) especially useful against gram-negative bacteria.

beta rhythm. Low voltage brain wave present when an individual is conscious and alert.

betatron *(bā'-tă-trŏn).* Circular electron accelerator that produces x-rays or high-energy electrons, used to inhibit cancer growth.

bezoar *(bē'-zŏr).* Hard mass of hair or other undigested material found in the stomachs of individuals who have eaten large quantities of such substances.

bicarbonate *(bī-kăr'-bō-nāt).* A salt of carbonic acid, found in all human blood. Too much bicarbonate is called *alkalosis*; too little, *acidosis*.

biceps *(bī'-sĕps).* Literally means a

muscle with two heads, it usually refers to the muscle in the front of the upper arm, which is used to flex the forearm.

bicuspid *(bī-kŭs'-pĭd).* Having two cusps; often used to refer to specific teeth and to the bicuspid valve on the left side of the heart.

b.i.d. Abbreviation for *bis in die,* "twice daily," often seen on prescriptions.

bifocal glasses *(bī'-fō-kăl).* Eyeglasses that combine two lenses, one for distant and one for near vision.

bifurcation *(bī-fŭr-kā'-shŭn).* Separation into two branches or parts.

bigeminal *(bī-jĕm'-ĭ-năl).* Occurring in pairs.

bigeminal pulse. Pulse in which two beats follow each other in rapid succession, followed by a longer pause.

bilateral *(bī-lăt'-ĕr-ăl).* Referring to or affecting two sides; for example, bilateral hearing loss is loss in both ears.

bilberry *(bĭl'-băr-ē).* A blue-black berry found on a shrub (*Vaccinium myrtillus*) that grows mainly in Europe, used by some people to treat and prevent vision problems.

bile *(bīl).* Thick secretion of the liver that travels from the bile duct of the liver into the duodenum. It is also stored in the gallbladder.

bile acids. Acids formed of bile from the liver and important in digestion of fats in the intestines.

bile ducts. Tubes that carry bile from the liver to other areas, such as the gallbladder and intestines.

biliary *(bĭl'-ē-ăr-ē).* Referring to bile and the bile system.

biliary colic. Severe pain caused by gallstones, especially when they attempt to pass through the bile ducts.

bilirubin *(bĭl-ĭ-rū'-bĭn).* The yellowish or orange pigment in bile; accumulation of bilirubin leads to jaundice.

bilirubinemia *(bĭl-ĭ-rū-bĭn-ē'-mē-ă).* Presence of bilirubin in blood.

biliuria *(bĭl-ĭ-ū'-rē-ă).* Presence of bile in the urine, indicating jaundice.

Billroth's operations. Surgeries that remove part of the stomach, frequently done when cancer or an ulcer is present.

bilocular *(bī-lŏk'-ū-lăr).* Having two compartments; for example, a bilocular cyst.

bimanual *(bī-măn'-ū-ăl).* Using both hands, as when doing a physical examination.

binaural *(bĭn-ăw'-răl).* Referring to both ears.

binge and purge syndrome *(bĭnj pŭrj sĭn'-drŏm). See* **bulimia.**

bioassay *(bī-ō-ăs'-ā).* Identification of the strength of a substance by testing it on germs, tissues, or animals.

bioavailability *(bī-ō-ă-văl-ă-bĭl'-ĭ-tē).* Rate and extent to which a drug or metabolite enters the body's general circulation for use by the body.

biochemic tissue salts therapy *(bī'-ō-kĕm-ĭk).* A self-help treatment based on the theory that cells need to maintain a certain balance of natural mineral salts to stay healthy.

biochemistry *(bī-ō-kĕm'-ĭs-trē).* Study of the chemistry of living tissue.

biodegradable. *(bī-ō-dē-grād'-ă-băl).* Organic substances that disintegrate naturally in or on the ground.

bioenergetics *(bī-ō-ĕn-ĕr-jĕ'-tĭks).* Type of psychotherapy in which people free blocked energy, which manifests as muscle tension, using exercise, breathing techniques, psychotherapy, and various body therapies such as massage and polarity therapy.

biofeedback *(bī-ō-fēd'-băk).* Procedure whereby individuals get information through visual or auditory means regarding one or more features of their physical state so they can gain some voluntary control over them.

bioflavonoid *(bī-ō-flă'-vĕ-noid).* A compound that regulates the permeability of capillary walls; citrus fruits are a good source.

biokinetics *(bī-ō-kĭ-nĕt'-ĭks).* Study of movement of living things.

biopsy *(bī'-ŏp-sē).* Removal and examination of tissue from a living body for the purpose of diagnosis.

endoscopic b. Biopsy performed with an endoscope equipped with an instrument to remove samples from the lining of a hollow organ.

excisional b. Removal of an entire lesion and the immediate surrounding area.

fine needle b. Collection of fluids or tissue from solid tumors, body cavities, bone marrow, or organs through a fine needle.

punch b. Removal of a plug of tissue using a special cutting instrument.

biorhythm *(bī-ō-rĭth'-ĕm).* Cyclic occurrence of certain biologic changes in a person's life; e.g., the sleep cycle. *See also* **circadian rhythm.**

biosynthesis *(bī-ō-sĭn'-thĭ-sĭs).* Manufacture of chemical compounds by or in a living organism.

bioterrorism *(bī-ō-tĕr'-ŏr-ĭzm).* Terrorism using biological agents, such as select viruses, bacteria, and toxins. The Centers for Disease Control and Prevention (CDC) maintains a list of these agents.

biotin *(bī'-ō-tĭn).* B-complex vitamin that helps with metabolism of carbohydrates and fat and body growth.

biparous *(bĭp'-ă-rŭs).* Giving birth to twins.

biped *(bī'-pĕd).* Any animal having two feet.

bipolar disorder *(bī-pō'-lăr).* Mental disorder characterized by episodes of depression and mania.

birth canal. The passage through which the fetus passes during childbirth (uterus and vagina).

birth control. *See* **contraceptive.**

birth defect. Abnormality present at birth, it can be inherited (genetic) or acquired during pregnancy or birth (congenital).

birthing room. Room, usually in a hospital or other health facility, dedicated for childbirth.

birthmark. Blemish or discoloration present at birth; *see also* **nevus.**

bisexual *(bī-sĕk'-shū-ăl).* Individual who engages in homosexual and heterosexual activities.

bismuth *(bĭz'-mŭth).* Chemical that, when combined with subcarbonate, protects the lining of the stomach and intestines.

bisphosphonate *(bĭs-fŏs'-fō-nāt).* Class of drugs used to strengthen bone.

black cohosh. Member of the buttercup and peony family, black cohosh (*Cimicifuga racemosa*) has long been used as an herbal remedy to treat gynecological problems, such as menstrual cramps, and to assist in childbirth.

blackhead. A plugged sweat gland in the skin.

bladder *(blăd'-dĕr).* Organ that can contain or store fluid, such as the urinary bladder or gallbladder.

neurogenic b. Any dysfunction of the urinary bladder caused by a lesion of the central or peripheral nervous system.

bland diet. Diet that is designed to be nonirritating; often prescribed for people who have peptic ulcer, colitis, or other intestinal disorders.

blastocyst *(blăs'-tō-sĭst).* Embryo at the time it implants into the uterine wall.

blastomycosis *(blăs-tō-mī-kō'-sĭs).* Chronic disease caused by inhalation of the *Blastomyces dermatitidis* fungus.

bleeding time. Time it takes for bleeding to stop after a blood stick is taken, normally 1 to 3 minutes.

bleomycin *(blē-ō-mī'-sĭn).* Any of a mixture of glycopeptide antibiotics; some are used to inhibit certain tumors, especially those of the lymph glands.

blepharectomy *(blĕf-ă-rĕk'-tō-mē).* Surgical removal of all or part of an eyelid.

blepharitis *(blĕf-ă-rī'-tĭs).* Inflammation of the eyelids.

blepharospasm *(blĕf'-ă-rō-spăz-m).* Spasmodic winking or contraction of the eyelid muscles.

blindness. Loss or lack of sight.

color b. Inability to distinguish differences between some colors.

flash b. Temporary loss of vision caused by exposure to intense light.

legal b. Legal definition: maximum correction of acuity of 20/200 or less in the better eye and diameter of visual field 20 degrees or less.

night b. Impaired vision in subdued light.

snow b. Temporary blindness caused by excessive exposure to sunlight reflected from snow.

blind spot. Center of the back of the eyeball where the optic nerve enters.

blind study. One in which the researchers cannot predict the outcome.

blister *(blĭs'-tĕr).* Collection of blood or serum just below the top layer of skin.

bloated *(blō'-tĕd).* State of being swollen, puffy; an abdomen distended with gas.

blood. Fluid, composed of red and white blood cells, platelets, and plasma, that circulates throughout the body in arteries, veins, and capillaries.

b. bank. Laboratory that collects donor blood for transfusions.

b. coagulation. Process by which liquid blood is changed into a blood clot (a semisolid mass).

b. count. Test to determine the number of red and white blood cells (and sometimes platelets) in a diluted 1 cubic millimeter blood sample.

b. doping. Administration of blood transfusions to athletes to enhance their performance.

b. gases. Pressures of oxygen and carbon dioxide in blood.

b. glucose. Main sugar that the body makes from food in the diet, it provides energy to the body's cells; also called *blood sugar.*

b. groups. Four groups—O, A, B, and AB—distinguishable by certain sensitivities (antigens) of the red blood cells in each.

b. pressure. Force of blood on the walls of the arteries resulting from the heart's pumping action.

b. sugar. Blood glucose.

b. test. Any of several techniques used to determine whether certain factors of the blood (e.g., amount of glucose, blood count) are within normal limits.

b. thinner. Common name for an anticoagulant agent that prevents formation of blood clots.

b. typing. Test to identify one's blood group.

b. urea nitrogen (BUN). Amount of nitrogenous material in the blood as urea, it indicates kidney function.

blood-brain barrier. Network of blood vessels and cells that filters blood flowing to the brain and protects the central nervous system from invading substances, such as viruses or some drugs.

Blount's disease *(blŏwntz).* Acquired disease of the tibial bone seen in young children.

blue baby. Infant born with heart and blood vessel defects that limit blood flow to the lungs, cause arterial and venous blood to mix, and cause the skin to be bluish.

board certified. A physician who has taken and passed a medical specialty examination in one or more of various areas, such as gastroenterology, internal medicine, or pediatrics.

body mass index (BMI). Index for relating a person's body weight to height, determined by dividing a person's weight in kilograms by height in meters squared.

body temperature. Level of heat produced and sustained by the body's processes; 98.6° is regarded as normal.

bodywork. A general term that refers to manipulative therapies, such as massage, shiatsu, rolfing, and others, that can help improve blood circulation, release muscle tension, and provide other health benefits.

boil. *See* **furuncle.**

bolus *(bō'-lŭs)*. 1. Soft mass of food in the mouth ready to be swallowed. 2. Concentrated mass of an injected drug.

bond. Force that holds adjacent atoms in place.

bonding. 1. Attachment that occurs between infants and their parents, especially the mother. 2. In chemistry, a chemical linkage.

bone. Dense, hard specialized form of connective tissue that forms most of the skeleton.

bone marrow. Specialized soft tissue within bone; blood cells are formed in the marrow.

bone marrow transfusion. Graft of bone marrow from one person to another.

bone mass (mineral) density. Measure of the amount of bone tissue in a certain volume of bone; can be determined using quantitative computed tomography.

bone scan. Process that allows tumors in bones to be seen long before they are visible on ordinary x-rays.

booster shot. Supplementary dose of a vaccine or other immunizing agent.

borborygmus *(bŏr-bō-rĭg'-mŭs)*. Rumbling noise produced by gas moving in the intestines.

boric acid *(bŏr'-ĭk)*. White, odorless powder sometimes used as a topical antiseptic and eyewash.

boron *(bŏr'-ŏn)*. Nonmetallic element used in antiseptics (e.g., boric acid).

Botox *(bō'-tŏx)*. Brand name of a highly purified preparation of botulinum toxin A, which is produced by the bacterium *Clostridium botulinum*, and used to block transmission of nerve signals to muscles, thus paralyzing them.

botulism *(bŏch'-ū-lĭz-m)*. Poisoning caused by the toxin of *Clostridium botulinum* found in improperly preserved/prepared foods.

bougie *(boo-zhē')*. Flexible rubber or silk instrument used to probe various body openings (e.g., urethra, esophagus).

bowel *(bŏw'-ĕl)*. Common name for the intestines, especially the large intestine.

bowel sounds. Sounds produced when contractions of the lower intestines move the contents forward.

Bowen technique. A form of bodywork in which vibrational movements are used on the muscles, connective tissues, and tendons to help heal various musculoskeletal problems.

bowlegs *(bō'-lĕgz)*. Outward curving of the legs, often associated with rickets.

brace *(brās)*. Device for supporting a body part.

brachial *(brā'-kē-ăl)*. Referring to the arm.

brachial artery. Main artery of the upper arm.

brachialgia *(brā-kē-ăl'-jē-ă).* Severe pain in the arm.

brachial plexus. Group of nerves that arise from the spinal cord in the neck and supply the arm, hand, and parts of the shoulder.

brachycephaly *(brā-kē-sĕ'-fă-lē).* Congenital deformity in which the skull is abnormally broad and short.

bradycardia *(brăd-ē-kăr'-dē-ă).* Abnormally slow heartbeat, less than 60 beats per minute.

bradypnea *(brăd-ĭp-nē'-ă).* Abnormally slow breathing.

brain *(brān).* Mass of nervous tissue in the skull, it is the main part of the central nervous system.

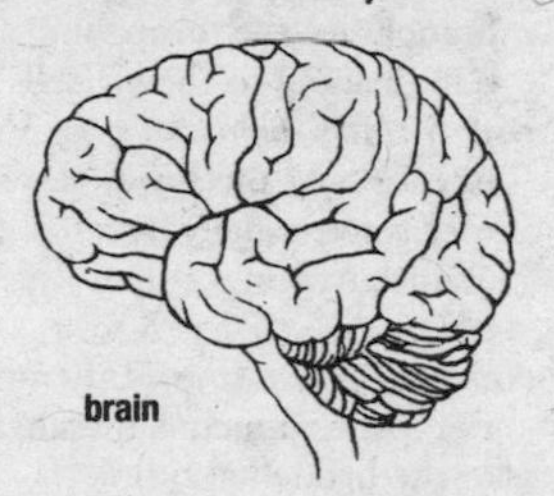

b. aneurysm. Abnormal widening of an artery or vein in the brain.

b. dead. Irreversible unconsciousness in which all brain waves have ceased.

b. scan. Painless diagnostic procedure to examine the brain and identify possible abnormalities.

b. stem. Portion of the brain that connects with the spinal cord.

branchial cyst *(brăng'-kē-ăl).* Deformity in the neck present since birth, caused by failure of the embryo's "gill slits" to close completely.

BRCA breast cancer gene. One of several genetic mutations associated with breast and ovarian cancer. By age 80, a woman with a BRCA mutation has about an 80 percent chance of developing breast cancer.

BRCA1. A gene that normally restrains cell growth in the breast but which predisposes the breast to cancer when the gene is mutated.

BRCA2. A gene that normally restrains cell growth in the breast and ovary but which may predispose the breast and ovaries to cancer. Mutations of this gene have also been linked to some early-onset prostate cancer.

breast *(brĕst).* 1. The upper anterior area of the thorax. 2. Either of two mammary glands which, in females, are capable of producing milk.

b. augmentation. Surgical procedure to enlarge the contour of the breast.

b. cancer. One of the most common malignancies among women.

b. pump. Suction device for extracting milk from the breast of a nursing woman.

b. reduction. Surgical procedure to reduce the size of the breast.

b. self-examination. Process by which a woman observes

and feels the breasts to detect any lumps or other changes.

breastbone. *See* **sternum.**

breath *(brĕth).* Air that is inhaled and exhaled during respiration.

breathing therapies. Techniques that help individuals focus on and be aware of their breathing as a way to improve overall health and energy levels, reduce stress, and overcome respiratory problems.

breech delivery *(brēch).* Birth that occurs with the infant presenting feet first.

Breema *(brē'-mă).* A bodywork technique similar to shiatsu and Thai massage, based on the concept that the body has three energy centers—mind, feelings, and physical—that must be in balance for health. The movements and postures of Breema are designed to achieve this goal.

brittle bones disease *(brĭt'-l bōnz dĭ-zēz').* *See* **osteogenesis imperfecta.**

brittle diabetes. Type of diabetes when a person's blood glucose level often and quickly fluctuates from high to low and vice versa.

Broca's area *(brō'-kăs).* Area in the brain that controls speech.

bromelain *(brŏ'-mĕ-lān).* An enzyme found in pineapple that helps in the digestion of protein and reportedly helps fight cancer and reduce inflammation. Available as a supplement.

bromocriptine *(brō-mō-krĭp'-tēn).* Substance that inhibits the production of prolactin by the pituitary gland.

bronchi *(brŏng'-kī)* (*sing.* bronchus). The main branches of the trachea that lead to the bronchioles and which convey air to and from the lungs.

bronchial *(brŏng'-kē-ăl).* Pertaining to the bronchi.

bronchial fistula. Abnormal opening or channel between a bronchial tube and the chest cavity.

bronchiectasis *(brŏng-kē-ĕk'-tĕ-sĭs).* Irreversible, abnormal widening of the bronchi, associated with chronic inflammation and lung infection.

bronchiole *(brŏn'-kē-ōl).* Small branch in the bronchial system, leading to the air cells.

bronchiolitis *(brŏn-kē-ō-lī'-tĭs).* Viral infection of the lower respiratory tract, usually occurring in children younger than 2 years old.

bronchitis *(brŏn-kī'-tĭs).* Inflammation of the mucous membrane of the bronchial tubes.

bronchoconstrictor *(brŏn-kō-kĕn-strĭk'-tĕr).* Any substance that causes the smooth muscle walls of the bronchi to narrow.

bronchodilator *(brŏn-kō-dī'-lā-tĕr).* Substance that causes the smooth muscle walls of the bronchi to relax.

bronchopneumonia *(brŏn-kō-nū-mōn'-yă).* Acute inflammation of the bronchi and alveoli of the lungs, characterized by fever, chills, cough, chest pain, and shallow breathing.

bronchoscopy *(brŏn-kŏs'-kĕ-pē).* Examination of the bronchi using a special device (bronchoscope).

bronchospasm *(brŏn'-kō-spăz-m).* Convulsion of the bronchi, causing them to narrow.

brucellosis *(brū-sĕ-lō'-sĭs).* Disease caused by infection with the bacterium (*Brucella* species) transmitted from livestock or their products.

Brudzinski's signs *(brū-jĭn'-skēz).* Specific neck and leg signs seen in meningitis.

bruit *(brūt).* Murmur or sound, especially an abnormal one heard when a stethoscope is placed over the heart or an artery.

bruxism *(brŭk'-sĭz-m).* Forceful clenching and grinding of the teeth, especially during sleep.

bryonia *(brī-ō'-nē-ă).* The roots of *Bryonia alba,* or wild hops, are used to make a homeopathic remedy that is given to individuals who have conditions that come on slowly and worsen with the slightest movement, which may include headache, chest pain, cough, and nausea.

bubo *(bū'-bō).* Swelling of a lymph gland, especially in the groin.

bubonic plague *(bū-bŏn'-ĭk).* Highly contagious, often fatal epidemic diseases caused by the toxin of the bacteria *Yersinia pestis,* transmitted by the bite of infected fleas.

buccal *(būk'-ăl).* Pertaining to the cheek.

Buerger's disease *(bĕr'-gĕrz).* Inflammation of the wall of a blood vessel with clot formation.

buffer *(bŭf'-ĕr).* Any substance that is added to another to prevent a radical change in concentration or pH in the second.

bulbourethral glands *(bŭl-bō-ū-rē'-thrăl).* Two glands, located one on each side of the prostate, that secrete a component of semen.

bulbous *(bŭl'-bŭs).* Swollen at the tip.

bulimia *(bū-lĭm'-ē-ă).* Disorder in which individuals (usually adolescents and young women) routinely overeat excessively and then self-induce vomiting; also called *binge and purge syndrome.*

bulla *(bū'-lă).* Blister or blisterlike elevation of the skin that contains fluid (serum).

BUN. *See* **blood urea nitrogen.**

bundle branch block *(bŭn'-dl brănch blŏk).* Defect in the electrical system of the heart, resulting in abnormal transmission of the heartbeat impulse.

bundle of His. Band of fibers in the heart through which the cardiac impulse is transmitted from the atrioventricular node to the ventricles.

bunion *(bŭn'-yŭn).* Painful swelling and thickening of the joint where the big toe joins the foot.

burn *(bĕrn).* Tissue injury caused by excessive exposure to heat, fire, electricity, radiation, or chemical.

brush b. One caused by friction of a rapidly moving object.

first-degree b. Reddening of the skin (outer layer—epidermis) without blistering.

radiation b. One caused by overexposure to ultraviolet rays, x-rays, or radium, for example.

second-degree b. Blistering of the skin, both epidermis and dermis involved.

third-degree b. Destruction of all layers of tissue; may include muscle, fat, and bone.

burp *(burp).* To bring up gas from the stomach through the mouth.

bursa *(bŭr'-să).* Membrane-lined, fluid-filled sac, usually between or near joints, which helps lubricate and protect joints, tendons, ligaments, and bone.

bursitis *(bŭr-sī'-tĭs).* Inflammation of a bursa.

butterfly rash *(bŭt'-ĕr-flī răsh).* Red, scaly eruption on both cheeks and across the nose, characteristic of lupus erythematosus and rosacea.

buttocks *(bŭt'-ŏks).* Part of the body behind the hips, consisting of the gluteal muscles.

bypass *(bī'-păs).* Surgically created channel or route around a body part, usually because that part has been damaged; e.g., a coronary bypass.

byssinosis *(bĭs-ĭ-nō'-sĭs).* Chronic inflammatory disease of the lungs caused by inhalation of dust in cotton, hemp, or flax.

C

CA125. Antigen often found in elevated levels in the serum of patients with various gynecological cancers (e.g., endometrial, ovarian, tubal, and endocervical).

cachexia *(kă-kēks'-ē-ă).* State of malnutrition, wasting, and weakness, which occurs in some cases of certain illnesses, including cancer, tuberculosis, cardiac disease, and AIDS.

cadmium sulfide *(kăd'-mē-ŭm).* Substance used to treat scalp dermatitis.

café au lait spot. Patchy pigmentation (usually light brown) of the skin, characteristic of neurofibromatosis.

Cafergot. Trade name of a medication that combines caffeine and ergot, used to relieve migraine attacks.

caffeine *(kă-fēn').* Central nervous system stimulant derived from coffee beans and dried tea leaves, used to counter mental fatigue and drowsiness.

caisson disease. *See* **bends.**

calamine lotion *(kăl'-ă-mīn).* Over-the-counter medication composed of zinc oxide and ferric oxide, used to treat itching and mild skin irritations.

calcaneus *(kăl-kā'-nē-ŭs).* Heel bone.

calcareous *(kăl-kā'-rē-ŭs).* Chalky; pertaining to calcium or limestone.

calcemia *(kăl-sē'-mē-ă).* Elevated level of calcium in the blood.

calciferol *(kăl-sĭf'-ĕr-ŏl).* Vitamin D2; a synthetic vitamin D, used to treat and prevent vitamin D deficiency and rickets.

calcification *(kăl-sĭ-fĭ-kā'-shŭn).* Process in which calcium accumulates in tissue.

calcinosis *(kăl-sĭ-nō'-sĭs).* Process in which calcium accumulates in the skin and subcutaneous tissues.

calcipenia *(kăl-sĭ-pē'-nē-ă).* Condition characterized by a deficiency of calcium in body fluids and tissues.

calcitonin *(kăl-sĭ-tō'-nĭn).* Hormone secreted by the thyroid gland, it plays a role in bone and calcium metabolism. In normal humans, its role in maintaining calcium blood levels is relatively minor, but high levels of calcitonin can be used to treat various disorders.

calcium *(kăl'-sē-ŭm).* Fifth most abundant element in the body, found mainly in bone, but also in teeth, fluids, and soft tissue.
c. carbonate. An antacid and astringent; often used as a calcium supplement.
c. citrate. Used as a calcium supplement.
c. gluconate. Used as a calcium supplement.
c. lactate. Used as a calcium supplement.

calcium channel blocker. Any drug that slows the flow of calcium within smooth muscle cells; used to treat angina, migraine, and hypertension.

calculus *(kăl'-kū-lŭs).* Stone, composed of mineral salts, that most often forms in ducts or hollow organs (e.g., gallbladder, kidney).

calendula *(kū-lĕnd'-yū-lă).* Homeopathic remedy (*Calendula officinalis*) prepared from the common marigold; reportedly has anti-inflammatory and antimicrobial properties.

calf *(kăf).* Fleshy, muscular back part of the leg below the knee.

caliber *(kăl'-ĭ-bĕr).* Diameter of a vessel, tube, or orifice.

calipers *(kăl'-ĭ-pĕrs).* Instrument for measuring the diameter or thickness of solids, such as body fat.

calisthenics *(kăl-ĭs-thĕn'-ĭks).* Exercises designed to develop range of motion, flexibility, muscle tone, and strength.

callus *(kăl'-ŭs).* 1. Hardened, thickened area of skin. 2. Tissue that forms around a bone fracture.

calorie *(kăl'-ŏ-rē).* Unit of heat. Most commonly, the fuel or energy value of food.

calorimeter *(kăl-ō-rĭm'-ĕ-tĕr).* Instrument for measuring the amount of heat given off by the body during a metabolic or chemical process.

Calve-Perthes disease *(kăl-vā' pĕr'-tās). See* **osteochondrosis.**

calx *(kălks)* (*pl.* calces). The heel.

camphor *(kăm'-fŏr).* Chemical, derived from the *Cinnamomum camphora* plant or man-made, used to treat some skin conditions.

canal *(kă-năl')*. Narrow tube or passageway for conducting materials other than blood or lymph; e.g., the alimentary canal (digestive tract).

canaliculus *(kăn-ă-lĭk'-ū-lŭs)*. Minute canal or channel.

cancellous *(kăn-sĕl'-ŭs)*. Pertaining to a spongy or latticelike structure; often used to describe a type of bone tissue.

cancer *(kăn'-sĕr)*. General term for any abnormal, malignant growth of cells that invade nearby tissues.

Candida *(kăn'-dĭ-dă)*. Genus of yeastlike fungi. Some species are part of the normal, healthy flora of the skin and mucous membranes, but overgrowth can also cause infections.

C. albicans. Species that normally lives in the intestinal tract of humans; it can cause infections, including thrush and vaginitis.

C. tropicalis. Species that can cause vaginitis, meningitis, onychomycosis, and bronchopulmonary infection.

candidiasis *(kăn-dĭ-dī'-ă-sĭs)*. Infection of the skin or mucous membranes with microorganisms of the genus *Candida.*

canker sore *(kăng'-kĕr)*. Ulceration often found on the gums or tongue.

cannula *(kăn'-ū-lă)*. Flexible tube inserted into a body opening to deliver or withdraw fluid or other materials.

canthoplasty *(kăn'-thō-plăs-tē)*. Cosmetic surgery of the canthus (upper eyelid) of the eye.

canthus *(kăn'-thŭs)*. Angle formed where the upper and lower eyelids meet; there is one inner and one outer canthus on each eye.

capillary *(kăp'-ĭ-lăr-ē)*. One of the minute blood vessels that connects venules and arterioles.

capsaicin *(kăp-sā'-ĕh-sĭn)*. Ingredient in cayenne (chili peppers) that helps reduce pain when used topically or taken orally.

capsule *(kăp'-sĕl)*. 1. Small, gelatinlike container used to enclose a dose of oral medication. 2. Fibrous or membranelike sac that surrounds an organ, body part, or tumor.

capsulotomy *(kăp-sū-lŏt'-ō-mē)*. Surgical repair or replacement of a joint capsule or other capsule.

captopril *(kăp'-tō-prĭl)*. One of the members of the angiotensin-converting enzyme inhibitor drugs, used to treat high blood pressure and congestive heart failure. Trade name: Capoten.

caput *(kā'-pŭt)*. 1. Head. 2. Main extremity of an organ.

carbohydrate *(kăr-bō-hī'-drāt)*. Any of a group of organic compounds composed of oxygen, hydrogen, and carbon, including starches and sugars.

carbo-loading. Practice of consuming large amounts of carbohydrates several days before a potentially physically exhausting event, done by some athletes.

carbon *(kăr'-bŏn)*. Nonmetallic element found in nearly all living things.

c. dioxide. Colorless, odorless gas given off by the lungs as a waste product of breathing.

c. monoxide Poisonous, odorless gas formed by the incomplete burning of organic substances, such as automotive fuel.

carbuncle *(kăr'-bŭng-kl).* Painful, extensive abscess or boil caused by infection with *Staphylococcus* bacteria and characterized by oozing pus, often through multiple openings.

carcinoembryonic antigen (CEA) *(kăr-sĭ-nō-ĕm-brē-ŏn'-ĭk).* Antigen found in minute amounts in normal tissues; elevated levels are looked for in a screening test for cancer.

carcinogen *(kăr-sĭn'-ō-jĭn).* Any substance that produces or increases the risk of developing cancer.

carcinoma *(kăr-sĭ-nō'-mă).* Malignant tumor that can develop in epithelial tissue of nearly any structure of the body and can spread to other body parts.

basal cell c. Skin malignancy that rarely metastasizes. It is the most common type of skin cancer.

c. in situ. Cancer that stays confined to where it originates.

epidermoid c. Tumor that forms on a surface, it can be of two types: wartlike and slow-growing, or flat and rapid-growing.

squamous cell c. Type of carcinoma that arises in squamous (scaly) cells; e.g., in skin, mouth, esophagus, cervix, or bronchi.

cardiac *(kăr'-dē-ăk).* Referring to the heart.

c. arrest. Sudden stoppage of the heart.

c. arrhythmia. Abnormal heart rate; *see also* **bradycardia, heart block, tachycardia** (types of cardiac arrhythmia).

c. conduction system. Specialized cardiac tissue that transports the signals that make the heart contract.

c. massage. Repeated, rhythmic compression of the heart, either as part of cardiopulmonary resuscitation, or during surgery.

c. monitor. Instrument that continuously monitors heart function.

c. output. Amount of blood (expressed in liters) released by the ventricles per minute; for a normal resting adult, this is 2.5 to 4.2 liters/minute/square meter.

c. tamponade. Condition in which accumulated fluid compresses the heart, rendering it unable to pump out sufficient amounts of blood.

cardiogram *(kăr'-dē-ō-grăm).* Electronic recording of the heart's rhythm, made with an electrocardiograph.

cardiography *(kăr-dē-ŏg'-ră-fē).* Process of recording the heart's activities using a cardiograph; also called *electrocardiography.*

cardiology *(kăr-dē-ŏl'-ō-jē).* Medical specialty concerned with the diagnosis and treatment of heart disease.

cardiomegaly *(kăr-dē-ō-mĕg'-ă-lē).* Enlargement of the heart.

cardiomyopathy *(kăr-dē-ō-mī-ŏp'-ĕ-thē).* Disease of the muscular wall of the heart (myocardium).

cardiopericarditis *(kăr-dē-ō-pĕr-ĭ-kăr-dī'-tĭs).* Inflammation of the myocardium and pericardium.

cardiopulmonary resuscitation (CPR) *(kăr-dē-ō-pŭl'-mō-nĕr-ē rē-sŭs-ĭ-tā'-shŭn).* An emergency technique performed in cases of cardiac arrest, involving simultaneous closed cardiac massage and mouth-to-mouth artificial respiration.

cardiospasm *(kăr'-dē-ō-spăz-m).* Muscular constriction of the valve between the distal end of the esophagus and the stomach, which prevents normal movement of food into the stomach. *See* **achalasia.**

cardiovascular system *(kăr-dē-ō-văs'-kū-lĕr).* System that includes the heart and blood vessels, which are involved in pumping blood and transporting nutrients throughout the body.

cardioversion *(kăr-dē-ō-vĕr'-zhŭn).* Use of electric shock to reestablish heart rhythm. *See* **defibrillation.**

carditis *(kăr-dī'-tĭs).* Inflammation of the heart, often due to infection.

Cardizem. *See* Appendix, Common Prescription and OTC Drugs: By Trade Name.

caries *(kăr'-ēz).* Decay and death of tooth tissue; also called dental caries or cavities.

carisoprodel. *See* Appendix, Common Prescription and OTC Drugs: By Generic Name.

carnitine *(kăr'-nĭ-tēn).* Naturally occurring derivative of the amino acid lysine, it is essential for the production of cell energy; also available as a dietary supplement.

carotene *(kăr'-ō-tēn).* One of four pigments, ranging in color from violet to red-yellow to yellow, found in fruits and vegetables. These pigments are converted into vitamin A in the liver.

carotenoids *(kă-rŏt'-ū-noyds).* Pigments synthesized by plants that provide the colors of fruits and vegetables, they are antioxidants and may have anticancer properties.

carotid artery *(kă-rŏt'-ĭd).* Either of two main arteries in the neck, it supplies blood to the neck and head.

carpal tunnel syndrome *(kăr'-păl tŭn'-ĕl).* Complex of symptoms that affect the wrist and hand, caused by compression of the median nerve in the wrist.

carrier *(kăr'-ē-ĕr).* 1. Individual who, although showing no signs of ill health, harbors organisms that can infect and cause disease in others. 2. In genetics, an individual capable of transmitting a variant gene even if the gene is not visibly expressed in the person.

car sickness. Nausea, vomiting,

and headache caused by riding in automobiles; *see also* **motion sickness.**

cartilage *(kăr'-tĭ-lĭj).* Connective tissue that protects and connects body parts, it covers some bones, especially joints, and is also found in body tubes (e.g., trachea).

cartilaginification *(kăr-tĭ-lă-jĭn-ĭ-fĭ-kā'-shŭn).* Abnormal formation of cartilage from other tissues.

caruncle *(kăr'-ŭng-kl).* Small fleshy growth.

lacrimal c. Growth seen at the medial junction of the eyelids.

urethral c. Growth seen near the entrance of the female urethra.

cascade *(kăs-kād').* Series of sequential events that, once started, continues until a final step is reached.

cascara sagrada *(kăs-kăr'-ă să-grăd'-ă).* Dried bark of a tree, *Rhamnus purshiana,* used as a laxative.

case-control study. Research technique in which information gathered from cases—individuals who have a disease or condition—are compared with information obtained from controls—people who do not have the disease.

case history. An individual's complete medical, family, and social history up to the time of presentation for treatment.

casein *(kā-sēn').* Main protein present in milk.

cast *(kăst).* 1. Rigid dressing, originally made of plaster of Paris and gauze, used to immobilize parts of the body. 2. Mold used to copy a body part (e.g., mold of teeth to make dentures).

castration *(kă-strā'-shĕn).* Surgical removal of the testes or ovaries.

CAT (CT) scan. *See* **computed tomography.**

catabolism *(kă-tăb'-ō-lĭzm).* Phase of metabolism in which chemical compounds are broken down into simpler forms.

catalepsy *(kăt'-ă-lĕp-sē).* Abnormal mental state characterized by a trance and loss of voluntary movement.

catalyst *(kăt'-ă-lĭst).* Substance that speeds up the rate of a chemical reaction, yet does not become permanently changed in the process.

cataphasia *(kăt-ă-fā'-zhă).* Speech disorder characterized by involuntary repetition of the same word or words.

cataplexy *(kăt'-ă-plĕk-sē).* Sudden and brief loss of muscle tone and reflexes, usually triggered by an emotional incident.

Catapres. *See* Appendix, Common Prescription and OTC Drugs: By Trade Name.

cataract *(kăt'-ă-răkt).* Condition in which there is loss of transparency of the lens of the eye, resulting in partial or complete blindness.

immature c. Cataract in which the lens is only slightly opaque.

juvenile c. Cataract in a child younger than nine years old, it is usually congenital or the result of trauma.

mature c. Cataract characterized by swelling and loss of transparency of the entire lens.

senile c. The most common type of cataract, it is associated solely with aging. It is painless and of unknown cause.

catatonia *(kăt-ă-tōn'-ē-ă).* Phase in schizophrenia characterized by a lack of reaction, movement, and speech.

catecholamine *(kăt-ĕ-kō'-lă-mēn).* Any one of several chemicals (e.g., dopamine, epinephrine, norepinephrine) derived from tyrosine and affecting many bodily functions by acting as neurotransmitters, hormones, or both.

catgut *(kăt'-gŭt).* Suturing material derived from sheep (not cats), used in surgery, and eventually absorbed by the body.

catharsis *(kă-thăr'-sĭs).* 1. Purging the body of toxins or other materials using a cathartic (laxative). 2. In psychology, purging the mind by recalling events, feelings, and emotions.

cathartic *(kă-thăr'-tĭk). See* **laxative.**

catheter *(kăth'-ĕ-tĕr).* Slender, usually flexible tube inserted into a body cavity to drain or deliver fluids.

Foley c. Catheter with an inflatable balloon near the tip which, when inflated, helps retain the catheter in place, usually in the bladder.

indwelling c. Catheter designed to be left within the body for a prolonged time.

Swan-Ganz c. A catheter inserted through one of the large veins that return blood to the heart. This catheter has a balloon, which carries it through the vena cava to the heart, through the right atrium and right ventricle, to the pulmonary artery.

catheterization *(kăth-ĕ-tĕr-ĭ-zā'-shŭn).* Introduction of a catheter into the body to withdraw fluid (e.g., urine, blood) or for diagnostic testing or treatment.

cardiac c. Insertion of a catheter into the chambers or blood vessels of the heart to gather information needed for diagnosis or treatment of heart disease.

cathexis *(kă-thĕk'-sĭs).* Investment of emotional energy in an idea, object, or person.

cat-scratch disease. Disease from the scratch or bite of a cat infected with the bacterium *Bartonella henselae*, characterized by swollen lymph nodes, fever, malaise.

cauda *(kaw'-dă).* Tail-like structure or tapered end of a structure.

c. equine. Collection of nerves that originate at the end of the spinal cord.

caudal *(kaw'-ăl).* Referring to the lower part or tail.

c. anesthesia. Type of regional anesthesia in which the anesthetic is injected into the lower end of the spinal canal to prevent pain in the areas served by the surrounding nerves.

caul *(kawl).* Fetal membranes that surround and adhere to the head of the fetus at birth.

cauliflower ear *(kaw'-lē-flŏ-ĕr).* Deformity of the external ear usually caused by repeated trauma, often seen in prizefighters.

causalgia *(kaw-zăl'-jē-ă).* Painful, burning feeling due to injury of nerve fibers.

caustic *(kaws'-tĭk).* 1. Capable of burning or destroying living tissue. 2. Substance that causes a burn or destroys tissue.

cauterize *(kaw'-tĕr-īz).* To burn or destroy tissue using heat, electric current, or chemicals for medical reasons.

cavernitis *(kăv-ĕr-nī'-tĭs).* Inflammation of the tissues that make up the shaft of the penis.

cavernous sinus thrombosis *(kăv'-ĕr-nŭs sī'-nŭs thrŏm-bō'-sĭs).* Blood clot, often due to infection, within the cavernous sinus in the skull, behind and above the eyes.

cavity *(kăv'-ĭ-tē).* 1. Hollow structure, as in a hollow organ. 2. General term for dental caries.

cayenne *(kī'-ăn). See* **capsaicin.**

CBC. Acronym for complete blood count.

CD4. A large protein molecule that serves as a receptor for HIV on the surface of T lymphocyte cells.

CDC. Acronym for Centers for Disease Control and Prevention, a government agency that investigates and helps control various diseases, especially communicable ones.

cecitis *(sē-sī'-tĭs).* Inflammation of the cecum.

cecostomy *(sē-kŏs'-tō-mē).* Surgical creation of an opening into the cecum through the abdominal wall, done to relieve intestinal obstruction.

cecum *(sē'-kŭm).* Wide, saclike portion of the large intestine where the appendix is attached.

cefadroxil *(sĕf-ă-drŏx'-ĭl).* One of the cephalosporin antibiotics, used to treat bacterial infections. Trade name: Duricef.

cefoperazone *(sĕf-ŏ-pĕr'-ă-zōn).* One of the cephalosporin antibiotics, used to treat severe infections. Trade name: Cefobid.

cefotaxime *(sĕf-ō-tăk'-sĕm).* One of the cephalosporin antibiotics, used to treat severe infections. Trade name: Claforan.

cefoxitin *(sĕ-fŏx'-ĭ-tĭn).* An antibiotic effective against a wide range of infections. Trade name: Mefoxin.

ceftazidime *(sĕf-tăz'-ĭ-dēm).* One of the cephalosporin antibiotics, used to treat moderate to severe infections. Trade names: Fortaz, Tazicef.

ceftriaxone *(sĕf-trī'-ă-zŏn).* An antibiotic effective in treating severe infections and sexually transmitted diseases. Trade name: Rocephin.

cefuroxime. One of the cephalosporin antibiotics, used to treat moderately severe infections. Trade name: Ceftin.

Celebrex. *See* Appendix, Common Prescription and OTC Drugs: By Trade Name.

celecoxib. *See* Appendix, Common Prescription and OTC Drugs: By Generic Name.

celiac *(sē'-lē-ăk).* Abdominal.

celiac disease. Disorder characterized by an intolerance to gluten (a protein in wheat and similar grains), leading to poor absorption of food, anemia, weight loss, and diarrhea.

cell *(sĕl).* Basic structure of tissues and organs, composed of an outer membrane, a nucleus, and cytoplasm, which contains various organelles. *See* individual types of cells by name.

cellulitis *(sĕl-yū-lī'-tĭs).* Inflammation of skin and deep subcutaneous tissue, characterized by pain, swelling, and redness.

cellulose *(sĕl'-yū-lōs).* Basic component of plant fiber, found in fruits, vegetables, and grains.

Celsius *(sĕl'-sē-ŭs).* Also called centigrade, it refers to the temperature scale in which 0° is the freezing point and 100° is the boiling point of water.

cementum *(sē-mĕn'-tŭm).* Specialized tissue that covers the root of a tooth.

centigrade *(sĕn'-tĭ-grād).* Having 100 degrees. A centrigrade thermometer is divided into 100 degrees, with 100 degrees being the boiling point and 0 degrees being the freezing point of water.

central nervous system. Central portion of the nervous system, including the brain and spinal cord, that coordinates all nervous system activity.

central venous catheter. Hollow tube inserted into an arm or neck vein and advanced until it reaches the vena cava in the chest.

centrifuge *(sĕn'-trĭ-fūj).* Device that spins test tubes at a very rapid speed for the purpose of separating substances of different densities.

cephalalgia *(sĕf-ă-lăl'-jē-ă).* *See* **headache.**

cephalexin *(sĕf-ă-lĕk'-sĭn).* *See* Appendix, Common Prescription and OTC Drugs: By Generic Name.

cephalhematoma *(sĕf-ăl-hē-mă-tō'-mă).* Accumulation of blood under the scalp of a newborn caused by use of foreceps during delivery.

cephalic *(sĕ-făl'-ĭk).* Referring to the head.

cephalic index. Measurement used to determine cranial capacity.

cephalosporin *(sĕf-ă-lō-spŏr'-ĭn).* Term for a group of antibiotics derived from cephalosporin C, which is obtained from the fungus *Cephalosporium,* and used to treat a wide variety of bacterial infections.

cephalothin *(sĕf'-ă-lō-thĭn).* One of the cephalosporin antibiotics, effective against aerobic organisms. Trade name: Keflin.

cerebellum *(sĕr-ĕ-bĕl'-ŭm).* Part of the brain, located behind the cerebrum and above the pons, responsible for muscle coordination.

cerebral *(sĕr-ē'-brĕl, sĕr'-ĕ-brĕl).* Referring to the cerebrum of the brain.

c. cortex. Thin layer on the surface of the cerebrum, where perception and higher mental functions are carried out.

c. edema. Excess fluid in the brain.

c. hemorrhage. Flow of blood from a ruptured blood vessel in the brain.

c. palsy. Deficiency or loss of muscle control caused by permanent, nonprogressive brain damage that occurs before or at birth.

c. thrombosis. Blood clot in a cerebral vessel.

cerebrospinal *(sĕr-ĕ-brō-spī'-năl).* Referring to the brain and spinal cord.

cerebrospinal fluid. Colorless fluid that protects the brain and spinal cord from physical trauma.

cerebrovascular accident *(sĕr-ĕ-brō-văs'-kū-lăr).* Bleeding or blockage of the blood vessels of the brain, resulting in lack of oxygen; also known as a *stroke.*

cerebrum *(sĕr'-ĕ-brŭm, sĕ-rē'-brŭm).* Largest portion of the brain, where conscious thoughts and actions are controlled.

certifiable disease *(sĕr-tĭ-fī'-ă-bl).* Infectious disease that, by law, health-care practitioners must report to health authorities when they identify it.

certify *(sĕr'-tĭ-fī).* To place someone in a mental-health facility, usually against his or her will.

cerumen *(sĕ-rū'-mĕn).* Soft, brown, waxy material secreted by the sebaceous glands of the external ear; earwax.

cervical *(sĕr'-vĭ-kăl).* 1. Referring to or in the region of the neck. 2. Referring to the cervix (neck) of an organ, including the cervix of the uterus.

c. cap. Contraceptive device, shaped like a small rubber cap, made to fit over the cervix to block entrance of sperm into the uterus.

c. disc syndrome. Condition caused by compression of cervical nerve roots, due to degenerative disease, trauma, or other factors, that results in neck pain.

c. smear. Small sample of secretions and cells of the cervix of the uterus, which is examined to detect abnormal cells.

c. spondylosis. Degenerative arthritis of the cervical vertebrae and related tissues.

c. vertebrae. First seven bones of the vertebral column, in the neck region.

cervicectomy *(sĕr-vĭ-sĕk'-tō-mē).* Surgical removal of the cervix of the uterus.

cervicitis *(sĕr-vĭ-sī'-tĭs).* Inflammation of the cervix of the uterus.

cervix *(sĕr'-vĭks).* Any necklike or constricted portion of an organ; typically used alone to denote the uterine cervix.

cesarean section *(sĭ-zăr'-ē-ăn).* Surgical procedure to deliver a newborn through an incision into the uterus, usually via the abdominal wall.

cesium *(sē'-zē-ŭm).* An element

that, in its radioactive form, is used to treat some cancers.

cetirizine. *See* Appendix, Common Prescription and OTC Drugs: By Generic Name.

CFS. Acronym for cerebral spinal fluid.

Chagas' disease *(chăg'-ăs).* Potentially serious infection with the protozoan parasite *Trypanosoma cruzi*, which can cause heart and intestinal damage.

chakras *(shŏk'-răs).* According to Eastern philosophy, the energy centers of the body (of which there are seven), said to absorb and release the life force, or chi.

chalazion *(kă-lā'-zē-ōn).* Inflamed benign cyst of the eyelid; also called *meibomian cyst.*

challenge *(chăl'-ĕnj).* Administration of a specific antigen (or physiologic stressor) to an individual suspected of having a sensitivity to the antigen (or other stressor) in order to provoke a response.

chamber *(chăm'-bĕr).* Enclosed area; often refers to the heart chambers (ventricles and auricles).

chamomile *(kăm'-ō-mēl).* Name for two different herbs (*Matricaria chamomilla* and *Anthemis nobilis*) not related botanically but with similar healing effects, including reported relief from digestive problems and menstrual cramps.

chancre *(shăng'-kĕr).* Hard, painless sore, typically the first sign of syphilis.

chancroid *(shăng'-kroid).* Contagious, pus-producing painful ulcer that usually appears on the genitalia and is caused by the bacterium *Haemophilus ducreyi.*

change of life. *See* **menopause.**

characteristic *(kăr-ăk-tĕr-ĭs'-tĭk).* Distinguishing feature of an organism; a trait.

Charcot-Marie-Tooth disease *(shăr-kōz').* Disease characterized by muscle atrophy and progressive weakness in the arms and legs.

charley horse *(chăr'-lē-hŏrs).* Colloquial term for the pain and tenderness in a calf or thigh, usually caused by muscle tear or strain.

CHD. Acronym for coronary heart disease.

cheek *(chēk).* Side of the face.

cheilitis *(kī-lī'-tĭs).* Inflammation and cracking of the lips due to vitamin deficiency, overexposure to sun, or reaction to cosmetics.

cheilosis *(kī-lō'-sĭs).* Disorder in which the lips are red and marked with splits, frequently caused by a riboflavin deficiency.

chelation *(kē-lā'-shŭn).* 1. Bonding interaction between a metal ion and two or more nonmetal binding sites in the same molecule forming a ring. 2. Chemical bonding technique used to remove substances (e.g., lead, iron) from the body's tissues.

chemistry *(kĕm'-ĭs-trē).* Science concerned with the atomic and molecular makeup of matter.

chemolysis *(kē-mŏl'-ĭ-sĭs).* Destruction by chemicals.

chemonucleolysis *(kē-mō-nū-klē-ŏl'-ĭ-sĭs).* Injection of an enzyme into a herniated disk to dissolve it.

chemoreceptor *(kē-mō-rĭ-sĕp'-tĕr).* Sense organ or sensory nerve ending that reacts to certain chemical stimuli.

chemosis *(kē-mō'-sĭs).* Inflammation of the mucous membrane that lines the eyelids (conjunctiva).

chemosurgery *(kē-mō-sŭr'-jĕr-ē).* Use of chemicals to destroy tissue.

chemotaxis *(kē-mō-tăk'-sĭs).* Movement of cells toward (positive) or away (negative) from a chemical stimulus.

chemotherapy *(kē-mō-thĕr'-ă-pē).* Use of chemical substances to treat disease; most commonly refers to use of drugs to treat cancer.

chest *(chĕst).* Thorax.

Cheyne-Stokes breathing *(chăn' stōks').* Abnormal respiration characterized by apnea followed by gradually increasing frequent, deep breathing.

chicken pox. Highly contagious disease caused by herpes varicella zoster virus and usually seen in childhood.

chilblain *(chĭl'-blān).* Condition characterized by blistering, redness, swelling, burning, and itching of the skin due to exposure to cold and dampness.

child abuse. Physical, emotional, and/or sexual maltreatment of a child.

childbirth. Process of delivering a fetus.

natural c. Delivery of a fetus without the use of anesthesia, analgesics, and sedatives, and during which women often use exercises and psychological conditioning to relax.

child neglect. Failure of parents or guardians to provide appropriate physical, nutritional, and emotional care, thus jeopardizing a child's health and development.

chill. Episode of shivering accompanied by coldness and pale skin, often occurring in the company of a fever or infection.

chiropractic *(kī-rō-prăk'-tĭk).* Method of diagnosis and treatment based on the belief that many diseases are caused by pressure on the nerves, caused by dislocation of the spinal column.

chlamydia *(klă-mĭd'-ē-ă).* General term for a group of microorganisms that cause various diseases, including genital infections.

C. pneumoniae. Species that causes bronchitis and pneumonia.

C. trachomatis. Common cause of sexually transmitted disease (e.g., vaginitis, urethritis) in men and women in the developed world.

chloasma *(klō-ăz'-mă).* Transient or permanent tan or yellowish-brown skin discoloration.

gravidarum c. Brownish pigmentation changes that often

occur in pregnancy, usually on the face.

idiopathic c. Skin discoloration caused by agents such as heat, sun, and x-rays.

chloramphenicol *(klŏr-ăm-fĕn'-ĭ-kŏl).* An antibiotic effective in treating typhoid and infections caused by salmonellae. Trade name: Chloromycetin.

chlorhydria *(klŏr-hī'-drē-ă).* Presence of excess hydrochloric acid in the stomach.

chloride *(klŏr'-īd).* Any compound of chlorine, levels are elevated in anemia, cardiac disease, and nephritis and decreased in diabetes, fever, and pneumonia.

chlorine *(klŏr'-ēn).* Gaseous element combined with sodium to form salt in the blood, it is also present in hydrochloric acid, which aids in digestion.

chlorophyll *(klŏr'-ĕ-fĭl).* Green pigment in plants that absorbs light during photosynthesis. In purified form, chlorophyll promotes wound healing.

chloroquine *(klŏr'-ĕ-kwĭn).* A medication used to treat malaria, systemic lupus erythematosis, and amebic dysentery. Trade name: Aralen.

chlorothiazide *(klŏr-ō-thī'-ă-zīd).* A medication prescribed to treat high blood pressure and edema. Trade name: Diuril.

chlorpromazine *(klŏr-prō'-mă-zēn).* A medication used to treat certain psychotic disorders and severe nausea and vomiting. Trade name: Thorazine.

chlorpropamide *(klŏr-prō'-pă-mīd).* A medication used to reduce blood sugar levels. Trade name: Diabinase.

chlorthalidone *(klŏr-thăl'-ĭ-dōn).* A medication used to treat high blood pressure. Trade name: Hygroton.

chocolate cyst *(chŏk'-ĕ-lăt).* Cyst, often on the ovary, that contains old blood that has turned brown.

cholangiography *(kō-lăn-jē-ŏg'-ră-fē).* X-ray of the bile ducts after injection of a contrast medium.

cholangiolitis *(kō-lăn-jē-ō-lī'-tĭs).* Inflammation of the small bile ducts located in the liver.

cholangitis *(kō-lăn-jī'-tĭs).* Inflammation of the bile ducts.

cholecystectomy *(kō-lē-sĭs-tĕk'-tō-mē).* Surgical removal of the gallbladder.

cholecystitis *(kō-lē-sĭ-tī'-tĭs).* Inflammation of the gallbladder.

cholecystoduodenostomy *(kō-lē-sĭs-tō-dū-ō-dē-nŏs'-tō-mē).* Surgical procedure in which a passage is created between the gallbladder and duodenum.

cholecystogastrostomy *(kō-lē-sĭs-tō-găs-trŏs'-tō-mē).* Surgical procedure in which a passage is created between the gallbladder and stomach.

cholecystography *(kō-lē-sĭs-tŏg'-rĕ-fē).* X-ray visualization of the gallbladder after oral administration or injection of a dye.

cholecystojejunostomy *(kō-lē-sĭs-tō-jĕ-jū-nŏs'-tō-mē).* Surgical procedure in which a passage is formed between the gallbladder and jejunum.

cholecystostomy *(kō-lē-sĭs-tŏs'-tō-mē).* Surgical procedure in which an incision is made through the abdominal wall to remove gallstones from the gallbladder.

choledochitis *(kō-ē-dō-kī'-tĭs).* Inflammation of the common bile duct.

choledocholithotomy *(kō-lĕd-ō-kō-lĭth-ŏt'-ō-mē).* Surgical incision made into the bile duct to remove a stone.

choledochoscope *(kō-lĕd'-ō-ōkŏ-skōp).* Flexible instrument used to inspect the inside of the bile duct.

choledochotomy *(kō-lĕd-ō-kŏt'-ō-mē).* Surgical incision made to remove stones from the bile duct.

cholelithiasis *(kō-lē-lĭ-thī'-ă-sĭs).* Presence of stones in the gallbladder.

cholemia *(kō-lē'-mē-ă).* Presence of bile or its pigments in the blood.

cholera *(kŏl'-ĕr-ă).* Acute infectious disease, caused by the toxin-producing bacterium *Vibrio cholerae,* and characterized by diarrhea, vomiting, and cramps with dehydration and shock in severe cases.

cholesteatoma *(kō-lē-stē-ă-tō'-mă).* Cyst filled with cellular debris, sometimes including cholesterol and most commonly found in the middle ear; can be congenital but is usually a complication of chronic ear infection.

cholesterol *(kō-lĕs'-tĕr-ŏl).* Sterol found in all animal fats and oils and throughout the body, especially the bile, brain, blood, and nerve sheaths. Excessive amounts in blood can contribute to the development of coronary artery disease.

cholesterosis *(kō-lĕs-tĕr-ō'-sĭs).* Excessive cholesterol deposits in an organ or in body tissues.

choline *(kō'-lēn).* A vitaminlike compound that is the basic constituent of lecithin and prevents fat from being deposited in the liver. It also plays a major role in the transmission of nerve impulses.

cholinergic *(kō-lĭ-nĕr'-jĭk).* 1. Nerve endings that free acetylcholine. 2. Substance that can produce the effect of acetylcholine. *See* **acetylcholine.**

choluria *(kō-lū'-rē-ă).* Presence of bile in the urine.

chondral calcinosis *(kŏn'-drăl kăl-sĭ-nō'-sĭs).* Often painful condition in which calcium deposits in cartilage within a joint; also called *pseudogout.*

chondritis *(kŏn-drī'-tĭs).* Inflammation of cartilage.

chondrofibroma *(kŏn-drō-fī-brō'-mă).* Nonmalignant tumor that contains cartilage and fibrous tissue.

chondroitin sulfate *(kŏn-droy'-tĭn).* Substance found naturally in the body that helps maintain the integrity of many of the body's tissues; available in supplement form.

chondroma *(kŏn-drō'-mă).* Benign,

slow-growing, painless tumor composed of cartilage.

chondromalacia *(kŏn-drō-mă-lā'-shă).* Softening of cartilage, usually of the kneecap.

chondrosarcoma *(kŏn-drō-săr-kō'-mă).* Malignant bone tumor composed of cartilage cells.

chord *(kŏrd).* Tendon, ligament, or band of fibrous tissue.

chorditis *(kŏr-dī'-tĭs).* Inflammation of a chord.

chordoma *(kŏr-dō'-mă).* A rare, slow-growing malignant tumor that can develop along the spine.

chorea *(kŏr-ē'-ă).* Any one of several disorders of the nervous system characterized by involuntary jerky movements of the limbs or facial muscles.

Huntington's c. Hereditary, progressive disease of the brain beginning in adulthood, characterized by involuntary movements usually of the trunk, lower limbs, and shoulders.

Sydenham's c. Type of chorea occurring in children, characterized by muscle weakness, lack of coordination, and involuntary movements; also called St. Vitus' dance.

choreoathetosis *(kŏr-ē-ō-ăth-ĕ-tō'-sĭs).* Condition characterized by abnormal involuntary movements such as twitching, facial contortions, heel walking, and writhing, frequently seen in cerebral palsy.

choriocarcinoma *(kŏr-ē-ō-kăr-sĭ-nō'-mă).* Rare, highly malignant tumor that develops most often in the uterus.

choriomeningitis *(kŏr-ē-ō-mĕn-ĭn-jī'-tĭs).* Inflammation of the membranes (meninges) that cover the brain and spinal cord.

chorion *(kō'-rē-ōn).* Outermost membrane of the sac that encloses the fetus within the womb.

chorioretinitis *(kŏ-rē-ō-rĕt-ĭ-nī'-tĭs).* Inflammation of a layer of cells (choroids) and the retina in the eye, resulting in blurred vision.

chorioretinopathy *(kŏr-ē-ō-rĕ-tĭ-nŏp'-ă-thē).* Eye disease that affects the retina and choroid.

choroid *(kō'-royd).* Middle blood vessel containing coat of the eye between the sclera and retina.

choroiditis *(kō-royd-ī'-tĭs).* Inflammation of the choroid.

Christmas disease. One form of hemophilia (hemophilia B); *see also* **hemophilia.**

chromatin *(krō'-mă-tĭn).* Gene-carrying part of the nucleus of a cell that stains easily when prepared for microscopic examination.

chromatosis *(krō-mă-tō'-sĭs).* Abnormal pigmentation in the tissues.

chromium *(krōm'-ē-ŭm).* Essential trace nutrient that plays a role in maintaining blood sugar levels; available in supplement form.

chromophobe *(krō'-mō-fōb).* Cells or tissues that do not stain well or not at all.

chromosome *(krō'-mō-sōm).* One of a group of threadlike structures, containing DNA, within the nucleus of a cell that transmits genetic information.

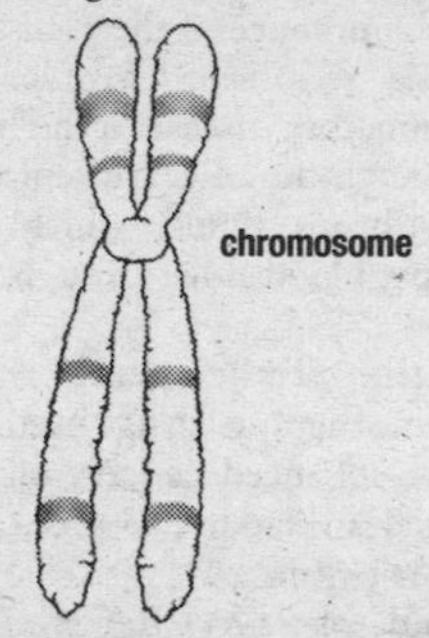

Philadelphia c. Abnormal chromosome seen in individuals who have chronic leukemia.
sex c. One of two chromosomes (X and Y chromosomes) responsible for determining sex: males have X and Y chromosomes; females have two X chromosomes.
X c. Chromosome that determines female sex characteristics.
XX c. The normal complement in females.
XXY c. An abnormal complement in males, characteristic of Klinefelter's syndrome.
XY c. The normal complement in males.
Y c. Chromosome that determines the male sex.

chronic *(krŏ'-nĭk).* Not acute; lasting a long time.

chronic bronchitis. Prolonged (more than three consecutive months in at least two successive years) narrowing and obstruction of the airways in the lungs.

chronic obstructive pulmonary disease (COPD). Disease process characterized by a decreased ability of the lungs to exchange air; may be caused by chronic bronchitis, pulmonary emphysema, chronic asthma, or chronic bronchiolitis.

chrysotherapy *(krĭs-ō-thĕr'-ă-pē).* Treatment with gold salts, used in cases of rheumatoid arthritis.

chyle *(kīl).* Milky fluid composed of digested fat and lymph which is transported from the intestines through lymph channels.

chylothorax *(kī-lō-thŏr'-ăks).* Accumulation of chyle in the chest cavity.

chylous ascites *(kī'-lŭs ă-sī'-tēz).* Large collection of lymph in the abdominal cavity.

chylous cyst. Lymph-containing sac, often located in the abdominal cavity near the small intestine.

chyluria *(kī-lūr'-ē-ă).* Presence of fat or chyle in the urine.

chyme *(kīm).* Combination of partially digested food and digestive secretions present in the stomach and duodenum during digestion.

chymotrypsin *(kī-mō-trĭp'-sĭn).* Enzyme present in the small intestines that helps digest proteins.

cilium *(sĭl'-ē-ŭm).* (*pl.* cilia) 1. Microscopic hairlike projection on the surface of a cell, capable of movement. 2. Eyelash.

cimetidine *(sĭ-mĕt'-ĭ-dēn).* Generic name of a medication that inhibits secretion of stomach acid; trade name: Tagamet.

cineangiography *(sĭn-ē-ăn-jē-ŏg'-ră-fē).* Continuous moving pictures showing the passage of an opaque dye through blood vessels.

cinefluorography *(sĭn-ē-flū-ŏr-ŏg'-ră-fē).* Continuous moving pictures of organs on a fluoroscopic X-ray screen.

cineplastic amputation *(sĭn-ē-plăs'-tĭk).* Surgical procedure in which the muscles and tendons of the amputation stump are formed to provide motion and direction to an artificial limb.

Cipro. *See* Appendix, Common Prescription and OTC Drugs: By Trade Name.

ciprofloxacin. *See* Appendix, Common Prescription and OTC Drugs: By Generic Name.

circadian rhythm *(sĭr-kā'-dē-ăn).* Rhythm or biological clock of an organism; for example, the natural sleep-wake cycle during a 24-hour period.

circle of Willis. Union of the cerebral arteries, forming a circle at the base of the brain.

circulation *(sĭr-kū-lā'-shŭn).* General term for the movement of blood, lymph, or other fluids throughout the body in a regular flow.

circulation time. Period of time it takes for a particle of blood to travel throughout the entire circulatory and pulmonary systems. Results can indicate various disease and cardiac problems.

circumcision *(sĭr-kŭm-sĭ'-zhŭn).* Surgical removal of part or all of the prepuce of the penis.

cirrhosis *(sĭ-rō'-sĭs).* Chronic inflammatory disease of the liver in which the cells are replaced by fibrous tissue, which restricts blood flow through the liver.

Cisplatin. Trade name of a chemotherapy drug used to treat advanced cancers of the testicles, bladder, ovaries, and other organs.

cisternal *(sĭs-tĕr'-năl).* Referring to any fluid-containing cavity or sac in the body.

citalopram. *See* Appendix, Common Prescription and OTC Drugs: By Generic Name.

citrated *(sĭt'-rāt-ĕd).* Referring to the addition of citric acid or citrate to a fluid.

citrate solution. Solution used to prevent blood from clotting.

citric acid *(sĭt'-rĭk).* Acid present in the juice of many fruits, especially citrus.

clairvoyance *(klăr-vŏy'-ăns).* Alleged ability to discern persons or objects or to be aware of events not perceptible to the senses.

clamp *(klămp).* Device used to compress, support, or grasp a tissue, organ, or other structure; typically used during surgery.

claudication *(klaw-dĭ-kā'-shĕn).* Lameness or limping, as well as cramplike pain in the legs caused by an inadequate blood supply to the muscles.

intermittent c. Severe pain in

the calf muscles that occurs when walking but subsides during rest.

claustrophobia *(klaw-strō-fō'-bē-ă).* Abnormal dread of being confined in closed or small spaces.

clavicle *(klăv'-ĭ-kl).* Bone that extends from the breastbone to the shoulder tip; also called the *collarbone.*

clawfoot *(klaw'foot).* Deformity of the foot characterized by an abnormally high arch and toes that are turned under.

clawhand *(klaw'-hănd).* Hand in which the fingers are hyperextended; usually the result of injury to the ulnar nerve.

clearance *(klĭr'-ĕns).* 1. Process of removing a substance. 2. Measure of the rate at which a substance is removed from the blood.

creatinine c. Clearance of endogenous creatinine by the kidneys.

renal c. Measure of the rate at which a substance is removed from the blood via the kidneys.

urea c. Clearance of urea from the blood, either by the kidneys or hemodialysis.

cleavage *(klē'-vĭj).* 1. The first stages of cell division after an egg has been fertilized. 2. Splitting a complex molecule into two.

cleft palate *(klĕft).* A common congenital defect characterized by a fissure in the roof of the mouth.

Cleocin. Trade name of an antibiotic effective against anaerobic bacteria; generic name: clindamycin.

click-murmur syndrome. A heart condition that affects 5–10 percent of the world's population, it is also known as mitral valve prolapse or Barlow's syndrome. Most patients are asymptomatic and do not need treatment, but some experience palpitations and/or fatigue.

climacteric *(klī-măk'-tĕr-ĭk).* In women, period marked by the cessation of menses, or menopause; in men, a corresponding period of declining sexual activity.

climax *(klī'-măks).* 1. Most intense stage of a disease. 2. Peak of sexual excitement.

clindamycin. An antibiotic effective against anaerobic bacteria. Trade name: Cleocin.

clinic *(klĭn'-ĭk).* Health-care facility designed to treat people on an outpatient basis.

clinical trial. Investigation of the effects of a drug given to human subjects, done for the purpose of identifying the effectiveness and effects of the substance.

clinician *(klĭ-nĭ'-shŭn).* Practicing physician or nurse who has expertise in clinical practice as distinguished from specializing in research.

clinicopathological *(klĭn-ĭ-kō-păth-ō-lŏj'-ĭk-ăl).* Referring to the signs and symptoms of a disease and the laboratory examination of specimens obtained from patients.

Clinoril. Trade name for a medication to treat rheumatoid arthritis; generic name: sulindac.

clitoridectomy *(klĭt-ō-rĭ-dĕk'-tō-mē).* Surgical removal of the clitoris.

clitoris *(klĭt'-ō-rĭs).* Erogenous part of the external female genitalia composed of erectile tissue, located between the labia minora; the female counterpart to the penis.

Clomid. Trade name of a drug used for infertility; generic name: clomiphene citrate.

clomiphene citrate. A drug used to stimulate ovulation in women who do not ovulate.

clonazepam. *See* Appendix, Common Prescription and OTC Drugs: By Generic Name.

clone *(klōn).* Genetically identical cells or organism that have been produced asexually from a single cell or organisms.

clonic *(klŏ'-nĭk).* Referring to jerky muscle spasms or contractions.

clonidine. *See* Appendix, Common Prescription and OTC Drugs: By Generic Name.

clopidogrel. *See* Appendix, Common Prescription and OTC Drugs: By Generic Name.

closed reduction. Setting a bone fracture by manipulating the bones.

Clostridium *(klŏs-trĭd'-ē-ŭm).* Genus of gram-positive, anaerobic, spore-bearing bacteria that can cause severe infections, such as gas gangrene and tetanus.

c. botulinum. Species that produces botulinum toxin, a cause of food poisoning.

c. difficile. Species that is normally found in infants and sometimes in adults. It can produce a toxin that causes pseudomembranous enterocolitis in individuals who take antibiotics.

c. perfringens. Species that is the main cause of gas gangrene.

c. tetani. Species responsible for causing tetanus.

closure *(klō'-zhŭr).* 1. Suturing or closing of a wound. 2. In psychotherapy, term used to represent resolution of an issue for a patient.

clot *(klŏt).* Clump of blood or lymph that has coagulated or solidified.

clotting time. Time required for blood to clot.

Clozapine. Trade name of a medication used to treat schizophrenia.

clubbed fingers. Rounded fingertips, sometimes seen in people who have chronic heart or lung disease.

clubfoot *(klŭb'foot).* Congenital condition in which the foot is shifted out of shape or position.

clumping *(klŭmp-ĭng).* Bacteria or other cells that cluster together into irregular masses; also called *agglutination.*

cluster headache *(klŭs'-tĕr).* Chronic headache caused by release of histamine from cells and characterized by a sharp pain on one side of the head, watery eyes, and runny nose.

CMF. Acronym for cytoxan, methotrexate, and 5FU, a drug combination used to treat breast cancer that has spread to other parts of the body.

CMV. Acronym for cytomegalovirus, often seen in people who have AIDS.

CO2 laser laparoscope. Laser beam directed through a laparoscope to destroy adhesions or other abnormalities, such as those seen in endometriosis.

coagulation *(kō-ăg-ū-lā'-shĕn).* Clotting process.

coagulopathy *(kō-ăg-ū-lŏp'-ă-thē).* Any disorder that affects the blood-clotting process.

coal tar derivatives. Chemicals and drugs derived from the breakdown of coal, such as benzene, phenol, and pitch.

coarctation *(kō-ărk-tā'-shĕn).* Narrowing or constriction of a blood vessel.

cobalt *(kō'-bawlt).* Metallic element, and an essential element in vitamin B12.

c. 60. Radioactive isotope (Co-60) of cobalt, used to treat cancer.

cocaine *(kō-kān').* White powder, derived from the coca plant or man-made, used both as an anesthetic and a drug of abuse.

crack c. Smokable form of cocaine, for illicit use.

c. hydrochloride. Form of cocaine used as a local anesthetic.

coccidioidomycosis *(kŏk-sĭd-ē-oi-dō-mī-kō'-sĭs).* Infection caused by the fungus *Coccidioides immitis,* occurring primarily in the southwestern United States and Central and South America; also known as *San Joaquin valley fever.*

coccus *(kŏk'-ŭs)* (*pl.* cocci). Bacteria whose cells are spherical. There are many species of cocci, some of which are associated with diseases such as sore throat, scarlet fever, pneumonia, and meningitis.

coccygectomy *(kŏk-sĭ-jĕk'-tō-mē).* Surgical removal of the coccyx.

coccygodynia *(kŏk-sĕ-gō-dĭn'-ē-ă).* Pain in the coccygeal (tailbone) region and surrounding area, often caused by a fall upon the buttocks.

coccyx *(kŏk'-sĭks).* Four fused vertebrae that form the tailbone in humans.

cochlea *(kŏk'-lē-ă).* Coiled, cone-shaped part of the inner ear where the receptor for hearing (organ of Corti) is located.

cochleal implant. Complex electronic device surgically placed within the inner ear to help people who have certain types of deafness to hear.

cochleitis *(kŏk-lē-ī'-tĭs).* Inflammation of the cochlea.

codeine *(kō'-dēn).* Derivative of opium or morphine used as a pain reliever and cough suppressant.

cod liver oil. Oil derived from the liver of codfish; very high in vitamins A and D, it is sometimes used as a supplement and to treat calcium and phosphorus deficiencies.

coenzyme *(kō-ĕn'-zīm).* Nonprotein substance that enhances the action of an enzyme, which is a protein that functions as a catalyst to speed up a chemical reaction.

c. Q10. Type of coenzyme whose main function is to help enzymes metabolize and utilize oxygen. Available as a supplement.

cognition *(kŏg-nĭsh'-ĕn).* Process of the mind by which individuals become aware of thoughts or perceptions; includes reasoning, judgment, intuition, and memory.

cognitive therapy *(kŏg'-nĭ-tĭv).* Form of psychotherapy based on the idea that emotional problems result from distorted attitudes and thinking patterns that can be corrected by learning how to change negative attitudes and perspectives.

cohesion *(kō-hē'-zhĕn).* Force that causes various components of a single substance to hold together.

cohort *(kō'-hŏrt).* In a clinical trial, a group of subjects that share a common experience as part of the trial and are then followed up to identify any new diseases or events that occur.

cohort study. A study in which a specific outcome (e.g., a seizure) is compared in groups of individuals who are alike in most ways but differ by a certain characteristic, such as alcohol use.

cohosh. *See* **black cohosh.**

coitus *(kō'-ĭ-tĕs).* Vaginal sexual union of a man and woman; sexual intercourse.

coitus interruptus. Contraceptive method in which the penis is withdrawn from the vagina before ejaculation to prevent sperm from entering the female's body.

colchicine *(kŏl'-chĭ-sĕn).* Name of a medication used to treat acute attacks of gout.

cold *(kōld).* General term for the common cold, a viral infection of the upper respiratory tract characterized by slight rise in temperature, inflammation of the nasal mucous membranes, sneezing, chills, and malaise.

cold sore. *See* **herpes simplex.**

colectomy *(kō-lĕk'-tō-mē).* Surgical removal of part or all of the colon.

colic *(kŏl'-ĭk).* Painful spasm in any tubular or hollow soft organ.

biliary c. Pain and other severe symptoms caused by movement of gallstones along the bile duct; also called *cholecystalgia.*

bilious c. Abdominal pain accompanied by vomiting of bile.

infantile c. Condition seen in infants younger than three months characterized by abdominal pain and frantic crying.

renal c. Flank pain associated with passage of a kidney stone down the ureter.

colitis *(kō-lī'-tĭs).* Inflammation of the colon.
cathartic c. Colitis caused by chronic use of laxatives.
infectious c. Colitis caused by an infectious agent.
ulcerative c. Chronic, recurrent ulceration in the colon, cause unknown. Characterized by crampy abdominal pain, rectal bleeding, loose discharges of pus, blood, and mucous with fecal matter.

collagen *(kŏl'-ă-jĕn).* Protein that makes up a major part of connective tissues (e.g., cartilage, bone, ligaments, skin).

collagen diseases (also called *collagen vascular diseases*). Group of diseases that involve inflammatory damage to connective tissues and blood vessels; examples include systemic lupus erythematosis, scleroderma and rheumatoid arthritis.

collarbone *(kŏl'-ăr-bōn).* Pair of bones that extend from the breastbone to the shoulder tip; also called *clavicle.*

collateral *(kō-lăt'-ĕr-ăl).* 1. Secondary; alternative. 2. Small side branch of a blood vessel or nerve.

Colles' fracture *(kŏl'ēz).* Common fracture of the wrist, in which a break occurs in the radius.

colloid *(kŏl'-oid).* Type of mixture in which microscopic particles, when dispersed in a solvent to the greatest degree possible, remain evenly distributed but don't form a true solution.

coloboma *(kŏl-ō-bō'-mă).* Defect or lesion of the eye that may include the eyelid; may be congenital or caused by trauma or disease.

colon *(kō'-lŏn).* Portion of the large intestine that extends from the cecum to the rectum.

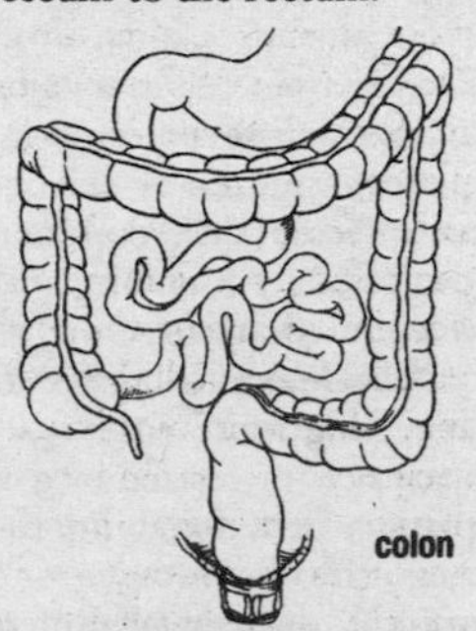
colon

ascending c. Portion of the colon between the cecum and the right colic (hepatic) flexure.
descending c. Portion of the colon between the left colic (splenic) flexure and the sigmoid colon.
sigmoid c. S-shaped terminal portion of the colon, located in the pelvis.
spastic c. *See* **irritable bowel syndrome.**
traverse c. Portion of the colon that runs across the upper part of the abdomen from the right to the left colic flexure.

colon hydrotherapy. Use of water to flush out accumulated waste and toxins from the upper and lower portions of the large intestine; also called *colonic irrigation.*

colonoscopy *(kō-lĕn-ŏs'-kĕ-pē).* Use of a flexible endoscope, inserted into the colon, to inspect the lining of the colon for polyps, tumors, and other abnormalities.

colony *(kŏl'-ĕ-nē).* Group of microorganisms grown from a single parent cell in a culture.

Colorado tick fever. Viral infection transmitted by the bite of an infected tick, common in the Rocky Mountain area of the United States.

color blindness. Inherited trait, occurring more commonly in males, characterized by a complete or partial inability to distinguish colors.

colorectal *(kō-lō-rĕk'-tăl).* Referring to the region of the body consisting of the lower large intestine and the rectum.

color therapy. A form of energy healing that involves the therapeutic use of different forms of color, including colored lights. The premise is that each color vibrates at its own frequency, and that proper use of color can influence body chemistry through these vibrations.

colostomy *(kĕ-lŏs'-tĕ-mē).* Surgical creation of an opening (stoma) in the abdominal wall to allow material to pass from the bowel through that opening rather than through the anus.

colostrum *(kĕ-lŏs'-trĕm).* Fluid released by the mother's breasts just before or after childbirth, it contains protective antibodies, proteins, and other important substances.

colposcopy *(kŏl-pŏs'-kō-pē).* Use of a special instrument (colposcope) to do a visual examination of the vagina and cervix.

coma *(kō'-mă).* Abnormal state of deep unconsciousness.

comedo *(kŏm'-ē-dō).* Accumulation of dead cells and a fatty substance (sebum) that plugs a hair follicle or oil gland; also called a *blackhead* (open comedo) or *whitehead* (closed comedo).

comedocarcinoma *(kō-mē-dō-kăr-sĭ-nō'-mă).* Type of breast cancer that generally is confined to the mammary ducts.

comfrey *(kŏm'-frē).* A prolific plant (*Symphytum officinale*) that has been used since ancient times to promote wound and bone healing when used as a poultice. This ability seems to be associated with a component called allantoin, which helps promote new cell growth.

comminuted fracture *(kŏm-ĭ-nūt'-ĕd).* Fracture in which the bone has been broken or crushed into several pieces.

Commission E Report. Report on herbal medication widely regarded to contain convincing evidence of the efficacy of herbal remedies; issued by the German Federal Institute for Drugs and Medical Devices.

commissure *(kŏm'-ĭ-shŭr).* 1. Place where two structures meet and join, such as the lips. 2. Band of nerve fibers that cross the midline of the brain.

commissurotomy *(kŏm-ĭ-shŭr-ŏt'-ō-mē).* Surgical procedure in which a commisure is cut.

communicable disease *(kŏ-mū'-nĭ-kă-bl).* Any infectious disease that may be transmitted from one person or animal to another, directly or indirectly.

compassionate use *(kŏm-pă'-shŭn-ĕt).* A term used in the US to describe the practice of providing an experimental therapy to patients before the Food and Drug Administration has approved it for human use. This procedure is limited to very ill individuals who have no other treatment options.

compatibility *(kŏm-păt-ĭ-bĭl'-ĭ-tē).* 1. Ability of two substances to exist together without unfavorable results. 2. Ability of blood from two different people to mix without interaction.

Compazine. Trade name of a medication used to control nausea and vomiting; generic name: prochlorperazine.

compensation case *(kŏm-pĕn-sā'-shŭn).* Disease or injury that occurs in relation to one's job, managed by the Workman's Compensation Board.

competence *(kŏm'-pĕ-tĕns).* Ability of a body part or organ to adequately perform its function.

complete blood count (CBC). Very common test used to determine the number of red and white blood cells and in some cases, platelets, in 1 cubic milliliter sample of blood.

complex *(kŏm'-plĕks).* 1. Collection or combination of various things or related factors (e.g., a complex of symptoms; *see also* **syndrome**). 2. All the emotions, ideas, and sensations associated with a subject.

complexion *(kŏm-plĕk'-shĕn).* Texture, color, and appearance of the facial skin.

complication *(kŏm-plĭ-kā'-shĕn).* Any unwanted condition that occurs during the course of or because of another physical disorder (e.g., development of bedsores because of being confined to bed).

compos mentis *(kŏm'-pŏs mĕn'-tĭs).* Latin term that means "of sound mind."

compound *(kŏm'-pound).* 1. Substance composed of two or more elements chemically combined in specific proportions by weight. 2. In pharmacology, a preparation that contains a mixture of medications.

compound fracture. *See* **fracture.**

compress *(kŏm'-prĕs).* Pad of gauze or other soft material applied with pressure to a wound or injured part to help control bleeding or relieve inflammation.

compression bandage. Strip of cloth wrapped around a body part to stop bleeding, immobilize the part, or prevent fluid from collecting in a limb.

compulsion *(kŏm-pŭl'-shŭn).* Irresistible urge to do something that is irrational or useless, done to relieve fear connected with an obsession.

computed tomography (CT) *(kŏm-pyū'-tĭd).* X-ray method for examining the body's soft tissues, in which a beam repeatedly passes through a body part and a computer calculates the level of absorption to assemble the image; formerly called computed axial tomography (CAT).

concave *(kŏn'-kāv).* Having a hollow surface.

concentration *(kŏn-sĕn-trā'-shŭn).* 1. The amount of a substance in a mixture or solution. 2. Increasing the strength of fluid by evaporation. 3. Ability to focus one's thoughts on one subject.

conception *(kŏn-sĕp'-shŭn).* Union of male sperm and the ovum of a female; fertilization.

concha *(kŏng'-kă).* Outer ear; the pinna.

concretion *(kŏn-krē'-shŭn).* Stone that forms as the result of the solidification and concentration of mineral salts.

concussion *(kŏn-kŭsh'-ŭn).* Injury that results from impact with an object, often associated with the brain.

condensation *(kŏn-dĕn-sā'-shŭn).* 1. Act of making something more compact. 2. Transformation of a gas into a liquid or a liquid into a solid.

conditioning *(kŏn-dĭsh'-ŭn-ĭng).* Process of training an individual or organism to respond to a specific stimulus in a specific way.

condom *(kŏn'-dŏm).* Sheath, usually made of thin rubber, used to cover the penis during sexual intercourse to prevent pregnancy and/or infection.

conduction *(kŏn-dŭk'-shŭn).* Transmission through a conductor of energy (heat, electricity, sound) or nerve signals from one point to another.

condyle *(kŏn'-dīl).* Rounded bump at the end of a bone where it forms a joint with another bone or bones.

condyloma *(kŏn-dĭ-lō'-mă).* Wart-like growth usually seen near the opening of the rectum or vulva.

confabulation *(kŏn-făb-ū-lā'-shŭn).* In psychology, a reaction to memory loss in which an individual fills in memory gaps with fabricated, inappropriate words or thoughts.

confidentiality *(kŏn-fĭ-dĕn-chē-ăl'-ĭ-tē).* Referring to information health-care practitioners obtain from patients that cannot be revealed to a third party without the patient's consent.

conflict *(kŏn'-flĭkt).* In psychiatry, mental/emotional struggle, conscious or unconscious, caused by the simultaneous presence of equally desirable or undesirable wishes.

confluent *(kŏn'-flū-ĕnt).* Merging or running together.

confusion *(kŏn-fyū'-zhŭn).* State of mind in which a person is disoriented or not aware with respect to time, place, or self; it can occur under severe stress due to an organic mental disorder.

congelation *(kŏn-jĕl-ā'-shŭn).* Frostbite or freezing.

congenital *(kŏn-jĕn'-ĭ-tl).* Present at birth.

c. defect. Abnormality present at birth; may be inherited, acquired during pregnancy, or result during the birth process.

congestion *(kŏn-jĕs'-chĕn).* Abnormal collection of blood or other fluid in a body part, as in the lungs.

congestive heart failure. Impaired heart function associated with retention of water and salt, resulting in shortness of breath, swollen feet and legs, and poor blood circulation.

conization *(kōn-ĭ-zā'-shŭn).* Surgical removal of a conical portion of tissue.

cervical c. Removal of a lesion and surrounding tissue from the cervix.

cold c. Conization done with a knife.

laser c. Conization accomplished with a laser beam.

conjugated estrogens. *See* Appendix, Common Prescription and OTC Drugs: By Generic Name.

conjunctivitis *(kŏn-jŭngk-tĕ-vī'-tĭs).* Inflammation of the mucous membrane that lines the eyelids and the front of the eye, caused by infection, allergy, or irritation.

connective tissue *(kŏ-nĕk'-tĭv).* Fibrous material that supports and binds other tissues and body parts; includes skin, ligaments, bone, and tendons.

conscious *(kŏn'-shĕs).* Being aware, awake; able to respond.

consent *(kŏn-sĕnt').* Voluntary authorization for a procedure or other event, such as surgery or to participate in a study.

consolidation *(kŏn-sŏl-ĭ-dā'-shŭn).* Solidification into a dense mass, especially when a lung affected with pneumonia becomes filled with fluid.

constipation *(kŏn-stĭ-pā'-shŭn).* Difficult or infrequent bowel movements due to loss of muscle tone in the intestines, poor diet, intestinal obstruction, or other causes.

constitution *(kŏn-stĭ-tū'-shŭn).* Psychological and physical makeup of an individual, including inherited traits.

constriction *(kŏn-strĭk'-shŭn).* Squeezing, narrowing, or tightening of a part.

consultation *(kŏn-sŭl-tā'-shŭn).* 1. Conference between two or more health-care practitioners to discuss the diagnosis and treatment of a specific patient. 2. Opinion provided by a specialist.

contact dermatitis *(kŏn'-tăkt).* Skin inflammation that results from exposure to an irritant or allergen.

contact lens. Flexible or rigid device that is fitted to an individual's eyeball to aid vision.

continuous wear c. Lens that can be worn for a week or longer without removing them from the eye.

contagion *(kŏn-tā'-jĕn).* Transmission of a disease from an affected person to others by either direct or indirect contact.

contamination *(kŏn-tăm-ĭ-nā'-shŭn).* Process by which something becomes infected or soiled, especially because of introduction of disease-causing microorganisms.

continence *(kŏn'-tĕ-nĕns).* 1. Ability to hold urine and feces and to release them voluntarily. 2. Self-restraint or moderation, as in eating or sexual activity.

contraceptive *(kŏn-tră-sĕp'-tĭv).* Any one of various means of preventing pregnancy by blocking conception.

contractility *(kŏn-trăc-tĭl'-ĭ-tē).* Ability to shorten or contract in response to a stimulus.

contraction *(kŏn-trăk'-shŭn).* Shortening in size, often used to refer to a muscle.

contracture *(kŏn-trăk'-chĕr).* Abnormal, permanent shortening of a muscle or tendon caused by atrophy of muscle fiber, scar tissue, or other reasons.

contraindication *(kŏn-tră-ĭn-dĭ-kā'-shŭn).* Any factor that prevents the use of a particular procedure or drug inadvisable for a certain patient because unwanted effects may result.

contralateral *(kŏn-tră-lăt'-ĕr-ăl).* Located on the opposite side (opposite: ipsilateral).

contrast medium. Radiopaque substance that is administered orally, intravenously, or via enema, which, when it is exposed to x-rays, allows tissues or organs to be visualized more clearly.

control *(kŏn-trōl').* In research or an experiment, the standard of comparison against which observations or conclusions may be checked for validity; for example, in an experiment one group of subjects receives a placebo (control group) and another receives a drug.

controlled drug or **substance.** Drug that a physician and pharmacist must record when it is dispensed; refers to drugs that are addictive or habit-forming as defined by the Controlled Substances Act of 1970.

contusion *(kŏn-tū'-shŭn).* Injury from a blow that does not break the skin; bruise.

convalescence *(kŏn-vĕ-lĕs'-ĕns).* Period of recovery after surgery, an episode of disease, or an injury.

conversion *(kŏn-vĕr'-shŭn).* 1. Act of changing from one state or position to another, e.g., changing the position of a baby to assist in delivery. 2. In psychiatry, unconscious defense mechanism by which emotional conflicts are suppressed and turned into physical symptoms that have no organic basis; also known as conversion disorder or reaction.

conversion hysteria. Emotional disorder in which emotional conflicts are transformed into physical symptoms, such as paralysis, blindness.

convex *(kŏn-vĕks').* Having an outwardly curved surface; opposite of concave.

convolution *(kŏn-vō-lū'-shŭn).* Twisting, folding, or coiling of a body part upon itself.

convulsion *(kŏn-vŭl'-shŭn).* Sudden, violent, involuntary muscular contraction, or a series of such contractions; seizure.

clonic c. Convulsion characterized by alternating contracting and relaxing of the muscles.

febrile c. Convulsion occurring in a child with a febrile illness.

tonic c. Convulsion in which the contractions are prolonged.

convulsive disorder. Any disease characterized by recurring convulsions; e.g., epilepsy.

coordination *(kō-ŏr-dĭn-ā'-shĕn).* Condition in which various parts or systems work together to perform a function.

COPD. Acronym for chronic obstructive pulmonary disease.

copper *(kŏp'-ĕr).* Metallic element that is essential to normal body functioning, as it is a component of various proteins.

copremesis *(kŏp-rĕm'-ĭ-sĭs).* Vomiting of feces, which can occur when the intestines are blocked.

coprophagy *(kŏp-rŏf'-ă-jē).* Eating excrement.

copulation *(kŏp-ū-lā'-shĕn).* Sexual intercourse.

cor *(kŏr).* Referring to the heart.

c. pulmonale. Enlargement or failure of the right ventricle, resulting from disorders of the lungs, pulmonary vessels, or chest wall.

Cordarone. Trade name of medication for treatment of irregularities of the heart; generic name: amiodarone.

cordectomy *(kŏr-dĕk'-tō-mē).* Surgical procedure in which all or part of a cord (e.g., vocal cord, spinal cord) is resected or removed.

cord presentation. Appearance of the umbilical cord at the vaginal opening during labor, an indication that delivery should be done quickly to save the child's life.

corectopia *(kŏr-ĕk-tō'-pē-ă).* Location of the pupil to one side of the center of the iris.

coreoplasty *(kŏr'-ē-ō-plăs-tē).* Surgical procedure to form an artificial pupil.

Corgard. Trade name of a medication for long-term treatment of angina pectoris; generic name: nadolol.

corium *(kō'-rē-ŭm).* Skin, specifically the dermis.

corn *(kŏrn).* Thickened, horny mass of skin cells that overlie a bone, usually on the toes, that is the result of chronic pressure.

cornea *(kŏr'-nē-ă).* Outer, transparent part of the eye overlying the iris, composed of five layers.

corneal transplant. Transplantation of a healthy cornea from a donor, typically someone who has recently died, to replace a damaged or diseased one in a recipient.

cornification *(kŏr-nĭ-fĭ-kā'-shŭn).* Process by which squamous epithelial cells are converted into horny tissue or keratin.

cornu *(kŏr'-nū).* Horn-shaped structure or projection of an organ; e.g., portion of the uterus near entry to the fallopian tubes.

cornual pregnancy. Pregnancy in which the embryo implants in the cornu of the uterus, which may cause the uterus to rupture as the embryo grows.

coronary *(kŏr'-ō-nă-rē).* Encircling as the blood vessels that supply the heart muscle; used to refer to the heart and to coronary heart disease.

c. arteries. Pair of blood vessels that branch from the aorta and supply the heart with blood.

c. artery bypass. Surgical procedure in which a blocked section of the coronary artery is bypassed by grafting a portion of a blood vessel around the blockage in order to improve blood supply to the heart.

c. artery disease. *See* **atherosclerosis.**

c. care unit (CCU). Intensive care unit dedicated to caring for individuals who have had a heart attack or other serious heart problem.

c. endarterectomy. Surgical procedure in which arteriosclerotic plaques are cleaned out of the coronary artery.

c. occlusion. Obstruction or closure of a coronary artery, caused by a blood clot or atherosclerosis; a heart attack.

c. thrombosis. Presence of a blood clot in any artery that supplies the heart muscle.

coroner *(kŏr'-ō-nĕr).* Government official empowered to investigate the cause of death. Depending on local laws, a coroner may or may not be a physician. *See* **medical examiner.**

corpulent *(kŏr'-pū-lĕnt).* Obese; fat.

corpus *(kŏr'-pŭs).* 1. Body. 2. The main or specialized portion of a structure that can be distinguished from the surrounding tissues.

c. cavernosum. Any erectile tissue, especially that of the penis and clitoris.

c. luteum. Structure in an ovary formed where an egg has developed and burst from the gland.

c. spongiosum. Middle column of erectile tissue of the penis through which the urethra passes.

corpuscle *(kŏr'-pŭs-ĕl).* Any small mass or body, such as a nerve ending or blood cell.

corrosive *(kŏ-rō'-sĭv).* Any agent that causes destruction of living tissue.

Cortef. Trade name for hydrocortisone, used to treat a wide variety of conditions, including allergic disorders, skin disorders, gastrointestinal disorders, respiratory disorders, rheumatic disorders, among others.

cortex *(kŏr'-tĕks).* 1. Outer layer of an organ; e.g., cerebral cortex (of the brain) or adrenal cortex (of the adrenal gland); *see* *also* **adrenal cortex, cerebral cortex.**

cortical *(kŏr'-tĭ-kăl).* Referring to the cortex.

corticoid *(kŏr'-tĭ-koid).* Any of various hormonal substances obtained from the cortex of the adrenal gland; corticosteroid.

corticosteroid *(kŏr-tĭ-kō-stĕr'-oid).* Any of a group of hormones (e.g., cortisol, aldosterone) produced by the adrenal cortex.

corticotropin. *See* **ACTH.**

cortisol *(kŏr'-tĭ-sŏl).* Hormone produced by the adrenal cortex, it is used to treat inflammatory conditions such as rheumatoid arthritis; also called *hydrocortisone.*

cortisone *(kŏr'-tĭ-sōn).* Hormone produced by the adrenal cortex, it regulates carbohydrate metabolism; as a drug it relieves inflammation.

Cortone. Trade name of a steroid medication used to treat skin disorders, allergic states, rheumatoid disorders and other conditions; generic name: cortisone acetate.

coruscation *(kō-rŭs-kā'-shŭn).* Sensation of flashes of light before the eyes.

cosmeceutical *(kŏs-mĕ-sū'-tĭ-kăl).* Cosmetic product claimed to have medicinal or druglike properties; the US Food and Drug Administration does not recognize these products as drugs.

cosmetic surgery. Surgery performed for the purpose of enhancing a person's appearance, not due to medical necessity.

costal cartilage. Cartilage that connects the end of the long ribs to the sternum (breastbone).

costochondral *(kŏs-tō-kŏn'-drăl).* Referring to a rib and its cartilage.

costochondritis *(kŏs-tō-kŏn-drī'-tĭs).* Inflammation of the cartilage of the chest wall, usually at the joint with the breastbone (sternum).

costovertebral junction *(kŏs-tō-vĕr'-tĕ-brăl).* Area in the back where the ribs join the spine.

cot, finger. Thin rubber covering worn on the finger, used by some physicians during examination.

cough *(kawf).* Forceful, sometimes violent, and sudden expulsion of air from the lungs.

Coumadin. Trade name for an anticoagulant medication, used to prevent and treat blood clots; generic name: warfarin.

counterirritant *(kŏwn-tĕr-ĭr'-ĭ-tănt).* Substance applied locally to produce a mild irritation for the purpose of alleviating an underlying inflammation.

Cowper's glands *(kow'-pĕrs).* Pair of pea-sized glands located under the bulb of the male urethra that excrete a mucous substance. Also known as *bulbourethral glands.*

coxa *(kŏk'-să).* Hip.

c. valga. Deformity produced when the angle of the femur head within the shaft is increased to more than 120 degrees.

c. vara. Deformity produced

by a decrease in the angle made by the head of the femur with the shaft.

coxalgia *(kŏk-săl'-jē-ă).* Pain in the hip.

coxsackie disease *(kŏk-săk'-ē).* Infectious disease that mimics infantile paralysis and meningitis, but which disappears after a few days.

CPR. Acronym for cardiopulmonary resuscitation.

cradle cap. Common skin disorder of newborns, characterized by a thick, yellow, greasy scalp.

cramps *(krămps).* Painful, frequently spasmodic tightening or contraction of a muscle, often in the uterus during menstruation or in the calf.

cranial *(krā'-nē-ăl).* Referring to the head, brain, or skull bones that encase the brain.

cranial nerves. Twelve pairs of nerves that originate in the brain, with each pair having specific motor and/or sensory functions.

craniectomy *(krā-nē-ĕk'-tō-mē).* Procedure to open the skull and remove a portion of it.

craniocleidedysostosis *(krā-nē-ō-klī-dō-dĭs-ŏs-tō'-sĭs).* Congenital condition in which the bones of the face, skull, and collarbone are improperly formed.

craniopharyngioma *(krā-nē-ō-fă-rĭn-jē-ō'-mă).* Tumor that develops at the base of the brain near the pituitary gland; also called *Rathke's pouch tumor.*

craniosacral *(krā-nē-ō-sā'-krăl).* Referring to the skull and the sacrum.

c. therapy. Alternative therapy in which practitioners apply gentle pressure to the craniosacral system (skull; face, spinal column) with the intention of releasing restricted flow of the cerebrospinal fluid and restore balance to the body.

craniosynostosis *(krā-nē-ō-sĭn-ŏs-tō'-sĭs).* Premature closure of the sutures of the skull in an infant.

craniotomy *(krā-nē-ŏt'-ĕ-mē).* Surgical procedure in which an opening is made into the skull to remove tumors, relieve pressure, or control bleeding.

cranium *(krā'-nē-ŭm).* Generally, the bones of the head; specifically, the bones that enclose the brain; also called the skull.

C-reactive protein. A plasma protein whose levels rise dramatically in the presence of certain conditions, such as inflammatory disorders, advanced cancer, trauma, surgery, burns, and infections; moderate or mild increases can be seen with psychological stress, childbirth, and strenuous exercise.

creatine *(krē'-ă-tĭn).* Amino acid produced in the body and found mainly in muscle and brain tissue.

creatinine *(krē-ăt'-ĭ-nĭn).* Product of creatine metabolism, it is removed from the body in the kidneys and excreted in urine. Levels of creatinine are used as an index of kidney function.

creatinuria *(krē-ă-tĭn-ū'-rē-ă).* Excessive concentration of creatinine in urine, seen especially in muscular dystrophy, polio, and other conditions.

cremation *(krē-mā'-shŭn).* Burning of a corpse.

crepitation *(krĕp-ĭ-tā'-shĕn).* 1. Grating sound, like that made by rubbing together the ends of a fractured bone. 2. Crackling sound heard in the chest with the stethoscope in conditions such as pneumonia.

CREST. Acronym for a syndrome characterized by calcinosis, Raynaud's phenomenon, esophageal involvement, sclerodactyly, and telangiactasia.

cretinism *(krēt'-ĭn-ĭzm).* Severe, congenital condition characterized by stunted growth, mental retardation, apathy, and protruding swollen tongue, due to inadequate production of thyroid hormones.

Creutzfeldt-Jakob disease *(kroit-s'fĕlt yăh'kōb).* Rare, untreatable, and fatal disease of the brain, caused by proteinlike particles called prions, characterized by progressive dementia, muscle wasting, and involuntary movements.

crib death. *See* **sudden infant death syndrome.**

crisis *(krī'-sĭs).* 1. Turning point in a disease, for better or worse, but usually for the better. 2. Sudden intensification of symptoms in the course of a disease.

crista *(krĭs'-tă)* (*pl.* cristae). Projection, sometimes branched, or a ridge, especially one on top of a bone or its border.

critical care specialist. Physician who specializes in treating critically ill patients.

Crohn's disease *(krōnz).* Chronic inflammatory condition that affects the colon and/or part of the small intestine and is characterized by episodes of diarrhea, abdominal pain, nausea, fever, weakness, and weight loss. Similar to but distinct from ulcerative colitis.

cromolyn sodium *(krō'-mŏ-lĭn).* A medication used to treat airway obstruction associated with asthma. Trade name: Intal.

crossmatching. Procedure used by blood banks to test whether a donor's blood is compatible with that of a potential transfusion recipient.

croup *(krūp).* Common term to describe a disease of infants and young children, characterized by hoarseness, fever, breathing difficulties, and a barking cough.

crush syndrome. Kidney failure and other consequences that occur as a result of a severe crushing injury to one or more limbs.

cryanesthesia *(krī-ăn-ĕs-thē'-zē-ă).* Loss of a sense of cold.

cryesthesia *(krī-ĕs-thē'-zē-ă).* Sensitivity to the cold.

crymodynia *(krī-mō-dĭn'-ē-ă).* Pain experienced when exposed to cold or damp weather. Common among people with arthritis.

cryobiology *(krī-ō-bī-ŏl'-ō-jē).* Study of the impact of cold on different biological systems.

cryoextraction *(krī-ō-ĕks-trăk'-shŭn).* Use of a cooling tube that is inserted into the eye to produce an ice ball, which is then removed. Used to treat cataracts.

cryogenics *(krī-ō-jĕn'-ĭks).* Science concerned with the production and effects of very low temperatures to treat human illness.

cryohypophysectomy *(krī-ō-hī-pō-fĭz-ĕk'-tō-mē).* Use of very low temperatures to destroy the pituitary gland.

cryoprobe *(krī'-ō-prōb).* Device that produces extreme cold, used to freeze tissue.

cryosurgery *(krī-ō-sŭr'-jĕr-ē).* Any surgical procedure in which tissues are exposed to extreme cold to destroy tissue, remove tumors, control bleeding, or produce lesions.

cryothalamectomy *(krī-ō-thăl-ă-mŏt'-ō-mē).* Use of extreme cold to destroy a portion of the thalamus.

cryotherapy *(krī-ō-thĕr'-ă-pē).* Use of very low temperatures to treat disease.

crypt *(krĭpt).* Small sac or cavity that extends to an epithelial surface.

cryptitis *(krĭp-tī'-tĭs).* Inflammation of a crypt or follicle.

cryptorchidism *(krĭp-tŏr'-kĭ-dĭz-ĕm).* Failure of one or both testes to descend into the scrotum as the male fetus develops.

crystalline lens *(krĭs'-tă-lĭn).* Lens of the eye in the capsule behind the pupil, it refracts light rays and brings them into focus on the retina.

crystalluria *(krĭs-tă-lū'-rē-ă).* Presence of crystals in the urine, which may occur after using sulfonamides.

CT scan. *See* **computed tomography.**

cubital *(kū'-bĭ-tăl).* Referring to the forearm or ulna.

cubitus *(kū'-bĭ-tŭs).* Forearm; elbow; ulna.

c. valgus. Deformity of the arm in which the elbow deviates away from the midline of the body when the arm is extended.

c. varus. Deformity of the arm in which the forearm deviates toward the midline of the body when the arm is extended.

culdoscope *(kŭl'-dō-skōp).* Special instrument (endoscope) used to visually examine female pelvic organs.

culdoscopy *(kŭl-dŏs'-kō-pē).* Examination of the internal female pelvic organs using an endoscope.

culture *(kŭl'-chĕr).* Intentional growing of microorganisms or living tissue cells in a medium.

culture medium. Substance on which microorganisms or living tissue may grow, such as agar, broth, or gelatin.

cupping *(kŭp'-ĭng).* Placement of a glass cup or other vessel from which air is withdrawn by heat, or use of a special suction device to attract blood to the surface.

curare *(kū-ră'-rē).* One of several extracts of South American trees, an active ingredient of which is used in a modified form to relax muscles during anesthesia.

curet *(kū-rĕt').* Spoon-shaped surgical instrument used to scrape the walls of a body cavity; also called *curette.*

curettage *(kū-rĕ-tăzh').* Scraping of a cavity (e.g., inside the uterus) to remove unwanted material (e.g., tumor) or to obtain a sample.

curvature of spine *(kŭr'-vĕ-chŭr).* General term referring to curves of the vertebral column—cervical, thoracic, lumbar, and sacral; also various misalignments of the spine, including kyphosis, lordosis, scoliosis, and others.

Cushing's disease *(koosh'-ĭngz).* Disorder characterized by excessive secretion of adrenocorticotropic hormone (ACTH) by the pituitary gland, which causes increased secretion of hormones by the adrenal cortex, resulting in fat deposits on the face, back, and chest, among other symptoms.

Cushing's syndrome. Disorder caused by excessive cortisol levels and characterized by moon face, high blood pressure, weight gain, emotional disturbances, and, in women, abnormal growth of body hair.

cusp *(kŭsp).* 1. Tapered point, especially on teeth. 2. Any of the flaps or segments of a heart valve.

cuspid *(kŭs'-pĭd).* Canine or dog tooth, it is the longest tooth in the mouth.

cutaneous *(kū-tā'-nē-ŭs).* Referring to the skin.

cutdown *(kŭt'-doun).* Surgical procedure in which a small opening is made over a vein or sometimes an artery so a needle or cannula can be placed into it.

cuticle *(kū'-tĭ-kl).* 1. Part of the skin overlying the base of a finger- or toenail. 2. Epidermis.

cutis *(kū'-tĭs).* Skin, consisting of the epidermis and the dermis (corium), which lay on top of subcutaneous tissue.

CVA. Acronym for cerebrovascular accident, or stroke.

cyanide poisoning *(sī'-ăn-īd).* Poisoning from ingestion or inhalation of cyanide, a dangerous compound found in smoke from fire and industrial chemicals, resulting in drowsiness, convulsions, and often death.

cyanocobalamin *(sī-ĕ-nō-kō-băl'-ĕ-mĭn).* Vitamin B12, essential for normal metabolism, blood formation, and nerve function.

cyanosis *(sī-ĕ-nō'-sĭs).* Bluish, grayish, or darkish discoloration of the skin and mucous membranes, which occurs when the amount of oxygen in the blood is greatly reduced.

cycle *(sī'-kĕl).* Series of steps or events that occur regularly.

cyclic therapy *(sī'-klĭk).* Hormone treatment given based on the phases of the menstrual cycle.

cyclitis *(sĭk-lī'-tĭs).* Inflammation of

the ciliary body (around the colored area) of the eye.

cyclobenzaprine. *See* Appendix, Common Prescription and OTC Drugs: By Generic Name.

cyclodialysis *(sī-klō-dī-ăl'-ĭ-sĭs).* Surgical procedure to create an opening in the eye to reduce pressure associated with glaucoma.

cyclophosphamide *(sī-klō-fŏs'-fă-mīd).* Antitumor agent, often used with other agents, for various cancers, including Hodgkin's disease, various leukemias, breast cancer, ovarian cancer, among others.

Cyclosporine A. Trade name of a drug used to fight organ rejection after organ transplantation.

cyst *(sĭst).* Closed, fluid-filled sac embedded in tissue.

cystadenocarcinoma *(sĭs-tăd-ĕ-nō-kăr-sĭ-nō'-mă).* Malignant tumor derived from gland tissue, usually seen in the ovaries, but also in the pancreas, appendix, and thyroid.

cystadenoma *(sĭs-tăd-ĕ-nō'-mă).* Benign tumor containing large cystic masses, usually found in the ovaries, salivary glands, and pancreas.

cystectomy *(sĭs-tĕk'-tō-mē).* 1. Removal of a cyst. 2. Removal of part or all of the urinary bladder.

cysteine *(sĭs-tē'-ēn).* Sulfur-containing nonessential amino acid.

cystic disease of the breast. *See* **fibrocystic change of breast.**

cystic duct *(sĭs'-tĭk).* Tube that connects the gallbladder to the common bile duct, a common site for gallstones.

cystic fibrosis. Inherited childhood disease in which exocrine glands malfunction, resulting in accumulation of mucus in the lungs, pancreas, and other organs.

cystine *(sĭs'-tēn).* A sulfur-containing amino acid and an important source of sulfur in metabolism.

cystinuria *(sĭs-tĭ-nū'-rē-ă).* 1. Presence of cystine in the urine. 2. Hereditary disorder characterized by excretion of large amounts of cystine and other amino acids, resulting in urinary stone formation.

cystitis *(sĭs-tī'-tĭs).* Inflammation of the urinary bladder and ureters, most commonly seen in women.

cystocele *(s ĭs'-tō-sēl).* Condition in which the urinary bladder pushes through the vaginal wall, sometimes occurring after childbirth; bladder hernia.

cystogram *(sĭs'-tō-grăm).* X-ray of the urinary bladder.

cystolithectomy *(sĭs-tō-lĭ-thĕk'-tō-mē).* Surgical removal of bladder stones; also called *cystolithotomy.*

cystolithiasis *(sĭs-tō-lĭ-thī'-ă-sĭs).* Presence of stones in the urinary bladder.

cystopyelitis *(sĭs-tō-pī-ĕ-lī'-tĭs).* Inflammation of the bladder and the pelvis of the kidney.

cystoscope *(sĭs'-tō-skōp).* Tubular

device with a light that allows doctors to examine the interior of the urinary bladder.

cystosteatoma *(sĭs-tō-stē-ă-tō'-mă).* Cyst of a sweat gland, located under the skin.

cystotomy *(sĭs-tŏt'-ō-mē).* Surgical incision of the urinary bladder.

cytoanalyzer *(sī-tō-ăn'-ă-lī-zĕr).* Instrument used to screen smears that contain cells suspected of being malignant.

cytology *(sī-tŏl'-ō-jē).* Study of the nature of cells.

cytolysis *(sī-tŏl'-ĭ-sĭs).* Destruction of living cells.

cytomegalovirus (CMV) *(sī-tō-mĕg-ă-lō-vī'-rĕs).* Any of a group of herpes viruses that cause disease in individuals who have an impaired or immature immune system, including infants and those with acquired immune deficiency syndrome.

Cytomel. Trade name for a medication that contains thyroid hormone, used to treat hypothyroidism; generic name: liothyronine.

cytopathology *(sī-tō-păth-ŏl'-ō-jē).* Study of cells in disease.

cytoplasm *(sī'-tō-plăz-ĕm).* All of the substance of a cell that is outside the nucleus.

cytoreductive surgery *(sī-tō-rē-dŭk'-tĭv).* Surgical procedure to reduce the size of a tumor that cannot be removed entirely.

cytotoxic *(sī-tō-tŏk'-sĭk).* Referring to the destruction of cells.

cytotoxic drug. Drug commonly used in the treatment of cancer to inhibit the growth and spread of cells.

cytotoxin *(sī'-tō-tŏk-sĭn).* Antibody that destroys or inhibits the function of cells.

D

Dacron *(dā'-krŏn).* Trade name for a plastic material that, in fiber form, is used as a suture material; in fabric form, for blood vessel grafts and catheters.

dacryocystitis *(dăk-rē-ō-sĭs-tī'-tĭs).* Inflammation of the lacrimal (tear) sac, usually seen in infants and menopausal women.

dacryolith *(dăk'-rē-ō-lĭth).* Stone in the tear-forming structures.

dacryopyosis *(dăk-rē-ō-pī-ō'-sĭs).* Formation or discharge of pus in the lacrimal (tear) duct.

dacryorrhea *(dăk-rē-ō-rē'-ă).* Excessive flow of tears.

dacryostenosis *(dăk-rē-ō-stĕ-nō'-sĭs).* Narrowing of any tear passageway.

dactylomegaly *(dăk-tē-lō-mĕg'-ĕ-lē).* Abnormally large fingers and/or toes.

dacyl *(dăk'-tĭl).* Digit; finger or toe.

daily reference values (DRV). Set of dietary references, typically seen on food labels, that reveal information about fat, saturated fat, cholesterol, carbohydrates, protein, fiber, sodium, and potassium values.

Dalmane. Trade name for a drug used to aid sleep; generic name: flurazepam.

daltonism. Inherited, sex-linked form of color blindness in which an individual cannot distinguish red from green.

dam. Barrier, often made of thin rubber, that prevents the flow of fluid; used in dentistry and surgery to isolate the operative area.

D and C. *See* **dilatation and curettage.**

dandelion *(dăn'-dē-līyon).* Plant valued by natural and some conventional health-care practitioners for its diuretic (urination-promoting) properties.

dander *(dăn'-dĕr).* Small scales from the hair or feathers of animals that can cause allergic reactions in sensitive individuals.

dandruff *(dăn'-rŭf).* Common name for a mild form of seborrheic dermatitis; *see* **seborrheic dermatitis.**

dark-field examination. Microscopic examination designed to identify the presence of the bacteria that cause syphilis.

Darvon. Trade name of a drug used to relieve pain; generic name: dextropropoxyphene.

DASH diet. DASH, an acronym for Dietary Approaches to Stop Hypertension, is an eating plan designed to reduce blood pressure. It is low in saturated and total fat and cholesterol, high in fiber and fruits and vegetables.

Datril. *See* Appendix, Common Prescription and OTC Drugs: By Generic Name, under "acetaminophen."

daughter cyst. Secondary cyst, sometimes rising from a previous one.

DCIS. Acronym for ductal carcinoma in situ.

deaf-mute *(dĕf'-mūt).* Individual who is unable to hear and speak.

deafness *(dĕf'-nĕs).* Partial or complete inability to hear.

deamination *(dē-ăm-ĭ-nā'-shŭn).* Process performed by the liver in which amino acids are broken down into sugars or fatty acids.

death *(dĕth).* Permanent cessation of all vital bodily functions. A proposed medical and legal definition is irreversible cessation in three areas: all brain function, spontaneous respiration, and spontaneous circulation.

brain d. Irreversible brain damage characterized by complete unresponsiveness to all stimuli, complete loss of spontaneous muscle activity, and flatline electroencephalogram for 30 minutes, all unrelated to hypothermia or poisoning by central nervous system depressants.

crib d. *See* **sudden infant death syndrome.**

fetal d. Death of the fetus in the uterus; also called *stillbirth.*

sudden cardiac d. Unexpected death associated with heart-related problems that occur rapidly after the onset of symptoms in an individual who has no known preexisting heart disease; time between onset of symptoms and death is less than 24 hours.

death rate. Number of deaths that occur in a population at risk in a specific area and within a specific time.

age-specific dr. Ratio of the number of deaths that occur in a specified age group in one year to the average or midyear population of that group.

cause-specific dr. Ratio of the number of deaths due to a specified cause in one year to the average or midyear total population.

crude dr. Ratio of the number of deaths in a specific area in one year divided by the average or midyear population in the area during the year.

death rattle. Gurgling noise sometimes heard in the throat of a dying person, caused by mucus accumulation and loss of the cough reflex.

death struggle. Twitching or convulsions sometimes experienced by an individual just before death.

debility *(dĭ-bĭl'-ĭ-tē).* Abnormal weakness or loss of body strength.

debridement *(dă-brēd'-mĕnt).* Removal of unhealthy tissue and foreign material from a wound in order to prevent infection and promote healing.

debulking operation *(dē-bŭlk'-ĭng).* Surgical procedure to remove portions of a large malignant tumor to reduce its size, provide oxygen to tumor tissues, and allow malignant cells to spread, which makes the tumor more susceptible to chemotherapy.

Decadron. Trade name for a cortisonelike drug used to treat inflammation; generic name: dexamethasone.

decalcification *(dē-kăl-sĭ-fĭ-kā'-shŭn).* Loss of calcium from bones or teeth.

decapsulation *(dē-kăp-sŭl-lā'-shŭn).* Removal of a capsule or covering membrane from an organ.

decibel *(dĕ'-ĭ-bĕl).* Unit for measuring the intensity or pressure of sound.

decidua *(dĕ-sĭj'-ū-ă).* Epithelial tissue of the lining of the uterus, which is shed during menstruation and after childbirth.

deciduitis *(dĕ-sĭd-ū-ī'-tĭs).* Acute inflammation of the deciduas (lining of a pregnant uterus).

decompensation *(dē-kŏm-pĕn-sā'-shŭn).* Failure of the heart to maintain adequate circulation, characterized by shortness of breath and edema.

decompression *(dē-kŏm-prĕ'-shŭn).* 1. Removal of excess pressure, as in removing gas from the intestinal tract. 2. Slow reduction or removal of pressure in deep-sea divers to prevent development of bends (nitrogen bubbles in tissue spaces and blood vessels).

decongestant *(dē-kŏn-jĕs'-tĕnt).* Substance that reduces congestion, especially nasal congestion.

decontamination *(dē-kŏn-tăm-ĭ-nā'-shŭn).* Process of making objects, people, or areas free of harmful substances.

decortication *(dē-kŏr-tĭ-kā'-shŭn).* Surgical removal of the cortex (outer portion) of an organ or structure.

decrepit *(dē-krĕp'-ĭt).* Weak; feeble.

decrudescence *(dĕ-krū-dĕs'-ĕns).* Reduction in the severity of symptoms.

decubitus *(dĕ-kū'-bĭ-tŭs).* 1. Act of lying down or the position itself. 2. Bedsore.

decubitus ulcer. Bedsore, sometimes appearing in people who have been confined to bed for a prolonged time.

deep sensibility. Sensibility to stimuli such as pain, pressure, and movement that activates receptors below the body surface, but not in the organs.

deep vein thrombosis (DVT). Blood clots in the veins of the leg, which can break off and be carried to the lung and potentially cause respiratory problems.

deerfly fever. *See* **tularemia.**

deet *(det).* N,N-diethyl-meta-toluamide, it is an active ingredient in most tick and insect repellants.

defecation *(dĕf-ĕ-kā'-shŭn).* Discharge or evacuation of feces from the bowels.

defect *(dē'-fĕkt).* Abnormality or malformation.

defense mechanism *(dē-fĕns').* Any reaction that helps protect against harmful events or circumstances.

defervescence *(dĕ-fĕr-vĕs'-ĕns).* Time that marks the reduction of fever to normal body temperature.

defibrillation *(dē-fĭb-rĭ-lā'-shŭn).* Technique—drug or mechanical—used to stop an irregular heartbeat and return it to normal.

defibrillator *(dē-fĭb'-rĭ-lā-tŏr).* Any device that delivers an electric shock to stop an irregular heartbeat and return it to normal.

deficiency disease *(dē-fĭsh'-ĕn-sē).* Any disease associated with a lack of essential nutrients, such as iron-deficiency anemia, scurvy, rickets, and others.

deformity *(dē-fŏr'-mĭ-tē).* Any disfigurement of the body.

degeneration *(dē-jĕn-ĕr-ā'-shŭn).* Deterioration of physical, mental, or moral characteristics, resulting in loss or decline in function.

degenerative disorder. Any of several conditions characterized by progressive loss of function (e.g., Alzheimer's disease, Parkinson's disease).

deglutition *(dē-glū-tĭsh'-ŭn).* Act of swallowing.

dehiscence *(dē-hĭs'-ĕns).* Process of splitting or breaking open, as when a closed surgical wound ruptures.

dehydration *(dē-hī-drā'-shĕn).* Extreme loss of water from the body's tissues caused by prolonged perspiration, diarrhea, and/or vomiting.

dehydrocholesterol-7 *(dē-hī-drō-kō-lĕs'-tĕr-ŏl).* Substance in the skin and other tissues that, after exposure to irradiation, forms a type of vitamin D (cholecalciferol) that the body can use.

dehydroepiadrosterone (DHEA) *(dē-hī-drō-ĕp-ē-ăn-drŏs'-tĕr-ōn).* Hormone produced by the adrenal gland and used by the body to make dozens of other hormones.

déjà vu *(dā'-zhă vū').* Sense that what one is experiencing has been encountered before, even though it has not; occurs in normal persons but also in some disorders, such as epilepsy.

delayed union. Failure of the ends of fractured bones to heal in the expected time period.

delinquency *(dē-lĭn'-kwĕn-sē).* Antisocial or criminal behavior, especially in a minor.

delirium *(dĭ-lĕr'-ē-ŭm).* Mental condition (usually brief) characterized by confused speech, hallucinations, restlessness, and incoherent excitement, usually the result of severe stress, high fever, exposure to toxins, nutritional deficiencies, or other causes.

delirium tremens. Acute mental disturbance caused by withdrawal from chronic alcohol use, characterized by sweating, tremor, anxiety, and hallucinations.

delivery *(dĭ-lĭv'-ĕ-rē).* Birth of a child.

delta waves *(dĕl'-tă).* Frequency of brain waves characteristic of dreamless, deep sleep; also called *delta rhythm.*

deltoid *(dĕl'-toyd).* Thick, large triangular muscle that covers the shoulder.

delusion *(dĭ-lū'-zhĕn).* A false belief that someone maintains even though it is not logical to do so.

dementia *(dĭ-mĕn'-shă).* Progressive decline in mental functioning, especially memory and judgment, due to organic factors, and which interferes with normal activities.

Demerol. Trade name for a drug used to treat moderate to severe pain; generic name: meperidine.

demineralization *(dē-mĭn-ĕr-ăl-ĭ-zā'-shŭn).* Loss of minerals, especially from bone, and other tissues.

demulcent *(dĭ-mŭl'-sĕnt).* Medication (oil, salve, ointment) used to soothe and relieve skin discomfort.

demyelinization *(dē-mī-ĕ-lĕ-nā'-shŭn).* Loss or destruction of the sheath (myelin) that normally covers nerve fibers, resulting in their impaired function.

denatured alcohol *(dē-nā'-chŭrd).* Alcohol that has been altered, making it unsuitable for drinking.

dendrites *(dĕn'-drīt).* Branches of a nerve cell that receive and transmit signals to the center of the cell. *See* **axon** for an illustration.

denervate *(dĕ-nĕr'-vāt).* To remove or cut the nerve supply to a body part, sometimes done to relieve severe pain.

dengue fever *(dĕng'-gă).* Viral, epidemic disease of tropical and subtropical regions transmitted by the bite of the *Aedes aegypti* and *Aedes albopictus* mosquitoes and characterized by fever, headache, muscle pain, and rash.

density *(dĕn'-sĭ-tē).* 1. Degree of compactness or relative weight of a substance compared with an established standard. 2. In radiology, the ability of a material to absorb x-rays.

dentalgia *(dĕn-tăl'-jă).* Toothache.

dental implant. Artificial tooth that is implanted into the soft tissues and bone of the jaw.

dentigerous cyst *(dĕn-tĭj'-ĕr-ŭs).* Fluid-containing cyst that forms around the crown of an unerupted tooth.

dentin *(dĕn'-tĭn).* Hard tissue that forms the main substance of teeth.

denture *(dĕn'-chur).* Artificial removable replacement tooth or teeth and surrounding tissue; also called a *plate.*

denudation *(dĕ-nū-dā'-shun).* Removal of a layer, usually of the skin, through surgery, trauma, or disease.

deossification *(dē-ŏs-ĭ-fĭ-kā'-shŭn).* Removal or loss of minerals from bone.

deoxygenation *(dē-ŏk-sĭ-jĕ-nā'-shŭn).* Process of removing or depriving oxygen.

deoxyribonucleic acid (DNA) *(dē-ō-sĕ-rī-bō-nū-klē'-ĭk).* Molecular basis of living tissue and of heredity.

Depakote. *See* Appendix, Common Prescription and OTC Drugs: By Trade Name.

dependence *(dē-pĕn'-dĕns).* Psychological and/or physical need for a substance, object, or person.

depersonalization *(dē-pĕr-sŏn-ăl-ĭ-zā'-shŭn).* Condition in which a person loses his/her sense of personal identity or feels as though his/her body is not real.

depilatory *(dĭ-pĭl'-ĕ-tŏr-ē).* Substance that removes or destroys hair from the body.

depletion *(dē-plē'-shŭn).* 1. Process of emptying. 2. Excessive loss of fluids, nutrients, tissue, or other elements from the body.

depot *(dē'-pō).* Organ or tissue in which drugs or other substances are deposited and stored by the body.

depression *(dē-prĕsh'-ŭn).* Mental state characterized by profound sadness accompanied by loss of interest in surroundings, feelings of hopelessness, and lack of energy.

major d. Disorder characterized by at least four of the following symptoms, lasting every day for at least 2 weeks: change in appetite with change in weight; insomnia; loss of interest and pleasure in surroundings; feelings of excessive guilt or worthlessness; fatigue; recurrent suicidal thoughts/attempts; agitation or motor retardation; indecisiveness.

postpartum d. Temporary mood disorder experienced by

some women after giving birth.

depressive reaction. Depressed stage of a manic-depressive reaction.

depressor *(dē-prĕs'-ŏr).* 1. Anything that depresses or reduces function, such as certain chemicals, muscles, or nerves. 2. Instrument for depressing a part.

derma *(dĕr'-mă).* Skin.

dermabrasion *(dĕr-mă-brā'-shŭn).* Procedure used to remove fine wrinkles, acne, scars, or tattoos, which involves scraping off surface layers of the skin with a high-speed rotary brush.

dermatitis *(dĕr-mă-tī'-tĭs).* Inflammation of the skin.

allergic contact d. Localized inflammation characterized by well-defined areas of redness and itchiness, resulting from contact with natural or manufactured substances to which the skin has been exposed and sensitized, especially cosmetics, topical medications, and jewelry.

atopic d. Skin reaction usually seen in people susceptible to hay fever and asthma, in which lesions occur mostly in front of the elbows and behind the knee.

contact d. Skin reaction caused by direct contact with a substance to which a person is hypersensitive.

d. medicamentosa. Rash caused by hypersensitivity to a medication taken internally.

seborrheic d. Condition that usually affects the scalp, but also behind the ears, on the chest and back, and characterized by redness, scaling, and itching.

dermatofibroma *(dĕr-mă-tō-fī-brō'-mă).* Benign, slow-growing skin tumor; also called *sclerosing hemangioma.*

dermatoglyphics *(dĕr-mă-tō-glĭf'-ĭks).* Study of the patterns of lines (whorls, loops, arches) in the fingertips and toes, which are unique to each person; helpful in the study of genetic disorders.

dermatologist *(dĕr-mă-tŏl'-ō-jĭst).* Specialist in skin disorders and related diseases.

dermatome *(dĕr'-mă-tōm).* Surgical instrument for cutting thin slices of tissue for a skin graft.

dermatomycosis *(dĕr-mă-tō-mī-kō'-sĭs).* Fungal infection of the skin, especially of the groin and feet; also called *dermatophytosis.*

dermatoneurosis *(dĕr-mă-tō-nū-rō'-sĭs).* Rash that erupts as a result of emotional stress.

dermatophytosis *(dĕr-mă-tō-fī-tō'-sĭs).* A superficial fungal infection of the hands and feet, caused by a dermatophyte; for example, athlete's foot.

dermoid cyst *(dĕr'-moid).* Common ovarian cyst, usually benign, that is lined with skin and contains pieces of hair, bone, and cartilage.

DES. Acronym for diethylstilbestrol; *see* **diethylstilbestrol.**

descending aorta *(dĭ-sĕn'-dĭng).* *See* **aorta.**

descending colon. *See* **colon.**

desensitization *(dē-sĕn-sĭ-tĭ-zā'-shŭn).* 1. Elimination of the sensation of pain by cutting a nerve. 2. Relief from an allergic reaction by administering desensitizing injections for hay fever. 3. In psychiatry, treatment for changing fear-induced behavior.

desert fever. *See* **coccidioidomycosis;** also known as *valley fever.*

desiccation *(dĕs-ĭ-kā'-shŭn).* Process of drying or making something moisture-free.

designer estrogen. Engineered drug that has some, but not all, of the actions of estrogen; *see also* **selective estrogen-receptor modulators (SERMs).**

desloratadine. *See* Appendix, Common Prescription and OTC Drugs: By Generic Name.

desmoid *(dĕz'-moyd).* Fibrous.

d. tumor. Nodule that results from the growth of fibrous tissue of muscle sheaths, especially in the abdominal wall.

desquamate *(dĕs'-kwă-māt).* To shed or cast off the outer layer of a surface, as the scaling off of the epidermis.

Desyrel. *See* Appendix, Common Prescription and OTC Drugs: By Trade Name.

detached retina. Separation of the retina from the choroids in the back of the eye, resulting from trauma or internal changes in the eye.

detoxification *(dē-tŏk-sĭ-fĭ-kā'-shŭn).* Process of neutralizing the toxic properties of a substance.

detumescence *(dē-tū-mĕs'-ĕns).* Return to a flaccid state or to normal size of a swollen organ or body part, or reduction in size of a tumor.

devascularization *(dē-văs-kū-lăr-ĭ-zā'-shŭn).* Interruption of blood circulation to a body part or organ, due to an obstruction or destruction of the blood vessels.

deviated septum *(dē'-vē-āt-ĕd).* A common abnormality in the wall-like part that separates the two portions of the nose.

deviation *(dē-vē-ā'-shŭn).* 1. Movement away from a normal course of activity. 2. Abnormal position of one or both eyes.

DEXA. Dual energy x-ray absorptometry, a method for scanning and measuring bone mineral density; also called DXA.

dexamethasone *(dĕx-ă-mĕth'-ă-zōn).* One in the group of corticosteroid drugs, used to treat allergic reactions. Trade name: Decadron.

Dexedrine. Trade name for an amphetamine used to inhibit appetite and provide mental stimulation; generic name: dextroamphetamine sulfate.

Dexon. Synthetic suture material absorbed by the body; a substitute for catgut and silk.

dextroamphetamine sulfate. A drug used for mental stimulation and to suppress appetite. Trade name: Dexedrine; street name: "speed."

dextrocardia *(děks-trō-kăr'-dē-ă).* Condition in which the heart is located toward the right side of the chest.

dextropropoxyphene. A drug used for pain relief. Trade name: Darvon.

dextrose *(děk'-strōs).* Simple sugar, also called glucose; *see* **glucose.**

diabetes *(dī-ĕ-bē'-tĭs).* General term for diseases characterized by excessive excretion of urine; usually refers to diabetes mellitus, although there are other types of diabetes.

d. insipidus. Rare form of diabetes caused by inadequate production of vasopressin by the pituitary.

d. mellitus. Chronic disorder of metabolism due to partial or total lack of insulin being secreted by the pancreas; includes both type I and type II diabetes.

gestational d. Form of diabetes mellitus that appears during pregnancy in women who did not have diabetes, and usually disappears after the baby is born.

labile d. Diabetes mellitus characterized by wide swings in glucose levels; also called *unstable diabetes.*

type I d. Autoimmune disease in which the pancreas makes too little or no insulin.

type II d. Type of diabetes in which the pancreas produces insulin but the body is unable to use it effectively.

diabetic neuropathy. Partial loss of sensation, including pain, especially in the limbs, seen in some people who have diabetes.

diabetic shock. Low blood sugar (hypoglycemia) associated with diabetes, characterized by fatigue, light-headedness or fainting, reddening of the skin, and a sweet odor to the breath.

Diabinase. Trade name of a drug used to lower blood sugar levels, used in diabetes mellitus.

diagnosis *(dī-ăg-nō'-sĭs).* Name of a disease or syndrome a person has or is believed to have.

clinical d. Diagnosis determined using a patient's history, laboratory study results, and symptoms.

differential d. Diagnosis determined by comparing symptoms of two similar diseases.

Diagnosis Related Group (DRG). Method used to evaluate average hospital stays for various medical conditions according to diagnosis.

Diagnostic and Statistical Manual of Mental Disorders (DSM-IV). Publication that delineates the standards for identifying emotional illnesses and is the standard reference for health-care practitioners.

dialysis *(dī-ăl'-ĭ-sĭs).* Procedure in which a membrane is used to separate smaller molecules from larger molecules.

renal d. Dialysis of blood to

remove liquid and chemicals that the kidneys would remove if they were functioning properly.

Diamox. Trade name of a drug used to stimulate urine output; generic name: acetazolamide.

diaper rash. Skin outbreak caused by irritation from a diaper or from substances used to clean the diaper.

diaphoresis *(dī-ĕ-fĕ-rē'-sĭs).* Excessive perspiration.

diaphragm *(d ī'-ă-frăm).* 1. Muscle layer that separates the chest and abdominal cavities. 2. Any membrane that divides.

contraceptive d. Flexible ring covered with rubber or other material that fits over the cervix of the uterus to prevent pregnancy.

diaphragm pacing. A method that assists breathing in patients who have spinal cord injuries. Pacing is accomplished by surgically implanting electrodes into the diaphragm, which is innervated by the phrenic nerve.

diaphysis *(dī-ăf'-ĭ-sĭs).* Shaft of a long cylindrical bone.

diarrhea *(dī-ă-rē'-ŭh).* Increase beyond normal in the looseness, fluidity, and/or frequency of bowel movements.

dysenteric d. Diarrhea with mucous and bloody stools.

fatty d. Diarrhea with stools that contain undigested fat.

inflammatory d. Diarrhea characterized by inflammation of the intestines caused by bacteria.

purulent d. Presence of pus in stools.

traveler's d. Diarrhea that occurs among travelers, especially those who visit tropical or subtropical areas where sanitation is poor.

diastase *(dī'-ă-stās).* Enzyme that converts starch into sugars.

diastole *(dī-ăs'-tĕ-lē).* 1. Period of relaxation, between two contractions of the heart, when the chambers widen and fill with blood. 2. In blood pressure readings, diastole is the second number.

diastolic pressure. Blood pressure level during the period of relaxation.

diathermy *(dī'-ă-thĕr-mē).* Application of heat to a body region with a machine that generates high frequency electrical currents, to increase blood flow to the area.

diazepam. *See* Appendix, Common Prescription and OTC Drugs: By Generic Name.

Dick test. Skin test used to discover a person's susceptibility to scarlet fever.

dicrotic *(dī-krŏt'-ĭk).* Double beat or split pulse, denoting a pulse with two beats for each heartbeat.

didelphic *(dī-dĕl'-fĭk).* Referring to a double uterus.

dienestrol *(dī-ĕn-ĕs'-trŏl).* Synthetic, nonsteroidal estrogen used for estrogen therapy.

diet *(dī'-ĕt).* 1. General term for all substances that are consumed as nourishment for the body. 2. To follow a specific eating plan for the purpose of reducing or increasing body weight.

dietetics *(dī-ĕ-tĕt'-ĭks).* Study of diet in relation to health and disease.

diethylstilbestrol (DES) *(dī-ĕth-ĭl-stĭl-bĕs'-trŏl).* Synthetic compound with properties of estrogen, once used to treat potential miscarriages but discontinued because it caused cancer in some daughters of women who took it while pregnant; now used to treat problems associated with menopause.

differential blood count *(dĭf-ĕ-rĕn'-shĕl).* Diagnostic tool in which the number or percentage of the specific types of white blood cells found in a given amount of blood are determined.

diffuse *(dĭ-fūs').* To spread out; not localized or limited.

Diflucan. *See* Appendix, Common Prescription and OTC Drugs: By Trade Name.

diflunisal. A nonsteroidal anti-inflammatory drug used to treat arthritis and other inflammatory conditions. Trade name: Dolobid.

digestion *(dī-jĕs'-chĕn).* Chemical and mechanical breakdown of food in the gastrointestinal tract and its conversion into substances that the body can use.

digestive system. Parts of the body involved in the digestion and absorption of food. These include the salivary glands, mouth, esophagus, stomach, small intestine, liver, gallbladder, pancreas, colon, rectum, and anus.

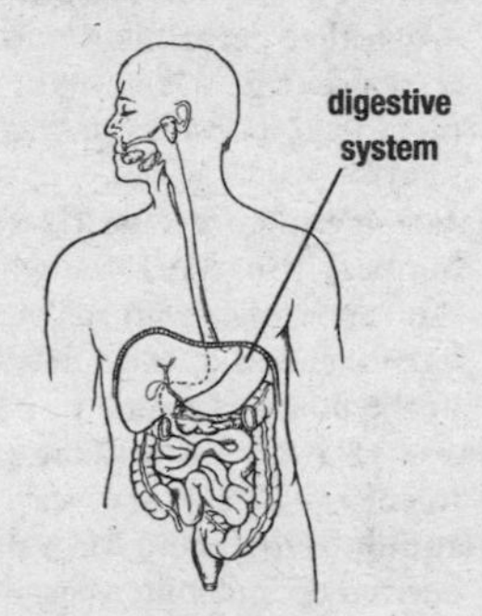

digit *(dĭj'-ĭt).* A finger or toe.

digitalis *(dĭj-ĭ-tăl'-ĭs).* Any of several drugs (e.g., digoxin, digitoxin) derived from foxglove plants and used to regulate and strengthen the heart.

digital rectal exam (DRE). Screening examination in which a doctor inserts a lubricated, gloved finger into the rectum to detect abnormalities of the prostate and within the rectum.

Digitek. *See* Appendix, Common Prescription and OTC Drugs: By Trade Name.

digitoxin *(dĭj-ĭ-tŏk'-sĭn).* Generic name of a digitalis drug used to treat congestive heart failure and cardiac arrhythmias; trade name: Crystodigin.

digoxin *(dĭj-ŏk'-sĭn).* *See* Appendix, Common Prescription and OTC: By Generic Name.

Dilantin. Trade name for a drug used to treat epileptic seizures; generic name: diphenylhydantoin.

dilatation *(dĭl-ĕ-tā'-shĕn).* Enlargement of an opening or an organ that occurs as a normal response (e.g., widening of the pupil in response to decreased light).

dilatation and curettage (D&C). Surgical procedure in which the cervix of the uterus is enlarged and the endometrium of the uterus is scraped.

dilator *(dī-lā'-tŏr).* Instrument used to enlarge a passage or cavity.

Dilaudid. Trade name for a drug derived from opium, used to treat pain; generic name: hydromorphone.

diltiazem *(dĭl-tī'-ĕ-zĕm). See* Appendix, Common Prescription and OTC Drugs: By Generic Name.

diluent *(dĭl'-ū-ĕnt).* Substance that reduces the concentration or strength of a solution to which it is added.

dilution *(dī-lū'-shŭn).* Process of reducing the concentration or strength of a solution.

Dimetapp. Trade name for a decongestant combination (phenylephrine hydrochloride and phenylpropanolamine hydrochloride) and an antihistamine (brompheniramine maleate); trade name: Dimetane.

dimethyl sulfoxide (DMSO) *(dī-mĕth'-ĕl sŭl-fŏk'-sīd).* Topical anti-inflammatory substance, used along with other drugs to aid their absorption through the skin.

dimpling *(dĭmp'-lĭng).* Depression(s) in the skin.

diopter *(dī-ŏp'-tĕr).* Unit of measurement for the lens of the eye.

dioptometer *(dī-ŏp-tŏm'-ĕ-tĕr).* Device used to measure the level of eye refraction.

Diovan. *See* Appendix, Common Prescription and OTC Drugs: By Trade Name.

dioxide *(dī-ŏx'-īd).* Any chemical that contains two atoms of oxygen plus one atom of another element.

dioxin *(dī-ŏks'-ĭn).* Highly toxic component once widely used in herbicides and preservatives and still present in the environment. Exposure is believed to cause birth defects, cancer, nervous system damage, and other effects.

diphenhydramine. *See* Appendix, Common Prescription and OTC Drugs: By Generic Name.

diphenlhydantoin *(dī-fĕn-ĭl-hī-dăn'-tō-ĭn).* A drug used to suppress and prevent convulsions. Trade name: Dilantin.

diphtheria *(dĭf-thĕr'-ē-ă).* Contagious disease caused by a bacillus, *Corynebacterium diphtheriae,* characterized by inflammation of the upper respiratory tract, fever, chills, and sore throat.

diplegia *(dī-plē'-jē-ă).* Paralysis of both arms or both legs; also called *bilateral paralysis.*

diplomate *(dĭp'-lō-māt).* Specialist who has been certified by one or more of the American Specialty Boards.

diplopia *(dĭ-plō'-pē-ă).* Double vision.

disarticulation *(dĭs-ăr-tĭk-ū-lā'-*

shŭn). Amputation or separation of a limb or appendage at the joint.

discharge *(dĭs'-chărj).* 1. Material that is secreted or excreted. 2. The process by which fluids, pus, feces, and so on flow away. 3. Release from a hospital or other type of care.

discission *(dĭ-sĭzh'-ŭn).* Surgical procedure in which the capsule of the lens of the eye is cut or punctured, typically done in individuals who have a cataract.

discoid lupus erythematosus (DLE) *(dĭs'-koid).* Chronic, recurrent disease characterized by a butterfly-shaped eruption of red lesions over the cheeks and nose.

disease *(dĭ-zēz').* Disruption of health, or a state of abnormal functioning, characterized by recognizable symptoms. *See* individual names of diseases.

disequilibrium *(dĭs-ē-kwĭ-lĭb'-rē-ŭm).* Lack of balance or stability.

disk *(dĭsk).* Flattened, rounded part, especially referring to the cartilage between the vertebrae (an intervertebral disk).

herniated d. An abnormal protrusion of the inner portion (either the nucleus pulposus or annulus fibrosus) of an intervertebral disk. Also called a *protruded disk.*

slipped d. Popular name for a herniated disk.

diskectomy *(dĭs-kĕk'-tō-mē).* Surgical removal of all or part of an intervertebral disk.

dislocation *(dĭs-lō-kā'-shŭn).* Displacement of a limb or body part, especially a bone.

disoriented *(dĭs-ŏr'-ē-ĕn-tĕd).* Mentally confused; loss of a sense of location or direction or of one's surroundings.

dispensary *(dĭ-spĕn'-sĕ-rē).* 1. Place where medications and other medical agents are dispensed to patients. 2. Free clinic for outpatients.

dispense *(dĭ-spĕns').* To give a prescription to a patient.

dissect *(dĭ-sĕkt').* To cut open or apart, especially in the study of anatomy.

dissemination *(dĭ-sĕm-ĭ-nā'-shŭn).* Spreading of disease-causing organisms.

dissimulation *(dĭ-sĭm-ū-lā'-shŭn).* Pretending one is sick when well or, vice versa, pretending one is well when sick.

dissociation *(dĭ-sō-sē-ā'-shŭn).* 1. In psychiatry, separation of ideas, thoughts, or emotions from the consciousness. 2. Transformation of a complex chemical compound into a simple one.

dissociative disorder. Mental condition in which repressed emotional conflict causes individuals to be confused about their identity; characterized by amnesia, multiple personality, or a dream state.

dissolution *(dĭs-ō-lū'-shŭn).* Breakup or decomposition of the normal physical state of part or parts of an organism, as of tissue.

dissolve *(dĭ-zŏlv').* To cause a substance to change from a solid to a dispersed state by placing it in a solution.

distal *(dĭs'-tăl).* 1. Farthest from a reference point. 2. In dentistry, a point most distant from the median line of the jaw.

distance healing. Healing in which people seek to help others using the power of the mind to try to alter the energy fields of the recipients.

distemper *(dĭs-tĕm'-pĕr).* In veterinary medicine, a general term for several viral diseases that affect dogs, cats, horses, and other animals.

distention *(dĭs-tĕn'-shŭn).* State of being stretched or distended.

distillation *(dĭs-tĭ-lā'-shŭn).* Vaporization of a liquid by heat followed by separation of its components by condensation of the vapor.

distilled water. Water cleansed of impurities by transforming it into steam and then condensing it, usually several times, as a liquid.

Diupres. Trade name of a combination of chlorothiazide and reserpine.

diuresis *(dī-ĕ-rē'-sĭs).* Increased excretion of urine usually caused by drinking large amounts of liquid, taking diuretics, or as a symptom of disease.

diuretic *(dī-ĕ-rĕt'-ĭk).* Drug that promotes the production and excretion of urine, often used to treat hypertension, congestive heart failure, and edema.

Diuril. Trade name for the diuretic chlorothiazide.

divalproex. *See* Appendix, Common Prescription and OTC Drugs: By Generic Name.

divergence *(dī-vĕr'-jĕns).* Spreading apart or going in opposite directions from a common point.

diverticulitis *(dī-vĕr-tĭk-ū-lī'-tĭs).* Inflammation and infection of one or more diverticulum (sac) in the intestinal tract, resulting in pain, fever, and nausea.

diverticulosis *(dī-vĕr-tĭk-ū-lō'-sĭs).* Formation of sacs in the mucous lining of the colon, which may be asymptomatic or cause intermittent pain.

diverticulum *(dī-vĕr-tĭk'-ū-lŭm).* Sac or pouch in the walls of an organ, especially the intestinal tract.

dizygotic twins *(dī-zī-gŏt'-ĭk).* Twins who develop from two different fertilized eggs; fraternal twins.

dizziness *(dĭz'-ē-nĕs).* Sensation of whirling or unsteadiness, or feeling a tendency to fall.

DNA. Acronym for deoxyribonucleic acid, the basic component of living matter; *see also* **deoxyribonucleic acid.**

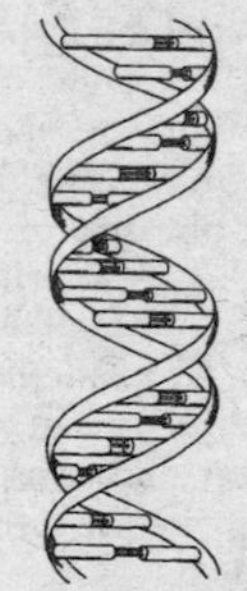

DNA

recombinant DNA. A DNA molecule that contains a new combination of genes that alters, to some extent, inherited characteristics.

DNR. Acronym for "do not resuscitate," an order from an individual who does not want to receive cardiopulmonary resuscitation if the heart stops or breathing ceases. *See* Appendix.

Dolobid. Trade name for the nonsteroidal anti-inflammatory drug diflunisal.

dominance *(dŏm'-ĕ-nĕns).* 1. Having a controlling influence. 2. When a specific genetic characteristic suppresses the expression of another.

Donnatal. Trade name of a combination drug that contains a sedative (generic name: phenobarbital) and several other substances, used to decrease gastrointestinal spasm.

donor *(dō'-nĕr).* Person who gives living tissue (e.g., blood, organ, sperm) to be used in another person.

donor insemination. Procedure in which a catheter is used to deposit a sperm sample from a man other than the woman's mate into the uterus, to achieve fertilization and pregnancy; also called *artificial insemination.*

dopamine *(dō'-pĕ-mēn).* Precursor of the hormone norepinephrine, it is found in the adrenal glands and in high concentrations in the brain. It acts as a neurotransmitter in the central nervous system.

doppler *(dŏp'-lĕr).* Instrument that emits ultrasonic beams into the body and is used to diagnose blood vessel and heart disorders.

dorsal *(dŏr'-săl).* Referring to the back or to the back of an organ.

dorsiflexion *(dŏr-sĕ-flĕk'-shŭn).* Bending backward of a body part, especially the foot.

dorsum *(dŏr'-sŭm).* 1. Back. 2. Back surface of any body part.

dosage *(dō'-sĭj).* Determination of the amount, frequency, and number of doses of a drug or other therapeutic approach (e.g., radiation) an individual should take.

dose *(dōs).* Amount of a medication or other substance to be given at one time.

booster d. Dose of an active immunizing substance, given to maintain immunity.

lethal d. Amount of an agent that will or may cause death.

maintenance d. Dose that is sufficient to maintain the influence of a drug that was achieved by a previous, greater amount given.

primary d. First and large dose administered to provide a high blood level quickly.

dosha *(dōsh'-ă).* According to ayurveda, the principle of constitution of the physical body based on a combination of three vital bioenergies (vata, pitta, kapha) responsible for a person's physical and emotional tendencies. The attributes of the doshas and their particular combination within each individual help determine a person's physical and mental characteristics, while an imbalance among the doshas causes disease.

dosimeter *(dŏ-sĭm'-ĕ-tĕr)*. Instrument that measures the amount of exposure to radiation to radiology personnel and patients.

double blind study. Experiment in which neither the investigator(s) nor the subject(s) knows which treatment the subject(s) receives.

double helix. Double strand structure that makes up DNA.

douche *(doosh)*. Current of water or vapor directed against a body part, especially to cleanse the vagina.

Down syndrome. Congenital defect caused by an extra chromosome, characterized by moderate to severe mental retardation and physical traits such as sloping forehead, flat nose or no bridge, and low-set ears, among others.

doxazosin. *See* Appendix, Common Prescription and OTC Drugs: By Generic Name.

doxepin. A tricyclic antidepressant used to treat depression. Trade names: Adapin, Sinequan.

doxycycline. *See* Appendix, Common Prescription and OTC Drugs: By Generic Name.

DPT. Acronym for the vaccination that contains diphtheria, whooping cough (pertussis), and tetanus vaccines.

drain *(drān)*. 1. Removal of blood, pus, or other liquid from an infected area of the body. 2. Tube or other device that allows fluids to leave an area.

Penrose d. A thin rubber tube with gauze at its center; it is inserted into a surgical wound to drain pus, blood, and other fluids.

sump d. A double-lumen drain that allows air to enter the drained area through the smaller lumen and to displace fluid into the larger luman.

dram *(drăm)*. A unit of weight that is equal to ⅛ ounce in the apothecaries' system and 1/16 ounce in the avoirdupois system.

Dramamine. Trade name for dimenhydrinate, a drug used to treat nausea, vomiting, and motion sickness.

drape *(drāp)*. 1. To cover an area around a part to be examined or operated upon. 2. Sterilized material used to cover such an area.

draw sheet. Sheet placed under the buttocks of a patient, which can be removed easily when soiled and which protects the underlying sheet and mattress from soilage.

dream *(drēm)*. Series of images, experienced during sleep, that can provoke various emotions, sensations and ideas.

day d. Mental fantasizing while awake.

wet d. *See* **nocturnal emission.**

dream therapy. Interpretation of

dreams as a way to get insight into a person's behavior, physical disorders, emotions, and creative potential.

dressing *(drĕs'-ĭng).* Material or substance applied to a wound or diseased area to prevent infection or to absorb excretions.

DRG. An acronym for Diagnosis Related Group.

dribble *(drĭb'-ĕl).* 1. To drool. 2. To pass urine in drops.

drop foot. Condition in which the foot droops or is flexed toward the sole and cannot voluntarily be flexed toward a normal position.

dropped beat. Skipped heartbeat.

drug *(drug).* Any substance that affects the function or processes of the body or mind.

d. abuse. Use or overuse of a drug for recreational or nontherapeutic purposes.

d. addiction. Condition caused by excessive or continued use of habit-forming drugs, such as heroin and barbiturates.

d. dependence. Condition in which people crave or depend on a specific drug that they have become accustomed to using.

d.-fast. Referring to microorganisms, such as bacteria, that resist the actions of a drug or chemical.

d. interaction. Interaction that can occur and the effect(s) on the body that may develop when one drug is taken along with another drug.

d. reaction. Negative, undesirable response to a substance that is taken for its medicinal effects.

dry eye syndrome. Condition characterized by lack of tears, increased sensitivity to light, and formation of thick mucous strands in the eye; also called *keratoconjunctivitis sicca.*

dry socket. Inflammation, pain, pus, and frequently infection that can occur at the site of an extracted tooth.

Duchenne's muscular dystrophy *(dū-shĕn'-ăr-ăn'). See* **muscular dystrophy.**

duct *(dŭkt).* Tube or channel usually used to transport secretions of a gland to other parts of the body.

ductal carcinoma in situ. Most common type of noninvasive breast cancer in women. The cells lining the milk ducts are cancerous but stay within the ducts without infiltrating surrounding breast tissue.

ductless glands. *See* **endocrine gland.**

ductus deferens *(dŭk'tŭs).* Duct that carries sperm from the epididymis to the ejaculatory duct; also called *vas deferens.*

Dulcolax. Trade name of an over-the-counter laxative; generic name: docosate sodium and bisacodyl.

duodenal ulcer *(dū-ō-dē'-năl).* Lesion of the mucous membrane of the duodenum.

duodenitis *(dū-ō-dē-nī'-tĭs).* Inflammation of the duodenum.

duodenojejunostomy *(dū-ō-dē-nō-*

jĕ-jū-nŏs'-tō-mē). Surgical procedure in which a passage is made between the duodenum and jejunum.

duodenoplasty *(dū-ō-dē-nō-plăs'-tē).* Surgical procedure in which the shape of the duodenum is altered, often to correct an obstruction.

duodenostomy *(dū-ōd-ē-nŏs'-tō-mē).* Surgical procedure in which a permanent opening is made into the duodenum through the abdominal wall.

duodenum *(dū-ō-dē'-nŭm).* First part of the small intestine, it receives material from the stomach and passes it to the jejunum.

Dupuytren's contracture *(dū-pwē-trăn').* Muscle contraction in which the little and ring fingers bend toward the palm and cannot be extended.

dura mater *(door'-ă).* Outer membrane that covers the spinal cord and brain.

D vitamin. *See* **vitamins.**

dwarfism *(dwŏr'-fĭz-ĕm).* Condition characterized by abnormally short stature and underdeveloped limbs; may be hereditary or caused by pituitary or thyroid dysfunction, kidney disease, or other causes.

dynamics *(dī-nă'-mĭks).* 1. Science of bodies in motion. 2. In psychiatry, the emotional forces that determine a person's patterns of behavior.

Dyazide *(dī'-ĕ-zīd).* *See* Appendix, Common Prescription and OTC Drugs: By Trade Name.

dysaphia *(dĭs-ā'-fē-ă).* Defect in the sense of touch.

dysarthria *(dĭs-ăr'-thrē-ă).* Difficulty in pronouncing words, usually because of poor control over muscles that control speech.

dyscalculia *(dĭs-kăl-kū'-lē-ă).* Inability to solve mathematical problems because of brain disease or injury.

dyscrasia *(dĭs-krā'-zhă).* General term for an abnormal condition.

dysentery *(dĭs'-ĕn-tĕr-ē).* Any one of various intestinal diseases characterized by inflammation of the intestines, abdominal pain, and frequent stools that contain mucus and blood.

dysfunction *(dĭs-fŭnk'-shŭn).* Abnormal or impaired functioning of a bodily system or an organ.

dysgenesis *(dĭs-jĕn'-ĭ-sĭs).* Abnormal development of a body part or organ, especially during the embryo stage.

dysgraphia *(dĭs-grăf'-ē-ă).* Impaired ability to write correctly, due to a brain or physical disorder.

dyshidrosis *(dĭs-hī-drō'-sĭs).* Disorder of the sweating mechanism.

dyskinesia *(dĭs-kĭ-nē'-sē-ă).* Impairment of voluntary movements, as in a tic or spasm.

dyslexia *(dĭs-lĕk'-sē-ă).* Condition in which people with normal vision have difficulty reading because of an impaired ability to identify and understand written symbols, with a tendency to reverse certain letters and words.

dysmenorrhea *(dĭs-mĕn-ĕ-rē'-ă).* Painful menstruation, usually characterized by cramplike pain in the lower abdomen and sometimes accompanied by nausea, vomiting, and intestinal discomfort just prior to and during menstruation.

dysostosis *(dĭs-ŏs-tō'-sĭs).* Defective bone formation.

dyspareunia *(dĭs-pĕ-rū'-nē-ă).* Abnormal condition in women in which sexual intercourse is painful, due to either a physical or psychological reason.

dyspepsia *(dĭs-pĕp'-shă).* Digestive disorder characterized by stomach discomfort, heartburn, and/or nausea.

dysphagia *(dĭs-fā'-jē).* Condition marked by painful or difficult swallowing.

dysphasia *(dĭs-fā'-zhĕ).* Speech impairment usually associated with a stroke, tumor, or brain injury.

dysphonia *(dĭs-fō'-nē-ă).* Difficulty speaking associated with an impairment of the voice.

dysplasia *(dĭs-plā'-zhĕ).* General term for any abnormal or impaired growth, often used when referring to abnormalities in cell shape and size.

dyspnea *(dĭsp-nē'-ă).* Labored, difficult breathing; shortness of breath.

dyssomnia *(dĭs-sŏm'-nē-ă).* Any of various sleep disorders, including narcolepsy, sleep apnea, restless leg syndrome, and jet-lag syndrome.

dyssynergia *(dĭs-sĭn-ĕr'-jē-ă).* Disturbance of muscular coordination.

dysthymia *(dĭs-thī'-mē-ĕ).* Chronic, mild depression with symptoms that are not disabling but which keep individuals from functioning fully. Some individuals with dysthmia also experience major depressive episodes.

dystocia *(dĭs-tō'-sē-ă).* Difficult labor, due to an unusually large fetus, abnormal position of the fetus, or other factors.

dystonia *(dĭs-tō'-nē-ă).* Abnormal muscle tone, especially sudden muscle spasms caused by a drug reaction or disease (e.g., dystonia musculorum deformans).

dystrophy *(dĭs'-trĕ-fē).* 1. Any disorder that develops from poor or faulty nutrition. 2. Defective development of tissue, especially muscle, which results in lost strength and decreased size.

dysuria *(dĭs-yū'-ē-ă).* Painful or difficulty urination, caused by inflammation of the bladder or urethra or other factors.

E

ear. Organ of hearing.

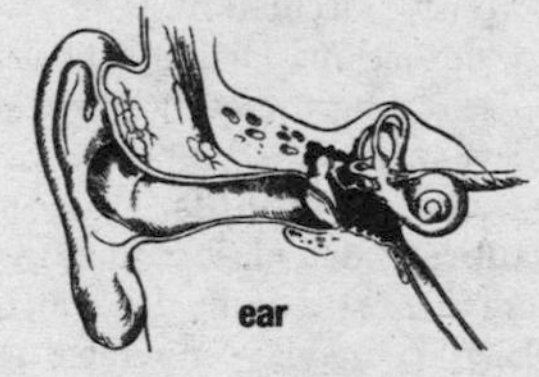

external e. Visible part of the ear that consists of the auricle (pinna) and the external auditory canal.

inner e. Innermost part of the ear, consisting of the semicircular canals, cochlea, and vestibule.

middle e. Portion just beyond the external ear, consisting of the tympanic chamber and the ossicles.

swimmer's e. Infection of the auditory canal usually associated with swimming.

earache *(ēr'-āk).* Pain in the ear.

ear coning. Alternative healing technique by which hollow candles are used to remove wax and other debris from the ears and to treat chronic ear infections in children.

eardrum *(ēr'-drŭm). See* **tympanic membrane.**

ear infection, middle (acute). Medically called acute otitis media, it is inflammation of the middle ear, characterized by a bulging eardrum usually accompanied by pain, or a perforated eardrum, with drainage of pus.

early ambulation. Practice of getting out of bed one to two days after surgery in an effort to prevent blood clots and promote circulation.

early detection. Diagnosing disease during the early stages of development.

ear rocks. *See* **otoliths.**

ear tubes. Small plastic tubes that are inserted into the eardrum to keep the middle ear aerated for a prolonged period of time (usually 6 months or longer) and to remove thickened secretions.

ear wax. Yellowish-brown waxlike material secreted by the glands that line the external ear canal; also called *cerumen.*

Ebola virus *(ē-bō'-lă).* Often fatal viral disease most often seen in African countries.

Ebstein's disease *(ĕb'-stīnz).* Congenital condition of the heart, characterized by fatigue, palpitations, and shortness of breath.

eccentric *(ĕk-sĕn'-trĭk).* Moving away from the center or located off-center.

ecchymosis *(ĕk-ĭ-mō'-sĭs).* Discoloration of the skin caused by bleeding beneath it; often begins as blue-black and changes to greenish brown or yellow; a bruise.

eccrine sweat glands *(ĕk'-rĭn).* Glands located over the entire skin surface and which secrete sweat and thus help regulate body temperature.

ECG. Acronym for electrocardiogram; also EKG. *See* **electrocardiogram.**

echinacea *(ĕk-ĭ-nā'-sē-ŭ).* Herb that is commonly used to fight infections, including colds and flu, and reduce inflammation.

echinococcus cyst *(ĕ-kī-nō-kŏk'-ŭs).* Cyst caused by infestation with a type of tapeworm often found in dogs; this cyst is usually seen in the liver of humans.

echocardiography *(ĕk-ō-kăr-dē-ŏg'-ră-fē).* Diagnostic method that uses ultrasound to see internal cardiac structures and their motions.

echolalia *(ĕk-ō-lā'-lē-ă).* Involuntary, echolike repetition of words spoken by other people; seen in catatonic schizophrenics.

ECHO virus *(ĕk'-ō).* Virus that belongs to a group of viruses (ECHO is an acronym for Enteric Cytopathogenic Human Orphan virus) associated with viral meningitis, acute respiratory infection, and myocarditis.

eclampsia *(ĕ-klămp'-sē-ă).* Potentially fatal (for both mother and infant) seizure disorder experienced by some women during pregnancy and/or up to six weeks after delivery.

E. coli. *See* ***Escherichia coli.***

ecology *(ē-kŏl'-ō-jē).* Study of the relationship between the environment and the organisms that live within it.

ecstacy *(ĕk'-stă-sē).* 1. Popular name for 3,4-methylenedioxymethamphetamine, a hallucinogenic drug that produces euphoria followed by difficulties with concentration 2. Emotional state characterized by extreme joy.

ectasia *(ĕk-tā'-zhă).* Dilatation of a hollow organ or tubular structure.

mammary e. Breast condition that affects women 50–60 who have had several children, it is characterized by thick secretions within some breast ducts and duct dilatation.

ectoderm *(ĕk'-tō-dĕrm).* Outermost of the three layers of the embryo, it eventually develops into the nervous system and the epidermis and related structures, such as hair.

ectogenous *(ĕk-tŏj'-ĕ-nŭs).* Having its origin outside of a structure of body; e.g., an infectious disease.

ectomorph *(ĕk'-tō-mŏrf).* Body type that is predominantly lean and generally nonmuscular; *See also* **endomorph** and **mesomorph.**

ectopia *(ĕk-tō'-pē-ă).* Congenital abnormal placement of a body part, especially at time of birth; also called *ectopy.*

ectopic pregnancy *(ĕk-tŏp'-ĭk).* Abnormal pregnancy in which the fertilized egg implants outside of the uterus, most often (90% of cases) in the fallopian tube, but also the ovary or abdominal cavity.

ectoplasm *(ĕk'-tō-plăz-m).* Outer layer of a cell.

ectropion *(ĕk-trō'-pē-ŏn).* Outward turning of an edge or margin, especially of an eyelid, usually the result of injury or loss of eye tissue.

eczema *(ĕk'-zĕ-mă).* Acute or chronic inflammation of the skin usually accompanied by itching and development of blisters.

edema *(ĕ-dē'-mă).* Swelling of any part of the body caused by fluid collection in the tissues; also known as *dropsy.*

EDTA. Acronym for ethylenediaminetetraacetic acid.

EEG. Acronym for electroencephalography.

effacement *(ĕ-fās'-mĕnt).* During labor, shortening of the cervix and thinning of its walls.

effector *(ē-fĕk'-tŏr).* 1. Agent that causes a specific effect. 2. Organ that produces an effect (e.g., secretion or contraction) in response to nerve stimulation.

efferent *(ĕf'-ĕr-ĕnt).* Carrying away from the center, as a nerve that carries signals from the brain to a muscle or other body part.

effervescent *(ĕf-ĕr-vĕs'-ĕnt).* Bubbling; giving off gas.

effeurage *(ĕf-loor-ăzh').* Firm, gentle, rhythmic stroking, especially when referring to massage.

Effexor. *See* Appendix, Common Prescription and OTC Drugs: By Trade Name.

effluent *(ĕ-flū'-ĕnt).* Outflow.

effluvium *(ĕ-flū'-vē-ŭm).* 1. Shedding of hair. 2. Outflow of a foul-smelling vapor or gas.

effusion *(ĕ-fū'-zhŭn).* Escape or flowing out of fluid into a body cavity.

egg. Female reproductive cell; also called *ovum.*

ego. Psychological term referring to awareness of the existence of the self as being different from others. It is one of three major divisions of the psyche; *see also* **id** and **superego.**

egocentric *(ē-gō-sĕn'-trĭk).* Condition characterized by constant or extreme preoccupation with one's own interests; self-centeredness.

egomania *(ē-gō-mā'-nē-ă).* Abnormal and excessive preoccupation with one's self.

Ehlers-Danlos syndrome *(ā'-lĕrz dăn-lōs').* Inherited disorder characterized by excessive elasticity of the skin and fragile blood vessels, which can result in excessive motion of the joints and discolored nodules of the skin.

ejaculation *(ē-jăk-ū-lā'-shŭn).* Emission of semen from the male urethra.

premature e. Emission of semen before or immediately upon engaging in sexual intercourse.

retrograde e. condition in which the semen is forced backward into the bladder.

ejaculatory ducts *(ē-jăk'-ū-lă-tŏr-ē).* Small tubes in the prostate gland through which semen travels into the urethra upon climax.

EKG. Acronym for electrocardiogram; also ECG.

elastin *(ē-lăs'-tĭn).* Yellowish protein, found in elastic fibers in tissue, that allows tissue to stretch.

Elavil. *See* Appendix, Common Prescription and OTC Drugs: By Trade Name.

elbow *(ĕl'-bō).* Joint between the arm and the forearm.

elderberry *(ĕl'-dĕr-băr-ē).* Herbal remedy that is high in vitamin C, carotenoids, amino acids, and flavonoids, and used to treat colds and flu, nervous disorders, and skin disorders.

elective *(ē-lĕk'-tĭv).* Not essential;

something that is decided upon by a patient and/or doctor, especially when referring to a surgical or other medical procedure.

electroacupuncture *(ē-lĕk-trō-ăk'-ū-pŭnk-chŭr).* Form of acupuncture in which the needles are stimulated with a weak electrical current.

electrocardiogram *(ē-lĕk-trō-kăr'-dē-ō-grăm).* Graphic recording of the electric current produced by the contraction of the heart, obtained with an electrocardiograph.

electrocautery *(ē-lĕk-trō-kăw'-tĕr-ē).* Use of a heated instrument to burn (cauterize) tissue for medical reasons (e.g., to remove warts).

electrocoagulation *(ē-lĕk-trō-kō-ăg-ū-lā'-shŭn).* Coagulation of diseased tissues using high-frequency electric currents.

electroconvulsive therapy (ECT) *(ē-lĕk-trō-kŏn-vŭl'-sĭv).* Type of therapy for emotional disorders in which an electric current is passed through the patient's head to produce convulsions.

electrode *(ē-lĕk'-trŏd).* Something that conducts electricity and allows it to enter or leave a surface or circuit; electrodes are often used to monitor heart rhythm.

electrodesiccation *(ē-lĕk-trō-dĕs-ĭ-kā'-shŭn).* Destruction of tissue by dehydration using an electrical current delivered through a special device.

electroencephalography *(ē-lĕk-trō-ĕn-sĕf-ă-lŏg'-ră-fē).* Recording of the electrical currents that are generated by brain activity, using an electroencephalograph.

electrolysis *(ē-lĕk-trŏl'-ĭ-sĭs).* 1. Use of electrical currents to destroy unwanted hair follicles. 2. Chemical decomposition of a compound in solution by passing an electrical current through it.

electrolyte imbalance *(ē-lĕk'-trō-līt).* Abnormal balance among the electrolytes chloride, potassium, and sodium.

electrolytes. Substances that, when in solution, can carry an electrical current. Those in the body include chloride, potassium, and sodium, among others.

electromyogram *(ē-lĕk-trō-mī'-ō-grăm).* Graphic record of the electric currents associated with muscle activity.

electromyography *(ē-lĕk-trō-mī-ŏg'-ră-fē).* Recording of the electric currents generated by muscular activity.

electronic fetal monitoring. *See* **fetal monitor.**

electron microscope *(ē-lĕk'-trŏn).* Microscope that uses a beam of electrons instead of light to scan surfaces and magnify images.

electrophoresis *(ē-lĕk-trō-fō-rē'-sĭs).* Movement of charged particles in liquid in response to changes in an electric field, a technique that is used to analyze certain substances (e.g., identifying proteins in serum).

electroretinography *(ē-lĕk-trō-rĕt-ĭn-ŏg'-ră-fē).* Procedure to measure and record the electrical activity of the retina, used to help diagnose retinal disease.

electroshock therapy. *See* **electroconvulsive therapy.**

electrosurgery *(ē-lĕk-trō-sŭr'-jĕr-ē).* Surgical procedure in which electrical devices (e.g., electric needles) are used.

electuary *(ē-lĕk-tū-ăr-ē).* Mixture of a medication with sugar or honey.

element *(ĕl'-ĕ-mĕnt).* Simplest chemical form composed of only one type of atom (e.g., oxygen, hydrogen).

elephantiasis *(ĕl-ĕ-fĕn-tī'-ĕ-sĭs).* Condition characterized by huge enlargement of the legs, scrotum, and sometimes other body parts, caused by obstruction of the lymph channels.

elevator *(ĕl'-ĕ-vā-tŏr).* 1. Instrument used to pry up a depressed bone fragment. 2. Instrument used to extract teeth and roots not reached with a forceps.

elimination diet *(ē-lĭm-ĭ-nā'-shŭn).* Systematic exclusion of specific foods from the diet in an effort to identify which ones trigger health problems, such as headache, diarrhea, bloating, flatulence, rash, and fatigue.

elixir *(ē-lĭk'-sĕr).* Clear, sweet solution of alcohol and water, used in some medicines taken by mouth.

elution *(ē-lū'-shŭn).* Separation of substances by washing.

emaciation *(ē-mā-shē-ā'-shŭn).* Excessive leanness, caused by disease or poor nutrition.

emanation *(ĕm-ă-nā'-shŭn).* 1. Exhalation. 2. Vapor, gas, or fluid that is given off by a chemical or other substance.

emasculation *(ē-măs-kū-lā'-shŭn).* *See* **castration.**

embalm *(ĕm-bălm').* To treat a dead body with preservatives to prevent decay.

embolectomy *(ĕm-bō-lĕk'-tō-mē).* Surgical removal of an embolus.

embolism *(ĕm'-bĕ-lĭz-m).* Blockage of a blood vessel, especially an artery, by a blood clot or other foreign material.

embolus *(ĕm'-bō-lŭs).* Blood clot (thrombus), piece of tissue, foreign object, gas, or air bubble that moves through the bloodstream until it lodges itself in a vessel.

embryo *(ĕm'-brē-ō).* 1. Stage in prenatal development of a mammal between the ovum and the fetus. 2. Young of any organism in its early stage of development.

embryonal cancer *(ĕm'-brē-ŏ-năl).* Highly malignant tumor that likely is derived from primitive embryonal cells.

emergency medical technician (EMT). Individual specially trained to provide prehospital care of injured or ill patients.

emergency medicine. Medical specialty concerned with the diagnosis and prompt treatment of trauma, sudden illness, or injury.

emesis *(ĕm'-ĭ-sĭs).* Vomiting.
emetic *(ĭ-mĕt'-ĭk).* Substance that induces vomiting, used to treat some cases of drug overdose and poisoning.
EMG. Acronym for electromyogram.
emission *(ĭ-mĭsh'-ĕn).* Discharge or release of something.
noctural e. Discharge of semen during sleep; also called *wet dream.*
emmenagogue *(ĭ-mĕn'-ă-gŏg).* Drug or measure that brings on menstruation.
emollient *(ē-mŏl'yĕnt).* Substance that softens and soothes the skin or mucous membranes.
emotion *(ē-mō'-shŭn).* Any intense feeling.
empathy *(ĕm'-pĕ-thē).* Ability to recognize and relate to the emotions of another person.
emphysema *(ĕm-fĭ-sē'-mă).* Abnormal condition of the lungs in which the air sacs (alveoli) overinflate, resulting in a breakdown of their walls.
empirical *(ĕm-pĭr'-ĭ-kăl).* Based on practical experience.
empyema *(ĕm-pī-ē'-mă).* Pus in the lung cavity or other body cavity, usually caused by a bacterial infection.
emulsion *(ē-mŭl'-shŭn).* Preparation composed of two liquids that do not mix.
Enalapril. *See* Appendix, Common Prescription and OTC Drugs: By Generic Name.
enamel *(ē-năm'-ĕl).* Hard, white substance that covers the crown of a tooth.
encanthis *(ĕn-kăn'-thĭs).* Small tumor at the inner angle (canthus) of the eye.
encapsulation *(ĕn-kăp-sū-lā'-shŭn).* Containing or having a covering, sheath, or capsule.
encephalitis *(ĕn-sĕf-ă-lī'-tĭs).* Inflammation of the brain, usually caused by a virus or a complication of another infection.
encephalogram *(ĕn-sĕf'-ă-lō-grăm).* X-ray of the brain made by withdrawing cerebrospinal fluid and replacing it with a gas.
encephalomalacia *(ĕn-sĕf-ă-lō-mă-lā'-shă).* Softening of the brain.
encephalomeningitis *(ĕn-sĕf-ă-lō-mĕn-ĭn-jī'-tĭs).* *See* **meningoencephalitis.**
encephalomyelitis *(ĕn-sĕf-ă-lō-mī-lī'-tĭs).* Inflammation of the brain and spinal cord.
encephalon *(ĕn-sĕf'-ă-lŏn).* Another name for the brain.
encephalopathy *(ĕn-sĕf-ă-lŏp'-ă-thē).* Any disease of the brain.
encephalosclerosis *(ĕn-sĕf-ă-lō-sklĕ-rō'-sĭs).* Hardening of the brain.
encephaloscopy *(ĕn-sĕf-ă-lŏsk'-ō-pē).* Technique in which a hole is drilled through the skull and an endoscope (hollow metal tube) is inserted to view the brain.
enchondroma *(ĕn-kŏn-drō'-mă).* Benign tumor composed of cartilaginous tissue that generally appears within a bone.
encopresis *(ĕn-kō-prē'-sĭs).* Condition characterized by constipation and retained hard stools, and by watery intes-

tinal contents that bypass the hard stools and exit the rectum, often mistaken for diarrhea.

endarterectomy *(ĕnd-ăr-tĕr-ĕk'-tŏ-mē).* Surgical removal of the lining of an artery.

endarteritis *(ĕnd-ăr-tĕr-ī'-tĭs).* Inflammation of the innermost coating of an artery caused by trauma, infection, or syphilis.

endemic *(ĕn-dĕm'-ĭk).* Disease that occurs continuously in a specific population or locality, but has a low mortality rate.

endocardiography *(ĕn-dō-kăr-dē-ŏg'-ră-fē).* Recording of the electric currents that cross the heart muscle.

endocarditis *(ĕn-dō-kăr-dī'-tĭs).* Inflammation of the lining membrane of the heart.

endocardium *(ĕn-dō-kăr'-dē-ŭm).* Serous lining membrane of the inner surface and cavities of the heart.

endocervical *(ĕn-dō-sĕr'-vĭ-kăl).* Referring to the uterine cervix.

endocervicitis *(ĕn-dō-sĕr-vĭ-sī'-tĭs).* Inflammation of the membrane that lines the cervix of the uterus.

endocrine gland *(ĕn'-dō-krĭn).* Ductless gland that produces and secretes a substance into the blood or lymph.

endocrinology *(ĕn-dō-krĭn-ŏl'-ō-jē).* Study of the endocrine glands and their functions.

endoderm *(ĕn'-dō-dĕrm).* Innermost of the three germ layers of the embryo.

endogenous *(ĕn-dŏj'-ĕ-nŭs).* Produced or originating from within a cell or organism.

endometrial cyst *(ĕn-dō-mē'-trē-ăl).* Cyst usually located in or near the ovary and composed of tissue from the endometrium.

endometriosis *(ĕn-dō-mē-trē-ō'-sĭs).* Abnormal condition in which the cells of the uterus invade other tissues in the area, such as the ovaries, bladder, or ureters.

endometritis *(ĕn-dō-mĕ-trī'-tĭs).* Inflammation of the inner lining of the uterus.

endometrium *(ĕn-dō-mē'-trē-ŭm).* Mucosal layer that lines the cavity of the uterus.

endomorph *(ĕn'-dō-mŏrf).* Person who has a body shape that tends to be soft and round, characterized by a prominent abdomen, *See also* **ectomorph** and **mesomorph.**

endomyocarditis *(ĕn-dō-mī-ō-kăr-dī'-tĭs).* Inflammation of the myocardium (heart muscle) and the endocardium (inner lining), due to infection or disease.

end organ. 1. Large, encapsulated ending of a sensory nerve. 2. Site of the ultimate damage by a disease process (e.g., kidney damage caused by hypertension).

endorphin *(ĕn-dŏr'-fĭn).* Any of several proteins in the brain that play a role in reducing or eliminating pain and enhancing pleasure.

endoscope *(ĕn'-dō-skōp).* Instrument used to examine the inside of a hollow organ or cavity.

endoscopy *(ĕn-dŏs'-kō-pē).* Use of an endoscope to inspect the inside of a canal or a food or air passage.

upper e. A common type of endoscopy in which a thin flexible instrument is advanced through the mouth to evaluate or treat problems of the esophagus, stomach, or the first part of the small intestine.

endoskeleton *(ĕn-dō-skĕl'-ĕ-tŏn).* Internal bony structure in vertebrates that supports the body from within.

endothelioma *(ĕn-dō-thē-lē-ō'-mă).* Any tumor, benign or malignant, that develops from cells that line blood vessels, lymphatic vessels, or serous membranes.

endothelium *(ĕn-dō-the'-lē-ŭm).* Thin layer of cells that line blood vessels, lymph vessels, and serous cavities.

endotoxic shock *(ĕn-dō-tŏk'-sĭk).* Condition brought on when poisons are freed by bacteria and enter the bloodstream; some symptoms include rapid pulse, shortness of breath, chills, fever.

endotoxin *(ĕn-dō-tŏk'-sĭn).* Poison produced and retained by bacteria and released when the cells are destroyed or die.

endotracheal anesthesia *(ĕn-dō-trā'-kē-ăl).* Anesthesia that is given through a rubber tube that is inserted through the mouth into the trachea (windpipe).

endotracheal tube. Catheter passed through the mouth or nose into the trachea (windpipe) to maintain an open airway.

endourologist *(ĕn-dō-ū-ŏl'-ĕ-jĭst).* A urologist who has special expertise in using endoscopic instruments and other devices inside the kidney, ureter, and bladder, and in treating diseases that affect these organs.

end plate. End part of a motor nerve fiber that sends nerve signals to the muscles or organs.

end product. The product that results after a series of chemical or physical reactions are completed.

end-stage renal disease (ESRD). Chronic irreversible renal failure.

enema *(ĕn'-ĕ-mă).* Injection of fluid into the rectum to remove feces, help diagnose a gastrointestinal disorder, give drugs, or provide nourishment.

energy *(ĕn'-ĕr-jē).* Exertion of power to cause physical change.

enervation *(ĕn-ĕr-vā'-shŭn).* Loss of strength or energy.

engagement *(ĕn-gāj'-mĕnt).* In obstetrics, entrance of the fetus's head in the maternal pelvis, which usually occurs in late pregnancy.

engorge *(ĕn-gŏrj').* Fill to excess or to the limit.

enophthalmos *(ĕn-ŏf-thăl'-mŏs).* Condition in which the eyeball is displaced back in the eye socket due to a defect or injury.

enteral nutrition *(ĕn'-tĕr-ăl).* Feeding through a tube that has been placed down the nose into the stomach.

enterectomy *(ĕn-tĕr-ĕk'-tō-mē).* Surgical removal of part of the intestine.

enteric-coated *(ĕn-tĕr'-ĭk).* Refers to pills or tablets that are covered with a film layer which allows them not to dissolve until they reach the small intestine.

enteritis *(ĕn-tĕ-rī'-tĭs).* Inflammation of the intestine, especially the small intestine.

Enterobacter *(ĕn-tĕr-ō-băk'-tĕr).* Genus of bacteria, widely found in nature, that occurs in the intestinal tract of humans and frequently causes nosocomial infections (those that originate in hospitals) associated with contaminated medical equipment and personnel.

Enterococcus *(ĕn-tĕr-ō-kŏk'-ŭs).* Genus of bacteria of the *Streptococcaceae* family, usually found in the intestines.

enterocolitis *(ĕn-tĕr-ō-kō-lī'-tĭs).* General term for inflammation of the small and large intestine.

enteroenterostomy *(ĕn-tĕr-ō-ĕn-tĕr-ŏs'-tō-mē).* Surgical procedure that connects any two noncontinuous segments of intestine.

enterolithiasis *(ĕn-tĕr-ō-lĭ-thī'-ă-sĭs).* Presence of stonelike substances in the intestines.

enteropathic *(ĕn-tĕr-ō-pă'-thĭk).* Capable of causing disease in the intestinal tract.

enteroptosis *(ĕn-tĕr-ŏp-tō'-sĭs).* Displacement of the intestines from their usual position downward into the abdominal cavity.

enterostomy *(ĕn-tĕr-ŏs'-tō-mē).* Surgical procedure to form a permanent opening into the intestine through the abdominal wall.

enterotoxin *(ĕn-tĕr-ō-tŏk'-sĭn).* Poison that is produced by or that originates in the intestinal contents, it can cause diarrhea and vomiting.

Enterovirus *(ĕn'-tĕr-ō-vī-rŭs).* Genus of viruses that infect the intestinal tract mainly, but also the muscles and other tissues.

entropion *(ĕn-trō'-pē-ŏn).* Abnormal inward turning of an edge, especially the eyelid toward the eyeball.

entropy *(ĕn'-trō-pē).* Portion of energy not available during a chemical reaction for performance of work because it has been used to increase the random motion of the atoms or molecules in a system.

enucleation *(ē-nū-klē-ā'-shŭn).* Surgical removal of an entire tumor or organ without rupture.

enuresis *(ĕn-ū-rē'-sĭs).* Involuntary release of urine, especially during sleep, after an age by which children should have bladder control.

enzyme *(ĕn'-zīm).* Protein, produced in cells, that can cause chemical changes while it remains unchanged in the process.

eosin *(ē'-ō-sĭn).* Synthetic dyes, produced from coal tar, used to stain cells to be studied under a microscope.

eosinophil *(ē-ŏ-sĭn'-ō-fĭl).* White blood cell that readily stains with eosin.

eosinophilia *(ē-ō-sĭn-ō-fĭl'-ē-ă).* Abnormally large number of eosinophils in the blood, often seen in allergic reactions and some inflammatory conditions.

ependyma *(ĕp-ĕn'-dĭ-mă).* Membrane that lines the ventricles in the brain and the central canal of the spinal cord.

ependymoma *(ĕp-ĕn-dĭ-mō'-mă).* A slow-growing, usually benign tumor that is composed of differentiated ependymal cells.

ephedra *(ĕ-fĕd'-ră).* Term used to describe any of several species of an herb that can be used as a decongestant.

e. sinica. Chinese herb, also known as ma huang, used as a stimulant and to help with weight loss. It can cause serious, even lethal, effects.

ephedrine *(ĕ-fĕd'-rĭn).* Adrenergic obtained from *Ephedra* species or made synthetically, it causes increased blood pressure and is a central nervous system stimulant. It is found in some supplements used to facilitate weight loss and increase energy.

epicanthus *(ĕp-ĭ-kăn'-thĕs).* Vertical fold of skin on both sides of the nose, it is a normal characteristic in some races and is a congenital abnormality in others.

epicardia *(ĕp-ĭ-kăr'-dē-ă).* Portion of the esophagus that extends from the diaphragm to the stomach.

epicondyle *(ĕp-ĭ-kŏn'-dīl).* Bony prominence located in the region of the condyle (rounded projection on a bone) near a joint.

epicondylitis *(ĕp-ĭ-kŏn-dĭ-lī'-tĭs).* Inflammation of the epicondyle of the humerus and the tissues that surround it.

external humeral e. Tennis elbow.

epicranium *(ĕp-ĭ-krā'-nē-ŭm).* Soft parts that cover the cranium.

epidemic *(ĕp-ĭ-dĕm'-ĭk).* Any disease, injury, or other health-related event that occurs suddenly in excessive numbers; usually associated with infectious diseases.

epidemiology *(ĕp-ĭ-dē-mē-ŏl'-ō-jē).* Study of the factors that determine and influence the frequency and distribution of disease and other health-related events and their causes in humans.

epidermal growth factor *(ĕp-ĭ-dĕr'-măl).* Genetically engineered protein that stimulates rapid healing of skin and the cornea.

epidermis *(ĕp-ĭ-dĕr'-mĭs).* The outermost layer of skin.

epidermoid tumor. Benign tumor that consists of abnormal epidermal cells.

epidermomycosis *(ĕp-ĭ-dĕr-mō-mī-kō'-sĭs).* Skin disease caused by a fungus.

epididymis *(ĕp-ĭ-dĭd'-ĭ-mĭs).* Long, cordlike structure located on and beside the posterior surface of the testes, it stores and transports spermatozoa.

epididymitis *(ĕp-ĭ-dĭd-ĭ-mī'-tĭs).* Inflammation of the epididymis.

epidural *(ĕp-ĭ-doo'-răl).* Located over or on top of the dura (outer membrane covering the spinal cord and brain).

epidural anesthesia. Loss of sensation caused by injection of a pain-killing agent into the epidural space.

epigastric *(ĕp-ĭ-găs'-trĭk).* Referring to the epigastrium, the area over the pit of the stomach.

epiglottis *(ĕp-ĭ-glŏt'-ĭs).* Thin, leaf-shaped structure that hangs over the entrance to the larynx and helps prevent food from entering the larynx and trachea while swallowing.

epiglottitis. *See* **supraglottitis.**

epilation *(ĕp-ĭ-lā'-shŭn).* 1. Extraction of hair. 2. Loss of hair due to exposure to ionizing radiation.

epilepsy *(ĕp'-ĭ-lĕp-sē).* Any of a group of syndromes characterized by recurrent seizures with impaired or loss of consciousness, abnormal motor activity, or sensory disturbances; *See also* **seizure.**

absence e. Form characterized by absence seizures (sudden transient break in consciousness of thought or activity).

grand mal e. Form often preceded by an aura and characterized by a loss of consciousness and tonic-clonic seizures.

idiopathic e. Epilepsy of unknown origin.

Jacksonian e. Form characterized by intense muscle spasms that occur on one side of the body; consciousness is usually maintained during episodes.

epileptic. 1. Referring to epilepsy. 2. Individual who has epilepsy.

epinephrine *(ĕp-ĭ-nĕf'-rĭn).* Neurotransmitter and hormone secreted by the adrenal medulla in response to certain stimuli, such as stress and hypoglycemia.

epineurium *(ĕp-ĭ-nū'-rē-ŭm).* Outermost layer of connective tissue of a peripheral nerve, it covers the entire nerve and contains blood vessels and lymphatics.

epiphora *(ĕ-pĭf'-ō-ră).* Abnormal overflow of tears down the cheeks due to excessive secretion of tears of an obstructed lacrimal duct.

epiphysis *(ĕ-pĭf'-ĭ-sĭs).* In infants and young children, a bone-forming site separated from a parent bone by cartilage. As children grow, it becomes a part of the larger (parent) bone.

epiphysitis *(ĕ-pĭf-ĭ-sī'-tĭs).* Inflammation of an epiphysis or of the cartilage that separates it from the parent bone.

epiploectomy *(ĕ-pĭp-lō-ĕk'-tē-mē).* Surgical procedure to remove the omentum; *see also* **omentum.**

epiploon *(ĕ-pĭp'-lō-ŏn).* The omentum.

episcleritis *(ĕp-ĭ-sklĕ-rī'-tĭs).* Inflammation of the tissues that overlay the sclera.

episiotomy *(ĕ-pĭs-ē-ŏt'-ō-mē).* Surgical incision into the perineum and vagina to prevent tearing during and to facilitate delivery of a fetus.

episode *(ĕp'-ĭ-sōd).* A noteworthy happening or series of happenings that occur in the course of continuous events; for example, an episode of illness.

epistaxis *(ĕp-ĭ-stăk'-sĭs).* Nosebleed.

epithelialization *(ĕp-ĭ-thē-lē-ĕl-ĭ-zā'-shŭn).* Healing by the growth of epithelium over a denuded surface.

epithelioma *(ĕp-ĭ-thē-lē-ō'-mă).* Malignant tumor that consists primarily of epithelial cells.

epithelium *(ĕp-ĭ-thē'-lē-ŭm).* The layer of cells that form the epidermis of the skin and top layer of mucous and serous membranes. Epithelium is classified into types based on the number of layers and shape of the cells.

attachment e. Area of soft tissue attached to the tooth.

glandular e. Epithelium consisting of cells that secrete.

pigmented e. Epithelium that contains pigment granules.

stratified e. Epithelium in which the cells are arranged in several layers.

epituberculosis *(ĕp-ĭ-tū-bĕr-kū-lō'-sĭs).* A type of primary tuberculosis that produces mild symptoms; seen in children.

eponychium *(ĕp-ō-nĭk'-ē-ŭm).* The narrow band of epidermis that extends from the nail wall onto the nail surface; also called *cuticle.*

Epsom salts *(ĕp'-sŏm).* Magnesium sulfate; it can be used as a laxative or, when dissolved in water, as an anti-inflammatory.

Epstein-Barr virus *(ĕp'-stēn-băr).* Virus that causes infectious mononucleosis and is associated with Burkitt's lymphoma and nasopharyngeal cancer; also called *human herpesvirus 4.*

epulis *(ĕp-ū'-lĭs).* General term applied to tumors and tumorlike growths that develop on the gingival.

equilibrium *(ē-kwĭ-lĭb'-rē-ŭm).* 1. State in which opposite forces counteract each other exactly; balance. 2. Postural balance of the body.

equine encephalitis. Type of encephalomyelitis that affects horses and mules, it can spread to humans by mosquitoes.

eradication *(ē-răd-ĭ-kā'-shŭn).* Complete elimination of a disease; usually refers to an epidemic or a disease that is endemic.

erasion *(ē-rā'-zhŭn).* Scraping away diseased tissue from a diseased body part.

Erb's paralysis *(ĕrbz).* Paralysis of the muscles in the upper arm and shoulder, causing the arm to hang limp and the hand to rotate inward.

ERCP. Acronym for endoscopic retrograde choledochopancreatography.

erectile tissue *(ē-rĕk'-tĭl).* Vascular tissue that becomes rigid or erect when filled with blood; usually refers to the clitoris, nipples, or penis.

erection *(ē-rĕk'-shŭn).* Enlargement and hardening of the penis, generally due to sexual excitement; can also refer, to a lesser extent, to a similar state occurring in the clitoris.

erg *(ĕrg).* The amount of work done when a force of 1 dyne acts through a distance of 1 cm.

ergograph *(ĕr'-gō-grăf).* Device for recording the contractions of muscles and measuring the amount of work done.

ergonomics *(ĕr-gō-nŏm'-ĭks).* Science concerned with finding ways to fit a job to people's anatomical, physiological, and psychological traits using methods that will enhance human efficiency and well-being.

ergosterol *(ĕr-gŏs'-tĕr-ŏl).* A sterol occurring mainly in yeast and forming vitamin D2 when exposed to ultraviolet light; also called *provitamin D2.*

ergot *(ĕr'-gŏt).* Fungus (*Claviceps purpurea*) derived from rye plants, it is the source of ergot alkaloids.

e. alkaloids. Group of related alkaloids derived from ergot or man-made; some have a medicinal use, others are poisonous.

ergotamine tartrate *(ĕr-gŏt'-ă-mēn).* A derivative of ergot, it stimulates the smooth muscles of blood vessels and the uterus; used to treat migraine. Trade names: Ergomar, Ergostat.

erogenous *(ēr-ŏj'-ĕ-nŭs).* Causing sexual excitement.

e. zone. Any part of the body that, when touched or stroked, causes sexual excitement.

erosion *(ē-rō'-shŭn).* Wearing or eating away of tissue.

erotic *(ē-rŏt'-ĭk).* Referring to sexual desire.

erotomania *(ē-rō-tō-mā'-nē-ă).* Excessive or exaggerated sexual behavior.

eructate *(ē-rŭk'-tāt).* Belch.

eruption *(ē-rŭp'-shŭn).* To break out or become visible; especially when referring to a skin lesion or rash.

erysipelas *(ĕr-ĭ-sĭp'-ĕ-lăs).* An acute febrile disease characterized by hot, bright red skin that also affects the subcutaneous tissue; it is caused by infection with streptococci.

erythema *(ĕr-ĭ-thē'-mă).* Form of macula characterized by diffused redness of the skin.

e. multiforme. A skin eruption characterized by dark red papules or tubercles, which may appear in rings or disk-shaped patches.

e. nodosum. Painful, red nodules on the legs, usually associated with rheumatism but may also be caused by food poisoning or certain drugs.

erythremia *(ĕr-ĭ-thrē'-mē-ă).* *See* **polycythemia.**

erythroblast *(ĕ-rĭth'-rō-blăst).* A nucleated, immature red blood cell.

erythroblastosis *(ĕ-rĭth-rō-blăs-tō'-sĭs).* Presence of an abnormally large number of erythroblasts in peripheral blood.

erythrocyte *(ĕ-rĭth'-rō-sīt).* A mature red blood cell.

erythrocyte sedimentation rate (ESR). The rate at which erythrocytes settle to the bottom of a well-mixed specimen of venous blood.

erythrolysis *(ĕr-ĭ-thrŏl'-ĭ-sĭs).* *See* **hemolysis.**

erythropoietin *(ĕ-rĭth-rō-poi'-ĕ-tĭn).* A hormone, produced mainly in the kidney, that stimulates red blood cell production.

escape mechanism. In psychology, a term that describes a mental effort to escape from a sense of responsibility or a sense of guilt.

eschar *(ĕs'-kăhr).* Scab or crust that forms over a raw surface.

Escherichia *(ĕsh-ĕ-rĭk'-ē-ă).* Genus of anaerobic, gram-negative, rod-shaped bacteria found in the large intestine of humans and other mammals.

E. coli. Main species of the genus that is almost constantly present in the alimentary canal of humans; it is responsible for various infections of the urinary tract, conjunctivitis, abscesses, and other infections.

esomeprazole. *See* Appendix, Common Prescription and OTC Drugs: By Generic Name.

esophagectomy *(ĕ-sŏf-ă-jĕk'-tō-mē).* Surgical removal of part or all of the esophagus.

esophagitis *(ĕ-sŏfĕ-jī'-tĭs).* Inflammation of the esophagus.

esophagogastrectomy *(ē-sŏf-ă-gō-găs-trĕk'-tĕ-mē).* Surgical removal of the esophagus and stomach, usually the distal portion of the esophagus and the proximal stomach.

esophagogastrostomy *(ē-sŏf-ă-gō-găs-trŏs'-tĕ-mē).* Surgical creation of a passageway between the stomach and esophagus.

esophagomyotomy *(ē-sŏf-ă-gō-mī-ŏt'-ĕ-mē).* Surgical incision through the muscular coat of the esophagus.

esophagoscopy *(ē-sŏf-ă-gŏs'-kō-pē).* Examination of the esophagus using an endoscope.

esophagospasm *(ē-sŏf'-ă-gō-spăzm).* Spasm of the esophagus.

esophagus *(ē-sŏf'-ă-gŭs).* A muscular passage that extends from the pharynx to the stomach.

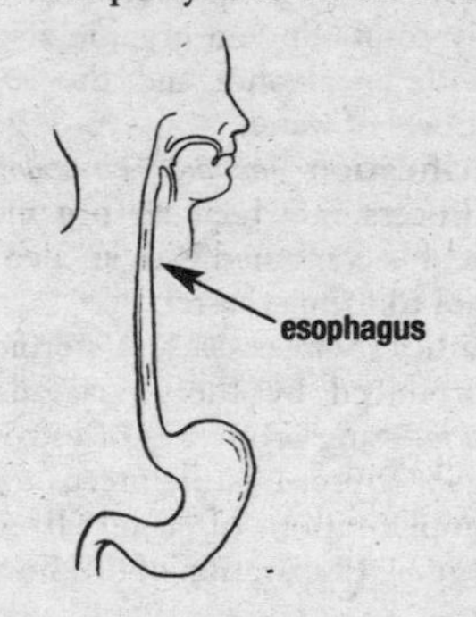

esotropia *(ĕs-ō-trō'-pē-ă).* Inward turning of an eye toward that of the other eye; cross-eye.

ESP. Acronym for extrasensory perception.

ESR. Acronym for erythrocyte sedimentation rate.

essence *(ĕs'-ĕns).* 1. The principle or property of something. 2. A solution of a volatile oil in alcohol.

flower e. Water extracts of fresh flowers selected for their effects on specific emotional or mental symptoms; *See also* **Bach flower remedy.**

essential amino acid. *See* **amino acids.**

essential fatty acids. *See* **fatty acid.**

essential hypertension. *See* **hypertension.**

essential oils. Extremely concentrated substances (amino acid precursors, vitamins, hormones, minerals, enzymes, nutrients) extracted from plants, used in aromatherapy; *See also* **aromatherapy.**

essential tremor. *See* **tremor.**

ester *(ĕs'-tĕr).* Compound formed by combining an organic acid with an alcohol and the removal of water.

esterification *(ĕs-tĕr-ĭ-fĭ-kā'-shŭn).* Process in which an organic acid is combined with an alcohol to form an ester.

estradiol *(ĕs-tră-dī'-ŏl).* 1. A steroid produced by the ovary and possessing properties of estrogen. It prepares the uterus for implantation of a fertilized egg. 2. Preparation of this hormone used in estrogen replacement therapy.

estrogen *(ĕs'-trĕ-jĕn).* General term for female sex hormones; estrogens (estradiol, estrone) are produced primarily in the ovary (in women) or testes (in men); *see also* **estradiol, estrone.**

conjugated e's. Mixture of estrone and equilin (derived from pregnant mares' urine or prepared synthetically), used to treat menopausal symptoms.

estrogen receptor test. Test to identify the impact of estrogen on the growth of breast cancer cells.

estrone *(ĕs'-trōn).* An estrogenic (producing the effects of an estrogen) hormone found in the urine of pregnant women and mares; it also can be prepared synthetically. Used to treat estrogen deficiencies.

estrus *(ĕs'-trŭs).* Cyclic period of sexual activity in nonhuman female mammals; also referred to as the animal being in "heat."

ESWL. Acronym for endoscopic shock wave lithotripsy; *see also* **lithotripsy.**

ethanol *(ĕth'-ă-nŏl).* Ethyl alcohol.

ether *(ē'-thĕr).* Organic compound with an oxygen atom bonded to two carbon atoms; once used as an anesthetic during surgery but now seldom used because it is highly flammable.

ethicist *(ĕth'-ĭ-sĭst).* Individual who specializes in the ethical aspects of a particular practice.

ethics *(ĕth'-ĭks).* System of moral standards or principles that govern one's conduct.

ethmoid *(ĕth'-moyd).* Sievelike.

e. bone. Sievelike spongy bone that forms a roof for the nasal fossae and part of the floor of the anterior fossa of the skull.

e. sinus. Air cells or space inside the ethmoid bone, opening into the nasal cavity.

ethmoidectomy *(ĕth-moy-dĕk'-tō-mē).* Excision of ethmoid cells that open into the nasal cavity.

ethmoiditis *(ĕth-moy-dī'-tĭs).* Acute or chronic inflammation of ethmoidal cells.

ethyl alcohol. Grain alcohol.

ethyl chloride. Volatile liquid that can be sprayed on the skin as a short-term anesthetic.

ethylenediaminetetraacetic acid (EDTA) *(ĕth-ĕ-lĕn-dī-ĕ-mĕn-tĕt-rē-ăs'-ĕ-tāt).* An agent that binds calcium and heavy metal ions; a chelator.

ethylestenol *(ĕth-ĕl-ĕs'-trĕ-nŏl).* An androgen and anabolic steroid.

etiology *(ē-tē-ŏl'-ĕ-jē).* Study of the factors that cause disease and how they are introduced to the host.

etodolac *(ē-tō-dō'-lăk).* A nonsteroidal anti-inflammatory drug used to relieve pain and inflammation.

eucalyptus *(ū-kĕ-lĭp'-tŭs).* An aromatic plant (*Eucalyptus globulus*) that yields an essential oil sometimes used as a decongestant and to treat sinusitis, flu, colds, allergies, and bronchitis.

eugenics *(ū-jĕn'-ĭks).* Improvement of a population by choosing the best specimens for breeding.

eunuch *(ū'-nĕk).* A boy or man who has had his testes or external genital organs removed.

euphoria *(ū-fŏr'-ē-ă).* An exaggerated feeling of physical and mental well-being, especially when it is not based in reality.

eustachian tube *(ū-stā'-kē-ăn).* Auditory tube that extends from the middle ear to the pharynx.

euthanasia *(ū-thă-nā'-zē-ă).* 1. An easy or painless death. 2. The act of willfully ending life in people who have an incurable condition; also called *mercy killing.*

euthenics *(ū-thĕn'-ĭks).* Science that attempts to improve a population by making changes to the environment.

euthyroid *(ū-thī'-royd).* Normal thyroid gland function.

evacuate *(ē-văk'-ū-āt).* To discharge or empty something, especially from the bowels.

evagination *(ē-văj-ĭ-nā'-shŭn).* Emergence from a sheath. 2. Bulging out of tissue to form a pouch or sac.

evanescent *(ĕv-ă-nĕs'-ănt).* Impermanent; passing away gradually.

evening primrose. Herbal remedy valued for its high levels of gamma-linolenic acid (GLA), which helps lower high blood pressure and high cholesterol levels.

eventration *(ē-vĕn-trā'-shŭn).* Partial protrusion of the abdominal contents through an opening in the abdominal wall; a hernia.

eversion *(ē-vĕr'-zhŭn).* 1. Turning outward. 2. Turning inside out; also called *ectropion.*

evidence-based medicine. The judicious use of the best current evidence when making health-care decisions for an individual patient.

evisceration *(ē-vĭs-ĕr-ā'-shŭn).* Partial protrusion of the contents of the abdomen through an opening in the abdominal wall, especially through a surgical incision.

Evista. *See* Appendix, Common Prescription and OTC Drugs: By Trade Name.

evoked potential *(ē-vōkt').* An electrical signal recorded from a nerve, muscle, sensory receptor, or site of the central nervous system that has been stimulated.

evolution *(ĕv-ō-lū'-shŭn).* Process of gradual and systematic development or change.

Ewing's tumor *(ū'-ĭngz).* Highly malignant, metastatic tumor of bone, usually occurring in the shafts of long bones, ribs, flat bones of children.

exacerbation *(ĕks-ăs-ĕr-bā'-shŭn).* Aggravation of or increase in the severity of symptoms of a disease.

exanthema subitum *(ĕks-ăn-thē'-mă sū'-bĭ-tŭm).* Acute disease of infants and young children, caused by human herpesvirus 6; also called *roseola.*

exanthropic *(ĕk-săn-thrŏp'-ĭk).* Referring to something, especially a disease, that originates outside the body.

excavation *(ĕks-kă-vā'-shŭn).* 1. The act of hollowing out or making a depression. 2. Formation of a cavity.

exchange transfusion. Repetitive withdrawal of small amounts of blood and replacement with a transfusion until blood volume is almost entirely exchanged.

excision *(ĕk-sĭzh'-ĕn).* Removal of a tumor, organ, and/or tissue by cutting.

excitant *(ĕk-sī'-ĕnt).* A stimulant.

excoriation *(ĕks-kŏr-ē-ā'-shĕn).* Abrasion or scratch of the skin or of the coating of an organ by trauma, chemicals, burns, or other causes.

excrement *(ĕks'-krĕ-mĕnt).* Material that the body eliminates as waste.

excrescence *(ĕks-krĕs'-ĕns).* Any abnormal outgrowth from a surface.

excrete *(ĕks-krēt').* To eliminate waste from the body, organs, or blood in a normal manner.

excursion *(ĕks-kŭr'-zhŭn).* Extent of movement of a body part from a normal or rest position.

exercise *(ĕks'-ĕe-sīz).* Physical activity performed to maintain or develop fitness and well-being.

exfoliation *(ĕks-fō-lē-ā'-shŭn).* The shedding, peeling, or scaling of skin.

exhalation *(ĕks-hă-lā'-shŭn).* Process of breathing out; the opposite of inhalation.

exhaustion *(ĕg-zaws'-chŭn).* State of extreme fatigue or weariness.

exhibitionism *(ĕg-zĭ-bĭsh'-ŭn-ĭzm).*

A psychoneurosis in which a person (usually a male) acts on an abnormal impulse to expose the genitals to unsuspecting individuals.

exocrine glands *(ĕks'-ō-krĭn).* Glands whose secretions reach the skin either directly or through a duct; sweat glands are an example.

exogenous *(ĕks-ŏj'-ĕ-nŭs).* Originating outside an organ or part; opposite of endogenous.

exomphalos *(ĕks-ŏm-fă-lŭs).* Protrusion of abdominal organs from the umbilical (navel) region; a hernia of the navel.

exophthalmos *(ĕks-ŏf-tăl'-mŏs).* Abnormal bulging of an eyeball, as may be seen in people who have an overactive thyroid gland, leukemia, tumor of the orbit, or other causes.

exoskeleton *(ĕk-sō-skĕl'-ĕ-tŏn).* The hard outer covering of certain invertebrates such as insects and crabs. In vertebrates, the term is used for structures produced by the epidermis, such as hair, nails, and teeth.

exostosis *(ĕks-ŏs-tō'-sĭs).* Bony growth that develops on the surface of a bone.

exoteric *(ĕks-ō-tĕr'-ĭk).* Referring to causes that develop outside the body.

exotoxin *(ĕks-ō-tŏks'-ĭn).* A toxin produced by a microorganism and excreted into its surrounding medium.

exotropia *(ĕks-ō-trō'-pē-ă).* Abnormal turning of one or both eyes outward.

expectorant *(ĕk-spĕk'-tō-rănt).* Substance that facilitates the removal of mucus or other fluids from the lungs, bronchi, and trachea.

expiration *(ĕx-pĭ-rā'-shŭn).* 1. Exhalation. 2. Death.

exploration *(ĕk-splō-rā'-shŭn).* Examination or investigation of an organ or other body parts for diagnostic reasons.

exsanguination *(ĕk-săn-gwĭn-ā'-shŭn).* Process of draining or expressing blood from a body part; to make bloodless.

exstrophy of the bladder *(ĕks'-trō-pē).* Congenital abnormality in which the lower part of the abdominal wall and anterior wall of the bladder are missing and the interior of the bladder is visible through the opening.

extended care facility. Health-care institution for patients who need long-term medical or custodial care, usually for chronic disease.

extended radical mastectomy. *See* **mastectomy.**

extended release. Form of medication designed to allow a twofold or greater reduction in frequency of administration compared with its conventional dosage form.

extension *(ĕk-stĕn'-shŭn).* Movement that straightens or increases the angle between the bones or parts of the body.

extensor muscle *(ĕk-stĕn'-sĕr).* Any muscle that extends a joint.

exteriorize *(ĕk-stĕr'-ē-ō-īz).* 1. To expose a part temporarily during surgery. 2. In psychiatry, to turn one's interest outside oneself.

extirpation *(ĕks-tĭr-pā'-shŭn).* Excision of tissue or organ; removing something by its roots.

extracapsular *(ĕks-tră-kăp'-sū-lăr).* Outside a capsule.

extracellular *(ĕks-tră-sĕl'-ū-lăr).* Outside the cell.

extracorporeal *(ĕks-tră-kŏr-pŏr'-ē-ăl).* Outside the body.

extracorporeal shock wave lithotripsy. *See* **lithotripsy.**

extract *(ĕks'-trăkt).* 1. To remove with force. 2. A concentrated preparation made from a plant or animal source, available in solid, powder, or semiliquid form.

extradural *(ĕks-tră-dū'-răl).* Located on the outer side of the dura mater.

extrahepatic *(ĕks-tră-hĕ-păt'-ĭk).* Located on the outside of the liver.

extramural *(ĕks-tră-mū'-răl).* Located outside the wall of an organ or vessel.

extrasensory perception *(ĕks-tră-sĕn'-sŏr-ē).* Ability to perceive external events through means other than the five senses.

extrasystole *(ĕks-trĕ-sĭs'-tō-lē).* A premature contraction of the heart that is separate from the normal rhythm and which occurs in response to an impulse in the impulse-conducting system.

extrauterine pregnancy *(ĕks-trĕ-ū'-tĕr-ĭn).* Development of the fertilized ovum outside of the uterus; also called *ectopic pregnancy.*

extravasation *(ĕks-trăv-ĕ-sā'-shŭn).* Escape or discharge of fluid, normally present in a tube or vessel, into the surrounding tissues.

extremity *(ĕks-trĕm'-ĭ-tē).* The end portion of anything. Usually refers to an upper or lower limb; also a hand or foot.

extrinsic *(ĕks-trĭn'-zĭk).* Originating from outside.

extroversion *(ĕks-trō-vĕr'-shĕn).* Turning inside out.

extrovert *(ĕks'-trō-vĕrt).* A type of personality in which an individual is interested mainly in external objects and actions; opposite of introvert.

extrusion *(ĕks-trū'-zhĕn).* Thrusting or pushing out by force.

extubation *(ĕks-tū-bā'-shŭn).* Removal of a previously inserted tube; e.g., a laryngeal tube.

exudate *(ĕks'-ū-dāt).* Material (e.g., cells, fluid, cell debris) that has escaped from blood vessels and has been deposited in tissues, usually because of inflammation.

eye. Organ of vision.

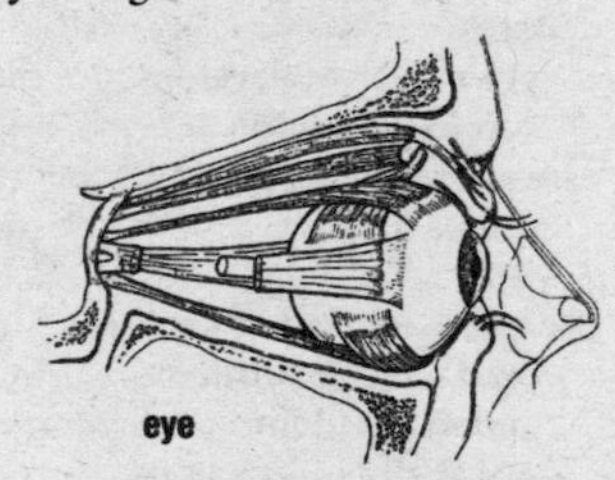

eye

dry e. *See* **keratoconjunctivitis sicca.**

lazy e. *See* **amblyopia.**

pink e. Another name for

acute contagious conjunctivitis.

eyebrow *(ī'-brou).* The arch over the eye and the hairs that cover it.

eyelash *(ī'-lăsh).* One of the hairs that grows at the edge of an eyelid; collectively the hairs are called *cilia.*

eye strain. Fatigue of the eye caused by overuse or from uncorrected visual problems.

eye tooth. Colloquial term for a canine tooth in the upper jaw.

eye wash. A solution used to bathe the eye.

F

Fabry's disease *(făh'-brēz).* An inherited disease in which a type of fat—glycolipid—accumulates in the tissues and organs in excessive amounts and impairs their function.

face. The front portion of the head from the forehead to the chin, including the eyes, nose, mouth, cheeks, and chin.

moon f. Rounded face seen in various disorders, such as Cushing's disease, or associated with use of corticosteroids.

face-lift. Popular term for rhytidectomy.

facet *(făs'-ĕt).* A small plane or rounded smooth surface on a hard body, as on a bone.

facial *(fā'-shĕl).* Referring to or directed toward the face.

facial nerve paralysis. Loss of voluntary movement of the muscles on one side of the face, caused by abnormal function of the seventh cranial nerve, which supplies those muscles; *see also* **Bell's palsy.**

facies *(fā'-shē-ēz)* (*pl.* facies). The outward appearance and expression of the face. Characteristic facial expressions are associated with certain diseases.

adenoid f. Characterized by a dull expression with an open mouth, usually seen in children who have adenoid growths.

f. hepatica. Thin face with sallow complexion, yellow eyes, and sunken eyeballs, seen in certain chronic liver disorders.

Parkinson's f. Lack of facial expression due to Parkinson's disease.

FACOG. Fellow of the American College of Obstetricians and Gynecologists.

FACP. Fellow of the American College of Physicians.

FACR. Fellow of the American College of Radiology.

FACS. Fellow of the American College of Surgeons.

factor *(făk'-tŏr).* 1. An element or agent that contributes to a process, result, or action. 2. A gene (hereditary factor).

coagulation f's. Substances in the blood that are necessary for proper blood clotting. There are more than a dozen such factors, each designated by Roman numerals.

growth f. A general term for various substances that promote normal or pathological growth of cells.

intrinsic f. A glycoprotein secreted by the gastric glands and necessary for absorption of vitamin B12; a deficiency of this factor can result in pernicious anemia.

platelet f's. Substances contained in or attached to platelets, they work with coagulation factors.

rheumatoid f. An antibody seen in the serum of the majority of people who have rheumatoid arthritis and in some individuals with other conditions.

Rh f. Any of several antigens that may be seen in red blood cells and which determine the Rh blood group system.

risk f. A clearly defined characteristic that is associated with an increased probability that a specific event or outcome (e.g., development of a disease) will occur.

tumor necrosis f. Either of two lymphokines that have the ability to cause the death of certain tumor cells without affecting normal cells. These factors have been used as experimental anticancer agents.

facultative *(făk'-ĕl-tā-tĭv).* 1. Optional. 2. In bacteriology, a bacterium that can grow with or without oxygen.

Fahrenheit *(fă'-ĕn-hīt).* A temperature scale on which 32 degrees is the freezing point of water and 212 degrees is the boiling point.

failure *(fāl'-yĕr).* 1. A state of being insufficient. 2. Cessation of normal functioning.

failure to thrive (FTT). Refers to a child whose physical development is significantly less than that of other children his/her age. During early infancy, this condition sometimes results in death; in older infancy or childhood it indicates an underlying disease.

faint *(fānt).* Transient loss of consciousness; syncope.

faith healing. Use of prayer to treat a disease or illness, a practice used by some Christian Scientists.

fallen womb. *See* **prolapsed uterus.**

fallopian tube *(fĕ-lō'-pē-ăn).* One of two tubes located on either side of the uterus, through which eggs are transported from the ovary to the uterus.

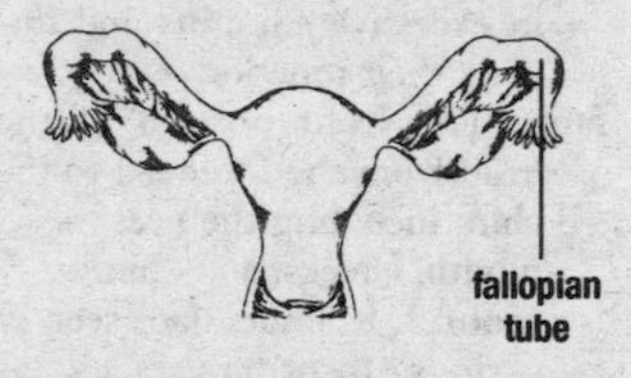

Fallot's tetralogy *(fă-lōz').* Congenital heart disease that involves four abnormalities: pulmonary stenosis, right ventricular hypertrophy, ventricular septal defect, and overriding of the aorta.

false-negative. A test or procedure that shows no evidence of disease when disease is actually present.

false-positive. A test or procedure that shows evidence of disease when disease is actually not present.

famciclovir *(făm-sī'-klō-vĕr).* Drug used to treat herpes zoster, herpes genitalis, and herpes simplex.

familial *(fă-mĭl'-yĕl).* Pertaining to or common to the same family; for example, a disease that occurs more frequently in a family than would be expected by chance.

familial tendency. Tendency for a disease to occur in the same family among members of several generations.

famotidine *(făm-ŏ'-tĭ-dīn).* Drug used to prevent and treat peptic ulcer, to relieve symptoms associated with hyperacidity, and to treat gastroesophageal reflux disease.

Fanconi syndrome *(făhn-kō'-nē).* A rare congenital disorder with a poor prognosis, characterized by various abnormalities of the musculoskeletal and genitourinary systems.

faradization *(făr-ĕ-dī-zā'-shŭn).* Treatment of nerves or muscles with an alternating electrical current, referred to as a faradic current.

farsightedness *(făhr-sīt'-ĕd-nĕs).* An error of refraction of light entering the eye, resulting in individuals being able to see distant objects clearly but cannot see near objects in clear focus.

fascia *(făsh'-ē-ă)* (*pl.* fasciae). Fibrous membrane found throughout the body that covers, supports, and separates muscles, and/or joins skin with underlying tissue. Fascia is found between muscles, around nerves and blood vessels, under the skin, and other places.

fasciculus *(fă-sĭk'-ū-lŭs).* A small bundle composed of muscle, tendon, or nerve fibers.

fasciectomy *(făsh-ē-ĕk'-tō-mē).* Excision of strips of fascia.

fasciotomy *(făsh-ē-ŏt'-ō-mē).* Surgical incision of fascia, often done to help relieve pressure in compartment syndrome (e.g., carpal tunnel syndrome).

fascitis *(făs-ē-ī'-tĭs).* Inflammation of fascia.

fasting. Abstinence from all food and drink except water for a specified amount of time.

fasting blood glucose. A method to detect how much glucose (sugar) is in a blood sample after an individual has fasted overnight. Commonly used to test for diabetes mellitus.

fat. White or yellowish tissue that serves as an energy reserve and helps to smooth and round out the body; also called *adipose tissue.*

fatal. Causing death; lethal.

fatigability *(făt-ĭ-gă-bĭl'-ĭ-tē).* Easily susceptible to exhaustion or fatigue.

fatigue *(fă-tēg').* Feeling of weariness or decreased efficiency resulting from excessive or prolonged exertion.

fat substitute. Compound that can replace conventional fat in foods and yet perform the same functions.

fatty acid. *See* **acid.**

fauces *(făw'-sēz).* The passageway from the mouth to the pharynx; also called *throat.*

faucitis *(făw-sī'-tĭs).* Inflammation of the fauces; also called *sore throat.*

FDA. Acronym for the Food and Drug Administration, a federal agency that monitors food and drug approvals and sales in the United States.

Fe. The chemical symbol for iron.

fear. An unpleasant emotional reaction to something one perceives as a threat, characterized by agitation, tension, heightened awareness, and various physical reactions.

febrile *(fē'-brĭl).* Referring to a fever; feverish.

fecal *(fē'-kăl).* Referring to feces.

fecal occult blood test. A test to check for hidden blood in stool. The source of bleeding may be anywhere in the gastrointestinal tract, and its presence may indicate any number of conditions.

feces *(fē'-sēz).* The waste or excrement discharged from the intestines; also called *stool.*

fecundability *(fē-kŭn-dĕ-bĭl'-ĭ-tē).* The probability that conception will occur in a specified population of couples during a named time period.

feedback *(fēd'-băk).* The return of some portion of the output to the place of origin by the system that receives it; *see also* **biofeedback.**

fee for service. In health care, a method of payment in which a health-care provider is paid for each individual service he/she renders to a patient.

Feingold diet *(fīn'-gōld).* Controversial dietary approach for hyperactive children in which foods that contain artificial colors, artificial flavors, preservatives, and salicylates are excluded.

felbamate *(fĕl'-bă-māt).* An anticonvulsant used to treat partial seizures in adults who have severe epilepsy.

Feldenkrais method *(fĕl'-dĕn-krīs).* A type of movement therapy that focuses on body functioning, the relationship between body movement and behavior, and how people learn behaviors.

fellatio *(fĕ-lā'-shē-ō).* Oral stimulation or manipulation of the penis.

felodipine *(fĕ-lō'-dī-pēn).* A calcium-channel-blocking drug used to treat hypertension.

felon *(fĕl'-ĕn).* An extremely painful abscess that develops on the palmar surface of the fingertips. Left untreated, the infection may spread to the bone of the finger.

Felty's syndrome *(fĕl'-tēz).* Syndrome characterized by an enlarged spleen, chronic rheumatoid arthritis, and pigmented spots on the lower extremities, occasionally anemia and thrombocytopenia (abnormally low level of platelets).

female *(fē'-māl).* Referring to the

sex that produces eggs or bears young.

feminization *(fĕm-ĭ-nĭ-zā'-shŭn).* 1. The normal development of primary and secondary sex characteristics in females. 2. Development of female characteristics by males.

femoral *(fĕm'-ŏr-ăl).* Referring to the femur, or to the thigh.

f. artery. The main artery that supplies blood to the thigh and leg.

f. nerve. One of the main nerves that supplies blood to the lower extremities.

femoral-popliteal bypass. Surgical procedure in which a graft is placed alongside the femoral artery, extending to the tibial artery below the knee.

femur *(fē'-mŭr).* The thighbone; it is the longest and largest bone in the body.

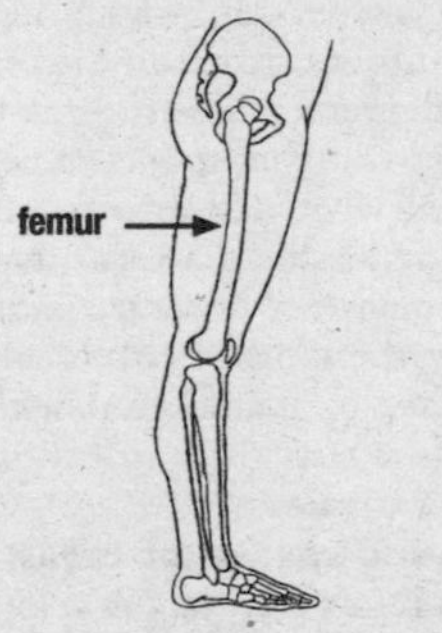

fenestra *(fĕ-nĕs'-tră).* 1. Window. 2. An opening in a cast or bandage.

fenestration *(fĕn-ĕs-trā'-shĕn).* The act of perforating.

f. operation. Surgical procedure in which a hole is drilled into the bone in the ear canal to relieve deafness.

feng shui *(fŭng'-shwā').* The Chinese practice of placing objects in an area based on the concept that patterns of arrangement influence the balance of yin and yang and the flow of qi, thus having an effect on health.

fenofenadine. *See* Appendix, Common Prescription and OTC Drugs: By Generic Name.

fenugreek *(fĕn'-ū-grēk).* A European herb (*Trigonella gruecum*) that reportedly can help reduce blood sugar levels, relieve menstrual symptoms, constipation, and diarrhea, and reduce inflammation.

fermentation *(fĕr-mĕn-tā'-shŭn).* The breakdown of organic compounds, especially carbohydrates, to simpler compounds with the use of enzymes.

ferric *(fĕr'-ĭk)* (*also* **ferrous**). Containing iron.

ferritin *(fĕr'-ĭ-tĭn).* One of the main forms in which iron is stored in the body.

fertility *(fĕr-tĭl'-ĭ-tē).* The ability to conceive or to bear children.

fertilization *(fĕr-tĭl-ĭ-zā'-shŭn).* The union of sperm with an egg (ovum).

fester *(fĕs'-tĕr).* To form pus or an abscess.

fetal *(fē'-tăl).* Referring to a fetus.

fetal alcohol syndrome. Condition seen in children born of alcoholic mothers, characterized by mental and growth retardation, defective facial features, and heart defects.

fetal monitor. An electronic apparatus that monitors fetal heartbeat and the mother's uterine contractions during childbirth.

fetishism *(fĕt'-ĭsh-ĭz-ĕm).* A paraphilia characterized by intense, recurrent sexual urges and fantasies concerning the use of inanimate objects (fetishes), usually women's clothing or accessories, as someone's preferred or necessary means for sexual arousal.

fetus *(fē'-tŭs).* The unborn offspring of any viviparous animal; in humans, from nine weeks after fertilization until birth.

fever *(fē'-vĕr).* Elevation of body temperature above the normal of 98.6°.

fever blister. A type of human herpesvirus (herpes simplex) that usually occurs along with a fever and commonly involves the lips and sometimes the nostrils; also called *cold sore.*

fiber. An elongated, threadlike structure.

fiber optics. The transmission of an image along flexible fibers.

fiberscope *(fī'-bĕr-skōp).* A flexible endoscope that uses fiber optics to transmit light and collect images.

fibrillation *(fī-brĭ-lā'-shŭn).* An involuntary contraction of muscle, not visible under the skin, that results from spontaneous activation of single muscle cells or muscle fibers whose nerve supply has been damaged.

atrial f. Irregular and rapid contractions of the atria (upper chambers of the heart).

ventricular f. Irregular and rapid contractions of the ventricles (lower chambers of the heart).

fibrin *(fī'-brĭn).* A fibrous protein derived from fibrinogen through the action of thrombin; it is the basic ingredient of a blood clot.

fibrinogen *(fī-brĭn'-ō-jĕn).* A protein found in blood plasma that is converted into delicate filaments (fibrin) by the action of the enzyme thrombin.

fibroadenoma *(fī-brō-ăd-ĕ-nō'-mă).* A benign tumor derived from glandular epithelium.

fibroblast *(fī'-brō-blăst).* A connective tissue cell, which forms the fibrous tissues in the body, including tendons.

fibroblastoma *(fī-brō-blăs-tō'-mă).* A tumor that develops from fibroblasts; there are two types, fibromas and fibrosarcomas.

fibrocystic *(fī-brō-sĭs'-tĭk).* Characterized by the development of cysts and an overgrowth of fibrous tissue.

fibrocystic change of breast. A condition that may develop in the breasts of women as they approach menopause, characterized by the development of cysts and associated with pain and tenderness.

fibroid *(fī'-broid).* Referring to a fibrous structure; a colloquial term for a fibroma.

fibroma *(fī-brō'-mă).* A benign, fibrous tumor that develops from connective tissue. Also called *fibroid* and *fibroid tumor.*

fibromatosis *(fī-brō-mă-tō'-sĭs).* Formation of fibrous, tumorlike nodules that develop from the deep fascia.

fibromyalgia *(fī-brō-mī-ăl'-ē).* Condition characterized by pain and stiffness in the muscles and joints, which may be diffuse or have multiple, specific trigger points.

fibroplasia *(fī-brō-plā'-zhē).* Formation of fibrous tissue, as occurs as wounds heal.

fibrosarcoma *(fī-brō-săr-kō'-mă).* A malignant tumor composed of cells and fibers derived from fibroblasts, which produce collagen.

fibrosis *(fī-brō'-sĭs).* The abnormal formation of fibrous tissue.

fibrositis *(fī-brō-sī'-tĭs).* Inflammation of muscle, characterized by pain and stiffness.

fibula *(fĭb'-ū-lă).* The outer and smaller of the two bones of the leg.

field *(fēld).* A specific area or open space, as a surgical field or visual field.

filament *(fĭl'-ĕ-mĕnt).* A tiny, delicate fiber or thread.

Filaria *(fĭ-lăr'-ē-ă)* (*pl.* filariae). A nematode worm of the superfamily Filarioidea.

filariasis *(fĭl-ă-rī'-ă-sĭs).* Chronic condition caused by infestation by filariae.

filling. Material inserted into a prepared tooth cavity to repair it, usually amalgam, gold, or a synthetic resin.

filter *(fĭl'-tĕr).* Any type of porous substance or device used to separate particulates or impurities from liquid or gas.

filtrate *(fĭl'-trāt).* A liquid or gas that has passed through a filter.

filum *(fī'-lĕm).* A threadlike structure.

fimbriated *(fĭm'-brē-āt-ĕd).* Fringed.

fine needle biopsy. The removal and examination of tissue from the body using a fine needle that is inserted into the affected tissue.

finger *(fĭng'-gĕr).* Any one of the five digits of the hand.

mallet f. Partial permanent flexion of the end phalanx of a finger caused by something striking the end or back of the finger, resulting in a rupture of the extensor tendon; also called *baseball finger, hammer finger.*

trigger f. A finger that can have a momentary spasm followed by a snapping into place. It may be caused by a nodule in the flexor tendon.

Fiorinal. Trade name of a pain-relieving drug composed of aspirin, butalbital, and caffeine.

first aid. Emergency care and treatment of an ill or injured individual before more sophisticated medical or surgical management can be initiated.

fission *(fĭsh'-ĕn).* 1. The process of splitting. 2. A form of asexual reproduction, seen primarily in unicellular organisms such as bacteria.

fissure *(fĭsh'-ĕr).* A groove, cleft, or deep fold.

fistula *(fĭs'-tū-lă).* An abnormal passage, usually between two internal organs or going from an organ to the surface of the body. A fistula may be surgically created.

fistulectomy *(fĭs-tū-lĕk'-tĕ-mē).* Surgical removal of a fistulous tract.

fit *(fĭt).* A seizure.

5-FU. *See* **fluorouracil.**

fixation *(fĭk-sā'-shŭn).* The act or process of suturing, fastening, or holding something in a fixed position.

elastic band f. A way to stabilize a fractured jaw using elastic bands applied to splints or appliances.

external pin f. A way to stabilize a fractured jaw by drilling pins into the bony parts and connecting them with metal bars.

skeletal f. A way to immobilize the end of a fractured bone using metal plates or wires applied directly to the bone or the body surface.

fixative *(fĭk'-sĕ-tĭv).* A fluid into which specimens are placed to preserve them in a close facsimile of their living state.

flaccid *(flăk'-sĭd).* Soft or weak.

flagellate *(flăj'-ĕ-lāt).* Any microorganism that has flagella.

flagellation *(flăj-ĕ-lā'-shĕn).* 1. The act of whipping or beating. 2. The formation of flagella.

flagellum *(flă-jĕl'-ŭm)* (*pl.* flagella). A long, whiplike projection from the surface of a cell which serves as a means of mobility.

Flagyl *(flăg'-ĕl).* Trade name for metronidazole, used to treat protozoal and bacterial infections, such as bacterial vaginosis and intestinal amebiasis.

flail chest. Chest that moves abnormally as the individual breathes, due to multiple fractures of the ribs.

flail joint. Any joint that shows excessive mobility, usually due to paralysis of the muscles that control it.

flank *(flănk).* The part of the body between the ribs and the upper edge of the ilium (one of the bones on each half of the pelvis); also refers to the outer side of the hip, thigh, and buttock.

flap. A piece of partially detached tissue used in cosmetic surgery of an adjacent area or to cover the end of a bone after resection.

flatfoot. A condition in which one or more of the arches of the foot have flattened out.

flatulence *(flăt'-ū-lĕns).* Excessive gas in the intestinal tract and stomach.

flatus *(flā'-tŭs).* 1. Excessive gas in the digestive tract. 2. Expelling of gas from the anus.

flavonoid *(flā'-vĕ-noid).* Any of a group of compounds found in plants and which act as antioxidants.

flaxseed *(flăks'-sēd).* Linseed.

Fleet enema. Trade name for sodium biphosphate.

flesh. Muscular tissue; skin.
proud f. Excessive amount of soft, edematous tissue that can develop in wounds before they have healed.

flesh-eating bacteria. Popular term for a type of streptococcal infection that quickly destroys tissue and, if not treated, causes death.

Flexeril. *See* Appendix, Common Prescription and OTC Drugs: By Trade Name.

flexible *(flĕk'-sĭ-bĕl).* Something that can be readily bent without breaking.

flexion *(flĕk'-shĕn).* The act of bending or the condition of being bent in contrast to extension.

flexor *(flĕk'-sŏr).* 1. Something that causes flexion. 2. Any muscle that flexes a joint.

floaters *(flō'-tĕrz).* Deposits in the vitreous (jellylike substance in the eyeball) of the eye that typically move about; often referred to as "spots before the eyes."

floating rib. A rib that does not attach to the sternum (breastbone) or to another rib; one of the last two ribs in the rib cage.

floccilation *(flŏk-sĭ-lā'-shŭn).* Aimless picking at bedclothes by individuals who have dementia, fever, delirium, or exhaustion.

Flomax. *See* Appendix, Common Prescription and OTC Drugs: By Trade Name.

Flonase. *See* Appendix, Common Prescription and OTC Drugs: By Trade Name.

floppy baby syndrome. A general term for an abnormal condition seen in newborns characterized by inadequate muscle tone. It can be caused by various neurological and muscle problems; *see also* **hypotonia.**

flora *(flŏr'-ă).* 1. The plant life, visible and invisible to the unaided eye, present in or characteristic of a specific area. 2. Bacteria and fungi, both normally and abnormally occurring in or on an organ.
intestinal f. The bacteria normally found within the intestinal tract, they usually facilitate digestion.

flotation therapy *(flō-tā'-shŭn).* Form of relaxation therapy in which individuals float in highly salted water that is enclosed in a flotation tank that allows in little or no external stimulation.

Flovent. *See* Appendix, Common Prescription and OTC Drugs: By Trade Name.

flow *(flō).* 1. The movement of a liquid or gas. 2. The rate at which a fluid moves through a body part or organ.

flu *(flū).* *See* **influenza.**

fluconazole *(flū-kŏn'-ĕ-zŏl).* *See* Appendix, Common Prescription and OTC Drugs: By Generic Name.

fluctuation *(flŭk-chū-ā'-shŭn).* 1. A variation from one course to another. 2. A wavelike motion produced by vibration of body fluid when the affected body area is palpated by hand.

fluid *(flū'-ĭd).* A nonsolid substance, liquid, or gas.

fluid balance. Regulation of the amount of water in the body, which can be upset by diarrhea, bleeding, vomiting, or dehydration.

fluidextract *(flū-ĭd-ĕk'-străkt).* A liquid preparation of a vegetable drug in which alcohol is used as the solvent and/or preservative, and of a strength such that each milliliter contains 1 gram of the standard drug it represents.

fluke *(flūk).* A worm of the *Trematoda* class, it can cause various diseases of the lungs, blood, intestines, and liver in humans.

fluorescein *(flū-rĕs'-ĕn).* A type of dye used as a diagnostic tool in evaluating corneal trauma and when fitting contact lenses.

fluoridation *(flŏr-ĭ-dā'-shŭn).* Treatment with fluoride, especially the addition of fluoride to the public water supply as a way to prevent dental caries.

fluoride *(flŏr'-īd).* A compound of fluorine.

stannous f. A compound applied to the teeth to prevent dental caries.

fluoroscope *(flŏr'-ō-skōp).* Instrument consisting of a screen (fluorescent screen) on which are projected the shadows of x-rays as they pass through the body.

fluoroscopy *(flŏr-ŏs'-kō-pē).* Examination that uses a fluoroscope.

fluorosis *(flū-ŏr-ō'-sĭs).* Condition in which excessive amount of fluorine or its compounds (especially fluoride), characterized by mottled enamel of the teeth.

fluorouracil *(floor-ō-ūr'-ă-sĭl).* 5-fluorouracil (5-FU), an anticancer drug administered intravenously, especially for tumors of the breast and gastrointestinal tract.

fluoxetine. *See* Appendix, Common Prescription and OTC Drugs: By Generic Name.

flush. 1. A temporary redness of the skin of the face and neck that can be caused by heat, emotional factors, physical exertion, certain diseases, or use of certain drugs.

flutamide *(flū'-tĕ-mīd).* A nonsteroidal antiandrogen used in the treatment of metastatic prostate cancer.

fluticasone. *See* Appendix, Common Prescription and OTC Drugs: By Generic Name.

flutter *(flŭt'-ĕr).* A rapid pulsation or vibration.

atrial f. Very rapid (250 to 350 contractions per minute) but regular heartbeat that originates in the atrial muscle.

ventricular f. Very rapid (approximately 250 contractions per minute) heartbeat that originates in the ventricular muscle; if untreated it usually progresses to ventricular fibrillation.

flux *(flŭks).* Excessive discharge or flow.

focus *(fō'-kŭs).* The main site of a disease process.

folate *(fō'-lāt).* The anionic (having a negative charge) form of folic acid.

fold. The doubling of a part upon itself.

Foley catheter. *See* **catheter.**

folic acid *(fō'-lĭk).* A member of the B vitamin complex; it is necessary for hematopoiesis and is found in green vegetables, liver, and yeast.

follicle *(fŏl'-ĭ-kĕl).* A sac or pouchlike depression or cavity.

hair f. A saclike depression of the skin from which the root of a hair develops.

lymph f. Small mass of lymphoid tissue present in the mucosa of the gut.

ovarian f. The egg and its surrounding cells, located in the cortex of the ovary.

sebaceous f. An oil gland of the skin; it opens into a hair follicle.

follicle-stimulating hormone. Hormone produced by the pituitary gland; it stimulates the growth of the follicle in the ovary and spermatogenesis in the testes.

folliculitis *(fō-lĭk-ū-lī'-tĭs).* Inflammation of hair follicles.

folliculosis *(fō-lĭk-ū-lō'-sĭs).* Presence of an abnormal number of lymph follicles.

follow-up study. Study in which subjects are exposed to a risk or given a specific preventive or therapeutic regimen and observed over time or at intervals to determine the outcome of the exposure or regimen.

fomentation *(fō-mĕn-tā'-shŭn).* Use of warm, moist applications as treatment.

fomite *(fō'-mīt).* An object, such as an article of clothing or an infant's rattle, that is not harmful itself, but which can harbor disease-causing microorganisms and thus transmit infection.

fontanelle *(fŏn-tĕ-nĕl').* A soft spot between the cranial bones of the skull of an infant.

food additives. One or more substances, other than the basic foodstuff, intentionally added to a food as part of the production, processing, storage, or packaging process.

food allergies. Allergic reactions that result from consumption of foods to which an individual is sensitive.

food exchange. A form of diet planning in which foods are grouped according to similarities in composition so they may be used interchangeably.

food poisoning. Intestinal symptoms caused by ingesting food contaminated with toxins or bacteria. Characteristic symptoms include nausea, vomiting, diarrhea, and abdominal cramps.

foot and mouth disease. A contagious viral disease of cattle, sheep, and swine, which rarely occurs in humans. When it does, it is characterized by fever and blisterlike eruptions on the palms and soles and in the mouth.

footdrop. Weakness or paralysis of the dorsiflexor muscles of the foot and ankle, which causes the toes to drag on the ground during walking.

foramen *(fŏr-ā'-mĕn).* A natural passage or opening through a bone or membrane.

forceps *(fŏr'-sĕps).* An instrument that resembles tongs, used to grasp, manipulate, compress, or extract tissue or specific structures.

forceps delivery. Childbirth during which the obstetrician applies forceps to either side of the infant's head to aid in delivery.

forearm *(fŏr'-ărm)* Part of the upper extremity between the elbow and wrist.

forefinger *(fō'-fĭng-ĕr).* Index finger.

forehead *(fŏr'-hĕd).* Front part of the head below the hairline and above the eyes.

forensic *(fŏr-ĕn'-sĭk).* Relating to or used in legal proceedings.

forensic medicine. Medicine as it relates to the law, as in autopsy proceedings, determining sanity, and the legal aspects of medical ethics.

foreplay *(fŏr'-plā).* The sexual stimulating activities of sex partners prior to intercourse.

foreskin *(fŏr'-skĭn).* The loose fold of skin that partly or completely covers the head of the penis; also called *prepuce.*

formaldehyde *(fŏr-măl'-dĕ-hīd).* A gas used as a preservative, disinfectant, and fixative; it is toxic if inhaled or absorbed through the skin and can cause cancer.

formative *(fŏr'-mă-tĭv).* Referring to the origin and development of an organism or tissue.

formication *(fŏr-mĭ-kā'-shĕn).* A hallucination characterized by the sensation of tiny insects crawling over the skin, most commonly seen in individuals who abuse cocaine or amphetamines.

formula *(fŏr'-mū-lă).* A specific statement, using numerals and other symbols, that provides the composition of or the directions for preparing a substance or following a procedure.

formulary *(fŏr'-mū-lăr-ē).* A collection of formulas and prescriptions for medications.

fornicate *(fŏr'-nĭ-kāt).* To have sexual intercourse with an individual to whom one is not married.

fornix *(fŏr'-nĭks).* Term for an archlike structure or the vaultlike space created by such a structure.

Fosamax. *See* Appendix, Common Prescription and OTC Drugs: By Trade Name.

fossa *(fŏs'-ă)* (*pl.* fossae). A channel, hollow place, or trench.

fovea *(fō'-vē-ă).* Pit or depression in the surface of a structure or organ.

foveation *(fō-vē-ā'-shŭn).* Pitting.

foxglove *(fŏks'-glŏv).* Common name for the plant *Digitalis purpurea,* from which the drug digitalis is derived.

fracture *(frăk'-chĕr)*. The breaking of a part, especially a bone.

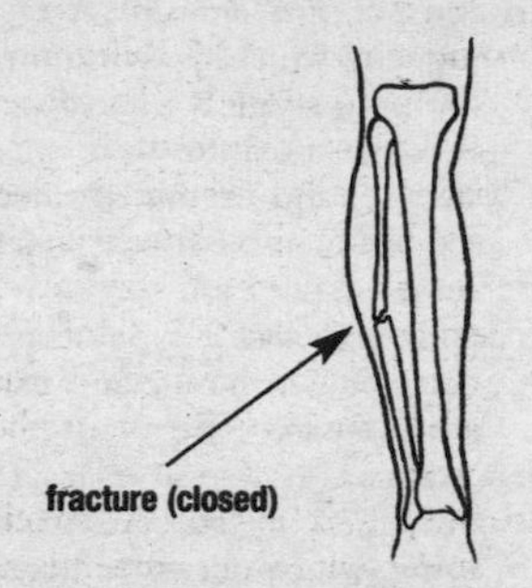

closed f. Break that does not result in an open wound in the skin; also called a *simple fracture*.

compound f. Break in which there is an open wound in the skin at the fracture site.

greenstick f. Fracture in which one side of a bone is broken while the other side is bent.

multiple f. One in which there are two or more lines of fracture of the same bone but they don't connect in any way.

stress f. Fracture caused by repeated or unusual stress on a bone.

fragile X syndrome *(fră'-jĭl)*. Inherited condition in which there is a defect of the X chromosome, resulting in mental retardation and large ears, chin, and testicles in males, and mild retardation in females. Twice as common in males than in females.

fragility *(fră-jĭl'-ĭ-tē)*. Tendency to break or disintegrate; brittleness.

Framingham study. A landmark study initiated in 1948 in which twelve thousand individuals living in Framingham, Massachusetts, underwent regular medical exams and provided important data, including DNA samples, from which invaluable health information has been extracted.

fraternal twins *(fră-tĕr'-năl)*. Nonidentical twins that result from the fertilization of two separate eggs rather than the splitting of one fertilized egg.

FRCP. Fellow of the Royal College of Physicians.

FRCS. Fellow of the Royal College of Surgeons.

free radical. A highly reactive chemical in the body that normally exists for a short time. Some are formed during oxidation and may be helpful (they kill infectious organisms) but they can also do extensive damage to cells and tissues.

fremitus *(frĕm'-ĭ-tŭs)*. A vibration perceptible on palpation.

frenulum *(frĕn'-ū-lŭm)* (*pl.* frenula). A small fold of mucous membrane that extends from a fixed to a movable part and limits the motion of the movable part

f. of clitoris. A fold that connects the underside of the clitoris with the labia minora.

f. of lips. Either one of the folds that extends from the gums to the midline of the upper or lower lips.

f. of tongue. Fold that extends from the midline of the underside of the tongue to the floor of the mouth.

frequency *(frē'-kwĕn-sē).* The number of regular recurrences of a given process or the number of members of a population.

Freudian *(froi'-dē-ăn).* Pertaining to Sigmund Freud or his psychological concepts and method of psychoanalysis.

friable *(frī'-ĕ-bĕl).* Easily crumbled, damaged, or pulverized.

friction *(frĭk'-shŭn).* The act of rubbing objects together.

Friderichsen-Waterhouse syndrome *(frĭd'-rĭk-sĕn wăw'-tĕr-hous).* The malignant or severe form of epidemic cerebrospinal meningitis, characterized by sudden onset, fever, coma, and collapse, hemorrhages of the skin and mucous membranes.

Friedreich's ataxia *(frĕd'-rĭks).* Hereditary disease characterized by paralysis of the lower limbs, speech defects, and significant curvature of the spine.

frigidity *(frĭ-jĭd'-ĭ-tē).* Coldness; former term for female sexual arousal disorder.

Froehlich's syndrome *(fră'-lĭks).* Dysfunction of the pituitary gland in children, resulting in obesity and underdeveloped genitals.

frontal *(frŏn'-tĕl).* Referring to the forehead.

frontal sinuses. *See* **sinus.**

frostbite *(frŏst'-bīt).* Freezing or the effect of freezing of a part of the body.

frozen section *(frō'-zĕn).* A technique in which tissue removed during a surgical procedure is frozen, cut into thin slices, stained, and examined microscopically immediately after its removal.

fructose *(frūk'-tōs).* A type of sugar occurring in honey and many sweet fruits.

fructosuria *(frūk-tō-sū'-rē-ă).* The presence of fructose in the urine, which occurs in hereditary fructose intolerance.

fruitarianism *(frū-tăr'-ē-ĕn-ĭzm).* A diet restricted to the consumption of fruits, honey, nuts, and olive oil.

frustration *(frŭs-trā'-shĕn).* Feeling of tension or uneasiness that results when one's purpose, desires, impulses, or actions are blocked.

fulguration *(fŭl-gū-rā'-shĕn).* Destruction of living tissue by electric sparks generated by a high frequency current; *see also* **electrocautery.**

fulminating *(fŭl'-mĭ-nāt-ĭng).* Occurring suddenly and with great severity, as in a medical condition that begins suddenly.

Fulvicin *(fŭl'-vĭ-sĭn).* Trade name for a preparation of griseofulvin, used to treat fungal infections of the skin.

functional disease. Disease associated with a functional problem, such as a headache, rather than one associated with a structural change.

functional food. Foods that contain modified ingredient(s) that may provide healthful benefits beyond the traditional nutrients it contains, such as those fortified with vitamins.

fundoplication *(fŭn-dō-plī-kā'-shŭn).* Suturing of the fundus of the stomach around the lower end of the esophagus, done to treat hiatal hernia.

fundus *(fŭn'-dĕs).* The bottom or base of an organ or the part of a hollow organ farthest from its mouth.

fungi *(fŭn'-jī).* A kingdom of living organisms that includes mushrooms, yeasts, and molds; these organisms lack chlorophyll and reproduce either sexually or asexually.

fungicide *(fŭn'-jĭ-sīd).* An agent that destroys fungi.

fungus *(fŭn'-gĕs)* (*pl.* fungi). An organism that belongs to the kingdom *Fungi.*

funiculitis *(fū-nĭk-ū-lī'-tĭs).* Inflammation of the spermatic cord.

funnel chest. Depression of the breastbone (sternum) caused by rickets.

funny bone. Site on the inner part of the elbow where the ulnar nerve is located.

FUO. Acronym for fever of unknown origin.

Furacin *(fū'-rĕ-sĭn).* Trade name for an antibiotic used to treat urinary and bowel infections; generic name: nitrofurazone.

Furadantin *(fŭr-ĕ-dăn'-tĭn).* Trade name for an antibacterial drug used to control infections of the bladder and kidneys; generic name: nitrofurnatoin.

furosemide *(fū-rō'-sĕ-mīd).* Generic name of a diuretic used to treat edema associated with congestive heart failure, hepatic or renal disease, or hypertension; trade name: Lasix.

furry tongue. A furlike appearance of the tongue that can develop after taking antibiotics; also called *hairy tongue.*

furuncle *(fū'-rŭng-kĕl).* Painful nodule caused by staphylococci, which enter through the hair follicles; also called *boil.*

furunculosis *(fū-rŭng-kū-lō'-sĭs).* Persistent occurrence of furuncles over a period of weeks or months.

fusion *(fū'-shŭn).* 1. The process or result of melting. 2. The merging or joining together of adjacent parts or bodies.

G

GABA. *See* **gamma-aminobutyric acid.**

gabapentin *(gă-bă-pĕn'-tĭn). See* Appendix, Common Prescription and OTC Drugs: By Generic Name.

gadolinium (Gd) *(găd-ō-lĭn'-ē-ŭm).* A rare element that is used in magnetic resonance imaging and in dual photon absorptiometry.

gag *(găg).* A device used to hold the mouth open. 2. To retch or cause to vomit.

gait *(gāt).* A manner or way of walking.

antalgic g. A limp adopted as a way to avoid pain on weight-bearing parts of the body.

ataxic g. An uncoordinated, unsteady walk in which individuals throw the feet outward, due to loss of muscle control.

myopathic g. Walk characterized by exaggerated lateral trunk movements and abnormal elevation of the hip, with the result being a gait of a penguin or duck.

stuttering g. Walk characterized by hesitancy that resembles stuttering; seen in some individuals who have neurological damage or those with schizophrenia.

galactagogue *(gă-lăk'-tă-gŏg).* A substance that promotes the flow of milk.

galactocele *(gă-lăk'-tō-sēl).* A cyst that develops in the breast resulting from obstruction of a milk duct.

galactography *(găl-ăk-tŏg'-rĕ-fē).* X-ray of the mammary ducts after injection with a radiopaque substance into the duct system.

galactorrhea *(gă-lăk-tō-rē'-ă).* Spontaneous or excessive flow of milk not associated with nursing.

galactose *(gă-lăk'-tōs).* A simple sugar produced in the body during digestion of lactose (milk sugar); it is then converted into glucose which the body uses for energy.

galactosemia *(gă-lăk-tō-sē'-mē-ă).* A genetic disorder that results from faulty galactose metabolism. It usually becomes apparent soon after birth and is characterized by feeding problems, mental retardation, enlarged liver and spleen, and elevated blood and urine galactose levels.

galactosuria *(gă-lăk-tō-sū'-rē-ă).* Presence of galactose in the urine.

gall *(gawl).* 1. Bile. 2. Localized swelling or skin sore caused by friction.

collar g. Pressure ulcer on a horse caused by repeated irritation from a poorly fitted collar or harness.

saddle g. Pressure ulcer on a horse caused by repeated irritation from a poorly fitted saddle.

gallbladder *(gawl'-blăd-ĕr).* A pear-shaped sac that stores bile; it is located under the liver.

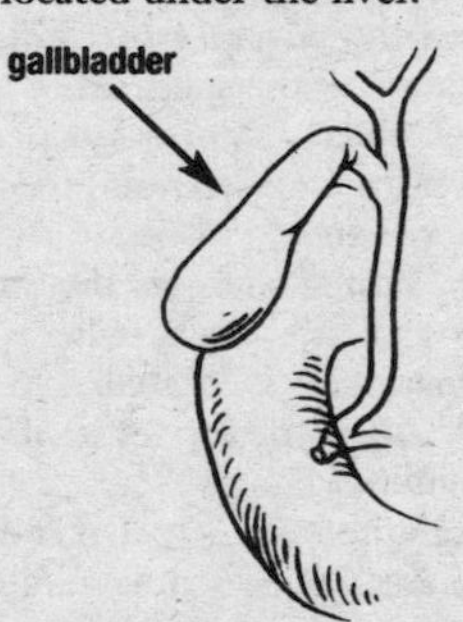

Courvoisier's g. A distended gallbladder due to an obstruction of the biliary tract.

fish-scale g. Gallbladder that

has a fish-scale appearance due to small cysts that have developed on the mucosa.

stasis g. A gallbladder that contracts sluggishly in response to a high-fat meal.

wandering g. Abnormal movement of the gallbladder.

gallop rhythm *(găl'-ŭp).* See **rhythm.**

gallstone *(gawl'stōn).* A concretion (stone) typically composed of cholesterol or a calcium salt, it forms in the gallbladder or bile ducts.

galvanic current *(găl-văn'-ĭk).* Direct electric current, usually from a battery, used therapeutically.

gamete *(găm'-ēt).* One of two reproductive cells (spermatozoon or ovum) that joins with another to form a zygote, from which a new organism develops.

gamma-aminobutyric acid *(găm'-ĕ-ĕ-mē-nō-bū-tĭr'-ĭk).* An amino acid neurotransmitter present in the brain, where it inhibits the transmission of nerve signals.

gamma globulin *(găm-ĕ glŏb'-ū-lĭn).* A protein that forms in the blood. The ability to resist infection is related to an individual's concentration of gamma globulin.

gamma linolenic acid (GLA). An essential polyunsaturated fatty acid present in some plant seed oils including evening primrose, black currant, and borage oils.

gamma ray *(găm'-ă rā).* Very short wavelength electromagnetic waves emitted by radioactive substances.

gamophobia *(găm-ō-fō'-bē-ă).* A morbid fear of marriage.

ganciclivir *(găn-sī'-klō-vĭr).* A drug that inhibits reproduction of herpesviruses, and is also used to treat cytomegalovirus infections.

ganglion *(găng'-glē-ŏn).* A knotlike mass; more specifically, the term for a group of nerve cell bodies located outside the central nervous system.

ganglionectomy *(găng-glē-ō-nĕk'-tō-mē).* Surgical removal of a ganglion.

ganglioneuroma *(găng-glē-ō-nū-rō'-mă).* A benign neoplasm made up of mature ganglion cells and nerve fibers.

ganglionitis *(găng-glē-ō-nī'-tĭs).* Inflammation of a ganglion.

gangrene *(găng'-grēn).* The death of tissue, usually associated with a loss of nutritional flow to the affected area and development of bacterial infection and decomposition.

Gardnerella *(găhrd-nĕr-ĕl'-ă).* A genus of gram-negative, rod-shaped bacteria found in a healthy female genital tract, but is also a cause of bacterial vaginitis.

gargle *(găhr'-gĕl).* A wash for rinsing or medicating the mouth and throat.

gas *(găs).* One of the basic forms of matter, it is characterized by molecules that are separated from each other and that move rapidly in all directions.

blood g. analysis. Use of laboratory tests to determine the oxygen and carbon dioxide concentrations and pressures with the pH of the blood.

coal g. A poisonous gas (it contains carbon monoxide) produced during the distillation of coal; it is used for cooking.

laughing g. Nitrous oxide.

tear g. A gas that causes severe tearing of the eyes.

gas gangrene. Gangrene caused by species of *Clostridium,* it often results from wounds that have gotten contaminated, allowing the muscles and subcutaneous tissues to become filled with gas.

gastrectomy *(găs-trĕk'-tō-mē).* Surgical removal of part or all of the stomach.

gastric analysis *(găs'-trĭk).* An analysis of the contents of the stomach, used to identify gastric bleeding, stomach cancer, or pernicious anemia.

gastric lavage. Procedure to wash out the stomach; may be used before stomach surgery or to empty poison or other irritants from the stomach.

gastric stapling. A surgical procedure in which the parts of the stomach are stapled together, which converts the upper part of the stomach into a small pouch and thus forces obese individuals to eat tiny portions yet feel full. Also called *gastric banding.*

gastrin *(găs'-trĭn).* A group of hormones released by the stomach during digestion; they increase the secretion of hydrochloric acid and, to a lesser degree, pepsinogen.

gastrinoma *(găs-trĭ-nō'-mă).* A tumor, usually of the pancreas, that produces gastrin and is associated with the Zollinger-Ellison syndrome.

gastritis *(găs-trī'-tĭs).* Inflammation of the stomach lining (mucosa).

gastrocnemius muscle *(găs-trŏk-nē'-mē-ŭs).* The large muscle in the back portion of the lower leg (calf); it extends the foot and helps to flex the knee.

gastrocolic *(găs-trō-kŏ'-lĭk).* Referring to the stomach and the colon.

gastrocolic reflex. The peristaltic motion in the colon initiated when food enters a fasting stomach.

gastroduodenal *(găs-trō-dū-ō-dēn'-ăl).* Pertaining to the stomach and the duodenum.

gastroduodenitis *(găs-trō-dū-ŏd-ĕn-ī'-tĭs).* Inflammation of the stomach and duodenum.

gastroduodenoscopy *(găs-trō-dū-ŏ-dē-nŏs'-kō-pē).* Use of a device called a gastroscope to visualize the interior of the stomach and duodenum.

gastroduodenostomy *(găs-trō-dū-ŏ-dĕn-ŏs'-tē-mē).* Surgical creation of a passageway between the stomach and the duodenum.

gastroenteritis *(găs-trō-ĕn-tĕr-ī'-tĭs).* Inflammation of the mucous membrane of the stomach and intestinal tract.

gastroenterologist *(găs-trō-ĕn-tĕr-*

ŏl'-ō-jĭst). Physician who specializes in the diagnosis and treatment of conditions that affect the stomach, intestines, and related structures (e.g., liver, esophagus, gallbladder, and pancreas).

gastroenterostomy *(găs-trō-ĕn-tĕr-ŏs'-tō-mē).* A surgical connection between the stomach and small bowel, typically necessary in patients who have stomach cancer.

gastroepiploic *(găs-trō-ĕp-ĭ-plō'-ĭk).* Pertaining to the stomach and greater omentum.

gastroesophageal reflux disease (GERD) *(găs-trō-ē-sŏf-ă-jē'-ăl).* The backflow of the contents of the stomach or duodenum through a faulty sphincter in the lower esophagus.

gastroesophagitis *(găs-trō-ē-sŏf-ă-jī'-tĭs).* Inflammation of the stomach and esophagus.

gastrogavage *(găs-trō-gă-văzh').* The introduction of nutritional substances into the stomach via a tube that is passed through the esophagus.

gastrohepatitis *(găs-trō-hĕp-ă-tī'-tĭs).* Inflammation of the stomach and the liver.

gastrointestinal *(găs-trō-ĭn-tĕs'-tĭ-năl).* Pertaining to the stomach and intestines.

gastrojejuostomy *(găs-trō-jĕ-jū-nŏs'-tō-mē).* Surgical connection between the stomach and the jejunum.

gastrologist *(găs-trŏl'-ĕ-jĭst).* A practitioner who specializes in diseases of the stomach.

gastromegaly *(găs-trō-mĕg'-ă-lē).* Abnormal enlargement of the stomach.

gastroparesis *(găs-trō-pă-rē'-sĭs).* Paralysis of the stomach; also called *gastroplegia.*

gastroplication *(găs-trō-plĭ-kā'-shŭn).* Surgical procedure in which a fold is stitched in the stomach as treatment of gastric dilatation.

gastroptosis *(găs-trŏp-tō'-sĭs).* Downward displacement of the stomach; it rarely causes symptoms.

gastroscope *(găs'-trō-skōp).* An endoscope used to inspect the interior of the stomach.

gastrosplenic *(găs-trō-splĕn'-ĭk).* Pertaining to the stomach and spleen.

gastrostomy *(găs-trŏs'-tĕ-mē).* Surgical creation of an artificial opening into the stomach.

Gaucher's disease *(gō-shăz').* A rare chronic congenital disorder that affects lipid metabolism and characterized by an enlarged spleen, increased skin pigmentation, and bone lesions.

Gauss' sign *(gows).* Unusual mobility of the uterus during the early weeks of pregnancy.

gauze *(găwz).* Thin, loosely woven material that is used for bandages and surgical sponges.

gavage *(gă-văzh').* Process in which an individual is fed a liquid or semiliquid food through a stomach tube or a tube that is passed through the nose and esophagus into the stomach.

Geiger counter *(gī'-gĕr).* An instrument used to detect ionizing radiation.

gel *(jĕl).* A colloid in which the solid dispersed phase forms a network in combination with the fluid.

gelatin *(jĕl'-ĕ-tĭn).* A product that is the result of partial hydrolysis of collagen derived from the tissue, bones, and skin of animals.

geld *(gĕld).* To castrate a male animal, especially a horse.

gemfibrozil *(jĕm-fī'-brō-zīl). See* Appendix, Common Prescription and OTC Drugs: By Generic Name.

gender *(jĕn'-dĕr).* The sex of an individual.

gene *(jēn).* A segment of a DNA molecule, located on a specific position on a chromosome, that contains hereditary information.

autosomal g. A gene located on any chromosome that is not a sex chromosome.

dominant g. Gene that will express its effect (e.g., a trait such as blue eyes) whether it is paired with an identical gene or a dissimilar one.

lethal g. A gene that causes the death of an individual or allows it to survive only under specific conditions.

recessive g. Gene that expresses its effect only when it is present in both chromosomes.

sex-linked g. Gene present within the X or Y chromosome.

gene mapping. A method used to determine the relative location of genes on a chromosome or another piece of DNA and the distances between them.

gene pool. The total number of genes, including all variations, that a specific species possesses at a specific time.

generalist *(jĕn'-ĕr-ăl-ĭst).* A physician who treats a wide range of diseases.

generation *(jĕn-ĕr-ā'-shŭn).* 1. The process of forming a new organism. 2. The average time between the birth of parents and the birth of their children.

generic drug *(jĕn-ĕr'-ĭk).* Drugs not protected by a trademark; also referred to as a *nonproprietary drug.*

gene splicing. The substitution of a portion of DNA is spliced into the DNA of another gene.

gene therapy. Insertion of a normal gene into an organism in an effort to correct a genetic defect.

genetic *(jĕn-ĕt'-ĭk).* Referring to reproduction or to genes.

genetic counseling. Advice dispensed by specialists in genetic medicine to prospective parents about the possibility their child will or will not be born with a hereditary disorder.

genetic engineering. Altering genes in living cells to create new or modified forms.

genetic screening. Testing a particular population to identify which individuals are at risk for a genetic disease or for transmitting it to others. Examples include screening blacks for the sickle cell gene,

Jews for the Tay-Sachs disease gene, and so on.

genetic testing. The study of an individual's DNA to detect genetic abnormalities that may be passed along to prospective children or to predict drug responses, diagnose genetic disease in children and adults, or identify future disease risks.

genicular *(jă-nĭk'-ū-lĕr).* Referring to the knee.

genioplasty *(jē'-nē-ō-plăs-tē).* Cosmetic surgery of the chin.

genital *(jĕn'-ĭ-tăl).* Referring to the reproductive organs.

genital herpes. Herpes simplex caused by human herpesvirus 2, mainly transmitted via sexual contact.

genitalia *(jĕn-ĭ-tāl'-ē-ă).* Referring to the genital organs.

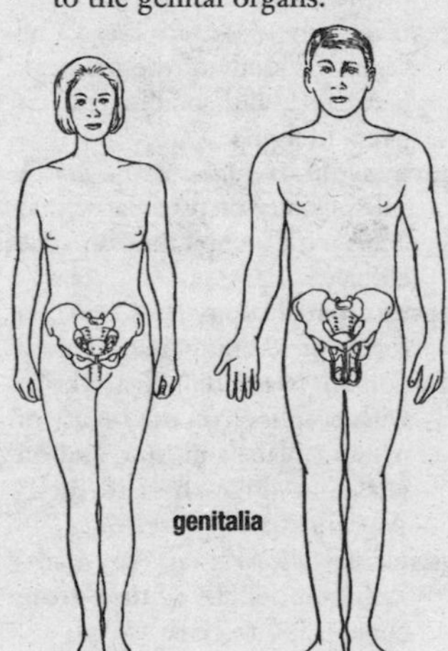

ambiguous g. Genital organs with characteristics typical of both male and female.

female g. Organs of reproduction in females, including the mons veneris, labia majora, labia minora, clitoris, fourchet, fossa navicularis, vestibule, vestibular bulb, Skene's glands, glands of bartholin, hymen, vaginal introitus, perineum, ovaries, fallopian tubes, uterus, and vagina.

male g. Organs of reproduction in males, including bulbourethral glands, ejaculatory ducts, testes, penis, urethra, seminal ducts, seminal vesicles, spermatic cords, scrotum, and prostate gland.

genitourinary *(jĕn-ĭ-tō-ūr'-ĭ-nār-ē).* Referring to the genital and urinary organs.

genius *(jēn'-yŭs).* 1. Exceptional ability or aptitude. 2. An individual who has exceptional mental, creative, and/or physical abilities.

genocide *(jĕn'-ō-sīd).* Premeditated murder of a specific ethnic or social group.

genome *(jē'-nōm).* The complete gene set of an organism, either haploid (the set from one parent) or diploid (from both parents).

genomics *(jē-nŏm'-ĭks).* Study of the function and structure of the genome.

genophobia *(jē-nō-fō'-bē-ă).* Morbid fear of sex.

genotype *(jē'-nō-tīp).* The complete genetic makeup of an individual or organism. Compare this with *phenotype.*

gentamicin *(jĕn-tă-mī'-sĭn).* An antibiotic derived from the fungi

Micromunospora, effective against many aerobic gram-negative bacilli and some gram-positive bacteria.

gentian violet *(jĕn'-shĕn).* A dye made from the dried rhizome and roots of *Gentiana lutea,* with antibacterial, anthelminthic and antifungal properties. It is used as a topical treatment of skin and mucous membrane infections.

genucubital position *(jĕn-ū-kū'-bĭ-tăl).* A position in which an individual is supported by knees and elbows with the chest raised above the table.

genus *(jē'-nĕs)* (*pl.* genera). A taxonomic category that is positioned between the species and the family.

geomedicine *(jē-ō-mĕd'-ĭ-sĭn).* Study of the influence of geography and climate on health.

geophagy *(jē-ŏf'-ă-jē).* A condition in which people eat inedible substances, such as dirt and clay.

GERD. *See* **gastroesophageal reflux disease.**

geriatric *(jĕr-ē-ăt'-rĭk).* Referring to elderly people or to the process of aging.

geriatrician *(jĕr-ē-ă-trĭ'-shŭn).* Physician who specializes in geriatrics.

geriatrics *(jĕr-ē-ăt'-rĭks).* The branch of medicine that deals with the diagnosis and treatment of conditions associated with aging and old age.

germ *(jĕrm).* A disease-causing microorganism.

German measles vaccine. Vaccine formerly developed to prevent rubella, or German measles, it is now combined with a vaccine known as MMR, which inoculates against measles, mumps, and rubella.

germ cell. Eggs and sperm; the reproductive cells. Each mature germ cell is haploid (has a single set of twenty-three chromosomes).

germicide *(jĕr'-mĭ-sīd).* A substance that destroys disease-causing microorganisms.

germinate *(jĕr'-mĭ-nāt).* Sprouting of a spore, seed, or plant embryo.

germinoma *(jĕr-mĭ-nō'-mă).* A type germ cell tumor usually found in the ovary, undescended testis, anterior mediastinum, or pineal gland.

gerontology *(jĕr-ŏn-tŏl'-ĕ-jē).* The scientific study of the medical, historical, and sociological aspects of aging.

geropsychiatry *(jĕr-ō-sī-kī'-ă-trē).* A subspecialty of psychiatry that deals with mental illness in the elderly.

gestalt psychology *(gĕs-tăwlt').* A form of therapy in which a person is treated as a whole, with emphasis on the reality of present place and time and on personal growth and self-awareness.

gestation *(jĕs-tā'-shŭn).* In mammals, the length of time from conception to birth.

gestational age. *See* **age.**

giant cell. Any very large cell, such as the megakaryocyte present in bone marrow.

giant cell tumor. A benign or malignant tumor that contains giant cells.

giantism *(jī'-ĕnt-ĭz-ĕm).* Excessive development of the size of a body or any of its parts.

Giardia *(jē-är'-dē-ă).* A genus of usually nonpathogenic intestinal protozoa.

G. lamblia. A species that can cause giardiasis in humans as well as domestic animals, including dogs.

Giardiasis *(jī-är-dī'-ĕ-sĭs).* A common infection of the small intestine in humans, caused by *Giardia lamblia* and spread via contaminated water or food or by direct human-to-human contact. Most infected individuals have no symptoms.

gibbosity *(gĭ-bŏs'-ĭ-tē).* The condition of having a humpback; kyphosis.

gigantism *(jī-'găn-tĭzm).* Abnormal, excessive growth in size and stature. Also called *giantism, hypersomia,* and *somatomegaly.*

gingival *(jĭn'jī-văl).* Referring to the gums.

gingivectomy *(jĭn-jī-vĕk'-tō-mē).* Surgical removal of diseased gum tissue.

gingivitis *(jĭn-jī-vī'-tĭs).* Inflammation of the gums characterized by swelling, redness, and a tendency for the gums to bleed.

ginkgo *(gĭng'-kō).* The dried leaves of *Ginkgo biloba,* used to improve blood flow to the brain and ringing in the ears; in traditional Chinese medicine it is used to treat asthma, ringing in the ears, and angina.

ginseng *(jĭn'-sĕng).* Any herb of the genus *Panax.*

Asian g. Dried roots of *Panax ginseng* used as a tonic and to improve fatigue.

Siberian g. Also known as *Eleutherococcus senticosuus,* used as a tonic.

glabella *(glă-bĕl'-ă).* The smooth surface of the bone that lies between the eyebrow arches.

gland *(glănd).* An organ that manufactures and secretes one or more chemicals that are used in other sites in the body. *See* individual names of glands.

glanders *(glăn'-dĕrz).* A contagious disease of horses caused by *Burkholderia mallei,* transmittable to humans.

glandular fever *(glăn'-dū-lăr).* Another name for infectious mononucleosis, characterized by fever, swollen lymph nodes, an enlarged spleen, atypical lymphocytes, and other symptoms.

glans *(glănz).* Term for a small, rounded glandlike structure.

g. clitoridis. The erectile tissue at the end of the clitoris.

g. penis. The cap-shaped formation at the end of the penis.

glare *(glăr).* Intense light that can cause temporary blurring of vision and possibly permanent damage to the retina.

glaucoma *(glaw-kō'-mă).* Eye disease characterized by increased pressure in the eye, which results in atrophy of the optic nerve and blindness.

absolute g. Form in which the affected eye is completely blind.

angle-closure g. Form that occurs suddenly when the outermost part of the iris is pushed against the inner boundary of the cornea, preventing the outflow of aqueous humor from the anterior chamber.

congenital g. Form characterized by an increase of intraocular fluid and enlargement of the eyeball and protrusion of the cornea.

narrow-angle g. *See* **angle-closure glaucoma.**

open-angle g. Chronic, slowly progressive form of glaucoma caused by a defect in the anterior chamber angle.

glioblastoma *(glī-ō-blăs-tō'-mă).* A general term for malignant forms of astrocytoma, the most common type of primary brain tumor; *see also* **astrocytoma.**

glioma *(glī-ō'-mă).* Term that describes all primary neoplasms that affect the brain and spinal cord.

gliosis *(glī-ō'-sĭs).* An excessive number of astrocytes in damaged areas of the central nervous system.

glipizide *(glĭp'-ĭ-zīd). See* Appendix, Common Prescription and OTC Drugs: By Generic Name.

Glisson's capsule *(glĭs'-ĕnz).* The outer fibrous membrane covering the liver.

globule *(glŏb'-yūl).* A small spherical body or mass; especially refers to a drop of semifluid or fluid.

globulin *(glŏb'-ū-lĭn).* Any one of a class of simple proteins found in blood and cerebrospinal fluid; globulins are soluble in salt solutions, insoluble in water, and coagulable by heat.

globulinuria *(glŏb-ū-lī-nū'-rē-ă).* Presence of globulin in the urine.

globus *(glō'-bŭs).* General term for a spherical structure.

glomerate *(glŏm'-ĕr-āt).* Forming into a ball.

glomerular *(glō-mĕr'-ū-lĕr).* Referring to a glomerulus.

glomerulonephritis *(glō-mĕr-ū-lō-nĕ-frī'-tĭs).* Inflammation of the kidney accompanied by inflammation of the capillary loops in the kidney's glomeruli.

glomerulus *(glō-mĕr'-ū-lŭs)* (*pl.* glomeruli). In anatomy, a general term for a cluster or tuft; often used to refer to a glomerulus of the kidney.

glossa *(glŏs'ă).* The tongue.

glossalgia *(glŏs-săl'-jē-ă).* Pain in the tongue.

glossectomy *(glŏs-ĕk'-tō-mē).* Surgical removal of part or all of the tongue.

glossitis *(glŏs-sī'-tĭs).* Inflammation of the tongue.

glossolalia *(glŏs-ō-lā'-lē-ă).* Meaningless speech or unintelligible chatter.

glossopharyngeal *(glŏs-ō-fĕ-rĭn-jē'-ăl).* Referring to the tongue and pharynx.

glottis *(glŏt'-ĭs).* The vocal structure of the larynx, it consists of the vocal cords and the opening that's between them.

glucagon *(glū'-kă-gŏn).* A polypeptide hormone, secreted by the pancreas, that increases the concentration of glucose in the blood.

glucocorticoid *(glū-kō-kŏrt'-ĭ-koid).* Any one of the steroid hormones produced by the adrenal cortex (or a synthetic steroid) that regulates fat, carbohydrate, and protein metabolism; cortisol is the main naturally occurring glucocorticoid.

glucokinase *(glū-kō-kī'-nās).* An enzyme present in the pancreas and liver. Along with ATP, it promotes the conversion of glucose to glucose 6-phosphate.

Glucophage. *See* Appendix, Common Prescription and OTC Drugs: By Trade Name.

glucosamine *(glū-kō'-să-mēn).* The amino sugar derivative of glucose.

g. sulfate. The sulfate salt of glucosamine, synthetically prepared as a nutritional supplement and commonly used to treat osteoarthritis.

glucose *(glū'-kōs).* Sugar; specifically in medicine, it refers to sugar dextrose or d-glucose. It is formed during digestion and is the most important carbohydrate in body metabolism.

g. tolerance test. A test to determine an individual's ability to metabolize glucose. It is done to diagnose diabetes mellitus, hypoglycemia, or malabsorption syndrome.

glucoside *(glū'-kō-sīd).* One of various substances present in plants that consist of glucose combined with an ether linkage.

glucosuria *(glū-kō-sū'-rē-ă).* Presence of an abnormal amount of sugar in the urine.

Glucotrol. *See* Appendix, Common Prescription and OTC Drugs: By Trade Name.

glutamate *(glū'-tă-māt).* A salt of glutamic acid.

glutamic acid *(glū-tăm'-ĭk).* A nonessential amino acid that occurs in proteins. It is also an excitatory neurotransmitter in the central nervous system.

glutamine *(glū'-tă-mēn).* An amino acid found in many plants and some animal tissue; it is essential in the hydrolysis of proteins.

glutathione *(glū-tă-thī'-ŏn).* A compound composed of glutamic acid, cysteine, and glycine that is important in cellular respiration.

gluteal *(glū'-tē-ăl).* Referring to the buttocks.

gluten *(glū'-tĭn).* The protein of wheat and other grains that is used as a flour substitute; it gives dough an elastic quality.

gluteus muscles *(glū'-tē-ŭs).* Any of the three buttock muscles.

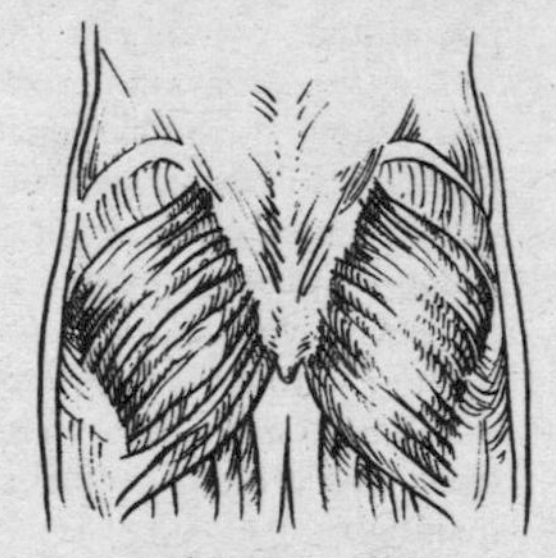

gluteus muscles

glyburide *(glī-bŭr'-īd).* See Appendix, Common Prescription and OTC Drugs: By Generic Name.

glycemia *(glī-sē'-mē-ă).* The presence of glucose (sugar) in the blood.

glyceride *(glĭs'-ĕr-īd).* An ester of glycerin compounded with an acid.

glycerin *(glĭs'-ĕr-ĭn).* A syrupy sweet liquid used as a sweetener, a lubricant, and a solvent for drugs.

glycine *(glī'-sēn).* A nonessential amino acid.

glycinuria *(glī-sĭ-nū'-rē-ă).* The presence of glycine in the urine.

glycogen *(glī-'kō-jĕn).* The form in which carbohydrates are stored in the body, especially in the muscles and liver.

glycogenesis *(glī-kō-jĕn'-ĭ-sĭs).* The formation of glycogen from glucose (sugar).

glycolysis *(glī-kŏl'-ĭ-sĭs).* The breakdown of sugar in the body into lactic acid.

glycoprotein *(glī-kō-prō'-tēn).* A compound consisting of a protein and a carbohydrate.

glycoside *(glī'-kō-sīd).* Any compound that contains a sugar and a nonsugar component and is named for the sugar it contains; for example, fructoside (fructose is the sugar).

glycosuria *(glī-kō-sū'-rē-ă).* The presence of sugar in the urine. Minute amounts may appear in normal urine, but greater levels may indicate diabetes mellitus.

glycyrrhiza *(glĭs-ĭ-rī'-ză).* The dried roots of licorice (*Glycyrrhiza glabra*) used in some drug preparations.

gnosia *(nō'-sē-ă).* The ability to perceive and recognize the significance of objects.

goblet cell *(gŏb'-lĕt).* A type of secretory cell present in the respiratory and intestinal tracts. Mucin droplets collect in one end of the cell and cause the cell to swell.

goiter *(goy'-tĕr).* An enlargement of the thyroid gland.

colloid g. Goiter in which the follicles of the thyroid are excessively extended with colloid.

cystic g. One in which the gland contains one or more cysts.

diffuse g. One in which the gland is diffusely enlarged, as seen in Grave's disease.

goldenseal *(gōl-dĕn-sēl').* A preparation made from the dried rhizome of *H. canadensis* that is widely used in herbal medicine and homeopathy to treat digestive problems, menstrual difficulties, and bronchitis.

golfer's cramp. Involuntary movements of the muscles in the hand and sometimes the forearm, seen in some people who play golf. Similar conditions (known as *dystonias*) are called writer's cramp, musician's cramp, and typist's cramp.

gonad *(gō'-năd).* A gland that produces gametes: for example, the ovary produces ova; the testis, sperm.

gonadotropin *(gōn-ă-dō-trōp'-ĭn).*

Any hormone that stimulates the gonads, especially the luteinizing hormone and the follicle-stimulating hormone.

goniotomy *(gō-nē-ŏt'-ō-mē).* Surgical procedure for glaucoma in which obstructions to the free flow of aqueous humor into the canal of Schlemm are removed.

gonococcemia *(gŏn-ō-kŏk-sē'-mē-ă).* The presence of gonococci in the blood.

gonococcus *(gŏn-ō-kŏk'-ŭs).* An individual microorganism of the species *Neisseria gonorrhoeae,* the organism that causes gonorrhea.

gonorrhea *(gŏn-ō-rē'-ă).* A sexually transmitted infection caused by *Neisseria gonorrhoeae,* characterized by painful urethritis and discharged pus in males, but usually asymptomatic in females.

gotu kola *(gŏ'-tū kō'-lă).* A preparation of the stems and leaves of *Centella asiatica* that is used topically to assist wound healing and to treat leprosy; also used in traditional Chinese medicine.

gout *(gout).* A metabolic disorder characterized by an excess of uric acid in the blood, painful inflamed joints (especially of the big toes), and deposits of sodium biurate in the cartilage of the inflamed joints and the kidney.

articular g. Gout that affects the joints.

polyarticular g. An unusual form that affects many joints.

secondary g. Gout that results from an acquired disorder such as chronic myelogenous leukemia or from an inborn metabolism malfunction.

Graafian follicle *(grăf'-ē-ăn).* A mature follicle of the ovary that, beginning with puberty and continuing until menopause, contains an ovum that is discharged in a process called ovulation.

gracilis muscle *(grăs'-ĭ-lĭs).* The long, slender muscle that extends from the groin down to the inner side of the knee.

graduate *(grăd'-ū-āt).* A container marked with lines used to measure liquids.

graft *(grăft).* Tissue removed from one part of the body and used to replace defective tissue in another part.

autograft. Living tissue that is transplanted from one site to another in the body of the same individual. Also called *autogenous graft, autologous graft.*

corneal g. A surgical procedure in which all or part of a defective cornea is removed and replaced with a healthy one; also known as *keratoplasty.*

fascial g. Use of fascia to repair defects in other tissues.

free g. Graft that is completely separated from its original site and transferred to another.

homologous g. *See* **allograft.**

mucosal g. A graft of mucosal tissue.

nerve g. A nerve segment used to repair part of a defective nerve.

skin g. Skin transplanted to replace a damaged or lost section of skin.

split-thickness g. A skin graft that consists of the epidermis and a portion of dermis; or a mucosal graft that consists of only a partial thickness of mucosa.

grain *(grān).* The smallest unit of weight in the apothecaries' and avoirdupois systems. In the apothecaries' system there are 480 grains in 1 ounce; in the avoirdupois system, there are 437.5 grains in 1 ounce.

gram *(grăm).* A unit of mass in the SI (Systeme International d'Unites) system. It is equal to one thousandth of a kilogram.

gram-negative bacteria. Bacteria that lose Gram's stain and take up the color of the red counterstain in Gram's method, making them look pink.

gram-negative sepsis. Blood poisoning caused by gram-negative bacteria.

gram-positive bacteria. Bacteria that retain the color of the gentian violet stain used in Gram's method.

Gram's stain. Staining of bacteria to identify those that stain pink (gram-negative) or purple (gram-positive) under a microscope.

grandiose *(grăn'-dē-ōs).* In psychiatry, referring to an exaggerated belief in one's self-worth, powers, wealth, or importance.

grand mal. A type of epileptic seizure characterized by a loss of consciousness and stiffness of the entire body; also called *generalized tonic-clonic epilepsy.*

granulated eyelids *(grăn'-ū-lā-tĕd).* Crusty formation on the eyelids caused by chronic inflammation.

granulation *(grăn-ū-lā'-shŭn).* 1. Process of dividing substances into small granules. 2. Formation of small, rounded, fleshy bodies on the surface of a healing wound.

granule *(grăn'-ūl).* 1. A small grain or particle. 2. A small pill.

granulocyte *(grăn'-ū-lō-sīt).* A mature granular white blood cell that develops in bone marrow.

granulocytopenia *(grăn-ū-lō-sī-tō-pēn'-ē-ă).* Condition characterized by a deficiency of granular white blood cells in the blood.

granuloma *(grăn-ū-lō'-mă).* A tumor or growth composed of granular tissue, seen in various infectious diseases such as syphilis and leprosy.

dental g. A mass of inflammatory tissue that develops at the root of a tooth; also called *periapical granuloma.*

g. inguinale. A slowly progressive granulomatous ulcerative disease in which the initial lesion usually appears in the genital area as a painless lesion.

pyogenic g. Small, red benign growth of granulation tissue on the skin or oral mucosa that develops as the result of trauma.

granulomatosis *(grăn-ū-lō-mă-tō'-*

sĭs). Any condition characterized by the presence of multiple granulomas.

graphology *(grăf-ŏl'-ō-jē).* Analysis of handwriting to evaluate the character of the writer.

gravel *(grăv'-ĕl).* Minute concretions, usually composed of calcium oxalate, uric acid, or phosphates, that develop in the bladder and kidney.

Grave's disease *(grāvz).* Thyroid disorder resulting from excessive production of thyroid hormone and characterized by an enlarged thyroid gland, muscle tremors, rapid pulse, weight loss, and bulging eyes.

gravity *(grăv'-ĭ-tē).* The phenomenon in which two objects having mass are attracted to each other.

specific g. The ratio of the density of a substance to that of a reference substance at a specific temperature.

gray matter. The cortex of the brain, which contains nerve cell bodies, and is in fact gray, in contrast to the white matter, the part of the brain that contains myelinated nerve fibers.

grippe *(grĭp). See* **influenza.**

griseofulvin *(grĭs-ē-ō-fŭl'-vĭn).* An antibiotic used as an antifungal in the treatment of skin and nail infections.

groin *(groin).* The inguinal area; the region around the crease that is formed at the junction of the thigh and trunk.

group therapy. A form of psychotherapy in which a group of people meet with a leader, usually a therapist, to gain insight into problems, provide emotional support, or help change troubled behavior, thoughts, or feelings.

growing pains. A general term used to describe aching sensations in the musculoskeletal system experienced by some young people. There is no evidence they are associated with growth.

growth factors. Polypeptides that have both positive and negative effects on the cell cycle.

growth hormone. Hormone secreted by the anterior lobe of the pituitary gland; it affects the rate of skeletal and organ growth and promotes the transport of fat.

guaiac test *(gwī'-ăk).* A test for occult blood; the presence of blood is indicated by a blue tint.

guided imagery. An alternative medicine technique in which individuals use their imagination to visualize better health or to eliminate a specific disease or symptom. It has been shown to be effective in some cancer centers and other medical facilities.

Gulf War syndrome. Controversial syndrome (experienced by 5,000 to 80,000 veterans) associated with Americans who served in the Gulf War. It is characterized by a wide variety of illnesses, physical and psychological, for which no specific cause has been found.

Guillain-Barre syndrome *(gē-yă—bă-rā').* A disorder characterized by progressive paralysis and loss of reflexes, usually beginning in the legs and proceeding from the end of an extremity toward the trunk of the body.

gullet *(gŭl'-ĭt).* The passage that leads from the mouth to the stomach, it includes the pharynx and the esophagus.

gumboil *(gŭm'boyl).* An abscess of the gum that typically drains itself by perforating the gum; also called *parulis.*

gum disease. Inflammation of the gingival accompanied by abnormal loss of bone that anchors the teeth. It is caused by toxins secreted by bacteria harbored in plaque. Early symptoms include painless bleeding of the gums, but more advanced disease is painful.

gumma *(gŭm'-mă).* A gummy infectious tumor that sometimes develops during the third stage of syphilis.

gums *(gŭmz).* The gingival.

gustatory *(gŭs'-tă-tŏr-ē).* Of or relating to the sense of taste.

gut *(gŭt).* The intestines.

gynatresia *(jĭn-ă-trē'-zhă).* Closing of a part of the female genital tract, usually the vagina.

gynecologist *(gī-nĕ-kŏl'-ō-jĭst).* Physician who specializes in gynecology.

gynecology *(gī-nĕ-kŏl'-ō-jē).* The medical specialty concerned with the diagnosis and treatment of disorders of the female reproductive organs, including the breasts.

gynecomastia *(jĭn-ĕ-kō-măs'-tē-ă).* Development of abnormally large mammary glands in males.

gynephobia *(jĭn-ĕ-fō'-bē-ă).* Morbid fear of or an abnormal aversion to women.

gyrus *(jī'-rŭs)* (*pl.* gyri). Any one of the convolutions of the cerebral hemispheres of the brain.

H

habit *(hă'-bĭt).* A fixed or constant behavior performed with constant or frequent repetition.

habituation *(hă-bĭch-ū-ā'-shŭn).* 1. The gradual adaptation to anything that you do frequently or to a stimulus in the environment. 2. In drug addiction, it is the mental equivalent to physical dependence on drugs.

habitus *(hăb'-ĭ-tŭs).* A physical appearance or posture that indicates a tendency for a person to have a certain disease or condition.

Haemophilus *(hē-mŏf'-ĭ-lĕs).* A genus of gram-negative aerobic or anaerobic bacteria that normally inhabit the upper respiratory tract. Also spelled *Hemophilus.*

H. haemolyticus. A nonpathogenic species found as a normal inhabitant of the upper respiratory tract.

H. influenzae. A species once thought to cause epidemic influenza in humans, now recognized as a cause of bacterial meningitis and pneumonia.

hair *(hār).* A threadlike outgrowth composed of keratin that grows out of the skin or scalp. Each hair consists of a shaft and a root, which are contained in a hair follicle.

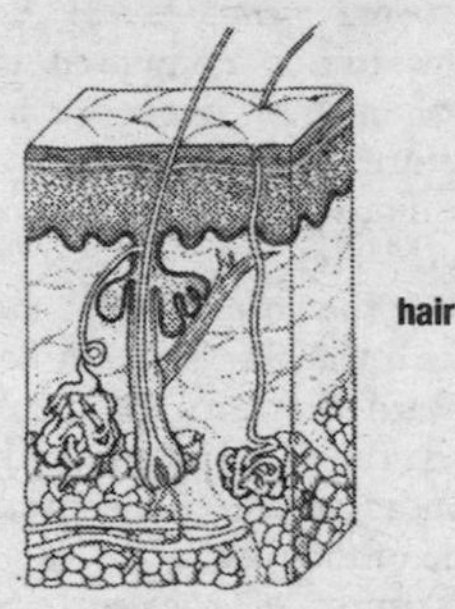

hair

burrowing h. A hair that grows horizontally under the skin, instead of emerging from the skin, which can cause an infection.

ingrown h. A hair that emerges from the skin but curves and reenters it, often causing an infection.

terminal h. Coarse hair that grows on various parts of the body during adulthood.

hair analysis. A test to determine the chemical composition of hair; used to study a person's level of exposure to toxins, nutritional status, or to monitor certain diseases.

hair cells. Type of cell found in the inner ear, they are receptors for the senses of hearing and positioning.

hair follicle. A tiny sac in the epidermis that holds a hair root and from which a hair emerges.

hairy tongue. Tongue that is covered with hairlike projections intertwined with threads from the fungus *Aspergillus niger* or *Candida albicans.* Can occur as a side effect of antibiotic therapy that inhibits the growth of bacteria normally found in the mouth.

Halcion *(hăl'-sē-ĕn).* Trade name for a preparation of triazolam, used to treat insomnia.

Haldol *(hăl'-dŏl).* Trade name for haloperidol, used to manage psychoses and control the tics of Tourette's syndrome.

half-life. Time necessary for half of the nuclei of a radioactive substance to lose their activity by undergoing radioactive decay.

drug h-l. Time required for the plasma level of a drug to decline to half of a certain measured level.

halfway house. A facility that houses individuals who have mental or addiction problems and who do not need complete hospitalization but who do require intermediate care to help them become established in the community.

halitosis *(hăl-ĭ-tō'-sĭs).* Offensive breath.

hallucination *(hă-lū-sĭ-nā'-shŭn).* In psychology, the perception of an external object when no such object exists in reality. A hallucination may be visual, auditory, tactile, gustatory, or olfactory.

auditory h. Imaginary perception of sounds.

gustatory h. Imaginary sense of tasting something that isn't present.

kinetic h. Imaginary sense of moving a part or all of the body.

olfactory h. Hallucination in which one smells something that isn't present.

tactile h. A false sense that one is touching something.

visual h. Sensation of seeing objects that are not real.

hallucinogen *(hă-lū'-sĭ-nō-jĕn).* A drug that produces hallucinations, such as LSD, mescaline, or peyote.

hallucinosis *(hă-lū-sĭ-nō'-sĭs).* A state characterized by the presence of hallucinations without any other impairment of consciousness.

hallux *(hăl'-ĕks).* The great or "big" toe; or the first digit of the foot.

h. malleus. Hammertoe of the great toe. Also called *hallux flexus.*

h. valgus. Condition in which the great toe points toward the midline of the body or away from the other toes.

halogen *(hăl'-ō-jĕn).* A substance that combines with hydrogen to form acids and with metals to form salts. Halogens include chlorine, bromine, iodine, and fluorine.

halothane *(hăl'-ō-thān).* A fluorinated hydrocarbon that is widely used as a general anesthetic.

Halsted's operation *(hăl'-stĕdz).* 1. A surgical procedure for inguinal hernia. 2. A radical mastectomy.

hamartoma *(hăm-ăr-tō'-mă).* A benign tumor composed of a spontaneous overgrowth of mature cells and tissues that are normally found in the affected site.

hammer toe *(hăm'-ĕr tō).* A toe in which the first bone is flexed toward the back of the foot and the second and third bones are flexed toward the sole of the foot.

hamstrings *(hăm'-strĭngz).* The three muscles on the backside of the thigh; they flex the leg and adduct and extend the thigh.

hand *(hănd).* The part of the body that is attached to the forearm at the wrist. It includes the carpus, metacarpus, and digits. The medical term for hand is *manus.*

handedness *(hăn'-dĕd-nĕs).* The tendency to use one hand in preference to the other.

handicap *(hăn'-dē-kăp).* Any mental or physical impairment, acquired or congenital, that prevents or restricts a person from performing everyday activities or engaging in work.

hangnail *(hăng'-nāl)*. Partly detached piece of skin that originates at the root or side of a finger or toenail.

hangover. A nonmedical term for describing a condition caused by ingesting an excessive amount of alcohol or another central nervous system depressant. Characteristics include headache, thirst, nausea, irritability, and fatigue.

Hanot's disease *(ăhn-ōz')*. Condition characterized by an enlarged, cirrhotic liver with jaundice.

Hansen's disease *(hăn'-sŏns)*. Leprosy.

hantavirus *(hăn'-tă vī'-rŭs)*. A genus of viruses that causes epidemic hemorrhagic fever or pneumonia and is believed to be transmitted to humans via direct or indirect contact with the feces of infected rodents.

haploid *(hăp'-lŏyd)*. Possessing half the diploid or a single set of chromosomes, as is seen in oocyte or spermatozoon. In humans, the haploid number is 23.

harelip *(hār'-lĭp)*. Cleft lip.

Harrington rods *(hăr'-ĭng-tŏn)*. A system of metal rods and hooks that are surgically inserted into the spine as treatment of scoliosis and other deformities.

Hartmann's pouch *(hărt'-măns)*. An abnormal sac on the neck of the gallbladder.

Hashimoto's disease *(hăh-shē-mō'-tōz)*. The most common cause of hypothyroidism, it is characterized by depression, fatigue, constipation, loss of sex drive, dry flaky skin, brittle nails, and intolerance to cold.

hashish *(hă-shēsh')*. The unadulterated resin gathered from the flowering tops of female hemp plants, it is chewed or smoked for its intoxicating effects.

haustra *(haws'-tră)*. Pouches or bulges in the wall of the colon.

hawthorn *(hăw'-thŏrn)*. A shrub or tree of the genus *Crataegus*, preparations of which are used to decrease output in congestive heart failure. It is also used in herbal and traditional Chinese medicine and homeopathy.

hay fever *(hay'-fē-vĕr)*. An allergic condition in which substances, such as pollen or dust, irritate the mucous membranes of the nose and upper air passages and cause inflammation, watery eyes, sneezing, and other allergic reactions.

Hayflick's limit *(hāy'-flĭks)*. The maximum number of divisions a cell undergoes before it dies, which is typically fifty to sixty for most human cells. Some malignant cells often exceed this limit.

HBV. Acronym for hepatitis B virus.

HCl. Acronym for hydrochloric acid.

HCV. Acronym for hepatitis C virus.

HDL. Acronym for high-density lipoprotein.

HDV. Acronym for hepatitis D virus.

headache *(hĕd'-āk).* Pain, acute or chronic, that is not confined to any specific part of the head. Also called *cephalalgia.*

cluster h. Head pain characterized by excruciating pain over one eye and the forehead, fever, and tearing eyes; attacks usually occur in clusters, sometimes a few times a day for several weeks, followed by months or years without recurrence.

lumbar puncture h. Headache that may occur after a lumbar puncture when a patient sits up but is relieved by lying down. Caused by lowering of intracranial pressure from leakage of cerebrospinal fluid through the puncture site.

migraine h. *See* **migraine.**

organic h. Head pain caused by intracranial disease or other organic disease.

tension h. Type of headache caused by prolonged physical or emotional stress or tension; it may be acute or chronic.

toxic h. Type of headache caused by poisoning or one associated with an illness.

vascular h. A classification of headache in which the proposed cause is abnormal functioning of the blood vessels of the brain's vascular system; types include migraine, cluster, and toxic.

healer *(hēl'-ĕr).* A person who heals, often used to refer to an individual who uses alternative or holistic medicine methods to heal.

health *(hĕlth).* A state of optimal physical, emotional, mental, and social well-being.

health maintenance organization (HMO). A general term that includes various health-care delivery systems that provide a prepaid program and utilize group practice for individuals who are under contract and within a specific geographic area.

hearing *(hĕr'-ĭng).* The ability to perceive sound.

heart *(hărt).* The muscular organ in the chest that maintains blood circulation for the body.

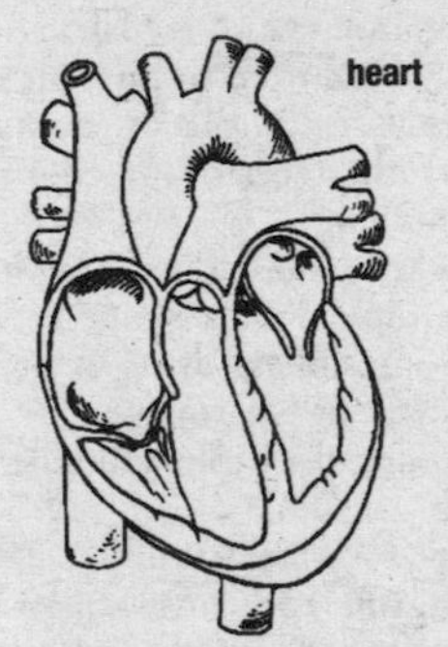
heart

athletic h. Enlargement or dilation of the heart not caused by disease or any disorder of circulation; often seen in athletes.

artificial h. A pumping device

that duplicates the rate, output, and blood pressure of a natural heart.

fatty h. Condition in which an excessive layer of fat is deposited about and in the heart muscle.

fibroid h. A heart in which fibrous tissue develops in the myocardium, sometimes seen in chronic myocarditis.

heart attack. A myocardial infarction.

heartbeat. A rhythmic contraction of the heart.

heart block. A condition in which the tissue that normally conducts impulses from the atrium to the ventricles fails to do so.

heartburn. A burning sensation that typically occurs under the sternum (breastbone), caused by the reflux of stomach acid into the lower esophagus.

heart failure. Condition in which the heart stops beating.

heart-lung machine. A device that maintains the functions of both the heart and lungs when they are unable to function, as during open-heart surgery.

heart murmurs. Sounds emitted by the heart other than those normally present. They typically are caused by blood passing over a rough valve or through a constricted valve, or may be due to a defective valve.

heart rate. The number of beats of the heart over a specific period of time, usually expressed by number per minute.

heartworm. A type of parasitic worm, *Dirofilaria immitis,* that can infect the heart of dogs, cats, and other animals, but rarely humans. The adult worms typically live in the animal's pulmonary artery and right side of the heart. The disease can be fatal.

heat *(hēt).* The sensation of an increase in temperature.

prickly h. Condition in which bacterial damage to sweat pores blocks sweat from escaping the skin and forces it to enter the epidermis, resulting in the development of tiny blisters that may become painful and itchy.

heat rash. Small red or pink spots on the skin, caused by inflammation of the sweat glands.

heat stroke. An acute and dangerous reaction to overexposure to heat in which an individual's body temperature is usually greater than 105° F; other symptoms include cessation of sweating, headache, confusion, and numbness, which can be followed by delirium or coma.

heaves *(hēvs).* Vomiting.

Heberden's nodes *(hē'-bĕr-dĕnz).* Hard nodules or enlarged lesions that develop on the last bones of the fingers, seen in osteoarthritis.

heel *(hēl).* Rounded posterior portion of the foot behind and under the ankle.

Heimlich maneuver *(hīm'-lĭk).* A method to dislodge food or other material from the throat of a choking individual.

Helicobacter *(hĕl-ĭ-kō-băk'-tĕr).* A genus of gram-negative bacteria; formerly classified as *Campylobacter.*

H. jejuni. A species that causes an acute disease characterized by diarrhea, abdominal pain, malaise, fever, nausea, and vomiting.

H. pylori. A species that causes gastritis and pyloric ulcers; it is also associated with gastric cancer.

heliosis *(hē-lē-ō'-sĭs).* Sunstroke.

helium *(hē'-lē-ŭm).* An odorless, tasteless, colorless gas, sometimes used with oxygen in the treatment of certain respiratory conditions.

helix *(hē'-lĭks).* 1. A spiral or coil. 2. Margin of the external ear.

Hellerwork *(hĕl'-ĕr-wĕrk).* A system of bodywork that uses deep massage, movement reeducation, and dialogue to improve posture, develop less stressful movement patterns, and increase awareness of the mind-body relationship.

helminth *(hĕl'-mĭnth).* A parasitic worm.

heloma *(hē-lō'-mă).* A callus or corn on the foot or hand.

helotomy *(hē-lŏt'-ĕ-mē).* Removal or paring of a corn or callus.

hemachrome *(hē'-mă-krōm).* A blood pigment that carries oxygen; for example, hemoglobin.

hemagglutination *(hēm-ă-glū-tĭ-nā'-shŭn).* Clumping together of red blood cells.

hemangiectasia *(hē-măn-jē-ĕk-tā'-shă).* Abnormal dilatation and sometimes lengthening of a blood or lymphatic vessel. Also called *angiectasis.*

hemangioma *(hē-măn-jē-ō'-mă).* A very common type of benign tumor composed of dilated blood vessels; it most often occurs in infancy and childhood.

capillary h. The most common type of hemangioma; it is usually composed of tightly packed capillaries.

cavernous h. A tumor composed of large dilated blood vessels, seen in the skin and some organs (e.g., liver, pancreas, spleen). Superficial lesions are usually bright to dark red, while deep lesions are bluish. Also called *strawberry mark.*

hemangiosarcoma *(hē-măn-jē-ō-săr-kō'-mă).* A rare, malignant tumor originating from blood vessels; it usually occurs in the skin, breast, liver, or soft tissues.

hemarthrosis *(hĕm-ăr-thrō'-sĭs).* Bloody leakage into the cavity of a joint.

hematemesis *(hĕm-ăt-ĕm'-ĕ-sĭs).* Vomiting of blood.

hematocele *(hē'-mă-tō-sēl).* Leakage of blood into a cavity.

hematocolpos *(hē-mă-tō-kŏl'-păs).* An accumulation of menstrual blood in the vagina.

hematocrit test *(hē-măt'-ō-krĭt).* Test to determine the proportion of the volume of a blood sample that is red blood cells.

hematologist *(hē-mă-tŏl'-ō-jĭst).* An individual who specializes in hematology.

hematoma *(hē-mă-tō'-mă).* A local-

ized accumulation of blood, usually caused by a leaking or ruptured blood vessel.
epidural h. Hematoma located above the dura mater.
intracerebral h. Hemorrhage localized in one area of the brain.
subdural h. Hematoma located beneath the dura; usually the result of head trauma.

hematometra *(hē-mă-tō-mē'-tră).* An accumulation of blood in the uterus.

hematopoiesis *(hē-mă-tō-poi-ē'-sĭs).* The formation and development of blood cells, usually in the bone marrow.

hematosalpinx *(hē-mă-tō-săl'-pĭnks).* Retained menstrual fluid in the fallopian tube.

hematoxylin *(hĕm-ă-tŏk'-sĭ-lĭn).* A colorless crystalline compound used to stain certain structures a deep blue, used commonly in microscopy.

hematuria *(hē-mă-tū'-rē-ă).* Presence of blood in the urine.

heme *(hēm).* The nonprotein, iron-containing component of hemoglobin.

hemianopia *(hĕm-ē-ă-nŏ'-pē-ă).* Defective vision or blindness in one-half of the visual field of one or both eyes.

hemic *(hē'-mĭk).* Referring to the blood.

hemicolectomy *(hĕm-ē-kō-lĕk'-tō-mē).* Surgical removal of approximately one-half of the colon.

hemicrania *(hĕm-ē-krā'-nē-ă).* Pain or ache in one side of the head.

hemigastrectomy *(hĕm-ē-găs-trĕk'-tō-mē).* Surgical removal of one-half of the stomach.

hemilaryngectomy *(hĕm-ē-lăr-ĭn-jĕk'-tō-mē).* Surgical removal of one-half of the larynx.

hemiparesis *(hĕm-ē-pă-rē'-sĭs).* Muscular weakness or partial paralysis that affects one side of the body.

hemiplegia *(hĕm-ē-plē'-jē-ă).* Paralysis of one side of the body.

hemisphere *(hĕm'-ĭ-sfēr).* Half of any spherical or roughly spherical object or structure.

hemithyroidectomy *(hĕm-ē-thī-roid-ĕk'-tō-mē).* Surgical removal of one lobe of the thyroid gland.

hemoagglutinin *(hē-mō-ă-glū'-tĭ-nĭn).* *See* **hemagglutinin.**

hemochromatosis *(hē-mō-krō-mă-tō'-sĭs).* A condition caused by the deposition of hemosiderin in the parenchymal cells, causing dysfunction of the liver, pancreas, heart, and pituitary.

hemoconcentration *(hē0mō-kŏn=sĕn-trā'-shŭn).* A decrease in the fluid content of the blood, resulting in an increase in concentration.

hemodialysis *(hē-mō-dī-ăl'-ĕ-sĭs).* Use of an apparatus (hemodialyzer) that filters the blood to remove certain elements when the kidneys are unable to do so. Also called *kidney dialysis, dialysis*, and *renal dialysis.*

hemodialyzer *(hē-mō-dī'-ă-lī-zĕr).* An apparatus used to perform hemodialysis. It allows blood to pass through a semipermeable membrane and make contact with a solution that removes certain elements from the blood.

hemodilution *(hē-mō-dī-lū'-shŭn).* An increase in the amount of fluid in the blood, resulting in a decrease in concentration of red blood cells.

hemodynamics *(hē-mō-dī-năm'-ĭks).* Study of the forces involved in blood circulation.

hemoglobin *(hē-mō-glō'-bĭn).* The iron-containing pigment found in red blood cells; it carries oxygen from the lungs to the body's tissues.

hemoglobinemia *(hē-mō-glō-bĭn-ē'-mē-ă).* Presence of hemoglobin in the blood plasma.

hemoglobinuria *(hē-mō-glō-bĭn-ū'-rē-ă).* Presence of hemoglobin in the urine.

hemolysin *(hē-mŏl'-ĭ-sĭn).* A substance or condition that disrupts red blood cells.

hemolysis *(hē-mŏl'-ĭ-sĭs).* The destruction of red blood cells, which frees hemoglobin to diffuse into the surrounding fluid.

hemolytic anemia. *See* **anemia.**

hemopericardium *(hē-mō-pĕr-ĭ-kăr'-dē-ŭm).* Accumulation of blood in the pericardial sac.

hemoperitoneum *(hē-mō-pĕr-ĭ-tō-nē'-ŭm).* Leakage of blood into the peritoneal cavity.

hemophilia *(hē-mō-fĭl'-ē-ă).* An inherited blood disease in which it takes the blood a prolonged time to clot, resulting in abnormal bleeding. The condition is caused by a deficiency of a specific factor in plasma.

hemophilia A. Form of hemophilia caused by a deficiency of blood coagulation factor VIII.

hemophilia B. Form of hemophilia caused by a deficiency of blood coagulation factor IX; *see also* **Christmas disease.**

hemopneumothorax *(hē-mō-nū-mō-thŏr'-ăks).* Presence of blood and air in the pleural cavity.

hemopoiesis *(hē-mō-poy-ē'-sĭs).* The formation of blood cells.

hemoptysis *(hē-mŏp'-tĭ-sĭs).* Coughing up and spitting of blood that arises from the mouth, larynx, trachea, bronchi, or lungs.

hemorrhage *(hĕm'-ĕ-rĭj).* Abnormal internal or external discharge of blood from the body.

cerebral h. Bleeding into the cerebrum. Also called *intracerebral hemorrhage.*

intracranial h. Bleeding within the cranium.

massive h. Loss of blood that is so rapid and profuse that the individual experiences shock unless adequate replacement blood is supplied promptly.

pulmonary h. Bleeding from the lungs.

subarachnoid h. Bleeding into the subarachnoid space.

subdural h. Bleeding into the subdural space.

hemorrhoidectomy *(hĕm-ĕ-roid-ĕk'-tō-mē).* Surgical removal of hemorrhoids.

hemorrhoids *(hĕm'-ĕ-roids).* A mass of dilated, twisted veins in the anorectal area that develop as the result of persistent, increasing pressure on the veins.

external h. Hemorrhoid that protrudes outside the rectum; it usually causes pain, burning, and itching.

internal h. Hemorrhoid found inside the rectum; it typically does not cause pain, burning, or itching.

strangulated h. Hemorrhoid that has had its blood supply cut off.

hemosiderin *(hē-mō-sĭd'-ĕr-ĭn).* An iron-containing pigment derived from hemoglobin, it is stored in cells until it is needed to make hemoglobin.

hemospermia *(hē-mō-spĕr'-mē-ă).* The presence of blood in the semen.

hemostasis *(hē-mō-stā'-sĭs).* Cessation of bleeding either by coagulation or surgical means.

hemostat *(hē'-mō-stăt).* A small surgical clamp used to close off a blood vessel and stop bleeding.

hemostatic *(hē-mō-stăt'-ĭk).* Something that causes hemostasis.

hemothorax *(hē-mō-thō'-răks).* Bloody fluid in the pleural cavity caused by ruptured blood vessels, inflammation of the lungs, or a malignancy.

hemotoxin *(hē-mō-tŏks'-ĭn).* A toxin that destroys red blood cells.

henna *(hĕn'-nă).* A coloring made from *Lawsonia inermis,* an Egyptian plant. It's been approved for use as a hair dye, but direct application to the skin can cause injury.

heparin lock. A solution of heparin sodium used specifically to maintain the patency of intravenous injection devices.

heparin sodium *(hĕp'-ă-rĭn).* An anticoagulant used to prevent and treat thrombosis and embolism; also to prevent clotting during blood transfusions and blood sampling.

hepatectomy *(hĕp-ă-tĕk'-tō-mē).* Surgical removal of part or all of the liver.

hepatic *(hĕ-păt'-ĭk).* Referring to the liver.

hepatitis *(hĕp-ă-tī'-tĭs).* Inflammation of the liver.

hepatitis A. Liver disease caused by hepatitis A virus, usually contracted via contaminated food, a contaminated blood transfusion, or other causes.

hepatitis B. Liver disease caused by hepatitis B virus, usually transmitted by blood transfusions, sharing of needles among drug users, or sexual contact; also from an infected mother to her infant.

hepatitis C. Liver disease caused by hepatitis C virus.

hepatitis D. Liver disease caused by hepatitis D virus, requiring simultaneous infection with hepatitis B virus, and having symptoms similar to hepatitis B.

hepatobiliary *(hĕp-ă-tō-bĭl'-ē-ăr-ē).* Referring to the liver and the bile or the bile ducts.

hepatocellular cancer *(hĕp-ă-tō-sĕl'-ū-lĕr).* A malignant tumor of the liver.

hepatolithiasis *(hĕp-ă-tō-lĭ-thī'-ă-

sĭs). The presence or formation of calculi in the liver.

hepatoma *(hĕp-ă-tō'-mă).* Tumor of the liver.

hepatomegaly *(hĕp-ă-tō-mĕg'-ă-lē).* Enlargement of the liver.

hepatonecrosis *(hĕp-ă-tō-nĕ-krō'-sĭs).* Gangrene of the liver.

hepatonephritis *(hĕp-ă-tō-nĕ-frī'-tĭs).* Inflammation of the liver and kidneys.

hepatoperitonitis *(hĕp-ă-tō-pĕr-ĭ-tō-nī'-tĭs).* Inflammation of the peritoneal covering of the liver.

hepatorenal syndrome *(hĕp-ă-tō-rē'-năl).* Failure of the liver and kidney after surgery; often fatal.

hepatosplenomegaly *(hep-a-to-sple-no-meg'-a-le).* Enlargement of the liver and spleen; seen in diseases such as Gaucher's disease.

hepatotoxic *(hĕp-ă-tō-tŏks'-ĭk).* Anything that is toxic to the liver.

Herceptin *(hĕr-sĕp'-tĭn).* Trade name for a preparation of trastuzumab, used to treat metastatic breast cancer.

heredity *(hĕ-rĕd'-ĭ-tē).* The genetic transmission of specific traits from parent to offspring.

hermaphrodite *(hĕr-măf'-rō-dīt).* An individual who has both male and female gonadal tissue.

hernia *(hĕr'-nē-ă).* The protrusion of tissues through an abnormal opening, caused by a rupture or weakness in a muscle.

abdominal h. Hernia through the abdominal wall.

congenital h. Hernia that is present at birth.

hiatal h. Protrusion of the stomach upward into the mediastinal cavity through the esophageal opening of the diaphragm.

sliding h. A type of hernia that may develop and then return to normal.

herniation *(hĕr-nē-ā'-shŭn).* An abnormal protrusion of an organ or other body part through a natural or defective opening in a membrane, muscle, or bone.

herniated disk. *See* **disk.**

hernioplasty *(hĕr-nē-ō-plăs'-tē).* Surgical repair of a hernia.

herniorrhaphy *(hĕr-nē-ŏr'-ĕ-fē).* Surgical repair of a hernia.

heroin *(hĕr'-ō-ĭn).* A narcotic derived from morphine.

herpes *(hĕr'-pēz).* Any inflammatory skin disease that is caused by a herpesvirus. When the term is used alone, it refers to herpes simplex or herpes zoster.

h. genitalis. Herpes simplex caused by human herpesvirus 2; usually transmitted via sexual contact and characterized by painful lesions in the genital region.

h. simplex. Infectious disease caused by herpes simplex virus type 1 and type 2; type 1 infections usually involve the mouth and other nongenital areas of the body; type 2 infections are primarily in the genital region.

h. zoster. An acute infectious disease caused by the varicella-zoster virus and characterized by painful lesions; also called *shingles.*

herpesvirus *(hĕr'-pēz-vī-rŭs).* Any virus that belongs to the family Herpesviridae.

heterogeneous *(hĕt-ĕr-ō-jē'-nē-ĕs).* Composed of dissimilar ingredients or elements.

heterograft *(hĕ'-tĕr-ō-grăft). See* **xenograft.**

heterometropia *(hĕt-ĕr-ō-mă-trō'-pē-ă).* A condition in which there are differences in the degree of refraction in the two eyes.

heterosexuality *(hĕt-ĕr-ō-sĕk-shū-ăl'-ĭ-tē).* Sexual attraction for one of the opposite sex.

heterotoxin *(hĕt-ĕr-ō-tŏk'-sĭn).* A toxin introduced to the body from an external source.

heterotransplant *(hĕt-ĕr-ō-trăns'-plănt). See* **xenograft.**

heterozygous *(hĕt-ĕr-ō-zī'-gŭs).* Possessing different alleles at a given site.

hexachlorophene *(hĕks-ă-klŏr'-ō-fēn).* A bactericidal compound included in soaps and emulsions used for preoperative cleansing of the skin and by surgeons and nurses before surgery.

hexadactylism *(hĕks-ă-dăk'-tĭl-ĭzm).* Possession of six fingers or six toes on one limb.

Hexadrol *(hĕk'-sĕ-drŏl).* Trade name for preparations of dexamethasone.

hiatus *(hī-ā'-tŭs).* An opening or gap.

hiccup *(hĭk'-ŭp).* An involuntary spasmodic contraction of the diaphragm that causes an inhalation that is suddenly stopped by closure of the glottis, resulting in a sharp, inspiratory cough.

hidradenitis *(hī-drăd-ĕ-nī'-tĭs).* Inflammation of sweat glands.

hidrosis *(hī-drō'-sĭs).* The formation and excretion of sweat; also refers to excessive sweating.

high blood pressure. *See* **hypertension.**

hilum *(hī'-lĕm).* Depression or pit at the entrance or exit of a duct into a gland or of nerves and blood vessels into an organ.

hip *(hĭp).* The upper part of the thigh, formed by the femur and the innominate bones.

hippocampus *(hĭp-pō-kăm'-pŭs).* Area deep in the forebrain that helps regulate emotion and memory. So called because its shape is like that of a seahorse (Greek "hippos" = horse; "kampos" = sea monster).

hippus *(hĭp'-ŭs).* Rapid, rhythmical dilatation and contraction of the pupils.

hip replacement. Surgical procedure during which the head of the femur and the acetabular portion of the hip joint are replaced, usually with a metal prosthesis.

hirsutism *(hŭr'-sūt-ĭzm).* Condition characterized by excessive growth of hair or the appearance of hair in unusual places, especially in women.

histamine *(hĭs'-tă-mēn).* A substance produced from the amino acid histidine. Some of its functions in the body include dilatation of capillaries, formation of bronchial smooth muscle, and increasing gastric secretions.

histidine *(hĭs'-tĭ-dēn).* An essential amino acid obtained by hydrolysis from proteins; it is necessary for tissue growth and repair.

histiocyte *(hĭs'-tē-ō-sīt).* *See* **macrophage.**

histiocytoma *(hĭs-tē-ō-sī-tō'-mă).* A tumor that contains histiocytes.

histiocytosis *(hĭs-tē-ō-sī-tō'-sĭs).* An excessive number of histocytes in the blood.

histocompatibility *(hĭs-tō-kŏm-păt-ĭ-bĭl'-ĭ-tē).* The ability of cells to survive without interference from the immune system.

histocompatible *(hĭs-tō-kŏm-păt'-ĭ-bĕl).* Referring to a tissue donor and recipient who share enough histocompatability antigens so the donated tissue will be accepted and functional.

histogram *(hĭs'-tō-grăm).* A graph that shows frequency distributions.

histology *(hĭs-tŏl'-ō-jē).* The study of the microscopic features of tissue.

histoplasmosis *(hĭs-tō-plăz-mō'-sĭs).* A systemic respiratory disease caused by the fungus *Histoplasma capsulatum.*

history *(hĭs'-tō-rē).* A systematic record of past events that related to a specific person, group of people, time period, or country.

family h. A record of the medical history and state of health of members in the immediate family of a patient.

medical h. A record of past medical conditions and illnesses and treatments that pertain to a specific patient.

HIV. Acronym for human immunodeficiency virus.

hives *(hīvz).* Eruption of intensely itchy wheals as an allergic response; often caused by ingesting a specific food or medication, or can be caused by sudden changes in climate. Also called *urticaria.*

HIV negative. A blood test result showing that an individual does not harbor the virus that causes AIDS.

HIV positive. A blood test result showing that an individual does harbor the virus that causes AIDS.

HLA test. Acronym for human leukocyte antigen test, in which blood samples from the mother, child, and alleged father are tested to determine paternity.

HMO. Acronym for health maintenance organization.

hoarseness *(hŏrs'-nĕs).* A rough or noisy quality of the voice.

Hodgkin's disease *(hŏj'-kĭns).* Disease of unknown cause characterized by enlarged lymphatic tissue in the neck, liver, and spleen. Other symptoms include fever, chills, loss of appetite, weight loss, and night sweats.

holistic medicine *(hō-lĭs'-tĭk).* An

approach to healing that considers all the needs of an individual—physical, emotional, spiritual, social, and economic.

holography *(hōl-ŏg'-ră-fē).* Recording images in three-dimensional form (called holograms) on photographic film by exposing the film to a laser beam reflected from the object.

holotropic breathwork *(hō-lō-trōp'-ĭk).* A technique that combines rapid deep breathing, bodywork, and music in an effort to help people open up their unconscious and unite their body and mind in a state of altered consciousness.

homatropine *(hō-măt'-rō-pēn).* A belladonna alkaloid that has antispasmodic and anticholinergic functions.

homeopathy *(hō-mē-ŏp'-ĕ-thē).* A therapeutic system in which conditions are treated with remedies capable of producing in healthy individuals symptoms like those of the condition being treated.

homeostasis *(hō-mē-ō-stā'-sĭs).* State of stability of the internal body, achieved through a system of control mechanisms activated by negative feedback.

homocysteine *(hō-mō-sĭs'-tēn).* An intermediary compound in the metabolism of the amino acid methionine. High levels have been shown to cause atherosclerosis and are suspected to damage neurons.

homogeneous *(hō-mō-jē'-nē-ĕs).* Anything that consists of similar elements or ingredients; opposite of heterogeneous.

homograft *(hō'-mō-grăft).* Transplant tissue that is obtained from the same species; *see also* **allograft.**

homolateral *(hō-mō-lăt'-ĕr-ăl).* Referring to or affecting the same side; ipsilateral.

homologous *(hō-mŏl'-ō-gĕs).* Referring to things that are similar in basic structure and origin but not necessarily in function; for example, a bird's wing and a man's arm are homologous.

homophobia *(hō-mō-fō'-bē-ă).* Abnormal fear of homosexuality.

homosexuality *(hō-mō-sĕks-ū-ăl'-ĭ-tē).* Sexual orientation that is directed toward people of one's own sex.

homotransplantation *(hō-mō-trăns-plăn-tā'-shŭn). See* **allograft.**

homozygote *(hō-mō-zī'-gōt).* An individual who has a pair of identical alleles at a given locus.

hookworm *(hook'wĕrm).* Any parasitic nematode in the *Ancylostomatidae* family.

hordeolum *(hŏr-dē'-ō-lăm).* An inflammatory staphylococcal infection of one or more sebaceous glands of the eyelids; also called a sty.

hormone *(hŏr'-mōn).* A chemical produced by a gland, organ, or part that has specific regulatory effects on the activity of one or more organs or glands. *See* names of individual hormones.

Horner's syndrome *(hŏr'-nĕrz).* Condition caused by paralysis of the cervical sympathetic nerve and characterized by a contracted pupil, drooping eyelid, facial flush, and displacement of the eye.

hornification *(hŏr-nĭ-fĭ-kā'-shŭn).* *See* **cornification.**

hospice *(hŏs'-pĭs).* A facility that provides an interdisciplinary program of palliative and supportive care for terminally ill patients and their families.

hospital *(hŏs'-pĭ-tăl).* An institution whose purpose is to treat the sick and injured.

general h. A facility that is equipped to provide a wide range of medical care.

private h. A facility that typically operates for profit and offers little or no uncompensated care.

teaching h. One that allocates a significant portion of its resources to formal medical education; it is often affiliated with a medical school.

voluntary h. A private, nonprofit facility that offers uncompensated care to the poor.

host *(hōst).* 1. An organism that provides nourishment for a parasite. 2. In transplantation of tissue, it is the individual who receives the graft.

hot flashes. A symptom complex usually associated with menopause. It involves a feeling of discomfort in the abdominal area, followed by a feeling of heat that moves toward the head, reddening of the face, and sweating.

house physician. A physician, usually an intern or resident, who cares for patients under the supervision of the medical and surgical staff.

house staff. The residents and interns in a hospital.

HTLV. Acronym for human T-lymphotropic virus.

human growth hormone. Any of several polypeptide hormones that affect protein, carbohydrate, and fat metabolism and the rate of skeletal and organ growth.

human insulin NPH. *See* Appendix, Common Prescription and OTC Drugs: By Generic Name.

humectant *(hū-mĕk'-tănt).* A moistening agent.

humerus *(hū'-mĕr-ŭs).* Bone of the upper arm that extends from the elbow to the shoulder, where it connects with the scapula.

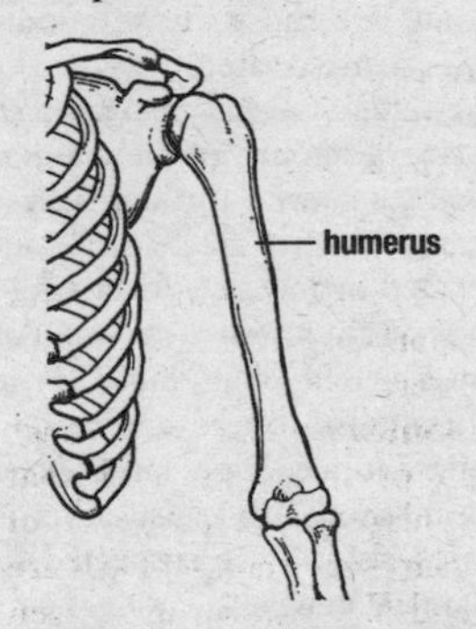

humor *(hū'-mŏr).* Any fluid or semifluid in the body.

humpback *(hŭmp'-băk).* Common term for kyphosis, a curvature of the spine to the extent that it results in a lump on the back.

Humulin. *See* Appendix, Common Prescription and OTC Drugs: By Trade Name.

hunger pains *(hŭn'-gĕr).* Dull or acute feeling felt in the upper abdomen caused by muscle contractions when a person is hungry.

Huntington's chorea *(hŭnt'-ĭng-tŏnz kŏ-rē'-ă).* An inherited disease of the central nervous system that usually appears between thirty and fifty years of age. Characterized by progressive dementia and involuntary movements of the limbs and face.

hyaline membrane disease *(hī'-ă-lĭn).* A respiratory disease of newborns; *see also* **respiratory distress syndrome.**

hyalinization *(hī-ă-lĭn-ĭ-zā'-shŭn).* The conversion into a substance that resembles glass.

hyaluronic acid *(hī-ĕ-lū-rŏn'-ĭk).* A substance found in connective tissue, where it is a binding and protective agent; also found in the synovial fluid, blood vessels, skin, vitreous humor, cartilage, and umbilical cord.

hyaluronidase *(hī-ĕ-lū-rŏn'-ĭ-dās).* Any of three enzymes that stimulate the breakdown of hyaluronic acid. They are found in testicular and spleen tissue.

hybrid *(hī'-brĭd).* An animal or plant that is produced from dissimilar parents, such as those from two different species, strains, or varieties.

hybridization *(hī-brĭd-ĭ-zā'-shŭn).* The process of producing hybrids; also called *crossbreeding.*

hydatid *(hī'-dĕ-tĭd).* Any cystlike structure.

hydatid of Morgagni. A benign cyst that can develop near the testicle in males and near the ovary in females.

hydralazine *(hī-drăl'-ĕ-zēn).* A drug that dilates blood vessels and is used as an antihypertensive.

hydramnios *(hī-drăm'-nē-ŏs).* Excessive amount of liquor amnii, secreted by the fetus, which surrounds the unborn child and overextends the uterus.

hydrarthrosis *(hī-drăr-thrō'-sĭs).* Accumulation of watery fluid in the cavity of a joint.

hydrase *(hī'-drās).* An enzyme that stimulates the addition or withdrawal of water from a compound.

hydrate *(hī'-drāt).* A crystalline substance that is formed by water when it combines with various compounds.

hydremia *(hī-drē'-mē-ă).* Excessive dilution of the blood, resulting in an abnormal proportion of serum to red blood cells; seen in splenomegaly and other conditions.

hydrocarbon *(hī-drō-kăr'-bŏn).* Organic compound composed only of hydrogen and carbon.

hydrocele *(hī'-drō-sēl).* The accumulation of serous fluid in a

saclike cavity, especially in the testes.
acute h. The most common type of hydrocele, it usually is the result of an inflamed testis or epididymis. Typically seen in children 2 to 5 years old.
chronic h. Type of hydrocele that usually occurs in middle age.
spinalis h. *See* **spina bifida.**

hydrocephalus *(hī-drō-sĕf'-ă-lŭs).* Accumulation of cerebrospinal fluid within the ventricles of the brain, usually due to obstructed circulation, which may result from infection, trauma, brain tumors, or developmental abnormalities.
congenital h. Chronic form of hydrocephalus that occurs in infancy.
normal pressure h. Type of hydrocephalus in which the ventricles enlarge without an increase in the spinal fluid pressure or any demonstrable block to the outflow of spinal fluid.

hydrochloric acid *(hī-drō-klŏr'-ĭk).* Gastric fluid produced by the cells that line the stomach; it aids in digestion.

hydrochlorothiazide *(hī-drō-klŏr-ō-thī'-ă-zīd). See* Appendix, Common Prescription and OTC Drugs: By Generic Name.

hydrocodone *(hī-drō-kō'-dōn). See* Appendix, Common Prescription and OTC Drugs: By Generic Name.

hydrocortisone *(hī-drō-kŏr'-tĭ-sōn).* The corticosteroid hormone that is produced by the adrenal cortex.

HydroDiuril *(hī-drō-dī'-ū-ĭl).* Trade name for hydrochlorothiazide, used to treat hypertension and edema.

hydrodynamics *(hī-drō-dī-năm'-ĭks).* The branch of science that deals with the mechanics of fluids and of solids contained in fluids.

hydrogen *(hī'-drō-jĕn).* An odorless, colorless, tasteless gas that is flammable and explosive when mixed with air. It is the lightest element and is found in nearly all organic compounds.

hydrogen peroxide. A cleansing, bleaching liquid typically diluted in water and used as a wash or spray.

hydrolysis *(hī-drŏl'-ĭ-sĭs).* Splitting of a compound into simpler elements by adding water.

hydronephrosis *(hī-drō-nĕ-frō'-sĭs).* Distention of the pelvis and calices of the kidney with urine as a result of an obstructed ureter.

hydrophobia *(hī-drō-fō'-bē-ă).* An irrational fear of water.

hydropneumothorax *(hī-drō-nū-mō-thŏr'-ăks).* An accumulation of gas and fluid within the pleural cavity.

hydrops *(hī'-drŏps).* Edema.

hydrotherapy *(hī-drō-thĕr'-ă-pē).* Therapeutic use of water in the treatment of disease.

hydruria *(hī-drū'-rē-ă).* Excessive production and discharge of urine.

hygiene *(hī'-jēn).* The study of

health and the methods used to achieve and preserve it.

hygroma *(hī-grō'-mă).* A cyst, bursa, or sac that contains fluid.

Hygroton *(hī'-grō-tŏn).* Trade name for a preparation of chlorthalidone, used to treat hypertension and edema.

hymen *(hī'-mĕn).* A fold of mucous membrane that partly or completely covers the opening of the vagina.

hymenotomy *(hī-mĕn-ŏt'-ō-mē).* Surgical incision of the hymen.

hyoid bone *(hī'-ŏyd).* U-shaped bone that lies at the base of the tongue.

hyoscine *(hī'-ō-sēn). See* **scopolamine.**

hypalgesia *(hī-păl-jē'-zē-ă).* A decreased sense of pain.

hyperacidity *(hī-pĕr-ă-sĭd'-ĭ-tē).* Having an excessive degree of acidity.

hyperadrenalism *(hī-pĕr-ă-drē'-năl-ĭz-ĕm).* Condition characterized by abnormally increased secretion of adrenal hormones.

hyperaldosteronism *(hī-pĕr-ăl-dŏs-tĕ-rō-nĭz'-ĕm).* Abnormal electrolyte metabolism, caused by an excessive secretion of aldosterone.

hyperalgesia *(hī-pĕr-ăl-jē'-zē-ă).* An abnormally increased sense of pain.

hyperalimentation *(hī-pĕr-ăl-ĭ-mĕn-tā'-shŭn).* Consumption of excessive quantities, especially of food.

hyperbaric chamber *(hī-pĕr-băr'-ĭk).* Chamber in which individuals are exposed to oxygen at a level greater than ordinary atmospheric conditions as a treatment method.

hyperbilirubinemia *(hī-pĕr-bĭl-ĭ-rū-bĕ-nē'-mē-ă).* Excessive bilirubin in the blood, which may result in jaundice.

hypercalcemia *(hī-pĕr-kăl-sē'-mē-ă).* Excessive calcium in the blood, causing muscle weakness, depression, nausea, constipation, anorexia, and fatigue.

hypercapnia *(hī-pĕr-kăp'-nē-ă).* Excessive carbon dioxide in the blood.

hypercatharsis *(hī-pĕr-kă-thăr'-sĭs).* Excessive bowel movements associated with the use of cathartics.

hyperchloremia *(hī-pĕr-klŏr-ē'-mē-ă).* Excessive chloride in the blood.

hypercholesterolemia *(hī-pĕr-kĕ-lĕs-tĕr-ŏl-ē'-mē-ă).* Excessive cholesterol in the blood.

familial h. An inherited form of the disease.

hyperchromatism *(hī-pĕr-krō'-mă-tĭz-ĕm).* Excessive pigmentation resulting from an excess of chromatin.

hypercyesis *(hī-pĕr-sī-ē'-sĭs).* Presence of more than one fetus in the uterus resulting from fertilization of a second egg within a short time of the first.

hyperemesis *(hī-pĕr-ĕm'-ĕ-sĭs).* Excessive vomiting.

hyperemia *(hī-pĕr-ē'-mē-ă).* An increase of blood in a body part.

hyperesthesia *(hī-pĕr-ĕs-thē'-zhă).*

An increased sensitivity to sensory stimulation, such as touch or sound.

hyperextension *(hī-pěr-ěks-těn'-shŭn).* Extreme or abnormal extension of a limb or body part.

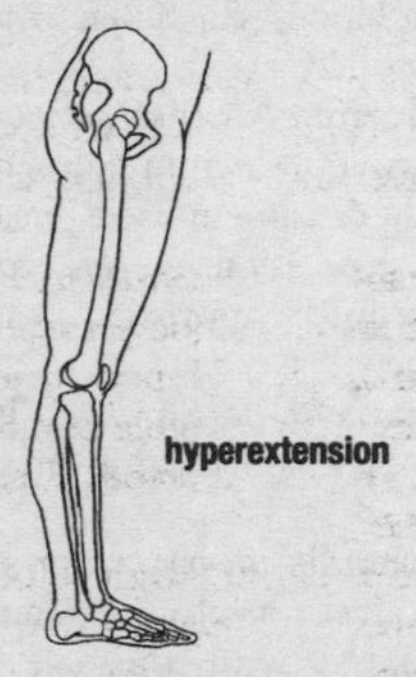

hyperflexion *(hī-pěr-flěks'-shŭn).* Forcible excessive extension of a limb or body part.

hyperglycemia *(hī-pěr-glī-sē'-mē-ă).* Abnormally elevated level of glucose in the blood.

hypergonadism *(hī-pěr-gō-năd'-ĭz-ěm).* A condition caused by excessive secretion of sex (gonadal) hormones.

hyperhidrosis *(hī-pěr-hī-drō'-sĭs).* Excessive sweating that is not warranted by the surrounding temperature.

hypericum *(hī-pěr'-ĭ-kŭm).* The traditional name for the homeopathic remedy known as St. John's wort (*Hypericum perforatum*), used mainly as a first aid remedy for injuries to the fingers, lips, nails, spine, or other areas rich in nerve endings.

hyperimmune *(hī-pěr-ĭm-mūn').* State in which an individual has large quantities of specific antibodies in the serum.

hyperinsulinism *(hī-pěr-ĭn'-sū-lĭn-ĭzm).* Excessive insulin in the blood.

hyperkalemia *(hī-pěr-kă-lē'-mē-ă).* Excessive potassium in the blood.

hyperkeratosis *(hī-pěr-kěr-ă-tō'-sĭs).* 1. Overgrowth of the outermost layer of the epidermis. 2. Any disease characterized by an overgrowth of the outer epidermis. 3. Overgrowth of the cornea.

hyperkinetic *(hī-pěr-kī-ně'-tĭk).* Hyperactive.

hyperlipidemia *(hī-pěr-lĭp-ĭ-dē'-mē-ă).* A general term for abnormally high concentrations of any of the lipids in plasma, such as hypertriglyceridemia and hypercholesterolemia.

hyperliposis *(hī-pěr-lī-pō'-sĭs).* Abnormal or excessive amount of fat.

hypermagnesemia *(hī-pěr-măg-ně-sē'-mē-ă).* An abnormally high level of magnesium in the blood, characterized by weakness, lethargy, and abnormal electrocardiographic readings.

hypermastasia *(hī-pěr-măs'-tē-ă).* Presence of one or more abnormally large mammary glands.

hypermotility *(hī-pěr-mō-tĭl'-ĭ-tē).* Excessive or abnormal level of motility, as can occur in the gastrointestinal tract.

hypernatremia *(hī-pěr-nă-trē'-mē-ă).* Excessive sodium in the blood.

hyperopia *(hī-pěr-ō'-pē-ă).* Farsightedness; when light rays entering the eye focus behind the retina because the eyeball is too short from the front to the back.

hyperparathyroidism *(hī-pĕr-păr-ă-thī'-royd-ĭz-ĕm).* A condition caused by an excessive amount of parathyroid hormone and characterized by calcium deposits in the kidney, osteoporosis, muscle weakness, and gastrointestinal symptoms.

hyperperistalsis *(hī-pĕr-pĕr-ĭ-stawl'-sĭs).* Excessively active peristalsis.

hyperpermeability *(hī-pĕr-pĕr-mē-ă-bĭl'-ĭ-tē).* Abnormal permeability, often referring to a cell membrane or a blood vessel wall.

hyperpituitarism *(hī-pĕr-pĭ-tū'-ĭ-tă-rĭz-ĕm).* A condition caused by excessive secretion of pituitary hormones, which can result in conditions such as acromegly, pituitary gigantism, or Cushing's disease.

hyperplasia *(hī-pĕr-plā'-zhă).* An abnormal increase in the number of normal cells in the normal tissue of an organ.

hyperpnea *(hī-pĕrp'-nē-ă).* An increased rate of respiration or breathing that is deeper than experienced during normal activity.

hyperprolactinemia *(hī-pĕr-prō-lăk-tĭ-nē'-mē-ă).* Elevated levels of prolactin in the blood. It is associated with amenorrhea in women and with impotence and hypogonadism in men.

hyperpyrexia *(hī-pĕr-pī-rĕk'-sē-ă).* *See* **hyperthermia.**

hyperreactive *(hī-pĕr-rē-ăk'-tĭv).* Characterized or referring to a greater than normal response to a stimulus.

hypersensitivity *(hī-pĕr-sĕn-sĭ-tĭv'-ĭ-tē).* A state of being in which the body reacts with an inappropriate or exaggerated immune response to what it perceives to be a foreign substance.

hypertension *(hī-pĕr-tĕn'-shŭn).* Elevated arterial blood pressure.

essential h. Hypertension that has no discernible organic cause; also called *primary h.* and *idiopathic h.*

gestational h. High blood pressure that develops during pregnancy and often disappears after delivery.

malignant h. A dangerous form of high blood pressure, often not responsive to antihypertensive medication.

portal h. Increased blood pressure due to partial obstruction of blood flow from the intestines through the liver, often seen in cirrhosis.

pulmonary h. High blood pressure in the pulmonary arteries that carry blood from the right ventricle of the heart to the lungs. This form of hypertension can cause irreparable damage to the lungs and right ventricle.

symptomatic h. Hypertension accompanied by symptoms such as headache or dizziness.

white coat h. A temporary rise in blood pressure triggered by seeing medical personnel in white coats (or other attire).

hyperthermia *(hī-pĕr-thĕr'-mē-ă).* 1. Elevation of the core body temperature to more than 99° F. 2. Raising the body temperature to greater than 107.6°F for therapeutic purposes.

hyperthyroidism *(hī-pĕr-thī'-royd-ĭzm).* A condition caused by excessive production of thyroid hormones. Characteristics include fatigue, nervousness, heat intolerance, sweating, weight loss, muscle weakness, goiter, and others.

hypertonic *(hī-pĕr-tŏn'-ĭk).* A solution that has a greater osmotic pressure than that of a compared solution.

hypertrichosis *(hī-pĕr-trī-kō'-sĭs).* Excessive growth of hair.

hypertrophy *(hī-pĕr'-trō-fē).* Enlargement of an organ or part caused by an increase in the size of its cells.

adaptive h. Increase in size in response to greater functional needs, as when the heart enlarges when one or more valves dysfunctions.

benign prostatic h. *See* **benign prostatic hypertrophy.**

hyperuricemia *(hī-pĕr-ū-rĭ-sē'-mē-ă).* Excessive uric acid or urates in the blood. It can lead to gout and renal disease.

hyperventilation *(hī-pĕr-vĕn-tĭ-lā'-shŭn).* A state in which there is an increase in the amount of air that enters the lungs during inspiration, which results in reduced carbon dioxide tension and may eventually cause alkalosis.

hypervolemia *(hī-pĕr-vō-lē'-mē-ă).* An abnormal increase in the volume of circulating blood.

hyphema *(hī-fē'-mă).* Bleeding within the anterior chamber of the eye, giving the eye a "bloodshot" look.

hypnoanalysis *(hĭp-nō-ă-năl'-ĕ-sĭs).* A method of psychotherapy that combines psychoanalysis and hypnosis.

hypnophobia *(hĭp-nō-fō'-bē-ă).* Irrational fear of falling asleep.

hypnosis *(hĭp-nō'-sĭs).* A state of altered consciousness, usually induced artificially, in which individuals have an abnormal sensibility to suggestions.

hypnotherapy *(hĭp-nō-thĕr'-ă-pē).* Therapy that involves use of hypnosis.

hypnotic *(hĭp-nŏt'-ĭk).* 1. Referring to hypnosis or sleep. 2. A substance that induces sleep or that dulls the senses.

hypoadrenalism *(hī-pō-ăd-rē'-năl-ĭsm).* *See* **adrenal insufficiency.**

hypo-allergenic *(hī-pō-ăl-ĕr-jĕn'-ĭk).* Something that has a reduced potential to cause an allergic reaction.

hypobaric *(hī-pō-băr'-ĭk).* Decreased atmospheric pressure.

hypobarism *(hī-pō-băr'-ĭz-ĕm).* A condition resulting from exposure to ambient gas pressure or atmospheric pressures that are less than those within the body.

hypocalcemia *(hī-pō-kăl-sē'-mē-ă).* Blood calcium levels below normal.

hypocapnia *(hī-pō-kăp'-nē-ă).* Deficiency of carbon dioxide in the blood, caused by hyperventilation.

hypochlorhydria *(hī-pō-klŏr-hī'-drē-ă).* A deficiency of hydrochloric acid in stomach acid.

hypocholesterolemia *(hī-pō-kĕ-lĕs-tĕr-ŏl-ē'-mē-ă).* Abnormally low levels of cholesterol in the blood.

hypochondriac *(hī-pō-kŏn'-drē-ăk).* Individuals who have an abnormal and excessive interest in and fear of disease, especially those who are healthy.

hypochondriasis *(hī-pō-kŏn-drī'-ă-sĭs).* Abnormal and excessive preoccupation with one's health.

hypochromia *(hī-pō-krō'-mē-ă).* An abnormal decrease in the hemoglobin content in red blood cells.

hypodermic *(hī-pō-dĕr'-mĭk).* Administered or inserted under the skin, as a hypodermic injection.

hypofibrinogenemia *(hī-pō-fī-brĭn-ō-jĕ-nē'-mē-ă).* Abnormally low level of fibrinogen in the blood.

hypogammaglobulinemia *(hī-pō-găm-mă-glŏb-ū-lī-nē'-mē-ă).* Abnormally low levels of all immunoglobulins in the blood.

hypogenitalism *(hī-pō-jĕn'-ĭ-tăl-ĭz-ĕm). See* **hypogonadism.**

hypoglossal *(hī-pō-glŏs'-ĕl).* Under the tongue; sublingual.

hypoglycemia *(hī-pō-glī-sē'-mē-ă).* An abnormally reduced concentration of glucose in the blood.

hypognathous *(hī-pŏg-nĕ-thŭs).* Having a lower jaw that is smaller than the upper jaw.

hypogonadism *(hī-pō-gō'-năd-ĭz-ĕm).* A condition that results from abnormally decreased gonadal function.

hypokalemia *(hī-pō-kă-lē'-mē-ă).* Abnormally low concentration of potassium in the blood.

hyponatremia *(hī-pō-nă-trē'-mē-ă).* A deficiency of sodium in the blood.

hypoparathyroidism *(hī-pō-păr-ă-thī'-roid-ĭz-ĕm).* A condition resulting from significantly reduced function of the parathyroid gland. It may be caused by removal of the glands, genetic factors, or an autoimmune disease.

hypophysectomy *(hī-pŏf-ĕ-sĕk'-tĕ-mē).* Surgical removal or destruction of the pituitary gland.

hypophysis *(hī-pŏf'-ĕ-sĭs).* A term for the pituitary gland.

hypoplasia *(hī-pō-plā'-zhă).* Underdevelopment of tissue or an organ.

hypoproteinemia *(hī-pō-prō-tē-nē'-mē-ă).* Abnormally low levels of protein in the blood.

hyporeflexia *(hī-pō-rē-flĕks'-sē-ă).* Reduced function of reflexes.

hyposensitive *(hī-pō-sĕn'-sĭ-tĭv).* Having a reduced ability to respond to stimuli.

hyposmia *(hī-pŏz'-mē-ă).* A diminished sensitivity to smell.

hypotension *(hī-pō-tĕn'-shĕn).* Abnormally low blood pressure.

hypothalamus *(hī-pō-thăl'-ă-mŭs).* The area of the brain below the thalamus, it is where secretions involved in metabolism, regulation of body temperature, and secretion of hormones are controlled.

hypothenar *(hī-pŏth'-ĕ-năr).* The fleshy part on the inner side of the palm next to the little finger.

hypothermia *(hī-pō-thĕr'-mē-ă).* 1. Having a body temperature below normal. 2. Lowering the body temperature to between 78° and 90° F for therapeutic reasons.

hypothesis *(hī-pŏth'-ĕ-sĭs).* An assumption that has not been proven scientifically.

hypothrombinemia *(hī-pō-thrŏm-bĭ-nē'-mē-ă).* A deficiency of thrombin in the blood.

hypothyroidism *(hī-pō-thī'-royd-ĭz-ĕm).* A condition caused by a deficiency of thyroid hormone secretion and characterized by a lowered basal metabolism and fatigue.

hypotonia *(hī-pō-tō'-nē-ă).* A condition of diminished muscle tone or intraocular pressure; *see also* **floppy baby syndrome.**

hypotonic solution *(hī-pō-tŏn'-ĭk).* A solution that has lower osmotic pressure than another.

hypoventilation *(hī-pō-vĕn-tĭ-lā'-shŭn).* A reduction in the rate and depth of breathing.

hypovolemia *(hī-pō-vō-lē'-mē-ă).* An abnormally low volume of circulating blood in the body.

hypoxia *(hī-pŏks'-ē-ă).* A deficiency of oxygen supply to body tissues.

hysterectomy *(hĭs-tĕr-ĕk'-tĕ-mē).* Surgical procedure to remove the uterus.

abdominal h. Surgical removal of the uterus through an incision in the abdominal wall.

subtotal h. Hysterectomy in which the cervix is not removed.

total h. Hysterectomy in which the uterus and cervix are completely removed.

vaginal h. Surgical removal of the uterus through the vagina.

hysteria *(hĭs-tĕr'-ē-ă).* A neurotic disorder whose definition has become nebulous because of the variety of meanings assigned to it over the centuries. Generally it is used to describe an extremely emotional state, typically seen among neurotics.

hysterotomy *(hĭs-tĕr-ŏt'-ĕ-mē).* Surgical incision of the uterus, usually to help deliver a fetus.

hysterotrachelectasia *(hĭs-tĕr-ō-trā-kĕl-ĕk-tā'-zhā).* Surgical dilation of the cervix and uterus.

Hytrin *(hī'-trĭn).* Trade name for a preparation of terazosin hydrochloride, used to treat hypertension.

I

-iatric. Suffix indicating a relation to medical treatment.

iatrogenic *(ī-ăt-rō-jĕn'-ĭk).* Term used to describe any adverse condition that happens to a patient as the result of treatment by a physician or surgeon.

ibuprofen *(ī-bū-prō'-fĕn). See* Appendix, Common Prescription and OTC Drugs: By Generic Name.

ICCU. Acronym for intensive cardiac care unit.

ichnogram *(ĭk'-nō-grăm).* A footprint, in ink on paper.

ichthyosis *(ĭk-thē-ō'-sĭs).* Any one of many skin disorders characterized by noninflammatory scaling of the skin.

icteric *(ĭk-tĕr'-ĭk).* Referring to or affected with jaundice.

icterus *(ĭk'-tĕr-ŭs).* Jaundice.

gravis neonatorum i. Severe jaundice that occurs in newborns.

praecox i. Mild jaundice that develops within the first twenty-four hours of life, due to incompatibility of the ABO blood group between mother and child.

ictus *(ĭk'-tŭs).* A stroke, seizure, or sudden attack.

ICU. Acronym for intensive care unit.

id. In psychiatry, the obscure, completely unconscious, primitive aspect of the personality that harbors instinctual drives.

IDDM. Acronym for insulin-dependent diabetes mellitus.

idealization *(ī-dē-ĕl-ĭ-zā'-shŭn).* A conscious or unconscious process by which an individual attributes excessive value to another person or object.

ideation *(ī-dē-ā'-shŭn).* The formation of a mental image, thought, or concept.

ideology *(ī-dē-ŏl'-ō-jē).* The science of the development of thoughts or ideas.

idiopathic *(ĭd-ē-ō-păth'-ĭk).* Having an unknown cause or spontaneous origin.

idiot savant *(ēd-jō să-vănt').* "Learned idiot," an individual who is severely mentally retarded but has a specific mental ability that is highly developed, such as math or music skills.

ignatia *(ĭg-nā'-shē-ă).* The dried ripe seed of *Strychnos ignatii* used in homeopathic remedies.

ileal pouch *(ĭl'-ē-ĕl).* A surgically created sac composed of a portion of the lower part of the small intestine. It is usually done for patients who have had a total colectomy, such as those with ulcerative colitis.

ileectomy *(ĭl-ē-ĕk'-tĕ-mē).* Surgical removal of the ileum.

ileitis *(ĭl-ē-ī'-tĭs).* Inflammation of the ileum.

ileocecal *(ĭl-ē-ō-sē'-kăl).* Pertaining to the ileum and the cecum.

ileocecal valve. Sphincter muscles that close the ileum where the small intestines open into the ascending colon, which prevents food from reentering the small intestines.

ileocolic *(ĭl-ē-ō-kŏl'-ĭk).* Pertaining to the ileum and the colon.

ileocolitis *(ĭl-ē-ō-kō-lī'-tĭs).* Inflammation of the ileum and the colon.

ileoproctostomy *(ĭl-ē-ō-prŏk-tŏs'-tō-mē).* Surgical procedure that establishes an opening between the ileum and the rectum.

ileosigmoidostomy *(ĭl-ē-ō-sĭg-moyd-ŏs'-tĕ-mē).* Surgical procedure that establishes an opening between the ileum and the sigmoid flexure of the colon.

ileostomy *(ĭl-ē-ŏs'-tĕ-mē).* Surgical procedure to create a passage through the abdominal wall into the ileum.

ileotransversostomy *(ĭl-ē-ō-trăns-vĕrs-ŏs'-tĕ-mē).* Surgical procedure to create an opening between the ileum and the transverse colon.

ileum *(ĭl'-ē-ŭm).* The lower three-fifths of the small intestines.

ileus *(ĭl'-ē-ŭs).* An obstruction of the intestinal tract.

ilium *(ĭl'-ē-ŭm).* The iliac bone; one of the bones on each half of the pelvis. Early in life it is a separate bone, until it fuses with adjacent pelvic bones.

illusion *(ĭl-ū'-shŭn).* A misinterpreted or false sensory image or impression.

image *(ĭm'-ĭj).* A picture or conception of an object or person.

body i. A three-dimensional concept an individual has of his or her physical self.

incidental i. The impression of an image that remains on the retina after the object is no longer there.

imaging *(ĭm'-ĭj-ĭng).* Production of a picture, shadow, or image that represents an object being studied. Imaging techniques include x-ray, ultrasound, and magnetic resonance imaging, among others.

imbalance *(ĭm-băl'-ăns).* An inability to stand upright or keep one's balance.

imbecile *(ĭm'-bĕ-sĭl).* Obsolete and offensive term for an individual who has moderate to severe mental retardation.

imbrication *(ĭm-brĭ-kā'-shŭn).* The overlapping of apposing surfaces.

Imferon *(ĭm-fĕr'-ŏn).* Trade name for iron dextran.

imipramine *(ĭ-mĭp'-rĕ-mēn).* Generic name of a tricyclic antidepressant drug; it was the first tricyclic antidepressant released to the market.

Imitrex *(ĭm'-ĕ-trĕks).* *See* Appendix, Common Prescription and OTC Drugs: By Trade Name.

immiscible *(ĭ-mĭs'-ĭ-bĕl).* Not susceptible to being mixed, like oil and water.

immobilize *(ĭ-mō'-bĭl-īz).* To render incapable of being moved.

immune *(ĭm-ūn').* Resistant to or protected from infectious disease due to the development of antibodies.

immune response. The reaction of the body to substances that are foreign or that it interprets as being foreign.

immune system. A complex system of components that produces antibodies and protects the body against harmful organisms and substances.

immunity *(ĭ-mū'-nĭ-tē).* State of

being protected from a disease, especially an infectious disease.

acquired i. Immunity that results from having the disease or by receiving a vaccination (active immunity) or transferred from mother to fetus in utero or through breast milk (passive immunity).

active i. *See* **acquired immunity,** above.

cell-mediated i. Immunity that results from the interaction of an antigen and T lymphocytes.

innate i. Immunity based on the genetic makeup of the individual; for example, humans have an innate immunity to canine distemper.

natural i. Immunity to disease that individuals are born with, which may be the result of race, the natural presence of immune bodies, or other factors.

passive i. *See* **acquired immunity,** above.

immunization *(ĭm-ū-nĭ-zā'-shŭn).* Becoming immune, or the process of making an individual immune.

immunoassay *(ĭm-ū-nō-ăs'-ā).* Any one of several tests used to measure the protein molecules involved in the reaction of an antigen with its antibody.

immunochemotherapy *(ĭm-ū-nō-kē-mō-thĕr'-ĕ-pē).* A combination of immunotherapy and chemotherapy.

immunocompromised *(ĭm-ū-nō-kŏm'-prĕ-mīzd).* State in which the immune response has been reduced, which may be the result of the use of immunosuppressive drugs, irradiation, malnutrition, or a disease.

immunodeficiency *(ĭm-ū-nō-dē-fĭsh'-ĕn-sē).* A reduced or compromised ability of the body to respond effectively (e.g., by producing antibodies) to assaults by antigens.

immunoglobulin *(ĭm-ū-nō-glŏb'-ū-lĭn).* One of a family of closely related proteins that can act as antibodies. There are five classes of immunoglobulins: IgM, IgG, IgA, IgD, and IgE.

IgA. The immunoglobulin that protects mucosal surfaces from invading microorganisms. It is found in respiratory and intestinal mucin, milk, saliva, and tears.

IgD. The immunoglobulin found on the surface of B cells. Little is known about its function.

IgE. This immunoglobulin is produced by cells of the lining of the intestinal and respiratory tracts and plays an important role in forming antibodies.

IgG. The main immunoglobulin in human serum. It moves through the placenta and produces immunity in the fetus.

IgM. The immunoglobulin produced in nearly all immune responses during the early stages of the reaction.

immunologist *(ĭm-ū-nŏl'-ō-jĭst).* An individual who studies and/or works in the field of immunology.

immunology *(ĭm-ū-nŏl'-ō-jē).* The branch of biomedical science that studies the structure and function of the immune system, immunization, and other aspects of immunity.

immunopathology *(ĭm-ū-nō-păth-ŏl'-ō-jē).* The branch of biomedical science that studies immune responses to disease, immunodeficiency diseases, and disease with an immunological origin or process.

immunosuppression *(ĭm-ū-nō-sū-prĕsh'-ĕn).* Prevention or reduction of an immune response by irradiation or administration of specific substances.

immunotherapy *(ĭm-ū-nō-thĕr'-ă-pē).* A general term for therapies that enhance immunity.

immunotransfusion *(ĭm-ū-nō-trăns-fū'-shŭn).* Transfusion of blood from donors who have been immunized previously by the organism or specific infection that has infected the receiving patient.

Imodium *(ĭ-mō'-dē-ĕm).* Trade name for preparations of loperamide hydrochloride, used to treat diarrheal diseases.

impacted *(ĭm-păk'-tĕd).* Firmly or closely lodged in position, as an impacted tooth.

impalpable *(ĭm-păl'-pĕ-bĕl).* Impossible to be detected by touch.

imperforate *(ĭm-pĕr'-fĕ-rāt).* Abnormally closed; having no opening.

impermeable *(ĭm-pĕr'-mē-ĕ-bĕl).* Not allowing passage, as of fluid.

impetigo *(ĭm-pĕ-tī'-gō).* A contagious, inflammatory skin disease characterized by crusty pustules that break and discharge fluid that dries and forms a crust. Usually caused by staphylococcus or streptococci.

implant *(ĭm'-plănt).* An object or material that is partially or totally inserted or grafted into the body for therapeutic, prosthetic, diagnostic, or experimental reasons.

impotence *(ĭm'-pĕ-tĕns).* The inability of the male to achieve or maintain an erection.

impregnate *(ĭm-prĕg'-nāt).* To make pregnant; to fertilize an egg (ovum).

impression *(ĭm-prĕsh'-ĕn).* 1. A slight indentation. 2. An effect on the body, mind, or senses by an external force.

dental i. An imprint of the teeth and/or area surrounding the teeth, made in a soft plastic material that hardens and is later filled with artificial stone or plaster of Paris.

impulse *(ĭm'-pŭls).* Act of pushing or driving onward with sudden force.

impulsion *(ĭm-pŭl'-shĕn).* Unquestioned obedience to internal urges without regard for what effect it has on others or pressure from the superego.

Imuran *(ĭm'-ū-răn).* Trade name for preparations of azathioprine, an immunosuppressive used to prevent transplant rejection in organ transplantation.

inanimate *(ĭn-ăn'-ĭ-măt).* Without life or animation.

inanition *(ĭn-ĕ-nĭsh'-ĕn).* Extreme weight loss, decreased metabolism, and significant weakness caused by prolonged starvation.

inarticulate *(ĭn-ăhr-tĭk'-ū-lĕt).* 1. Without joints. 2. An inability to pronounce distinct syllables or express oneself intelligibly.

inassimilable *(ĭn-ă-sĭm'-ĭ-lă-bl).* Unable to be utilized as a nutrient.

inborn *(ĭn'-bŏrn).* Congenital; genetically determined and present at birth.

inborn error of metabolism. An inheritable biochemical disorder, such as albinism, cystinuria, sensitivity to sunlight, thyroid disease, and phenylketonuria, among hundreds of others. In infants inborn errors often cause symptoms such as lethargy, fast breathing, recurrent vomiting, and poor feeding.

incest *(ĭn'-sĕst).* Sexual intercourse or other sexual behavior between blood relatives or individuals so closely related that marriage between them would be legally or culturally prohibited.

incidence *(ĭn'-sĭ-dĕns).* The rate or frequency at which an event or conditions occurs; for example, the number of new cases of a specific disease that occurs during a given time period and certain population.

incipient *(ĭn-sĭp'-ē-ĕnt).* Coming into existence.

incision *(ĭn-sĭzh'-ĕn).* A cut made with a knife, especially during surgery.

incisor *(ĭn-sī'-zŏr).* In dentistry, either of the two most frontal teeth in each jaw in adults; also referred to as the cutting teeth.

inclination *(ĭn-klĭ-nā'-shŭn).* Deviation or leaning away from the normal position.

inclusion *(ĭn-klū'-zhĕn).* The act of enclosing something.

dental i. A tooth that is surrounded with so much bony material it cannot erupt through the gums.

incoherent *(ĭn-kō-hĕr'-ĕnt).* 1. Without proper sequence. 2. Not understandable.

incompatible *(ĭn-kŏm-păt'-ĭ-bĕl).* State of not being suitable for mixing or combining with something or someone else.

incompetent *(ĭn-kŏm'-pĕ-tĕnt).* An individual who is legally unable to execute a contract.

incontinence *(ĭn-kŏn'-tĭ-nĕns).* An inability to retain urine, feces, or semen due to a loss of sphincter control or because of cerebral or spinal lesions.

fecal i. Involuntary passage of feces due to failure of control of the anal sphincters.

stress i. Involuntary release of urine under stressful conditions, frequently seen among women of middle age or older.

incoordination *(ĭn-kō-ŏr-dĭ-nā'-shŭn).* An inability to produce coordinated, rhythmic muscular movement that is not due to weakness.

incrustation *(ĭn-krĕs-tā'-shŭn).* The formation of scabs or crust.

incubation *(ĭn-kū-bā'-shŭn).* 1. The time between exposure to an infectious organism and the appearance of the first symptoms of the disease or illness. 2. Maintenance of a controlled environment (temperature, humidity, oxygen) for an infant, usually one that is premature.

incubator *(ĭn'-kū-bā-tŏr).* An apparatus that maintains optimal temperature, humidity, and oxygen level to maintain a premature infant.

incus *(ĭng'-kĕs).* The middle of the three small bones of the ear; along with the stapes and malleus, it helps to conduct vibrations from the tympanic membrane to the inner ear.

indentation *(ĭn-dĕn-tā'-shŭn).* A pit, depression, or notch.

Inderal *(ĭn'-dĕr-ĕl).* Trade name for preparations of propranolol hydrochloride, used to treat high blood pressure and heart conditions.

indication *(ĭn-dĭ-kā'-shŭn).* A sign or situation that indicates the cause, treatment, pathology, or issue of an attack of a disease.

indicator *(ĭn'-dĭ-kā-tĕr).* 1. The index finger. 2. In chemical analysis, a substance that is used to determine the pH level.

indigenous *(ĭn-dĭj'-ĕ-nĕs).* Something that is native to a country or region.

indigestion *(ĭn-dĭ-jĕs'-chĕn).* Incomplete digestion of food, usually accompanied by one or more symptoms, such as pain, nausea, vomiting, gas, belching, acid regurgitation, and heartburn.

indinavir *(ĭn-dĭn'-ĕ-vĭr).* Generic name for a drug used to treat human immunodeficiency virus infection and acquired immunodeficiency syndrome.

Indocin *(ĭn'-dō-sĭn).* Trade name for preparations of indomethacin, used to treat inflammatory conditions such as rheumatoid arthritis and osteoarthritis.

indole *(ĭn'-dōl).* A byproduct of the decomposition of tryptophan, it is found in feces and is largely responsible for their odor.

indolent *(ĭn'-dō-lĕnt).* 1. Causing little pain. 2. Slow growing.

induction *(ĭn-dŭk'-shŭn).* The process of causing something to occur.

Indur *(ĭn'-dŭr).* *See* Appendix, Common Prescription and OTC Drugs: By Trade Name.

induration *(ĭn-dūr-ă'-shŭn).* The quality of being hard.

indwelling catheter. *See* **catheter.**

inebriation *(ĭn-ē-brē-ā'-shŭn).* Intoxication with alcohol.

inertia *(ĭn-ĕr'-shă).* The tendency for a body to remain motionless until an outside force acts upon it.

in extremis *(ĭn ĕks-trē'-mĭs).* At the point of death.

infant *(ĭn'-fănt).* A child from birth to 12 months of age.

immature i. Infant that usually weighs less than 2,500

grams (5.5 lbs) at birth and is not physiologically well developed.

low birth weight i. An infant that weighs less than 2,500 grams (5.5 lbs) at birth.

premature i. Infant usually born after the twentieth completed week and before full term, arbitrarily defined as weighing 500 to 2,499 grams (2.2 to 5.5 lbs).

infanticide *(ĭn-făn'-tĭ-sīd).* Taking the life of an infant.

infantilism *(ĭn'-fĕn-tĭ-lĭz-ĕm).* Persistence of childhood traits into adulthood. The condition is characterized by mental retardation and underdeveloped sexual organs; dwarfism is also sometimes present.

infarct *(ĭn'-făhrkt).* An area of tissue in the body in which the cells die due to cessation of the blood supply, which is usually caused by a blood clot (a thrombus or embolus).

infarction *(ĭn-făhrkt'-shŭn).* Formation of an infarct; the infarct itself.

acute myocardial i. Infarction that occurs when circulation to a region of the heart is obstructed and cell death occurs. Usually characterized by pain, dizziness, nausea, paleness, and perspiration.

cerebral i. A condition in which blood flow to a portion of the brain has been obstructed, resulting in cell death.

myocardial i. Gross cell death of the myocardium due to cessation of the blood supply to the area. It is usually caused by atherosclerosis of the coronary arteries.

infection *(ĭn-fĕk'-shĕn).* A condition in which a part or the entire body is invaded by disease-causing agents that, if conditions are favorable, multiply and cause injury to the cells.

airborne i. An infection that is contracted by inhaling microorganisms that are in the air on dust particles or water droplets, such as occurs with a sneeze or cough.

mixed i. Infection of an organ or tissue by more than one microorganism. This occurs most frequently in wound infections, pneumonia, and abscesses.

opportunistic i. Infection by an organism that does not usually cause disease but does so under certain conditions, especially a compromised immune system.

secondary i. Infection by a microorganism following an infection by another type of microorganism.

subclinical i. Infection in which signs and symptoms are not apparent either on examination or laboratory tests. May occur during the very early stages of an infection.

infectious disease *(ĭn-fĕk'-shĕs).* A disease capable of being transmitted or communicated, with or without contact, to other organisms.

infectious hepatitis. *See* **hepatitis.**

inferiority complex. In psychology, a state of mind in which an individual feels inferior to others.

infertility *(ĭn-fĕr-tĭl'-ĭ-tē).* A diminished or absent capacity in either a male or female to produce offspring.

primary i. Infertility in an individual who has never conceived.

secondary i. Infertility in an individual who has conceived previously.

infestation *(ĭn-fĕs-tā'-shŭn).* An attack by parasites, such as mites, insects, or ticks, on the skin.

infiltration *(ĭn-fĭl-trā'-shŭn).* The accumulation in cells or tissues of substances not normally found there, or in abnormally excessive amounts.

infirmity *(ĭn-fĭr'-mĭ-tē).* A weakened or feeble state of the mind or body.

inflammation *(ĭn-flă-mā'-shŭn).* A localized response by tissue to an injury or destruction in which various body processes work to destroy, reduce, or isolate both the injured tissue and the substance that caused the injury so healing can proceed.

acute i. Inflammation that comes on quickly and usually lasts a short time.

chronic i. Inflammation that progresses slowly, lasts a long time, and is usually accompanied by the formation of scar tissue.

suppurative i. Inflammation accompanied by the formation of pus.

influenza *(ĭn-flū-ĕn'-ză).* An acute, contagious, viral infection of the respiratory tract, characterized by fever, chills, headache, muscle ache, and inflammation of the nasal cavity.

informed consent. Voluntary, competent permission given by an individual for a medical procedure or test to be performed, or for medications to be administered. Such consent is based on the individual understanding the risks, benefits, and options surrounding the procedure, test, or medication.

infraclavicular *(ĭn-fră-klă-vĭk'-ū-lăr).* The area below the clavicle.

infracostal *(ĭn-fră-kōs'-tăl).* The area below the rib.

infrared rays *(ĭn-fră-rĕd').* Invisible heat rays that are beyond the red end of the spectrum. They are the basis of thermography.

Infusaid pump *(ĭn-fū'-săd).* A device that provides continuous administration of chemotherapy.

infusion *(ĭn-fū'-zhĕn).* 1. Steeping a substance in water until it reaches medicinal potency. 2. Introduction of a fluid other than blood into a vein for therapeutic purposes.

saline i. Administration of a saline (salt) solution, either intravenously or subcutaneously (under the skin).

ingest *(ĭn-jĕst').* To eat.

ingrown toenail *(ĭn'-grōwn).* A nail

that has grown into the nailbed, usually causing a painful infection.

inguinal *(ĭng'-gwĭ-nĕl).* Referring to the groin or the groin area.

inhalant *(ĭn-hă'-lĕnt).* Something that is meant to be inhaled.

inhalation *(ĭn-hă-lā'-shĕn).* 1. The intake of air into the lungs; also called *inspiration.* 2. The intake of an aerosolized drug into the lungs with the breath.

inhaler *(ĭn-hăl'-ĕr).* A device used to administer vapor or processed medications by inhalation.

dry powder i. An inhaler, usually breath activated, that dispenses dry powdered medication.

metered dose i. An inhaler that delivers aerosolized medications in fixed doses to patients who have respiratory disease.

inherent *(ĭn-hĕr'-ĕnt).* Innate; belonging to anything naturally and existing as a basic characteristic of something.

inherited *(ĭn-hĕr'-ĭ-tĕd).* Body characteristics and genetic makeup received through heredity (genes).

inhibition *(ĭn-hĭ-bĭsh'-ĕn).* Repression or restraint of a process or function.

inhibitor *(ĭn-hĭb'-ĭ-tŏr).* Any substance that interferes with biological activity; for example, a chemical that stops the actions of an enzyme.

injectable *(ĭn-jĕk'-tă-bĕl).* Any substance that is capable of being injected.

injection *(ĭn-jĕk'-shĕn).* The act of forcing a liquid into a vessel, cavity, or under the skin.

intramuscular i. An injection made into the substance of a muscle.

intravenous i. An injection made into a vein.

Ringer's i. A sterile solution of sodium chloride, potassium chloride, and calcium chloride in water, used as a vehicle for injecting medications, a fluid and electrolyte replenisher, or to wash the skin or wounds.

subcutaneous i. An injection made beneath the skin.

inlay *(ĭn'-lā).* 1. Material, such as skin or bone, that is inserted into a tissue defect. 2. A solid filling made to fit a cavity of a tooth and then cemented into the tooth.

innervation *(ĭn-ĕr-vā'-shĕn).* The distribution or supply of nerves to a body part.

innocuous *(ĭ-nŏk'-ū-ĕs).* Harmless.

inoculation *(ĭ-nŏk-ū-lā'-shĕn).* 1. The process of introducing a substance such as microorganisms or serum into tissues of plants and animals or into culture media. 2. The process of introducing a disease agent (e.g., a vaccine virus) into a healthy individual to produce immunity.

inoperable *(ĭn-ŏp'-ĕr-ĕ-bĕl).* Not suitable to undergo a surgical procedure.

inorganic *(ĭn-ŏr-găn'-ĭk).* Not of organic origin.

inositol *(ĭn-ŏs'-ĭ-tŏl).* A sugarlike substance found in the body and in most plants; it is also part of the vitamin B family.

inotropic *(ĭn-ō-trō'-pĭk).* Affecting the energy or force of muscle contraction. An inotropic drug affects the force with which a muscle contracts.

inpatient *(ĭn'-pā-shĕnt).* A patient who goes to a hospital or other health-care facility for medical services that require an overnight stay.

inquest *(ĭn'-kwĕst).* A legal inquiry into the cause and circumstances surrounding a sudden, unexplained, or violent death.

insanity *(ĭn-săn'-ĭ-tē).* A legal term used to describe a mental condition due to which a person lacks criminal responsibility for a crime and thus cannot be convicted of it for "reasons of insanity."

insemination *(ĭn-sĕm-ĭ-nā'-shĕn).* The deposit of semen or seminal fluid into the vagina or cervix.

artificial i. Introduction of semen into the vagina or cervix by artificial means.

donor i. Artificial insemination in which the donated semen used is from a man other than the recipient's husband. Also called *artificial insemination by donor.*

homologous i. Artificial insemination in which the donated semen is from the recipient's husband. Also called *artificial insemination by husband.*

insensible *(ĭn-sĕn'-sĭ-bĕl).* Not perceptible to the senses.

insidious *(ĭn-sĭd'-ē-ŭs).* Something that occurs or develops in a gradual or subtle manner.

in situ *(ĭn sī'-tū).* Confined to the original site without interference or invasion from adjacent tissues.

insoluble *(ĭn-sŏl'-ū-bĕl).* Incapable of being dissolved.

insomnia *(ĭn-sŏm'-nē-ă).* Inadequate or poor-quality sleep due to one or more of the following reasons: difficulty falling asleep, waking up often with difficulty returning to sleep, waking up too early in the morning, or unrefreshing sleep.

chronic i. Insomnia that occurs on most nights and lasts a month or longer.

transient i. Insomnia that occurs from time to time.

inspiration *(ĭn-spĭ-rā'-shŭn).* Inhalation.

inspiratory capacity. The volume of gas the lungs can hold when a full breath (full inhalation) is taken.

inspissated *(ĭn-spĭs'-āt-ĕd).* To become thickened or less fluid due to evaporation, absorption, or dehydration.

instability *(ĭn-stă-bĭl'-ĭ-tē).* A lack of steadiness.

instep *(ĭn'-stĕp).* The arched midportion of the bottom of the foot.

instillation *(ĭn-stĭ-lā'-shŭn).* Administration of a liquid drop by drop.

instinct *(ĭn'-stĭnkt).* The inherited, unlearned responses to certain environmental stimuli and conditions that are characteristic of a species.

insufficiency *(ĭn-sŭ-fĭsh'-ĕn-sē).* The state of being inadequate for a given situation or purpose.

adrenal i. Reduced or abnormally low production of hormones by the adrenal glands.

cardiac i. An inability of the heart to function normally.

gastric i. An inability of the stomach to eliminate its contents.

hepatic i. An inability of the liver to function properly.

renal i. An inability of the kidney to eliminate waste materials from the blood at its normal rate.

thyroid i. *See* **hypothyroidism.**

venous i. Failure of the valves of the veins to function, which interferes with the return of blood to the heart.

insufflation *(ĭn-sū-flā'-shŭn).* The process of blowing a powder, gas, air, or vapor into a body cavity.

perirenal i. The injection of air around the kidneys to facilitate radiographic visualization of the adrenal glands.

insulin *(ĭn'-sū-lĭn).* A hormone secreted by the beta cells of the pancreas; it is necessary for the proper metabolism of glucose (blood sugar).

insulinoma *(ĭn-sĕ-lĭn-ō'-mă).* An islet cell tumor of the beta cells of the pancreas, it is usually benign and can cause the body to produce excess insulin.

insulin pump. A pump that delivers insulin into the body through a thin plastic tube and soft needle (cannula), usually inserted in the abdomen. The pump helps maintain blood sugar control and allows users more lifestyle flexibility.

insulin shock. A condition that results from taking an overdose of insulin, which causes a decline in blood sugar levels to below normal (hypoglycemia).

insusceptibility *(ĭn-sĕ-sĕp-tĭ-bĭl'-ĭ-tē).* Immunity; the quality of not being susceptible.

integration *(ĭn-tĕ-grā'-shŭn).* Bringing together different functions or parts so they can cooperate toward a common end.

integument *(ĭn-tĕg'-ū-mĕnt).* A covering.

intelligence *(ĭn-tĕl'-ĭ-jĕns).* The capacity to comprehend or understand relationships, to solve problems, and to adjust to new situations.

intelligence quotient. An index of relative intelligence, determined by how an individual answers certain questions, and then the score is compared with those of others within the same age group.

intemperance *(ĭn-tĕm'-pĕr-ĕns).* Lack of self-control with respect to consumption of food and drink, especially alcoholic beverages.

intensity *(ĭn-tĕn'-sĭ-tē).* The extent or degree of force, strength, activity, or power.

intensive care unit. A separate

unit in a hospital where extremely sick patients, some of whom may require life-support equipment, are cared for.

intensivist *(ĭn-tĕn'-sĭ-vĭst).* A physician who specializes in providing care for patients in an intensive care unit.

interaction *(ĭn-tĕr-ăk'-shŭn).* The process of two or more things acting on each other.

interarticular *(ĭn-tĕr-ăr-tĭk'-ū-lăr).* Situated between joints.

intercellular *(ĭn-tĕr-sĕl'-ū-lăr).* Positioned between the cells of a structure.

intercostal *(ĭn-tĕr-kŏs'-tăl).* Situated between the ribs.

intercourse *(ĭn'-tĕr-kŏrs).* Mutual exchange between people or groups.

sexual i. Coitus; any physical contact between two persons that involves genital stimulation of at least one of the individuals.

interdigitate *(ĭn-tĕr-dĭj'-ĭ-tāt).* To interlock, as the fingers of clasped hands.

interdiscipline *(ĭn-tĕr-dĭs'-ĭ-plĭn).* The overlapping of various branches of medicine or health care.

interferon *(ĭn-tĕr-fĕr'-ŏn).* A type of protein that may form when cells are exposed to viruses, parasites, bacteria, or protozoa.

interleukin *(ĭn-tĕr-lū'-kĭn).* A general term for a group of cytokines capable of performing many functions and that have an effect on the lymphatic system.

interleukin-1. An interleukin produced by various cells, it raises body temperature, stimulates production of interferon, promotes growth of disease-fighting cells, among other functions.

interleukin-2. An interleukin that is produced by T cells; it stimulates the growth of T lymphocytes and is used as an anticancer drug in the treatment of various solid malignant tumors.

intermittent *(ĭn-tĕr-mĭt'-ĕnt).* Activity that occurs at separated intervals.

intermural *(ĭn-tĕr-mū'-rĕl).* Positioned between the walls or organs.

intermuscular *(ĭn-tĕr-mŭs'-kū-lăr).* Situated between muscles.

intern *(ĭn'-tĕrn).* A graduate of a medical school who is training in a hospital prior to being eligible to be licensed to practice medicine or dentistry.

internal *(ĭn-tĕr'-năl).* Within the body.

internal medicine. A branch of medicine that treats diseases of the internal organs using methods other than surgery.

international unit (IU). An internationally accepted amount of a substance; it is usually used to indicate the quantity of fat-soluble vitamins (e.g., vitamin E), and some hormones, enzymes, and other biological substances.

internist *(ĭn-tĕr'-nĭst).* A physician who specializes in internal medicine.

intersex *(ĭn'-tĕr-sĕks).* An individual who has both male and female characteristics; *see also* **hermaphrodite.**

interstice *(ĭn-tĕr'-stĭs).* A small gap or space in a tissue or structure.

interstitial *(ĭn-tĕr-stĭsh'-ĕl).* Referring to or situated between parts or in the interstices of a tissue.

intertrigo *(ĭn-tĕr-trī'-gō).* A superficial skin inflammation that occurs on apposed skin surfaces, such as the creases of the neck, between toes, and beneath heavy breasts.

intervention *(ĭn-tĕr-vĕn'-shĕn).* Taking action for the purpose of modifying an effect; specifically, any action done to improve health or change the course of a disease.

interventricular *(ĭn-tĕr-vĕn-trĭk'-ū-lĕr).* Positioned between ventricles.

intestinal *(ĭn-tĕs'-tĭ-năl).* Referring to the intestines.

intestinal bypass surgery. Procedure in which part of the small intestine is removed as a means to treat morbid obesity.

intestinal flora. The bacteria, both favorable and unfavorable, present in the intestinal tract.

intestinal obstruction. Blockage of the lumen of the intestines.

intestine *(ĭn-tĕs'-tĭn).* The portion of the alimentary canal that extends from the pyloric opening of the stomach to the anus.
large i. The portion of the intestines that receives material from the small intestine and then completes the digestive process and eliminates stool from the body. The large intestine has four parts: cecum, appendix, colon, and rectum.
small i. The portion of the intestines that receives digested material from the stomach, adds digestive enzymes and bile to it, and moves it along to the large intestine. The small intestine has three parts: the duodenum, jejunum, and ileum.

intima *(ĭn'-tĭ-mă).* The innermost lining of a structure, as a blood vessel.

intolerance *(ĭn-tŏl'-ĕr-ăns).* An inability to withstand or endure something, or an incapacity to bear pain or the effects of a drug or other substance.
drug i. A state in which an individual reacts to the normal pharmacologic doses of a drug with symptoms of overdose.
exercise i. A limited ability to do exercise or work at generally accepted levels.
lactose i. An inability to digest lactose, a sugar component of milk and most other dairy foods, due to a lack of an enzyme—lactase—necessary for such digestion to occur.

intoxication *(ĭn-tŏks-ĭ-kā'-shŭn).* State of being poisoned by a substance; often refers to alcohol intoxication.

intra-. Prefix meaning inside or within.

intra-abdominal *(ĭn-tră-ăb-dŏm'-ĭ-năl).* Within the abdomen.

intra-arterial *(ĭn-tră-ăhr-tĕr'-ē-ăl).* Within one or more arteries; also called *endarterial.*

intra-atrial *(ĭn-tră-ā'-trē-ăl).* Within one or both atria of the heart.

intracapsular *(ĭn-tră-kăp'-sū-lăr).* Within a capsule.

intracardiac catheter *(ĭn-tră-kăr'-dē-ăk).* A catheter that is placed within the heart.

intracellular *(ĭn-tră-sĕl'-ū-lăr).* Within a cell.

intracranial *(ĭn-tră-krā'-nē-ăl).* Within the cranium.

intractable *(ĭn-trăk'-tĕ-bĕl).* Resistant to therapy, control, cure, or relief.

intracutaneous *(ĭn-tră-kū-tā'-nē-ŭs).* Within the substance of the skin. An intracutaneous injection is placed into the skin.

intraductal papilloma. *See* **papilloma.**

intrahepatic *(ĭn-tră-hĕ-păt'-ĭk).* Within the liver.

intraluminal *(ĭn-tră-lū'-mĭ-năl).* Within the lumen of a tube, such as a blood vessel.

intramuscular *(ĭn-tră-mŭs'-kū-lăr).* Within the muscle.

intranasal *(ĭn-tră-nā'-săl).* Within the nasal cavity.

intraocular *(ĭn-tră-ŏk'-ū-lăr).* Within the eyeball.

intraocular implant. A plastic lens inserted into the eye to replace the natural lens, usually removed due to a cataract.

intraocular pressure (IOP). Pressure created within the eyeball by the continual renewal of fluids. IOP is increased in conditions such as glaucoma.

intraoperative *(ĭn-tră-ŏp'-ĕr-ă-tĭv).* The period during a surgical procedure.

intraoral *(ĭn-tră-ŏr'-ăl).* Within the mouth.

intraorbital *(ĭn-tră-ŏr'-bĭ-tăl).* Within the eye.

intraperitoneal *(ĭn-tră-pĕr-ĭ-tō-nē'-ăl).* Within the peritoneal cavity.

intraspinal *(ĭn-tră-spī'-năl).* Occurring or positioned within the spinal column.

intrauterine *(ĭn-tră-ū'-tĕr-ĭn).* Within the uterus.

intravenous *(ĭn-tră-vē'-nŭs).* Within or into a vein.

intravenous feeding. Providing complete nutritional needs intravenously, usually through a major vein, such as the jugular or subclavian.

intravenous pyelography (IVP). *See* **pyelography.**

intraventricular *(ĭn-tră-vĕn-trĭk'-ū-lăr).* Within a ventricle.

intrinsic *(ĭn-trĭn'-sĭk).* 1. Something that is essential and natural. 2. Situated entirely within or referring solely to a part.

intrinsic factor. A substance present in the gastric juice of humans that facilitates the absorption of vitamin B12.

introitus *(ĭn-trō'-ĭ-tŭs).* An opening or entryway into a body cavity or canal.

introjection *(ĭn-trō-jĕk'-shŭn).* In psychoanalysis, an unconscious defense mechanism in which a person identifies him- or herself with another person and assumes the feelings of the other personality.

intromission *(ĭn-trō-mĭsh'-ŭn).* To

place or insert one part into another; usually refers to insertion of the penis into the vagina.

introspection *(ĭn-trō-spĕk'-shŭn).* Contemplation of one's own mind, thoughts, and feelings.

introvert *(ĭn'-trō-vĕrt).* An individual who turns his or her interests inward.

intubation *(ĭn-tū-bā'-shŭn).* Insertion of a tube into a body cavity or canal.

endotracheal i. Insertion of a tube into the trachea to assist breathing.

nasal i. Insertion of a tube through the nose.

oral i. Insertion of a tube through the mouth.

intumescence *(ĭn-tū-mĕs'-ĕns).* A swelling, either normal or abnormal.

intussusception *(ĭn-tŭ-sŭ-sĕp'-shĕn).* The slipping of one part of the intestine into another part adjoining it. This occurs most often in children.

inunction *(ĭn-ŭngk'-shŭn).* A medicated substance, such as an ointment, rubbed into the skin.

in utero *(ĭn ū'-tĕr-ō).* Inside the uterus.

invaginate *(ĭn-văg'-ĭ-nāt).* To ensheath or to insert one part of a structure within a part of the same structure.

invalid *(ĭn'-vă-lĭd).* An individual who is disabled by illness or disease, especially one who is confined to bed or a wheelchair.

invasive *(ĭn-vā'-sĭv).* Something that has the tendency to spread, especially when referring to a malignancy.

invasive procedure. Procedure that involves puncturing or making an incision of the skin or inserting an instrument or foreign material into the body.

inversion *(ĭn-vĕr'-shŭn).* Reversal of a normal relationship, such as turning something inside out, upside down, or turned inward.

invertebrate *(ĭn-vĕr'-tĕ-brāt).* Any animal that does not have a spinal column.

investment *(ĭn-vĕst'-mĕnt).* Any tissue that covers other tissues or body parts.

in vitro *(ĭn vē'-trō).* Occurring within a glass, such as a test tube, or in an artificial environment.

in vitro fertilization. Fertilization that takes place within a test tube rather than inside the human body.

in vivo *(ĭn vē'-vō).* Occurring within a living body or organism.

involuntary *(ĭn-vŏl'-ĕn-tār-ē).* Something that is performed independently of the will.

involution *(ĭn-vō-lū'-shŭn).* Rolling or turning inward.

iodide *(ī'-ō-dīd).* A compound of iodine that contains another element.

iodine *(ī'-ō-dīn).* A nonmetallic element, it is essential in nutrition, especially for the synthesis of thyroid hormones, thyroxine, and triiodothyronine.

ion *(ī'-ŏn).* An atom or radical that has a positive or negative charge. Substances that form ions are called *electrolytes.*

ionization *(ī-ŏn-ĭ-zā'-shŭn).* A process during which neutral atoms or radicals become electrically charged ions.

ipecac *(ĭp'-ĕ-kăk).* The dried root and rhizome of *Cephaelis ipecacuanba* used in syrup to induce vomiting, especially in cases of poisoning.

ipsilateral *(ĭp-sĭ-lăt'-ĕr-ăl).* Pertaining to or affecting the same side.

IQ. Acronym for intelligence quotient.

iridectomy *(ĭr-ĭ-dĕk'-tĕ-mē).* Surgical removal of a portion of the iris.

iridencleisis *(ĭr-ĭ-dĕn-klī'-sĭs).* A surgical procedure to relieve elevated intraocular pressure in the eye, as seen in glaucoma.

iridocapsulitis *(ĭr-ĭ-dō-kăp-sū-lī'-tĭs).* Inflammation of the iris and the capsule of the lens of the eye.

iridochoroiditis *(ĭr-ĭ-dō-kō-roi-dī'-tĭs).* Inflammation of the iris and the choroid.

iridocyclitis *(ĭr-ĭ-dō-sī-klī'-tĭs).* Inflammation of the iris and the ciliary body.

iridocystectomy *(ĭr-ĭ-dō-sĭs-tĕk'-tĕ-mē).* Surgical procedure to remove a cyst from the iris.

iridokinesis *(ĭr-ĭ-dō-kĭ-nē'-sĭs).* The expanding and contracting movements of the iris.

iridology *(ĭr-ĭ-dŏl'-ĕ-jē).* A method of diagnosing disease by examining the iris of the eye. It is not considered to be a scientific approach to medicine.

iridotomy *(ĭr-ĭ-dŏt'-ō-mē).* Surgical incision of the iris.

iris *(ī'-rĭs).* The circular pigmented membrane suspended between the lens and the cornea of the eye. It contracts and dilates to regulate the entrance of light into the eye.

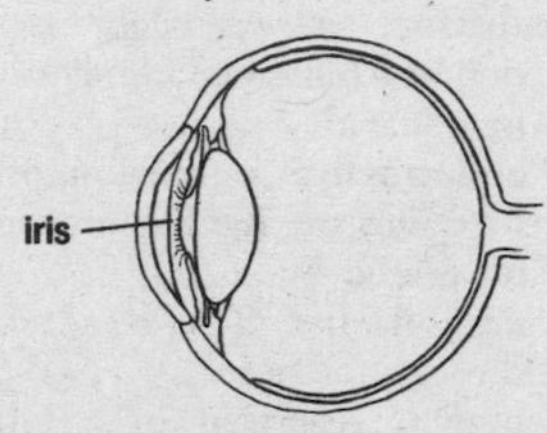

iritis *(ī-rī'-tĭs).* Inflammation of the iris.

iron *(ī'-ĕrn).* A metallic element and an essential constituent of hemoglobin and myoglobin; also present in enzymes that aid cellular respiration.

iron-deficiency anemia. *See* **anemia.**

irradiation *(ĭ-rā-dē-ā'-shĕn).* The therapeutic use of x-rays, radium rays, ultraviolet rays, or other radiation.

irrational *(ĭr-ră'-shŭn-ĕl).* Contrary to what is logical or reasonable.

irreducible *(ĭr-ē-dūs'-ĭ-bĕl).* Not capable of being reduced or made smaller, as a fracture or dislocation.

irreversible *(ĭr-ē-vĕr'-sĭ-bĕl).* Not capable of being reversed.

irrigate *(ĭr'-ĭ-gāt).* To wash out a wound with a fluid.

irritable bowel syndrome. A

chronic noninflammatory disease characterized by abdominal pain and diarrhea and/or constipation; also called *spastic colon.*

irritant *(ĭr'-ĭ-tănt).* An agent that causes irritation, such as inflammation or redness.

ischemia *(ĭs-kē'-mē-ă).* Local and temporary deficiency of blood supply to an organ or body part due to obstruction of circulation.

myocardial i. A deficiency of blood flow to the heart muscle, caused by an obstruction or constriction of the coronary arteries.

silent i. Cardiac ischemia without pain or other symptoms.

ischialgia *(ĭs-kē-ăl'-jē-ă).* Neuralgic pain in the hip; also see **sciatica.**

ischium *(ĭs'-kē-ŭm).* Lower portion of the hip bone.

island *(ī'-lĕnd).* A cluster of cells or an isolated portion of tissue; an islet.

islet *(ĭs'-lĕt).* A cluster of cells or an isolated portion of tissue; an island.

islets of Langerhans. Clusters of four types of cells in the pancreas; destruction of one type—beta cells—is a major cause of type I diabetes.

iso-. Prefix indicating equal, alike, or uniform.

isoflavone *(ĭs-ō-flā'-vōn).* A type of plant estrogen (phytoestrogen) found primarily in soybeans and foods made from soybeans.

isograft *(ī'-sō-grăft). See* **syngraft.**

isoimmunization *(ī-sō-ĭm-ū-nĭ-zā'-shŭn).* The development of antibodies against an antigen that comes from a genetically dissimilar member of the same species.

isolation *(ĭs-ō-lā'-shŭn).* Separation of infected individuals from uninfected individuals for a period of time, usually done in cases of communicable disease.

isomer *(ī'-sō-mĕr).* One of two or more chemical substances that have the same molecular formula but different chemical and physical properties.

isometric *(ī-sō-mĕt'-rĭk).* Maintaining equal dimensions.

isometric exercise. Contraction of a muscle that is not accompanied by movement of the joints that would usually move because of that muscle's action.

isometropia *(ī-ōs-mĕ-trō'-pē-ă).* Equal refraction in both eyes.

isoniazid *(ī-sō-nī'-ă-zīd).* An antibacterial drug used to treat tuberculosis.

isopia *(ī-sō'-pē-ă).* Equal vision in both eyes.

isopropyl alcohol *(ī-sō-prō'-pĕl).* Rubbing alcohol, which is a preparation of acetone, methyl isobutyl ketone, and ethanol.

Isordil *(ī'-sŏr-dĭl).* Trade name for preparations of isosorbide dinitrate, used to reduce intraocular pressure.

isosorbide dinitrate *(ī-sō-sŏr'-bīd).* Generic name for an antianginal drug that is given orally and sublingually; trade name, Isordil.

isosorbide mononitrate. An active metabolite of isosorbide dinitrate, it has the same qualities.

isotonic *(ī-sō-tŏn'-ĭk).* Having the same tone or tension.

isotonic solution. A solution that has the same tonicity as another solution with which it is compared.

isotope *(ī'-sō-tōp).* A chemical element that has the same number of nuclear protons (e.g., the same atomic weight) but a different number of nuclear neutrons (e.g., a different atomic mass). Many isotopes are radioactive.

isthmus *(ĭs'-mŭs).* A narrow structure of passage that connects two cavities or two parts.

-itis. Suffix meaning "inflammation of."

IUD. Acronym for intrauterine device.

IV. Abbreviation for intravenous.

IVP. Acronym for intravenous pyelogram. *See* **pyelogram.**

J

jacket *(jăk'-ĕt).* A structure or garment used to cover part of the body, usually the upper portion or the trunk.

Jacksonian epilepsy *(jăk-sō'-nē-ăn). See* **epilepsy.**

jaundice *(jawn'dĭs).* A syndrome characterized by excessive levels of bilirubin in the blood and accumulation of bile pigment in the skin, mucous membranes, and sclera.

cholestatic j. Jaundice resulting from an abnormal flow of bile.

hemolytic j. Jaundice caused by increased production of bilirubin caused by significant destruction of red blood cells and release of large amounts of hemoglobin.

hepatocellular j. Jaundice caused by trauma to or destruction of the liver cells.

infectious j. Jaundice caused by an infectious organism attacking the liver.

nonhemolytic j. Jaundice caused by an abnormality in the metabolism of bilirubin, resulting in elevated levels of bilirubin in the blood.

obstructive j. Form of jaundice caused by obstruction in the flow of bile from the liver to the duodenum.

regurgitation j. Form of jaundice caused by bile that has escaped from the bile canaliculi into the bloodstream.

retention j. Jaundice due to an inability of the liver to eliminate bilirubin provided by circulating blood.

jaw *(jaw).* Either or both of the maxillary and mandibular bones in the head that bear the teeth.

jejunectomy *(jĕ-jū-nĕk'-tō-mē).* Surgical removal of all or part of the jejunum.

jejunitis *(jĕ-jū-nī'-tĭs).* Inflammation of the jejunum.

jejunoileal bypass *(jĕ-jū-nō-īl'-ē-ăl).* Surgical procedure used to cause weight loss in morbidly obese individuals, in which the upper end of the small intestine is stitched to the lower end, which diverts food from much of the small intestine.

jejunum *(jĕ-jū'-nŭm).* The portion of the small intestine that extends from the duodenum to the ileum. It is about 8 feet long.

jelly *(jĕ'-ē).* A semisolid, gelatinous mass.

contraceptive j. A nongreasy jelly or cream inserted into the vagina or a diaphragm to prevent conception.

jerk *(jĕrk).* A sudden muscle reflex or involuntary movement.

jet lag. A condition caused by rapid travel across several time zones and characterized by insomnia, fatigue, and disturbances in body function that can last several days.

jing *(jĭng).* In traditional Chinese medicine, it is one of the basic substances that pervade the body; also referred to as "essence."

Johnny *(jŏh'-nē).* The short, collarless hospital gown worn by patients who are being examined or treated in a doctor's office or hospital.

joint *(joint).* The place where two or more bones are joined, such as the elbow and knee joints.

jugular veins *(jŭg'-ū-lăr).* Two pairs of veins in the neck: the external jugular veins receive blood from the exterior of the cranium and deep parts of the face; the internal veins receive blood from the brain and superficial parts of the neck and face.

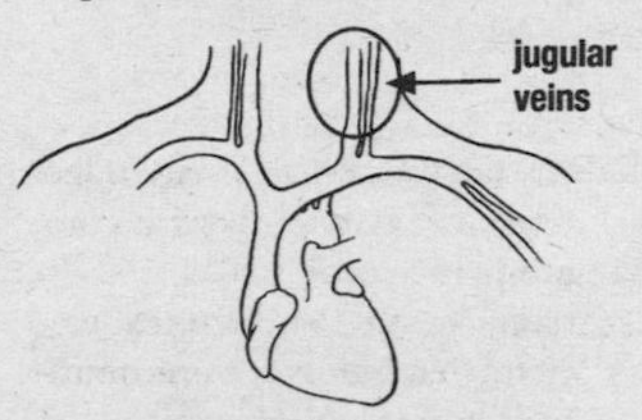

juice *(jūs).* Liquid that is excreted, secreted, or expressed from an organism.

appetite j. Gastric juice secreted while eating, the properties of which vary with the appetite for the food being eaten.

gastric j. Secretions of the stomach, consisting of salts, water, free hydrochloric acid, and pepsin.

pancreatic j. Clear digestive juice secreted by the pancreas and poured into the duodenum.

junction *(jŭnk'shŭn).* The place where two parts come together.

juniper *(jū'-nĭ-pĕr).* An herbal and homeopathic remedy prepared from the berries of *Juniperus communis* or the oil extracted from them; used for indigestion and loss of appetite.

juvenile *(jū'-vĕ-nīl).* Referring to childhood or youth.

juvenile diabetes. *See* **diabetes, type I.**

juvenile rheumatoid arthritis. *See* **arthritis.**

juxtaposition *(jŭks-tă-pō-zĭ'-shŭn).* Position that is side by side or adjacent.

K

kala-azar *(kă'-lă ă-zăr').* An infectious, often fatal disease common in rural areas in the tropics and subtropics and characterized by lesions on the liver and spleen.

kalemia *(kă-lē'-mē-ă).* Presence of potassium in the blood.

kallikrein *(kăl-ĭ-krē'-ĭn).* An enzyme found in blood plasma, urine, and body tissue that, when activated, forms kinin and is a very potent vasodilator.

kanamycin *(kăn-ĕ-mī'-sĭn).* Generic name for an antibiotic effective against aerobic gram-negative bacteria and some gram-positive bacteria. Trade name: Kantrim.

Kanner's syndrome *(kăh'-nĕrz).* An autistic disorder, named after child psychiatrist Leo Kanner.

kaolin *(kā'-ō-lĭn).* A type of aluminum silicate that is purified, pulverized, and used in the treatment of diarrhea, frequently along with pectin.

Kaopectate *(kā-ō-pĕk'-tāt).* Trade name for a drug that contains kaolin and is used to treat diarrhea.

kapha *(kăh'-făh).* In ayurvedic medicine, one of the three doshas (principles of constitution of the body). Kapha is the principle of stabilizing energy and regulates growth.

Kaposi's sarcoma *(kăh'-pō-sēz).* Multiple, bluish-red, malignant lesions that usually appear on the lower limbs but which slowly increase in size and number and spread to proximal locations. One form is seen in immunocompromised individuals, especially those with HIV/AIDS.

karyocyte *(kăr'-ō-sīt).* A nucleated cell.

karyogamy *(kăr-ē-ŏg'-ĕ-mē).* The union of nuclei during fertilization.

kava kava *(kă'-vă).* An herbal remedy prepared from the rhizome of *Piper methsticum,* the kava plant, used as a muscle relaxant, anticonvulsant, and sedative.

Kawasaki disease *(kăh-wăh-săh'-kē).* Mucocutaneous lymph node syndrome.

K-Dur. Trade name for a preparation of potassium chloride.

Keflin *(kĕf'-lĭn).* Trade name for a preparation of cephalothin sodium, an antibiotic.

Kegel exercises *(kē'-gĕl).* Exercises done to strengthen the pubococcygeal muscles, which in turn can help control or prevent stress incontinence, improve sexual response, and relieve discomfort in pregnancy.

keloid *(kē'-lŏyd).* An enlarged, elevated, irregularly shaped scar caused by the formation of an abnormal amount of collagen during connective tissue repair.

Kelvin scale *(kĕl'-vĭn).* Temperature scale in which absolute zero is equal to minus 273° on the Celsius scale.

Kenalog *(kĕ'-ă-lŏg).* Trade name for preparations of triamcinolone acetonide, an anti-inflammatory and immunosuppressant.

keratectomy *(kĕr-ă-tĕk'-tō-mē).* Surgical removal of a portion of the cornea.

keratitis *(kĕr-ă-tī'-tĭs).* Inflammation of the cornea.

keratoacanthoma *(kĕr-ă-tō-ak-ăn-thō'-mă).* A benign tumor that closely resembles squamous cell carcinoma. It usually originates in a hair follicle and grows quickly. Exposure to sunlight appears to play a role in its development.

keratoconjunctivitis *(kĕr-ă-tō-kŏn-jŭnk-tĭ-vī'-tĭs).* Inflammation of the cornea and the conjunctiva.

keratodermatitis *(kĕr-ă-tō-dĕr-mă-tī'-tĭs).* Inflammation and overgrowth of the horny layer of the skin.

keratoiritis *(kĕr-ĕ-tō-ī-rī'-tĭs).* Inflammation of the cornea and iris.

keratoma *(kĕr-ă-tō'-mă).* A callous or horny growth on the skin.

keratomileusis *(kĕr-ă-tō-mĭ-lū-sĭs).* Cosmetic surgery of the cornea in which a section is removed, frozen, the curvature reshaped, and then reattached to the cornea.

keratomycosis *(kĕr-ă-tō-mī-kō'-sĭs).* Fungal growth on the cornea.

keratoplasty *(kĕr'-ă-tō-plăs-tē).* Cosmetic surgery on the cornea.

lamellar k. A transplant of the anterior (front) half of the cornea while allowing the anterior chamber to remain intact.

optic k. Transplantation of corneal tissue to replace scar tissue that is obstructing vision.

refractive k. Procedure in which a portion of the cornea is removed, the segment is shaped to the required curvature, and it is reinserted.

keratoscleritis *(kĕr-ĕ-tō-sklĕ-rī'-tĭs).* Inflammation of the cornea and sclera.

keratosis *(kĕr-ĕ-tō'-sĭs).* Any horny growth on the skin.

actinic k. A common keratosis caused by excessive exposure to the sun that usually affects middle-aged and older individuals.

seborrheic k. A common benign, noninvasive tumor that usually occurs in middle life. It occurs most often on the face, trunk, and extremities and appears as soft lesions with slight to significant pigmentation.

keratotomy *(kĕr-ĕ-tŏt'-ĕ-mē).* Surgical incision of the cornea.

kernicterus *(kĕr-nĭk'-tĕr-ŭs).* A condition that affects newborns, in which bilirubin infiltrates certain areas of the brain and spinal cord. Prognosis is poor if left untreated.

ketogenesis *(kē-tō-jĕn'-ĕ-sĭs).* The production of ketone bodies.

ketolysis *(kē-tŏl'-ĕ-sĭs).* The destruction of ketone bodies.

ketone bodies *(kē'-tōn).* Compounds produced during the oxidation of fatty acids; namely acetone, beta-hydroxybutyric acid, and acetoacetic acid.

ketonuria *(kē-tō-nū'-rē-ă).* Presence of ketone bodies in the urine.

ketosis *(kē-tō'-sĭs).* The accumulation in the body of ketone bodies, the result of faulty metabolism of fatty acids.

kidney *(kĭd'-nē).* Either of the two organs located at the back of the abdominal cavity, whose function is to filter blood, excrete byproducts of metabolism as urine, and regulate the concentrations of certain substances, such as hydrogen and potassium.

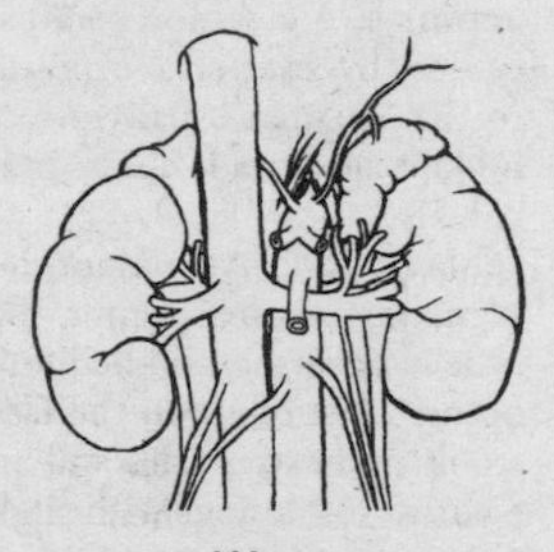

kidney

artificial k. Device used to remove elements from the blood while it is circulated outside the body.

cystic k. Kidney that contains one or more cysts.

floating k. Kidney that is displaced and movable.

polycystic k. Kidney that contains many cysts.

kidney stone. A calcification in the kidney or lower in the urinary tract, which develops because of decreased urine volume of increased excretion of substances that form stones (e.g., calcium, oxalate, cystine, urate, xanthine, phosphate).

kilocalorie *(kĭl-ō-kăl'-ĕ-rē).* The amount of heat needed to raise the temperature of 1 kilogram of water 1 degree Celsius; it is the calorie used to express the energy value of food.

kilogram *(kĭl'-ō-grăm).* A unit of weight that is equal to one thousand grams or 2.2 pounds in the avoirdupois system of weight.

kinescope *(kī'-nō-skōp).* A device used to test the refraction of the eye.

kinesia *(kĭ-nē'-sē-ă).* Any type of motion sickness, such as airsickness and seasickness.

kinesiology *(kĭ-nē-sē-ŏl'-ĕ-jē).* The study of muscles and body movement.

kinesitherapy *(kĭ-nē-sĭ-thĕr'-ĕ-pē).* The treatment of disease using exercise or movement.

kinesthesia *(kĭn-ĕs'-thē'-zhĕ).* The awareness or knowledge of one's movements, weight, tension, and body position, which depends on input from sensors in the body (e.g., joint and muscle receptors and hair cells).

kinetics *(kĭ-nĕt'-ĭks).* The science of motion.

kinetoscopy *(kĭ-nĕ-tŏs'-kĕ-pē).* A series of photographs that show the movement of the limbs, used in the diagnosis of gait disorders.

Klebsiella *(klĕb-sē-ĕl'-ă).* A genus of gram-negative bacteria widely distributed in nature and common in the human intestinal tract.

K. oxytoca. A cause of urinary tract infections.

K. pneumoniae. A cause of acute bacterial pneumonia and urinary tract infections.

kleptomania *(klĕp-tō-mā'-nē-ă).* An uncontrollable impulse to steal objects one has no personal or monetary use for, and the taking of the objects is preceded by tension and followed by relief or pleasure.

Klinefelter's syndrome *(klīn'-fĕl-tĕrs).* A congenital condition in males characterized by small testes, enlarged breasts, lack of live sperm, and long legs. There are several forms, which are determined by chromosomal abnormalities.

Klippel-Feil sequence *(klĭ-pĕl').* Condition characterized by a short neck, low hairline at the nape of the neck, and restricted head movement, all due to a defect in the early development of the vertebrae in the neck.

knee *(nē).* The front portion of the leg at the junction of the femur and tibia and the junction itself, which is covered with the patella (kneecap).

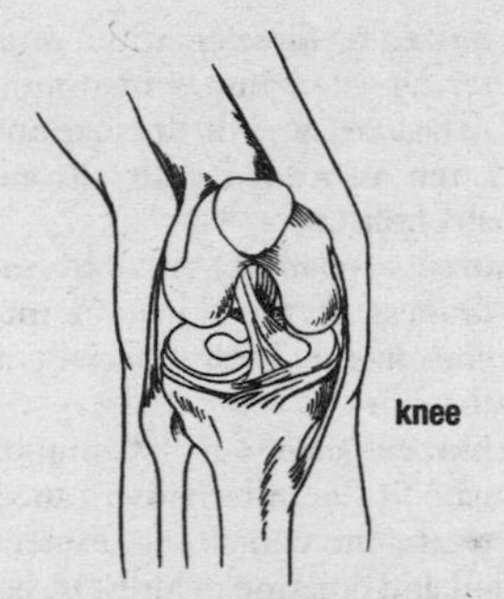

knee

housemaid's k. Inflammation of the bursa in the front of the knee with an accumulation of fluid. May occur in individuals who kneel frequently.

knock k. Condition in which the knees come together while the ankles are far apart.

locked k. Condition in which the leg cannot be extended.

trick k. Popular term for a knee joint that frequently locks in position.

kneecap *(nē'-kăp).* The patella.

knee jerk. The reflex produced when the tendon below the kneecap is struck sharply while the leg is hanging loosely flexed at a right angle.

knit *(nĭt).* To heal, as when a broken bone mends.

knuckles *(nŭk'-ĕlz).* The joints in the hand, especially when the fist is clenched.

Koplick spots *(kŏp'-lĭk).* Small red spots with blue-white centers that appear in the mouth before the rash of measles erupts.

Korsakoff's syndrome *(kŏr'-sĕ-kŏfs).* Mental disorder characterized by disorientation, muttering delirium, insomnia, delusions, and hallucinations; often associated with chronic alcoholism.

kraurosis *(krŏ-rō'-sĭs).* Atrophy and dryness of the skin and mucous membranes, especially of the vulva.

krebiozen *(krĕ-bī'-ō-zĕn).* Common name of an alternative cancer treatment containing mineral oil and creatine. The FDA has determined that krebiozen has no use as a cancer treatment and can cause dangerous side effects.

Kupfer cells *(koop'-fĕr).* Large star-shaped or pyramidal cells with a large oval nucleus; they are found in the liver.

kwashiorkor *(kwăsh-ē-ŏr'-kŏr).* A severe protein-deficiency condition seen in children.

K-Y jelly. Trade name of a lubricating jelly used during physical examinations and to facilitate sexual intercourse.

kyphosis *(kī-fō'-sĭs).* Curvature of the thoracic spine (upper back); humpback.

Kytril *(kī'-trĭl).* Trade name for preparations of granisetron hydrocholoride, used to treat vomiting associated with cancer chemotherapy.

L

labia *(lā'-bē-ă).* Plural of labium.

L. majora. The two folds of fatty tissue that lie on either side of the vaginal opening; they form the side borders of the vulva.

L. minora. The two thin folds of skin that lie inside the vestibule of the vagina and between the labia majora and hymen.

labile *(lā'-bē-ăl).* Unsteady; rapidly shifting or changing.

labor *(lā'-bŏr).* The process by which a female expels the fetus from the uterus to the outside world.

false l. Uterine contractions that occur before it is time for actual labor to begin. These contractions eventually cease.

induced l. Use of mechanical means or other methods, usually intravenous oxytocin, to stimulate onset of labor.

premature l. Delivery of a viable fetus before the normal end of gestation; usually refers to delivery that occurs between the twentieth and thirty-seventh week after the onset of the last menstrual period.

labyrinth *(lăb'-ĭ-rĭnth).* Term for the interconnecting structures of the inner ear, consisting of the semicircular canals, utricle, saccule, and cochlea.

labyrinthitis *(lăb-ĭ-rĭn-thī'-tĭs).* Inflammation of the labyrinth, which may be accompanied by extreme dizziness or hearing loss.

laceration *(lăs-ĕr-ā'-shŭn).* A ragged or torn wound.

lacrimal *(lăk'-rĭm-ăl).* Referring to the tears.

lacrimal duct. One of two ducts

that carries tears from the lacrimal lake to the lacrimal sac.

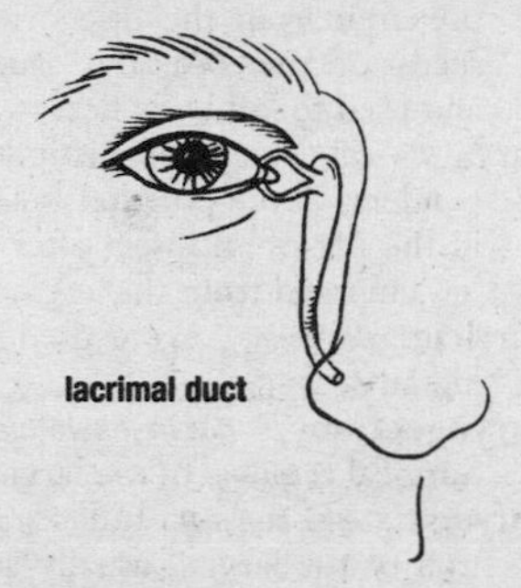

lacrimal gland. Gland that secretes tears.

lacrimal punctum. The opening in the inner corner of the lower eyelid where tears exit into the nasolacrimal duct (the passageway leading to the nose).

lacrimation *(lăk-rĭ-mā'-shĕn).* The secretion and discharge of tears.

lactalbumin *(lăk-tăl-bū'-mĭn).* A simple, soluble protein present in milk and cheese.

lactase *(lăk'-tās).* An enzyme that digests the sugar present in milk.

lactation *(lăk-tā'-shĕn).* The secretion of milk from a mother's breast.

lacteals *(lăk'-tē-ăl).* Referring to milk.

lactic acid *(lăk'-tĭk).* An acid formed as an end-product of sugar metabolism, it helps digest milk.

lactobacillus *(lăk-tō-bĕ-sĭl'-ĕs).* A genus of bacteria that are found widely in nature and in the human vagina, intestinal tract, and mouth.

l. acidophilus. An organism that produces lactic acid by fermenting milk sugars. Preparations of this organism are also available in supplements, used to treat digestive difficulties and help prevent infections after antibiotic therapy.

lactose *(lăk'-tōs).* A major ingredient of mammalian milk; *see also* **lactose intolerance.**

lactose intolerance. An inability to withstand ingestion of lactose, usually due to an inherited deficiency of lactase (which digests lactose) in the intestinal tract.

lactosuria *(lăk-tō-sū'-rē-ă).* Presence of lactose (milk sugar) in the urine.

lacto-ovo-vegetarian *(lăk-tō-ō-vō-vĕg-ĕ-tăr'-ē-ăn).* An individual whose animal product consumption includes eggs and dairy products but does not include meat, poultry, or fish.

laetrile *(lā'-ĕ-trĭl).* Natural substance derived from the crushed pits of certain fruits, especially apricots, and alleged to have anticancer abilities.

Lamaze method *(lă-măz').* A method for expectant parents to prepare for and ease the discomfort of childbirth; it includes education regarding the physiology of pregnancy, and techniques for breathing and bearing down during labor.

lame *(lām).* An inability to ambulate (walk) normally.

lamella *(lă-mĕl'-ă).* A thin plate.

lamina *(lăm'-ĭ-nă).* 1. A thin layer or membrane. 2. Term often used to refer to the flattened part of either side of the arch of a vertebra.

laminectomy *(lăm-ĭ-nĕk'-tō-mē).* Surgical procedure to remove the posterior arch of a vertebra.

lance *(lăns).* 1. To make an incision with a lancet. 2. A two-edged surgical knife.

lancet *(lăn'-sĕt).* A pointed, two-edged surgical knife.

Lanoxin *(lă-nŏk'-sĭn). See* Appendix, Common Prescription and OTC Drugs: By Trade Name.

lansoprazole *(lăn-sō'-prĕ-zōl).* Generic name for a proton pump inhibitor, used to treat conditions related to excess stomach acid; trade name: Prevacid.

lanugo *(lĕ-nū'-gō).* The fine hair present on the skin of a fetus.

laparoscope *(lăp'-ĕ-rō-skōp).* An instrument, similar to an endoscope, that is inserted into the peritoneal cavity to perform a visual inspection.

laparoscopic surgery *(lăp-ĕ-rō-skŏp'-ĭk).* A surgical procedure that utilizes a laparoscope, which is inserted into the body through a small incision.

laparoscopy *(lăp-ĕ-rŏs'-kĕ-pē).* Examination of the interior of the abdomen using a laparoscope.

laparotomy *(lăp-ĕ-rŏt'-ĕ-mē).* Surgical incision made through the flank (area between the ribs and pelvis on the side of the body).

larkspur *(lăhrk'-spĕr).* A medicine prepared from the dried ripe seeds of the *Delphimiu ajacis*, and used to kill body lice.

larva *(lăr'-vă)* (*pl.* larvae). An independent developmental stage in the life of an insect after it has emerged from the egg.

larvicide *(lăr'-vĭ-sīd).* Agent used to kill larvae.

laryngectomy *(lăr-ĭn-jĕk'-tō-mē).* Surgical removal of the larynx.

laryngitis *(lăr-ĭn-jī'-tĭs).* Inflammation of the larynx, usually accompanied by a sore, dry throat and hoarseness, cough, and temporary loss of voice.

laryngopharyngeal *(lă-rĭng-gō-fă-rĭn'-jē-ăl).* Referring to the larynx and pharynx.

laryngoplasty *(lă-rĭng'-gō-plăs-tē).* Reconstructive surgery of the larynx.

laryngoscope *(lăr-ĭn-gŏs'-kō-pē).* Examination of the interior of the larynx.

laryngospasm *(lă-rĭng'-gō-spăz-ĕm).* Spasm of the laryngeal muscles, which causes the larynx to close.

larynx *(lăr'-ĭnks).* The enlarged upper end of the trachea below the root of the tongue, consisting of nine cartilages connected by ligaments and muscles. It guards the opening into the trachea and is the organ of the voice.

laser *(lā'-zĕr).* Acronym for light amplication by stimulated emission of radiation. A device that converts light frequencies into one small and extremely intense, focused hot beam of one wavelength.

argon l. A laser with ionized argon, used for photocoagulation.

carbon dioxide l. A laser whose active medium is carbon dioxide, used to cut and remove tissue.

Nd:YAG. A laser whose active medium is yttrium, aluminum, garnet, and neodymium; used for photocoagulation and photoablation.

laser coagulation. Clotting tissue using a laser. A coagulation laser produces light that is absorbed by hemoglobin to seal off bleeding blood vessels.

laser conization. Removal of a section of the cervix of the uterus using a laser knife.

LASIK. Acronym for laser-assisted in-sit keratomileusis. Surgical procedure of the cornea in which laser and a keratome (knife for cutting the cornea) are utilized to correct vision.

Lasix *(lā'-sĭks). See* Appendix, Common Prescription and OTC Drugs: By Trade Name.

Lasser's paste *(lăhs'-ĕrz).* An ointment composed of zinc oxide, starch, and petroleum jelly, used to treat certain skin conditions, especially eczema.

lassitude *(lăs'-ĭ-tūd).* Weariness; tiredness.

latanoprost *(lă-tăn'-ō-prŏst). See* Appendix, Common Prescription and OTC Drugs: By Generic Name.

latency *(lā'-tĕn-sē).* A state of seeming inactivity.

motor l. The time between initiation of a stimulus and the beginning of a response from the muscles.

sleep l. The period between the time a person lies down to rest and the onset of sleep.

latent *(lā'-tĕnt).* Concealed.

latent period. The time it takes for a reaction to take effect after a stimulus.

lateral *(lăt'-ĕr-ĕl).* Referring to a position farther from the median plane

latex *(lā'-tĕks).* Any one of various white sticky fluids secreted by some plants.

latissimus *(lă-tĭs'-ĭ-mĕs).* A general term referring to a broad structure, as a muscle.

laughing gas *(lăf'-ĭng).* Nitrous oxide, a gas used as a general anesthetic along with other anesthetic agents.

lavage *(lăh-văhzh').* To wash out or irrigate an organ.

lavender *(lăv'-ĕn-dĕr).* Any plant of the genus *Lavandula.* The essential oil is used medicinally.

lavender oil. A volatile oil distilled from the flowers of *Lavandula angustifolia* and used internally for loss of appetite, nervousness, and insomnia, and externally for circulatory disorders.

laxative *(lăk'-să-tĭv).* A substance that promotes evacuation of the bowel.

laxity *(lăk'-sĭ-tē).* Displacement or slackness in the movement of a joint.

lazy eye. *See* **eye.**

LD. Acronym for lethal dose.

LDL. Acronym for low-density lipoprotein.

L-dopa. *See* **levodopa.**

lead *(lĕd).* A soft, gray-blue metal that is poisonous if ingested.

lead poisoning, acute. Ingestion or inhalation of enough lead to cause vomiting, diarrhea, headache, stupor, convulsions, and coma.

lead poisoning, chronic. Chronic ingestion or inhalation of lead, which damages the nervous system, the gastrointestinal tract, and blood-forming organs.

learning disability. An inability to learn or a defect in the ability to learn, often referring to basic skills such as math, reading, and writing.

lecithin *(lĕs'-ĭ-thĭn).* Phosphatidylcholine.

leech *(lēch).* A blood-sucking water worm that secretes hirudin, an anticoagulant, which makes it useful for certain bleeding situations. *See* **leech therapy.**

leech therapy. The use of leeches in cosmetic and reconstructive surgery, as in the reattachment of severed body parts, such as a finger, toe, ear, or nose.

Lee-White test. Test that identifies the time it takes blood to clot.

leg. In common terms, the entire lower limb. However, in medical terms, it refers to the part between the knee and ankle.

bow l. Condition in which the legs curve outward at the knees.

scissor l. Deformity in which the legs cross while an individuals is walking, caused by spasticity in the thigh muscles.

legal blindness. Blindness as defined by law. In most states, the criteria is 20/200 or less in the better eye.

Legg-Calve-Perthes disease *(lĕg-kăhl'-vā—pĕr'tĕz).* A hip disorder that occurs in children and is caused by an interruption of the blood supply to the femur (the ball in the ball-and-socket hip joint), which leads to its deterioration.

legionella *(lē-jē-nĕl'-ă).* Any microorganism that belongs to the genus *Legionella.*

Legionnaires' disease *(lē-jĕn-ărz').* A bacterial disease caused by infection with *Legionella pneumophila* and characterized by pneumonia, high fever, gastrointestinal pain, headache, and sometimes involvement of the nervous system, liver, or kidneys.

leiomyoma *(lī-ō-mī-ō'-mă).* A benign tumor consisting primarily of smooth muscle tissue, most commonly of the uterus. Also called a *fibroid tumor.*

leiomyosarcoma *(lī-ō-mī-ō-săhr-kō'-mă).* A sarcoma that contains spindle cells of smooth muscle; it usually develops in the uterus, retroperitoneal area, or the extremities.

leishmaniasis *(lĕsh-mē-nī'-ĕ-sĭs).* An infection caused by species of *Leishmania;* it usually appears on the skin, in the nasal cavities, or on the pharynx.

Lennox syndrome *(lĕn'-ĕks).* A severe form of epilepsy, characterized by frequent seizures of several types, a specific brain wave pattern, developmental delay, and poor social skills.

lens *(lĕnz).* A transparent piece of glass or other substance shaped to allow rays of light to converge or scatter.

lens implant. The insertion of an artificial lens to replace the natural lens in the eye; usually the replaced lens has been damaged by a cataract.

lentigo *(lĕn-tī'-gō).* A small, flat, tan to dark brown or black pigmented spot on the skin that resembles a freckle but is different in that it contains an increased number of melanocytes.

leprosy *(lĕp'-rĕ-sē).* A chronic, slowly progressive infectious disease caused by *Mycobacterium leprae* and characterized by lesions of the skin, mucous membranes, nerves, bones, and organs.

leptomeninges *(lĕp-tō-mĕ-nĭn'-jēs).* Two of the three membranes (pia mater and arachnoid) that cover the brain and spinal cord.

leptomeningitis *(lĕp-tō-mĕn-ĭn-jī'-tĭs).* Inflammation of the leptomeninges.

Leptospira *(lĕp-tō-spī'-rā).* Genus of bacteria of the family Leptospiraceae, consisting of single, coiled, aerobic cells.

L. interrogans. The species that contains all the strains of the genus that cause leptospirosis.

Leptospirosis *(lep-to-spi'-ro'-sis).* Any one of several febrile illnesses that affect both humans and other animals, caused by infection with one of the serovars of *Leptospira interrogans.*

lesbian *(lĕz'-bē-ăn).* 1. A female homosexual. 2. Referring to homosexuality between females.

lesion *(lē'-zhĕn).* A change in tissue structure due to disease or trauma. Examples include boil, tumor, and abscess.

coin l. A round or nodular shadow seen on chest x-rays, caused by a disease process.

gross l. One that is visible to the naked eye.

peripheral l. A lesion that develops on the nerve endings.

structural l. A lesion that causes an obvious change in a tissue.

lethal *(lē'-thăl).* Fatal.

lethal gene. A gene that causes the death of the organism.

lethargy *(lĕth'-ĕr-jē).* A state of reduced energy, characterized by significant drowsiness, apathy, and listlessness.

leucine *(lū'-sēn).* An essential amino acid that is necessary for optimal growth in infants and nitrogen balance in adults.

leukemia *(lū-kē'-mē-ă).* A progressive, malignant disease of the blood-forming organs, characterized by unrestrained growth of white blood cells (leukocytes). It is classified according to the severity of the disease and the prominent cell type.

acute l. A form of leukemia that lasts only a few weeks and is characterized by rapid onset, severe anemia, hemorrhage, and severe infections, resulting in death.

aleukemic l. Form of the disease in which the white blood cells do not increase in number.

chronic l. Form of leukemia in which those affected live with the disease for many years; some for 20 or longer.

feline l. A general term for various leukemias that occur in domestic cats. The disease usually affects the liver, spleen, thymus, and gastrointestinal tract.

hairy cell l. Form of chronic leukemia characterized by an enlarged spleen and a large amount of abnormal large mononuclear cells covered by hairlike projections in the bone marrow, spleen, liver, and peripheral blood.

lymphocytic l. Form associated with an increased number of malignant lymphocytes and lymphoblasts.

leukemoid *(lū-kē'-moid).* Having blood and sometimes clinical findings that resemble those of leukemia.

leukocidin *(lū-kō-sī'-dĭn).* Any substance that is toxic to leukocytes (white blood cells); specifically, an exotoxin produced by disease-causing streptococci and staphylococci that kill leukocytes by destroying the cytoplasmic granules.

leukocytes *(lū'-kō-sīts).* White blood cells. They are classified into two main groups: granular (basophils, eosinophils, and neutrophils) and nongranular (lymphocytes and monocytes).

leukocytosis *(lū-kō-sī-tō'-sĭs).* A temporary increase in the number of leukocytes in the blood, which often occurs with vigorous exercise and is accompanied by fever, infection, hemorrhage, or inflammation.

leukoderma *(lū-kō-dĕr'-mă).* An acquired type of skin depigmentation caused by a specific substance or dermatosis.

leukopenia *(lū-kō-pē'-nē-ă).* Reduction in the number of leukocytes in the blood, less than 5,000 per cubic milliliter. Types of leukopenia are named for the type of cell that is in low supply.

leukoplakia *(lū-kō-plā'-kē-ă).* Formation of thickened, white patches on a mucous membrane, usually the tongue, cheek, or vulva.

l. buccalis. Leukoplakia of the cheek.

l. lingualis. Leukoplakia of the tongue.

l. vulvae. Leukoplakia of the female external genitalia, usually seen in older women.

leukorrhea *(lū-kō-rē'-ă).* A white, sometimes yellow, sticky discharge from the vagina.

Levaquin *(lē'-vĕ-kwĭn).* *See* Appendix, Common Prescription and OTC Drugs: By Trade Name.

levator *(lē-vā'-tŏr).* A muscle that elevates the organ or structure into which it is inserted.

levitation *(lĕv-ĭ-tā'-shŭn).* A hallucinatory sensation of floating or rising in the air.

levodopa *(lē-vō-dō'-pă).* Generic name for a drug used in the treatment of Parkinson's disease. Trade names: Bendopa, Dopar.

levofloxacin *(lē-vō-flŏk'-sĕ-sĭn). See* Appendix, Common Prescription and OTC Drugs: By Generic Name.

levothyroxine *(lē-vō-thī-rŏk'-sĕn). See* Appendix, Common Prescription and OTC Drugs: By Generic Name.

Levoxyl *(lĕ-vŏk'-sĭl). See* Appendix, Common Prescription and OTC Drugs: By Trade Name.

libidinous *(lĭ-bĭd'-ĭ-nĕs).* Erotic.

libido *(lĭ-bē'-dō).* Sexual drive or desire.

Librium *(lĭb'-rē-ŭm). See* Appendix, Common Prescription and OTC Drugs: By Trade Name.

lice *(līs). See* **louse.**

lichenification *(lī-kĕn-ĭ-fĭ-kā'-shŭn).* Hypertrophy of the skin, resulting in thickening and exaggeration of normal skin markings, giving the skin a leathery appearance.

lichen planus *(lī'-kĕn).* An itchy, inflammatory skin disease that usually affects the mucous membranes of the mouth and genital area as well as the nails, and is characterized by violet, flat, scaly papules.

lidocaine *(lī'-dō-kān).* A drug used as a local, topical anesthetic for the skin and mucous membranes.

lifetime risk. The risk of developing a disease during a person's lifetime or dying of the disease. For example, the lifetime risk of dying from prostate cancer is 3.4% for American men; for individuals born in 2000 in the United States, the lifetime risk of developing diabetes is 32.8% for males and 38.5% for females.

ligament *(lĭg'-ă-mĕnt).* A band of tissue that connects bones or supports organs.

ligature *(lĭg'-ă-chŭr).* Any substance, such as surgical gut, wire, silk, or cotton, used to tie a vessel or tie off a part of the body.

absorbable l. Ligature made of a substance that is eventually absorbed into the body.

nonabsorbable l. Ligature made of a substance that the body does not absorb.

lightening *(līt'-ĕn-ĭng).* The sensation of decreasing abdominal distension produced as the uterus descends into the pelvic cavity, usually occurring two to three weeks before the first stage of labor begins.

light therapy. Treatment of physical conditions (e.g., psoriasis, herpes simplex) or mental disorders (e.g., seasonal affective disorder) by exposure to light of various concentrations or wavelengths.

limb *(lĭmb)*. An extremity; an arm or leg.

limbus *(lĭm'-bŭs)*. The margins or edge of a part.

limbus of the eye. The circular area where the colored portion of the eye meets the white of the eye.

liminal *(lĭm'-ĭ-năl)*. Barely perceptible to the senses.

lincomycin *(lĭn-kō-mī'-sĭn)*. An antibiotic that is used primarily against gram-positive bacteria.

linea *(lĭn'-ē-ă)*. An anatomical line.
l. alba. The white line of connective tissue in the middle of the abdomen that runs from the sternum to the pubic area.
l. albicantes. Whitish lines that may appear on the abdomen, breasts, or buttocks due to pregnancy, obesity, weight loss, or rapid growth during adolescence. They are due to weakening of the elastic tissues.

linear *(lĭn'-ē-ĕr)*. Referring to or resembling a line.

liniment *(lĭn'-ĭ-mĕnt)*. An oily substance containing medication used on the skin.

linseed oil *(lĭn'-sēd)*. Oil derived from the seeds of common flax, *Linum usitatissimum,* and used to smooth and soothe the skin.

lipase *(lī'-pās)*. A fat-splitting enzyme found in the blood, tissues, and pancreatic secretions.

lipectomy *(lī-pĕk'-tō-mē)*. Removal of fatty tissue.

lipemia *(lī-pē'-mē-ă)*. An abnormal amount of fat in the blood.

lipid *(lĭp'-ĭd)*. Any one of a group of fats or fatlike substances. Lipids include fatty acids, neutral fats, steroids, and waxes.

Lipitor *(lĭp'-ĭ-tŏr)*. *See* Appendix, Common Prescription and OTC Drugs: By Trade Name.

lipoblast *(lĭp'-ō-blăst)*. An immature fat cell.

lipodystrophy *(lĭp-ō-dĭs'-trō-fē)*. Any disruption of fat metabolism.

lipoma *(lĭp-ō'-mă)*. A benign, soft, encapsulated tumor consisting of fatty tissue.

lipometabolism *(lĭp-ō-mĕ'-tăb'-ōl-ĭzm)*. Fat metabolism.

lipoprotein *(lī-pō-prō'-tēn)*. Compounds consisting of simple proteins combined with lipid (fat) components; examples include cholesterol, triglycerides, and phospholipids.

liposarcoma *(lĭp-ō-săr-kō'-mă)*. A malignant tumor that develops from embryonal lipoblastic cells.

liposuctioning *(lĭp-ō-sŭk'-shŭn)*. Surgical removal of fat deposits through use of a high pressure vacuum. The fat is sucked through a thin flexible tube that is inserted through a small incision.

lipping *(lĭp'-ĭng)*. Development of a bony overgrowth seen in osteoarthritis.

lippitude *(lĭp'-ĭ-tūd)*. Ulcerations of the edges of the eyelids.

lip reading. Interpreting what an individual is saying by watching the speaker's lip movements, a method used by deaf people.

lipuria *(lĭ-pū'-rē-ă).* Presence of fat in the urine.

liquefaction *(lĭk-wĕ-făk'-shŭn).* Conversion of a solid into a liquid.

lisinopril *(lī-sĭn'-ō-prĭl). See* Appendix, Common Prescription and OTC Drugs: By Generic Name.

liter *(lē'-tĕr).* A metric measure of capacity. One liter equals 1.057 liquid quarts.

lithectasy *(lĭth-ĕk'-tă-sē).* Removal of a stone from the bladder through a dilated urethra.

lithiasis *(lĭ-thī'-ĕ-sĭs).* The formation or presence of stones.

gallbladder l. Stones in the gallbladder; cholecystolithiasis.

pancreatic l. Stones in the pancreas; pancreatolithiasis.

renal l. Stones in the kidney; nephrolithiasis.

urinary l. Stones in the urinary tract; urolithiasis.

lithium carbonate *(lĭth'-ē-ŭm).* A generic drug used in the treatment of manic phases in bipolar disorder and in maintenance therapy to reduce the severity and frequency of manic episodes.

lithotomy *(lĭ-thŏt'-ō-mē).* Incision of a duct or organ to remove a stone.

transurethral l. A surgical procedure in which stones are removed from the ureter by introducing an instrument through the urethra and bladder into the ureter.

lithotripsy *(lĭth'-ō-trĭp-sē).* The crushing of a urinary stone or gallstone while it is in the body, followed immediately by washing out the fragments.

extracorporeal shock wave l. A procedure for treating urinary stones and gallstones in which the patient is placed in contact with water and a high-energy shock wave is focused on the stone, which disintegrates it so it can be eliminated from the body.

laser l. Use of a pulsed dye laser to disintegrate stones in the urinary tract.

litmus *(lĭt'-mĕs).* A blue pigment prepared from coarsely ground lichens and ammonia.

litmus paper. Paper prepared with litmus and used as a test for alkalinity and acidity: if the blue paper turns red, it indicates an acidic solution; if it remains blue, an alkaline solution.

Little's disease. A form of cerebral palsy and a congenital condition characterized by spastic stiffness of the limbs, muscle weakness, convulsions, and mental deficiency.

"live-flesh." A nonmedical term used to describe muscle twitchings that often occur in the muscles of the eyelids. They are caused by excessive use or tiredness.

liver *(lĭv'-ĕr).* The largest organ in

the body, it is located in the upper right and part of the upper left side of the abdominal cavity. It performs various metabolic and detoxification functions and is essential for life.

fatty l. A liver that has fatty deposits, usually from alcohol abuse, jejunoileal bypass surgery, or diabetes. The liver is enlarged but patients are usually asymptomatic. It can progress to cirrhosis or hepatitis if the underlying cause is not addressed.

liver profile. A series of tests performed on blood samples to determine the state of liver functioning.

livid *(lĭv'-ĭd).* Discolored due to congestion or a contusion.

living will. *See* Appendix, Advance Directives: What Everyone Should Know.

lobar pneumonia. *See* **pneumonia.**

lobe *(lōb).* A defined portion of any organ, especially the brain, lungs, or a gland, that is demarcated by fissures, connective tissue, shape, and sulci.

lobectomy *(lō-bĕk'-tĕ-mē).* Removal of a lobe, as from the lung, brain, thyroid, or liver.

lobotomy *(lō-bŏt'-ĕ-mē).* Incision into a lobe. The term is also used to refer to a surgical procedure in which the fibers in a lobe of the brain are cut to relieve mental disturbances.

lobular neoplasia *(lŏb'-ū-lăr).* A precancerous neoplasia found in the lobules of mammary glands and identified only upon microscopic examination.

lobule *(lŏb'-ūl).* A small lobe.

localized *(lō'-kĕl-īzd).* Limited to a specific region or to one or more areas.

lochia *(lō'-kē-ă).* The vaginal discharge that occurs the first week or two after childbirth.

lockjaw *(lŏk'-jăw).* Spasms of the jaw muscles, especially the masticatory muscles, which make it difficult to open the mouth. It is an early symptom of tetanus.

locomotion *(lō-kō-mō'-shŭn).* Movement or the ability to move from one location to another.

locus *(lō'-kŭs).* 1. An anatomical site. 2. In genetics, the site of a gene on a chromosome.

logaphasia *(lŏg-ă-fā'-zē-ă).* Aphasia in which there is an impairment of the ability to write or speak, due to a lesion in the motor speech region of the brain.

loin *(loyn).* Lower part of the back and sides between the ribs and pelvis.

Lomotil *(lō'-mō-tĭl).* Trade name for combination preparations of diphenoxylate hydrochloride and atropine sulfate, used to treat diarrhea.

longevity *(lŏn-jĕv'-ĭ-tē).* The quality or condition of living a long life.

longitudinal *(lŏn-jĭ-tū'-dĭ-nĕl).* Lengthwise; parallel to the long axis of the body.

long QT syndrome. An inherited defect in heart rhythm that causes individuals to faint without warning and to experience dizziness, palpitations, seizures, and sudden death. The name refers to the QT segment seen on electrocardiogram (ECG) tracings.

Lopid *(lō'-pĭd). See* Appendix, Common Prescription and OTC Drugs: By Trade Name.

Lopressor *(lō-prĕs'-sĕr). See* Appendix, Common Prescription and OTC Drugs: By Trade Name.

loratadine *(lĕ-răt'-ĕ-dēn). See* Appendix, Common Prescription and OTC Drugs: By Generic Name.

lorazepam *(lŏr-ăz'-ĕ-păm). See* Appendix, Common Prescription and OTC Drugs: By Generic Name.

lordosis *(lŏr-dō'-sĭs).* Excessive, abnormal arching of the back, causing the abdomen to be thrust forward and the lower back to be concave; also called *swayback.*

losartan *(lō-săhr'-tăn). See* Appendix, Common Prescription and OTC Drugs: By Generic Name.

Lotensin *(lō-tĕn'-sĭn). See* Appendix, Common Prescription and OTC Drugs: By Trade Name.

lotion *(lō'-shĕn).* Liquid medicinal preparation for application to the skin or to use in bathing.

Lotrel *(lō'-trĕl). See* Appendix, Common Prescription and OTC Drugs: By Trade Name.

louse *(lous)* (*pl.* lice). A small wingless insect that lives as a parasite on mammals and birds. Human lice transmit diseases such as typhus, trench fever, and plague.

body l. *Pediculus humanus corporis,* lives primarily in clothing.

crab l. *Phthirus pubis,* lives mainly in pubic hair, but also in beards, eyelashes, and eyebrows.

head l. *Pediculus humanus capitis,* lives mainly in head hair.

lovastatin *(lō'-vĕ-stăt-ĭn).* Generic name of a drug used to control cholesterol levels in the blood; trade name: Mevacor.

LSD. Acronym for lysergic acid diethylamide, a compound that causes hallucinations and can result in permanent brain damage with uncontrolled use.

lucidity *(lū-sĭd'-ĭ-tē).* Quality of clarity or brightness, especially when referring to mental conditions.

lumbago *(lŭm-bā'-gō).* A general term for dull, aching pain in the lower (lumbar) part of the back.

lumbar *(lŭm'-băhr).* The area of the back between the thorax and the pelvis.

lumbarization *(lŭm-băr-ĭ-zā'-shĕn).* Surgical procedure in which the first sacral vertebra is fused with the last lumbar vertebra.

lumbar puncture. A puncture made by placing a needle into the subarachnoid space in the spinal column for the purpose of removing spinal fluid for analysis or to inject an anesthetic.

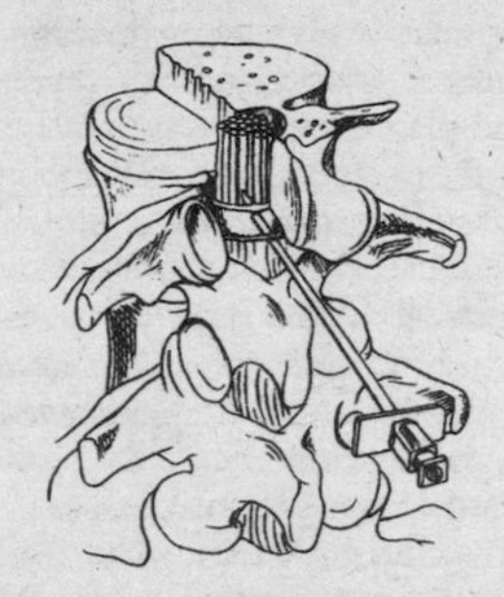

lumbar puncture

lumbodorsal *(lŭm-bō-dŏr'-sĕl).* Referring to the lumbar and thoracic regions.

lumbosacral *(lŭm-bō-să'-krĕl).* Referring to the lumbar and sacrum.

lumen *(lū'-mĕn).* The space within a tube or tubular structure, such as a blood vessel or the intestines.

luminal *(lū'-mĭ-nĕl).* Pertaining to the lumen of a tubular structure.

lumpectomy *(lŭm-pĕk'-tĕ-mē).* Surgical removal of a tumor from the breast without also removing surrounding tissue or lymph nodes.

lumpy jaw. *See* **actinomycosis.**

lung *(lŭng).* Either one of the pair of organs used for breathing, located in the pleural cavity in the chest. The right lung has three lobes and the left has two.

Lupron *(lū'-prŏn).* Trade name for preparations of leuprolide acetate, used to treat advanced prostate cancer.

lupus erythematosus *(lū'-pŭs ĕr-ĕ-thēm-ă-tō'-sŭs). See* **systemic lupus erythematosus.**

luteal *(lū'-tē-ĕl).* Referring to the corpus luteum, its hormone, or its cells.

luteinization *(lū-tē-ĭn-ī-zā'-shĕn).* Process of development of the corpus luteum.

lutin *(lū'-tĭn).* A hormone of the corpus luteum that helps prepare the endometrium for a fertilized egg.

luxation *(lĕk-sā'-shĕn).* Dislocation.

lycopene *(lī'-kō-pēn).* A red carotenoid pigment found in tomatoes and some berries and fruits. Some studies suggest it has cancer-fighting abilities.

Lyme disease *(līm).* An inflammatory disorder transmitted to humans by a spirochete, *Borrelia burgdorferi,* which is carried by a tick. The disease is characterized by distinctive skin lesions and polyarthritis and affects the heart and nervous system.

lymph *(līmf).* A transparent, alkaline fluid present in the lymphatic system. Lymph is produced in tissue spaces throughout the body and is transported in small vessels.

lymphadenectomy *(līm-făd-ĕ-nĕk'-tĕ-mē).* Surgical removal of a lymph node.

lymphadenitis *(līm-făd-ĕ-nī'-tĭs).* Inflammation of one or more lymph nodes, usually due to an infection elsewhere in the body.

lymphadenoma *(lĭm-făd-ĕ-nō'-mă).* See **lymphoma.**

lymphadenopathy *(lĭm-făd-ĕ-nŏp'-ĕ-thē).* Disease of the lymph nodes.

lymphangiogram *(lĭm-făn'-jē-ō-grăm).* A radiograph of the lymphatic vessels.

lymphangioma *(lĭm-făn-jē-ō'-mă).* A benign tumor composed of lymphatic vessels.

cavernous l. Dilated lymph vessels filled with lymph, always occurring in the neck or armpit.

simple l. A lymphangioma composed of small lymphatic channels that usually occur in the head and neck or armpit.

lymphangitis *(lĭm-făn-jī'-tĭs).* Inflammation of one or more lymphatic vessels.

lymphatic *(lĭm-făt'-ĭk).* Referring to lymph.

lymphatic vessels. Thin-walled channels that carry lymph from the tissues.

lymphedema *(lĭm-fĕ-dē'-mă).* Edema caused by obstruction of the lymphatic vessels.

lymphoblast *(lĭm'-fō-blăst).* A cell that develops into a lymphocyte.

lymphoblastoma *(lĭm-fō-blăs-tō'-mă).* A tumor composed of lymphocytes.

lymphocyte *(lĭm'-fō-sīt).* A lymph cell or a white blood cell that has no cytoplasmic granules. They are classified into two groups: B-lymphocytes and T-lymphocytes.

lymphocytosis *(lĭm-fō-sī-tō'-sĭs).* An excessive amount of normal lymphocytes in the blood of any effusion.

lymphoepithelioma *(lĭm-fō-ĕp-ĭ-thē-lē-ō'-mă).* A type of squamous cell cancer that involves the lymphoid tissue of the tonsils and nasopharynx.

lymphogranuloma *(lĭm-fō-grăn-ū-lō'-mă).* Hodgkin's disease.

lymphokine *(lĭm'-fō-kīn).* A substance released by sensitized lymphocytes when they encounter specific antigens.

lymphoma *(lĭm-fō'-mă).* Any new and abnormal growth disorder of lymphoid tissue.

Burkitt l. A type of non-Hodgkin's lymphoma that usually occurs in people ages 12 to 30. The disease most often (90% of the time) involves a rapidly growing tumor in the abdomen; other locations include the testes, sinuses, bone, lymph nodes, skin, bone marrow, and central nervous system.

low-grade l. A lymphoma that grows and spreads slowly, in contrast to intermediate- and high-grade lymphomas, which usually grow and spread more quickly.

non-Hodgkin's l. Malignant tumor that arises in the lymphatic system. Symptoms may include swollen but not painful lymph nodes, gastrointestinal distress, skin problems, night sweats, unexplained weight loss, fever, and itching.

lymphosarcoma *(lĭm-fō-săhr-kō'-mă).* A malignant tumor of lymphatic tissue.

lyophilization *(lī-ŏf-ĭ-lī-zā'-shĕn).* A process by which material is rapidly frozen and then dehydrated under high vacuum.

lysine *(lī'-sĕn).* An essential amino acid that is necessary for optimal growth in infants and maintenance of nitrogen balance in adults.

lysis *(lī'-sĭs).* Destruction. For example, hemolysis is the destruction of red blood cells; bacteriolysis is the destruction of bacteria.

M

Maalox *(mā'-lŏkx).* Trade name for an over-the-counter preparation of magnesia and alumina, for treatment of indigestion.

macerate *(măs'-ĕr-āt).* To soften by soaking or steeping in water.

macrobiotics *(măk-rō-bī-ŏ'-tĭks).* A dietary approach based on ayurvedic principles in which foods are combined in specific ways. The diet is based mainly on brown rice and vegetables.

macrocyte *(măk'-rō-sīt).* An abnormally large red blood cell (larger than 10 microns in diameter), also called *megalocyte.*

macrodontia *(măk-rō-dŏn'-shă).* A developmental disorder characterized by an increase in the size of one or more teeth.

macroglossia *(măk-rō-glŏs'-ē-ă).* Having an abnormally large tongue.

macromolecule *(măk-rō-mŏl'-ĕ-kūl).* A very large molecule; examples include proteins and polysaccharides.

macrophage *(măk'-rō-fāj).* Any one of the many types of mononuclear phagocytes present in tissues.

macroscopic *(măk-rō-skŏp'-ĭk).* Something that is visible with the unaided eye or without a microscope.

macrostomia *(măk-rō-stō'-mē-ă).* An abnormally wide mouth due to the failure of the maxillary and mandibular prominences to come together properly.

macula *(măk'-ū-lă).* A small spot, stain, or thickened area distinguishable from the surrounding area because of color or another characteristic.

m. folliculi. Point on the ovarian follicle where it ruptures.

m. lutea retinae. An irregular yellow depression on the retina and the site where short wavelengths of light are absorbed. Differing size, shape, and color of this area may be associated with variant types of color vision.

macular degeneration. An eye disease in which the macula is progressively destroyed, impairing central vision and hindering an individual's ability to see straight ahead and to read, drive, and perform many other everyday tasks.

macule *(măk'-ūl).* A discolored, nonelevated lesion on the skin.

mad cow disease. An infectious,

fatal disease that affects the brain. It is seen in cattle and other animals and can be transmitted to humans.

magnesium *(măg-nē'-sē-ŭm).* A mineral element found in soft tissue, bones, muscles and in minute amounts in body fluids.

m. carbonate. Available as an oral supplement to neutralize stomach acidity.

m. citrate. Form used as a laxative before diagnostic procedures or surgery of the colon; also used to prevent and treat hypomagnesemia (magnesium deficiency).

m. gluconate. Available as an oral supplement to help prevent hypomagnesemia.

m. lactate. Available as an oral supplement to help prevent hypomagnesemia.

m. stearate. A compound of magnesium, used as a lubricant in tablets during drug manufacturing.

m. sulfate. A salt used to prevent and treat seizures associated with toxemia in pregnancy; also used as a laxative and to treat or prevent hypomagnesemia.

magnetic resonance imaging (MRI) *(măg-nĕ'-tĭk).* A method for viewing internal organs using magnetism, radio waves, and a computer that produces the body images on a screen. It is painless and avoids exposure to X-rays.

magnet therapy. An alternative therapy in which magnetic fields (as magnet, magnetic field-generating devices, or magnetic blankets or other objects) are applied to parts of the body. This therapy reportedly helps treat arthritis, circulatory disorders, fibromyalgia, infections, inflammation, insomnia, pain, and other conditions.

magnification *(măg-nĭ-fĭ-kā'-shŭn).* Process of increasing the apparent size of an object, as under a microscope.

maidenhead *(mā'-dĕn-hĕd).* Lay term for the thin fold of mucous membrane that covers or surrounds the vaginal opening; *see also* **hymen.**

mainstreaming *(mān'-strēm-ĭng).* Term referring to the practice of taking disabled children or adults from institutionalized settings and attempting to integrate them into normal environments.

maintenance drug therapy. A course of drug therapy that is sufficient to prevent recurrence of symptoms.

maitake *(mĕ-tăh'-kē).* A medicinal mushroom (*Grifola frondosa*) that has been used by the Japanese for centuries to improve overall health and strengthen the body. There is some evidence that use of maitake can reduce blood pressure.

malabsorption syndrome *(măl-ăb-zŏrp'-shŭn).* Any condition in which there is inadequate or disordered absorption of nutrients from the intestinal tract. May be associated with conditions such as celiac, lactase deficiency, pancreatic insufficiency, among others.

malacia *(mă-lā'-shē-ă).* Abnormal softening of tissues or of an organ.

maladjusted *(măl-ă-jŭs'-tĕd).* Poorly adjusted.

malady *(măl'-ĕ-dē).* Disease.

malaise *(măh-lāz').* A vague feeling of fatigue and overall physical discomfort.

malalignment *(măl-ĕ-līn'-mĕnt).* Displacement of structures that should be in line; especially refers to teeth.

malar bone *(mā'-lĕr).* A bone on each side of the face; cheekbone.

malaria *(mă-lār'-ē-ă).* An infectious disease endemic in warm regions, caused by protozoa of the genus *Plasmodium,* usually transmitted by infected mosquitoes.

malemission *(măl-ĕ-mĭsh'-ŭn).* Failure of semen to be ejaculated during intercourse.

maleruption *(măl-ē-rŭp'-shŭn).* Incorrect eruption of a tooth, resulting in it being out of its normal position.

malformation *(măl-fŏr-mā'-shŭn).* Abnormal shape or structure, especially congenital, of an organ or larger region of the body.

malignancy *(mă-lĭg'-năn-sē).* A cancerous tumor or neoplasm.

malingerer *(mă-lĭng'-gĕr-ĕr).* An individual who pretends to be ill or to have a nonexistent disorder to arouse sympathy.

malleable *(măl-ē-ă-bl).* Susceptible to being shaped by pressure.

malleolus *(măl-ē'-ō-lŭs).* The protuberance on either side of the ankle joint.

malnutrition *(măl-nū-trĭ'-shŭn).* Any disorder of nutrition caused by an unbalanced or insufficient diet or an inability to properly assimilate or utilize nutrients.

malocclusion *(măl-ō-klū'-zhĕn).* Malposition of the maxillary and mandibular teeth to a degree that it interferes with chewing.

malposition *(măl-pĕ-zĭsh'-ĕn).* An abnormal position of an organ or body part.

malpractice *(măl-prăk'-tĭs).* An improper or damaging practice, especially referring to faulty medical or surgical treatment.

malrotation *(măl-rō-tā'-shŭn).* Failure of the intestine to rotate normally during development in the womb. Malrotation is usually not apparent until the intestine becomes obstructed, characterized by vomiting of bile, abdominal pain, bloody stools, and a swollen abdomen.

maltase *(măwl'-tās).* An enzyme, secreted by the salivary glands and the pancreas, that converts maltose into glucose.

maltose *(măwl'-tōs).* The basic structural component of glycogen and starch, it is used as a sweetener and a nutrient. It is converted into glucose by the enzyme maltase.

malunion *(măl-ūn'-yŭn).* Condition

in which the fragments of a fractured bone grow in a faulty position, forming an imperfect union.

mammalgia *(măm-ăl'-jē-ă).* *See* **mastalgia.**

mammaplasty *(măm'-ĕ-plăs-tē).* Cosmetic surgery of the breast, to either augment or decrease its size.

mammary glands *(măm'-ĕr-ē).* Glands in the female breast that can secrete milk.

mammography *(mă-mŏg'-ră-fē).* Radiography of the mammary gland.

managed care. A system of medical care in which a third party, such as a governmental agency, a corporation, a partnership, or an insurance company, regulates the criteria under which doctors can practice.

mandol *(măn'-dŏl).* Trade name for a preparation of cefamandole, an antibiotic effective against many gram-positive and some gram-negative bacteria.

mange *(mānj).* A communicable skin disease, caused by various mites, that affects cats, dogs, and other domestic animals.

mania *(mā'-nē-ă).* Mental disorder characterized by excessive excitement.

manic-depressive *(măn'-ĭk dē-prĕs'-ĭv).* Alternating between episodes of mania and depression, as seen in bipolar disorder.

manipulation *(mĕ-nĭp-ū-lā'-shĕn).* 1. In physical therapy, the forceful passive movement of a joint beyond its active limit of motion. 2. Skillful treatment, usually by hand.

mannitol *(măn'-ĭ-tŏl).* A carbohydrate obtained from plant sources, it is used as a diuretic to prevent and treat acute renal failure, excrete toxic substances, reduce cerebral edema, and other uses.

manometer *(mĕ-nŏm'-ĕ-tĕr).* An instrument for measuring the pressure or tension of gases or liquids, especially of the blood.

Mantoux test *(măhn-tū').* A test for tuberculosis to determine whether exposure to or infection with *Mycobacterium tuberculosis* has occurred.

manual lymph drainage *(măn'-ū-ĕl).* A massage technique in which practitioners use slow, gentle, repetitive strokes to improve circulation of lymph to heal various conditions, stimulate the nervous system, and facilitate recovery.

manus *(mă'-nĕs).* Hand.

MAO inhibitors. *See* **monoamine oxidase inhibitor.**

maple sugar urine disease. A fatal, inherited metabolic disease of infants, in which amino acid metabolism is defective. The disease gets its name from the fact that the urine and sweat of affected infants have the odor of maple syrup.

marasmus *(măr-ăz'-mŭs).* A form of protein and energy malnutrition mainly caused by prolonged calorie deprivation, usually during the first year of life, and characterized by progressive wasting of muscle and subcutaneous fat.

marble bones *(măr'-bĕl).* Bones that are abnormally hard and dense due to excessive calcification; *see also* **osteopetrosis.**

Marfan's disease *(măhr-făhnz').* An inherited disease that affects the connective tissue, bones, ligaments, skeletal structures, and muscles. Characteristics include an irregular gait, abnormal joint flexibility, flat feet, stooped shoulders, and a dislocated optic lens.

margination *(măr-jĭ-nā'-shŭn).* The adhesion of red blood cells to blood vessel walls during the early stages of inflammation.

Marie's disease *(mă-rēz').* Chronic condition characterized by enlargement of bones and soft tissues of the hands, feet, and face; *see also* **acromegaly.**

marijuana *(măr-ĭ-wă'-nă).* Dried flowing tops of the *Cannabis sativa* plant, usually rolled into cigarettes and smoked for its euphoric effects.

marrow *(măr'-ō).* The soft tissue present in the cavities of long bones (yellow marrow) and in the spaces of spongy bones (red marrow).

marrow transplant. Transplantation of bone marrow from a donor to a recipient who has a blood disease, such as leukemia.

marsupialization *(măr-sū-pē-ăl-ĭ-zā'-shŭn).* Surgical procedure in which the borders of an evacuated tumor sac are raised and stitched to the edges of the abdominal wound to form a pouch.

masculinization *(măs-kū-līn-ĭ-zā'-shŭn).* The normal development of primary or secondary sex characteristics in males.

masking of symptoms. Concealment of a medical condition that may result when an individual takes a narcotic or other type of medication.

masochism *(măs'-ō-kĭz-ĕm).* The act of deriving pleasure from experiencing physical or psychological pain; it usually refers to sexual masochism.

mass *(măs).* 1. A body or lump composed of cells that adhere to each together. 2. The characteristic that gives matter the quality of not being able to move spontaneously.

massage *(mă-săhzh').* A systematic therapeutic method that involves stroking, kneading, and in other ways applying friction to the body, primarily using the hands but, in some forms of massage, also using the forearms and elbows.

cardiac m. The use of rhythmic pressure applied over the sternum or directly to the heart to reinstitute and maintain circulation.

Swedish m. The most commonly used type of classical Western massage, used especially for relief of muscular tension, improvement of circulation and range of motion, and relaxation.

masseter muscle *(măs-sē'-tĕr).* The muscle that closes the mouth; also the main muscle involved in chewing.

mastalgia *(măs-tăl'-jē-ă).* Pain in the breast.

mast cell. A connective tissue cell whose normal function is not known but which releases histamine and other chemicals when it is damaged in allergic reactions. Also known as a *mastocyte.*

mastectomy *(măs-tĕk'-tŏ-mē).* Surgical removal of a breast or part of the breast.

extended radical m. Radical mastectomy along with removal of the ipsilateral half of the sternum, part of the ribs two through five, and internal mammary lymph nodes.

modified radical m. Simple mastectomy and removal of the axillary lymph nodes, but not the pectoral muscles.

radical m. Removal of the breast, pectoral muscles, axillary lymph nodes, and relevant associated skin and subcutaneous tissue.

simple m. Removal of only the breast tissue and nipple and a minimal portion of the overlying skin.

masticate *(măs'-tĭ-kāt).* To chew food in preparation for swallowing and digestion.

mastitis *(măs-tī'-tĭs).* Inflammation of the mammary gland or breast.

mastoid portion of temporal bone *(măs'-toid).* Portion of the temporal bone in back of the ear; it is filled with air cells and can become infected after a bout of severe ear infection.

mastoidectomy *(măs-toid-ĕk'-tĕ-mē).* Surgical removal of the mastoid air cells.

mastoiditis *(măs-toid-ī'-tĭs).* Inflammation of the mastoid portion and the air cells, often as a result of an ear infection.

masturbation *(măs-tĕr-bā'-shŭn).* Self-stimulation of the genitals for sexual pleasure.

maternal *(mă-tĕr'-năl).* Referring to the mother.

matrix *(mā'-trĭks).* The intracellular structure of a tissue or the tissue from which a structure develops.

maturation *(măch-ū-rā'-shŭn).* The process or stage of becoming mature or fully developed.

maturity *(mă-chūr'-ĭ-tē).* The period when maximal development has been reached.

maxillary *(măk'-sĭ-lăr-ē).* Referring to the upper jaw.

maxillofacial surgeons *(măks-ĭl-ō-fā'-shăl).* Physicians who concern themselves with surgical treatment of conditions affecting the maxilla, head, and neck.

McBurney's point *(mĕk-bĕr'-nē).* A point on the abdomen that corresponds with the normal position of the base of the appendix; it is especially tender in acute appendicitis.

mean *(mēn).* The average.

measles *(mē'-zĕls).* A highly contagious viral disease caused by a paramyxovirus and characterized by fever, sneezing, general malaise, nasal congestion, cough, spotted cheeks, conjunctivitis, and a body rash.

meatitis *(mē-ĕ-tī'-tĭs).* Inflammation

of the urinary meatus, most often seen in males as a complication of circumcision.

meatotomy *(mē-ă-tŏt'-ō-mē).* Surgical incision of the urinary meatus to enlarge the opening.

meatus *(mē-ā'-tŭs).* An opening or passage.

meconium *(mē-kō'-nē-ŭm).* 1. The first feces of a newborn infant, often green-black to light brown and almost odorless.

media *(mē'-dē-ă)* (*sing.* medium). Middle.

medial *(mē'-dē-ăl).* Referring to the middle or to the midline of a body or structure.

median *(mē'-dē-ăn).* 1. Middle. 2. In statistics, a number obtained by arranging a series of numbers in order of magnitude and then taking the middle number.

mediastinitis *(mē-dē-ăs-tĭ-nī'-tĭs).* Inflammation of the tissue of the mediastinum.

mediastinotomy *(mē-dē-ăs-tĭ-nŏt'-ō-mē).* Surgical removal of the mediastinum.

mediastinum *(mē-dē-ăs-tī'-nŭm).* A partition (septum) or cavity between two main areas of an organ.

Medicaid *(mĕd'-ĭ-kād).* A state and federally funded government program that provides medical care for the poor.

medical *(mĕd'-ĭ-kăl).* Referring to medicine or to the treatment of disease or other health-related conditions.

medical examiner. A physician who is qualified to investigate the cause of death and the circumstances under which death occurred.

medical group. A number of physicians of different specialties who agree to work together in the treatment of patients.

medicamentosus *(mĕd-ĭ-kĕ-mĕn-tō'-sĕs).* Referring to, used in, or caused by one or more medications.

Medicare *(mĕd'-ĭ-kăr).* The US government's health insurance program for people 65 and older and individuals with certain disabilities. Part A covers inpatient hospital stays; part B covers doctor and outpatient services.

medication *(mĕd-ĭ-kā'-shŭn).* A drug or medicine.

medicinal *(mĕd-dĭs'-ĭn-ăl).* Possessing healing qualities.

medicine *(mĕd'-ĭ-sĭn).* Any drug or remedy.

alternative m. A system of techniques that are used instead of or in place of conventional medical remedies. For example, in a cancer patient, use of a special diet rather than chemotherapy, radiation, or other conventional techniques to treat the disease.

aviation m. Branch of medicine that deals with health problems related to aviation.

Chinese herbal m. A branch of traditional Chinese medicine that focuses on the use of medicinal herbs to both support the body's healing powers and to treat disease.

complementary m. A system

of techniques that are used along with conventional medicine to achieve a goal. For example, the use of aromatherapy after surgery to reduce a patient's pain and discomfort.
environmental m. A branch of medicine that looks at the impact of the environment on humans, including factors such as population growth, pollution, radiation, and temperature changes.
forensic m. A branch of medicine that involves the application of medical knowledge to the law.
holistic m. A system of medicine that considers each human being as a whole, functioning unit rather than looking at an individual's disease or symptoms as being separate from the person.
integrative m. According to the National Center for Complementary and Alternative Medicine, it is a medicine that combines conventional medical techniques and complementary/alternative medicine therapies for which there is some significant scientific evidence of effectiveness and safety.
internal m. A medical specialty concerned with the diagnosis and medical treatment of diseases and disorders that affect the internal structures.
mind-body m. A holistic approach that considers the effect of one's thoughts, emotions, and feelings on the physical body, especially the immune system.
nuclear m. A branch of medicine that involves the use of radionuclides in the diagnosis and treatment of disease.
occupational m. A branch of medicine that involves the study, prevention, and treatment of diseases and injuries that are related to the workplace
orthomolecular m. A system for the prevention and treatment of disease based on the concept that an individual's biochemical makeup is unique and genetically determined. Treatment consists of supplementation with substances that are natural to the body, including vitamins, minerals, and amino acids.
socialized m. A system of medical care that is controlled by the government and for which individuals pay no direct cost or a nominal fee, as care is covered by tax monies collected from the population.
social m. Branch that deals with environmental and community factors as they impact health and disease.
sports m. A field of medicine that involves the prevention, diagnosis, and treatment of injuries sustained during athletic activities.
traditional Chinese m. A complex system of medical theory and practice composed of four branches—herbal med-

icine, acupuncture and moxibustion, qi gong, and tui na—and based on several basic characteristics, including the principle of yin/yang and the idea that basic substances pervade the body.

veterinary m. A medical specialty consisting of the diagnosis and treatment of diseases of animals.

medicolegal *(mĕd-ĭ-kō-lē'-gĕl).* Referring to medicine and law; *see* **forensic medicine.**

medigap *(mĕd'-ĭ-găp).* In the US, health insurance sold by private insurance companies that is designed to supplement Medicare and to fill the gaps in health-care coverage.

meditation *(mĕd-ĭ-tā'-shŭn).* A self-directed practice for calming the mind and relaxing the body. Studies show that regular meditation can improve quality of life, reduce pain, high blood pressure, anxiety, and stress, and increase longevity.

medium *(mē'-dē-ŭm).* 1. A substance that transmits impulses. 2. A preparation used to cultivate microorganisms or cellular tissue.

Medline *(mĕd'-līn).* A computerized medical bibliography individuals can access online.

Medrol *(mĕd'-rŏl).* Trade name for preparations of methylprednisolone, used for treatment of inflammatory conditions and as replacement therapy for adrenocorticol insufficiency.

medroxyprogesterone *(mĕd-rŏk-sē-prō-jĕs'tĕr-ŏn). See* Appendix, Common Prescription and OTC Drugs: By Generic Name.

medulla *(mĕ-dŭl'-ă).* Term for the most interior part of a structure or organ.

medulla oblongata. The lower portion of the brain stem. It consists of critical nerve cells that are involved with vital functions such as respiration and circulation.

medulloblastoma *(mĕ-dŭl-ō-blăs-tō'-mă).* A soft malignant tumor that develops on the roof of the fourth ventricle and cerebellum, often invading the meninges. Most occur in children.

megacardia *(mĕg-ă-kăr'-dē-ă). See* **cardiomegaly.**

megacolon *(mĕg-ă-kō'-lŏn).* An abnormally large or dilated colon.

megakaryocyte *(mĕg-ă-kăr'-ē-ō-sīt).* A giant cell with a large or multiple nuclei, seen in bone marrow.

megaloblast *(mĕg'-ă-lō-blăst).* A large, nucleated abnormal red blood cell with a slightly irregular shape, found in pernicious anemia.

megalocyte *(mĕg'-ă-lō-sīt).* An abnormally large red blood cell. Also called *macrocyte.*

megalodactyly *(mĕg-ă-lō-dăk'-tĕ-lē).* Abnormally large fingers or toes.

megalomania *(mĕg-ă-lō-mā'-nē-ă).* An unreasonable conviction that one is extremely important or has tremendous power or influence. The ideas are referred to as *delusions of grandeur.*

megavitamin therapy *(mĕg-ă-vī'-tĕ-mĭn). See* **orthomolecular medicine.**

meibomium glands *(mī-bō'-mē-ĕn).* Minute glands in the eyelids. When the exit points become blocked, a cyst can form.

meibomium stye. Stye or cyst that forms in the eyelid due to a blockage of fluid from the meibomium glands; *see also* **chalazion.**

meiosis *(mī-ō'-sĭs).* A method of cell division among sex cells in which two successive divisions of the nucleus produce cells that contain half the number of chromosomes present in somatic (body) cells.

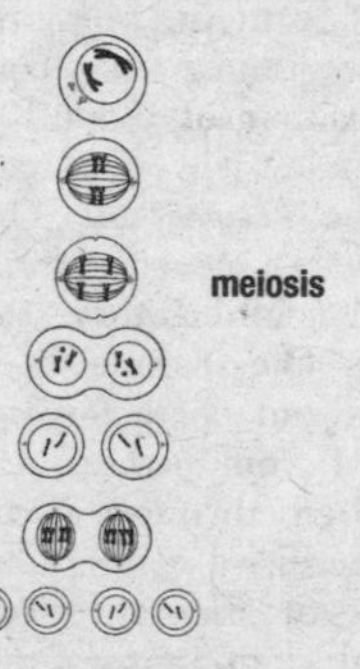

melancholia *(mĕl-ăn-kŏl'-lē-ă).* In current psychiatric terms, a term used to refer to severe forms of major depressive disorder.

melanin *(mĕl'-ă-nĭn).* Any of several dark pigments of the skin, hair, various tumors, the substantia nigra of the brain, and the choroids of the eye.

melanoblastoma *(mĕl-ă-nō-blăs-tō'-mă).* A tumor that contains melanin.

melanocyte *(mĕl-ăn'-ō-sīt).* A melanin-forming cell.

melanoma *(mĕl-ă-nō'-mă).* A pigmented tumor or mole that may or may not be malignant.

acral-lentiginous m. The most common type of melanoma in nonwhite individuals, it most often appears on the soles and palms.

lentigo maligna m. A malignant skin melanoma that occurs most often on sun-exposed areas of the skin.

superficial spreading m. The most common type of malignant melanoma, it occurs most often on the lower leg or back and appears as a small pigmented macule or a slightly palpable flat lesion with an irregular outline.

melanosis *(mĕl-ĕ-nō'-sĭs).* Excessive pigmentation of a part of the body due to a disorder in melanin metabolism.

melanuria *(mĕl-ă-nū'-rē-ă).* Excretion of urine that is darkly stained or that turns dark when left standing.

melatonin *(mĕl-ă-tō'-nĭn).* Hormone produced by the pineal gland in mammals. In humans it is involved in the regulation of sleep, mood, ovarian cycles, and puberty.

melena *(mĕl'-ĕ-nă).* Black, tarry

stools stained with blood pigments or altered blood, seen in some newborns.

melitis *(mĕ-lī'-tĭs).* Inflammation of the cheek.

Mellaril *(mĕl'-ĕ-rĭl).* Trade name for preparations of thioridazine, used to treat schizophrenia and acute psychotic episodes.

membrana tympani *(mĕm-brā'-nă).* *See* **eardrum.**

membrane *(mĕm'-brān).* A thin layer of tissue that covers a surface, divides a space or organ, or lines a cavity.

elastic m. A membrane composed mainly of elastic fibers.

fetal m. One of several membranous structures that protects and supports the embryo. They include the allantois, amnion, chorion, deciduas, placenta, and yolk sac.

mucous m. Any membrane that lines cavities or canals that are in contact with the air and are kept moist by secretion of mucus.

nictitating m. A third eyelid present in lower vertebrates (e.g., a cat).

permeable m. Any membrane that allows water and certain substances in solution to pass through it.

placental m. Membrane of the placenta that separates the maternal blood from fetal blood.

serous m. A membrane that is composed of mesothelium lying on a thin layer of connective tissue that lines closed body cavities (e.g., peritoneal).

synovial m. A membrane that lines a joint and secretes synovial fluid.

memory *(mĕm'-ŏr-ē).* The mental registration, retention, and recall of past experiences, ideas, sensations, thoughts, and knowledge.

anterograde m. The ability to remember events that occurred in the remote past but an inability to remember recent events.

immunologic m. The ability of the immune system to respond more quickly and effectively to subsequent antigenic challenges than to the first exposure.

long-term m. Memory that is retained over long periods of time.

short-term m. Memory that is lost within a few seconds to a maximum of about 30 minutes unless it is reinforced.

menarche *(mĕ-năhr'-kē).* The beginning of menstruation.

Mendelian inheritance *(mĕn-dē'-lē-ĕn).* The manner in which genes and characteristics are passed from parents to their children through dominant and recessive genes. The four modes of inheritance are autosomal dominant, autosomal recessive, X-linked dominant, and X-linked recessive.

Meniere's disease *(mān-ē-ārz').* A recurrent, usually progressive group of symptoms that include severe dizziness, progressive deafness, ringing in the ears, and a sensation of fullness or pressure in the ears.

meningeal *(mĕ-nĭn'-jē-ĕl).* Of or referring to the meninges.

meninges *(mĕ-nĭn'-jēz).* The three membranes that cover the brain and spinal cord: pia mater, dura mater, and arachnoid.

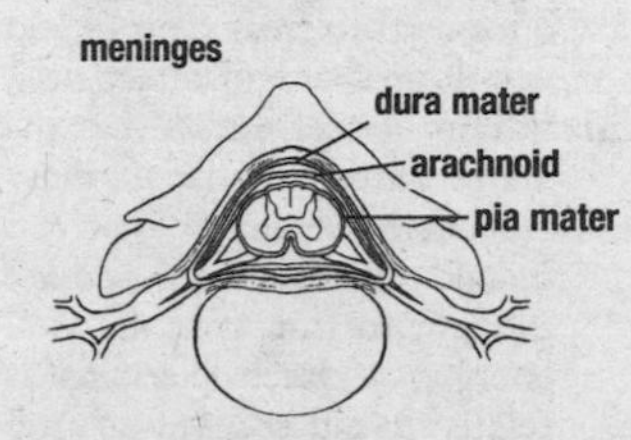

meningioma *(mĕ-nĭn-jē-ō'-mă).* A benign, slow-growing tumor that affects the meninges, it may cause increased intracranial pressure or other problems.

meningitis *(mĕn-ĭn-jī'-tĭs).* Inflammation of the meninges, usually caused by a bacterium or a virus.

aseptic m. Any of several mild types of meningitis, most of which are caused by viruses.

bacterial m. Meningitis caused by bacteria, most often including *Haemophilus influenzae, Neisseria meningitides, Streptococcus pneumoniae,* and *Mycobacterium tuberculosis.*

viral m. Meningitis due to any of various viruses, including coxsackieviruses, the virus of lymphocytic choriomeningitis, and the mumps virus.

meningocele *(mĕn-ĭn'-gō-sēl).* Congenital hernia in which the meninges protrude through an opening of the spinal column or the skull.

meningococcus *(mĕn-ĭn-gō-kŏk'-ŭs).* A microorganism of *Neisseria meningitides* and the cause of epidemic cerebral meningitis.

meningoencephalitis *(mĕn-ĭn-gō-ĕn-sĕf-ă-lī'-tĭs).* Inflammation of the brain and its meninges, usually caused by bacteria but occasionally due to free-living amebae.

meniscectomy *(mĕn-ĭ-sĕk'-tō-mē).* Removal of meniscus cartilage of the knee.

meniscus *(mĕn-ĭs'-kŭs).* Generally, any crescent-shaped structure in the body. Specifically, one of the crescent-shaped disks of fibrocartilage in the knee.

menopause *(mĕn'-ō-păws).* The period that marks the permanent cessation of menstruation. It usually occurs between the ages of 35 and 58.

menorrhagia *(mĕn-ō-rā'-jē-ă).* Excessive bleeding during a menstrual period, either in terms of the number of days and/or the amount of blood expelled.

menorrhea *(mĕn-ō-rē'-ă).* 1. Normal menstruation. 2. Profuse menstruation.

Mensendieck system *(mĕn-sĕn'-dē-ĕk).* A type of movement therapy designed to help improve the body's structure and function and eliminate any accompanying aches and pains. The system consists of more than 200 exercises and is taught by trained teachers.

menstrual cycle *(mĕn'-strū-ăl).* The

recurrent preparation of the uterus to receive a fertilized egg and the expulsion of the uterine lining (menstruation) when no egg reaches the uterus. This occurs at roughly 28-day intervals.

menstruation *(mĕn-strū-ā'-shŭn).* The periodic discharge of blood and other fluids from the uterus, occurring at roughly 28-day cycles during the life of women from puberty to menopause.

mentality *(mĕn-tăl'-ĭ-tē).* Mental power or ability.

mental retardation. A mental disorder characterized by subnormal intellectual functioning associated with impaired adaptive behavior and which becomes evident during the developmental stage.

menthol *(mĕn'-thŏl).* An alcohol derived from the oils of numerous species of *Mentha* or made synthetically and used topically to treat itching or as an inhalant for treatment of upper respiratory conditions.

meperidine *(mĕ-pĕr'-ĭ-dēn).* Generic name of a synthetic opioid analgesic, used to relieve moderate to severe pain.

meprobamate *(mĕ-prō'-bĕ-māt).* Generic name for a drug used to relieve anxiety and tension and to promote sleep; trade names: Equanil, Miltown.

mercurial *(mĕr-kū'-rē-ăl).* 1. Referring to mercury. 2. A substance that contains mercury.

mercury *(mĕr'-kū-rē).* A metallic element that is liquid at room temperature and toxic when it enters the body.
ammoniated m. A topical antiseptic used to treat certain skin diseases.
bichloride m. Extremely poisonous compound that was once used to treat syphilis and now is used as a disinfectant.

meridians *(mĕ-rĭd'-ē-ăns).* According to Eastern medicine, these are energy channels that run up and down the body and are thought to transport the vital energy, qi. Each meridian is related to either a yin or yang organ. *See* **yin and yang.**

meropia *(mĕr-ō'-pē-ă).* Partial blindness.

merosmia *(mĕr-ŏs'-mē-ă).* Partial loss of the sense of smell, with certain odors not being perceived.

mescaline *(mĕs'-kă-lēn).* A poisonous hallucinogenic substance found in mescal buttons (mescal cactus).

mesencephalon *(mĕz-ĕn-sĕf'-e-lŏn).* The midbrain.

mesenchyma *(mĕz-ĕng'-kĭ-mĕ).* The meshwork of cells that form the embryonic mesoderm and develop into connective tissues, blood, blood vessels, the lymphatic system, and the reticuloendothelial system.

mesenchymoma *(mĕs-ĕn-kĭ-mō'-mă).* A neoplasm that contains mesenchymal and fibrous tissue.

mesenteric arteries *(mĕs-ĕn-tĕr'-*

īk). Two large blood vessels that originate in the abdominal aorta and supply blood to the intestines.

mesentery *(mĕz'-ĕn-tĕr-ē).* A membranous fold that attaches an organ to the body wall.

mesial *(mē'-zē-ăl).* 1. Toward the middle point. 2. In dentistry, on the side toward the center line of the dental arch.

mesoappendix *(mĕz-ō-ă-pĕn'-dĭx).* The peritoneal fold that attaches the appendix to the membranous fold of the ileum.

mesoderm *(mĕz'-ō-dĕrm).* The middle layer of the three main germ layers of the embryo, from which are derived connective tissue, bone, cartilage, muscle, blood and blood vessels, lymphatics, lymphoid organs, pericardium, peritoneum, kidney, and gonads.

mesomorph *(mĕz'-ō-mŏrf).* An individual with a body type in which the tissues are derived mainly from the mesoderm, characterized by a heavy, hard physique with a rectangular shape.

mesothelioma *(mĕz-ō-thē-lē-ō'-mă).* A tumor derived from the mesothelial tissue (pleura, peritoneum, pericardium), it can be either benign or malignant.

mesothelium *(mĕz-ō-thē'-lē-ŭm).* The layer of fat cells, derived from the mesoderm, that line the coelom or body cavity of the embryo and, in adults, it forms the epithelium that covers serous membranes.

metabiosis *(mĕt-ă-bī-ō'-sĭs).* The dependence of an organism upon another for its existence.

metabolism *(mĕ-tăb'-ō-lĭz-ĕm).* The sum of all the physical and chemical changes that occur within an organism, including all the energy and material transformations, such as transforming food into elements to be used by the body.

basal m. The minimal amount of energy expended to maintain respiration, circulation, body temperature, muscle tone, and other functions concerned with growth and nutrition.

drug m. The biotransformation of drugs in the body.

metacarpal bones *(mĕt-ă-kăr'-pĕl).* The bones in the hand between the wrist and the fingers.

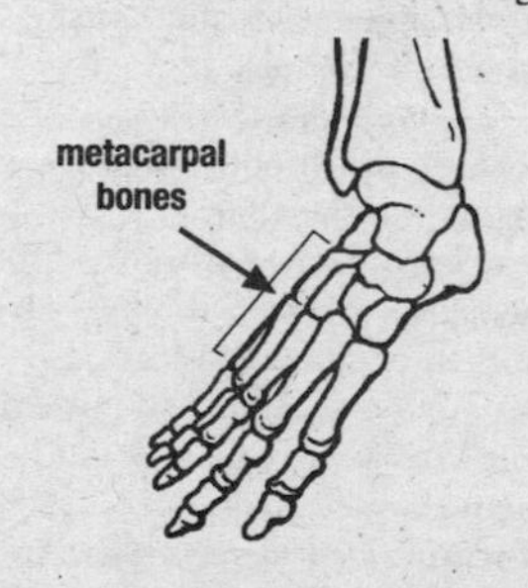

metachronous *(mĕ-tăk'-rĕ-nĕs).* Occurring at different times.

metamorphosis *(mĕt-ă-mŏr'-fĕ-sĭs).* A change in shape, especially when something changes from one developmental stage to another, as from an egg to a larva.

Metamucil *(mĕt-ĕ-mū'-sĭl).* Trade

name for a preparation of psyllium, used as a bulk-forming laxative.

metaplasia *(mĕt-ă-plā'-sē-ă).* A change in the type of adult cells in a tissue to a form that is abnormal for that tissue.

metastasis *(mĕ-tăs'-tĕ-sĭs).* The spread of disease from one organ or part of an organ to another not directly connected with it.

metastasize *(mĕ-tăs'-tĕ-sīz).* To form a new foci of disease in another part of the body by metastasis.

metatarsal bones *(mĕt-ĕ-tăhr'-sĕl).* Referring to the metatarsus.

metatarsalgia *(mĕt-ĕ-tăhr-săl'-jĕ).* Tenderness and pain in the metatarsal area.

metatarsophalangeal joints *(mĕt-ă-tăhr-sō-fĕ-lăn'-jē-ăl).* Referring to the metatarsus and the phalanges of the toes.

metatarsus *(mĕt-ĕ-tăhr'-sĕs).* The part of the foot between the tarsus and the toes.

meteorism *(mē'-tē-ŏr-ĭzm).* The presence of gas in the abdomen or intestinal tract.

meter *(mē'-tĕr).* Unit of linear measure approximately equivalent to 39.37 inches.

metformin *(mĕt'-fŏr-mĭn). See* Appendix, Common Prescription and OTC Drugs: By Generic Name.

meth *(mĕth).* A common abbreviation for methamphetamine.

methadone *(mĕth'-ă-dōn).* A synthetic opioid painkiller that has actions similar to those of morphine and heroin. It is used in the treatment of heroin addiction.

methamphetamine *(mĕth-ăm-fĕt'-ă-mēn).* A substance closely related chemically to amphetamine and ephedrine, with actions similar to those of amphetamine.

methemoglobin *(mĕt-hē'-mō-glō-bĭn).* A brown pigment formed from hemoglobin by oxidation of ferrous iron to ferric iron, which may be due to poisoning by certain chemicals or heredity.

methicillin *(mĕth-ĭ-sĭl'-ĭn).* A semisynthetic penicillin used to treat resistant staphylococcal infections.

methionine *(mĕ-thī'-ō-nīn).* An essential amino acid that is necessary for normal metabolism.

methocarbamol *(mĕth-ō-kăhr-bĕ-mŏl).* Generic name of a drug used to treat painful musculoskeletal conditions.

methotrexate *(mĕth-ō-trĕk'-sāt).* Generic name for a drug used to treat various malignancies, as well as severe rheumatoid and psoriatic arthritis.

methylcellulose *(mĕth-ĕl-sĕl'-ū-lōs).* A tasteless powder that is used as a bulk substance in foods and laxatives, and also as an adhesive or emulsifier.

methyldopa *(mĕth-ĕl-dō'-pă).* Generic name for a drug used to treat hypertension; it is a derivative of the amino acid phenylalanine.

methylene blue *(mĕth'-ĭ-lēn).* A dark green crystalline powder that is used as an antidote for carbon monoxide and cyanide poisoning.

methylmercury *(mĕth-ĕl-mĕr'-kĕr-ē).* An organic form of mercury that is highly toxic and causes many adverse effects and readily affects the brain. Those at highest risk are children of women who consume large amounts of certain fish during pregnancy, as methylmercury accumulates in predator fish.

metoclopromide *(mĕt-ō-klō'-prĕ-mīd).* Generic name for a drug used to stimulate gastric motility and to treat gastroesophageal reflux (GERD) and gastroparesis. Trade name: Reglan.

me-too drug. A drug that is structurally very similar to drugs that are already on the market and thus may provide little or no medicinal benefit but may help create competition and force prices down.

metopic *(mē-tŏp'-ĭk).* Frontal.

metoprolol *(mĕt-ō-prō'-lŏl). See* Appendix, Common Prescription and OTC Drugs: By Generic Name.

metric system *(mĕt'-rĭk).* A system of weights and measurements based on the meter (measurement), gram (weight), and liter (volume).

metritis *(mĕ-trī'tĭs).* Inflammation of the uterus.

metrocolpocele *(mĕ-trō-kō'-pō-sēl).* A hernia of the vagina and uterus.

metrodynia *(me-tro-din-e-a).* Pain in the uterus.

metronidazole *(mĕt-rō-nī'-dĕ-zōl).* Generic name for an antibacterial, antiprotozoal drug, used to treat bacterial vaginosis and intestinal amebiasis.

metrophlebitis *(mĕ-trō-flĕ-bī'-tĭs).* Inflammation of the veins of the uterus.

metroplasty *(mĕ-trō-plăs'-tē).* Plastic surgery on the uterus.

metroptosis *(mĕ-trō-tō'-sĭs).* Prolapse of the uterus.

metrorrhagia *(mĕ-trō-rā'-jē-ă).* Uterine bleeding that occurs at irregular but frequent intervals, with a variable amount of flow, although sometimes prolonged.

miconazole *(mī-kŏn'-ă-zŏl).* A broad-spectrum antifungal medication used topically to treat tinea and cutaneous candidiasis and intravenously to treat systemic fungal infections.

microaerophilic *(mī-krō-ār'-ō-fĭl-ĭk).* The ability to grow at concentrations of oxygen that are lower than those present in the atmosphere. Bacteria are referred to as being microaerophilic.

microanalysis *(mī-krō-ĕ-năl'-ĕ-sĭs).* The chemical analysis of minute particles of matter.

microanastomosis *(mī-krō-ĕn-ăs-tĕ-mō'-sĭs).* Anastomosis between very small tubular structures.

microangiopathy *(mī-krō-ăn-jē-ŏp'-ĕ-thē).* Disease of the small blood vessels.

diabetic m. A complication of diabetes mellitus in which there is a thickening of the capillaries in many vascular beds.

thrombotic m. Development of thrombi in the capillaries and arterioles.

microbes *(mī'-krōbs).* One-celled life forms, especially those that can cause disease, such as fungi, bacteria, and protozoa. Also called *microorganisms.*

microbiology *(mī-krō-bī-ŏl'-ĕ-jē).* The study of microorganisms.

microbiotic *(mī-krō-bī-ŏ'-tĭk).* Referring to microscopic living organisms.

microcalcifications *(mī-krō-kăl-sĭ-fĭ-kā'-shŭn).* A minute area of calcification in the tissue.

microcephaly *(mī-krō-sĕf'-ă-lē).* Congenital condition in which the head is abnormally small; often seen in mental retardation.

microchemistry *(mī-krō-kĕm'-ĭs-trē).* The study of chemical reactions using microscopic quantities.

Micronase *(mī'-krō-nās). See* Appendix, Common Prescription and OTC Drugs: By Trade Name.

micropigmentation *(mī-krō-pĭg-mĕn-tā'-shŭn). See* **permanent makeup.**

microscope *(mī'-krō-skōp).* An instrument used to see enlarged images of minute substances and objects and reveal details not observable with the naked eye.

binocular m. A microscope that has two eyepieces, which allows simultaneous viewing with both eyes.

compound m. A microscope that permits simultaneous viewing of portions of images of two different specimens.

electron m. A microscope that uses an electron beam instead of light to form an image for viewing.

operating m. A microscope used to perform delicate microsurgery, such as those on small blood vessels or the middle ear.

microsurgery *(mī-krō-sĕr'-jĕr-ē).* Dissection of minute structures using an operating microscope and handheld instruments.

microtome *(mī'-krō-tōm).* A device for cutting thin slices of tissue to be studied under a microscope.

micturition *(mĭk-tū-rĭ'-shĕn).* Urination.

midriff *(mĭd'rĭf).* The mid region of the torso.

midwife *(mĭd'wīf).* An individual trained to assist women during childbirth, which may occur in the mother's home, in a hospital, or a special birthing center, depending on local law.

migraine *(mī'-grān).* A vascular headache that typically occurs on one side of the head, and usually accompanied by nausea, vomiting, irritability, diarrhea or constipation, and frequently visual effects.

abdominal m. Migraine in which nausea and vomiting are most prominent, rather than the head pain.

classic m. Migraine that is

preceded by neurologic symptoms (an aura), such as seeing zigzags or spots of light, sensitivity to light and sound, and nausea.

common m. Migraine without an aura.

migration *(mī-grā'-shŭn).* A spontaneous change of place, as when cells move from one position to another or symptoms move.

Mikulicz's disease *(mē'-koo-lĭch).* A chronic, benign, usually painless inflammatory swelling of the salivary and lacrimal glands.

Miles' operation *(mīlz).* A surgical treatment for cancer of the lower sigmoid and rectum, which involves removal of the pelvic colon, mesocolon, and adjacent lymph nodes, and a permanent colostomy.

miliaria *(mĭl-ē-ăr'-ē-ă).* A syndrome of skin changes caused by retention of sweat.

milk *(mĭlk).* The liquid secretion of the mammary gland.

acidophilus m. Milk that is fermented with *Lactobacillus acidophilus* and used to treat gastrointestinal disorders and balance the bacterial flora of the intestinal tract.

homogenized m. Milk in which the fat particles are emulsified to the extent that the cream does not separate.

pasteurized m. Milk that has been heat treated to kill bacteria.

skim m. Milk from which some or all the cream has been removed.

milk thistle. A thorny plant (*Silybum marianum*) used to make an extract called silymarin, which is available as a supplement and has the ability to heal liver problems.

millicurie *(mĭl-ĭ-kū'rē).* A unit of radioactivity, equal to one thousandth of a curie.

milliequivalent (mEq) *(mĭl-ē-ē-kwĭv'-ĕ-lĕnt).* The weight of a substance found in 1 cc (cubic centimeter) of a normal solution.

milligram *(mĭl'-ĭ-grăm).* One one-thousandth of a gram.

milliliter *(mĭl'-ĭ-lē-tĕr).* One one-thousandth of a liter.

millimeter *(mĭl'-ĭ-mē-tĕr).* One one-thousandth of a meter.

mimetic *(mī-mĕt'-ĭk).* Referring to simulation of another bodily process or disease; imitative.

mind *(mīnd).* Integration of the functions of the brain by which an individual becomes aware of his or her surroundings and experiences them.

mindfulness *(mīnd'-fŭl-nĕs).* A Buddhist meditation technique in which individuals strive to be silent, nonjudgmental witnesses to their thoughts, actions, feelings, and the world around them and be intensely aware of the present moment.

mineral *(mĭn'-ĕr-ăl).* A nonorganic element or compound that occurs in nature.

mineralocorticoid *(mĭn-ĕr-ĕl-ō-kăr'-tĭ-koid).* A type of corticosteroid that is involved in the regulation of electrolytes and water balance.

minimally invasive surgery. A surgical procedure performed through an endoscope or laparoscope, resulting in very small incisions.

minipill *(mĭn'-ē-pĭl).* A type of oral contraceptive that contains only the hormone progestin and no estrogen. It is a less effective contraceptive than the combination pill, but is useful for women who cannot take estrogen.

minocycline *(mĭ-nō-sī'-klēn).* Generic name for a broad-spectrum antibiotic; trade name: Minocin.

minoxidil *(mĭ-nŏk'-sĭ-dĭl).* Generic name for a drug used internally as an antihypertensive and topically for treatment of hair loss.

miosis *(mī-ō-'sĭs).* Contraction of the pupil.

miscarriage *(mĭs'-kăr-ĕj).* Common term for *spontaneous abortion.*

misogyny *(mĭs-ŏj'-ĭn-ē).* An aversion to or hatred of women.

misoncism *(mī-sō-nē'-ĭzm).* An aversion to new ideas or new things.

mites *(mīts).* Any arthropod of the order Acarina except ticks. Some are parasitic and cause conditions such as scabies and mange; others carry disease-causing organisms.

mitochondria *(mī-tō-kŏn'-drē-ă).* Rod-shaped organelles found in the cytoplasm of cells. They are the main sites of energy production in cells and perform other tasks as well.

mitomycin *(mī-tō-mī'-sĭn).* Generic name for any one of a group of antitumor antibiotics produced by *Streptomyces caespitosus.* Because it is highly toxic, it is reserved for individuals who have not responded to other treatment.

mitosis *(mī-tō'-sĭs).* A type of cell division in four phases: prophase, metaphase, anaphase, and telophase. It is the process by which the body grows and somatic cells are replaced.

mitral *(mī'-trăl).* Referring to the bicuspid or mitral valve.

mitral stenosis. Narrowing opening of the mitral valve that blocks the free flow of blood from the atrium to the ventricle.

mitral valve. The valve on the left side of the heart between the atrium and the ventricle.

mitral valve prolapse. A common condition in which the cusp or cusps of the mitral valve prolapse (fall) into the left atrium during systole (when the heart contracts and expels blood).

mixed infection. *See* infection.

mixture *(mĭks'-chĕr).* A combination of different ingredients or drugs.

mm Hg. Millimeter of mercury; a unit of pressure equal to that exerted by a column of mercury at 0° C one millimeter high at standard gravity.

MMR. Acronym for the vaccine for measles, mumps, and rubella.

mobility *(mō-bĭl'-ĭ-tē).* The quality of being mobile or capable of being moved or flowing freely.

mobilization *(mō-bĭ-lĭ-zā'-shĕn).* The process of making a fixed part or stored substance mobile.

modality *(mō-dăl'-ĭ-tē).* A method of applying or utilizing any therapeutic agent, usually physical agents.

molar *(mō'-lĕr).* In dentistry, a back or grinding tooth, one of three on each side of each jaw.

sixth-year m. The permanent first molar tooth, which usually erupts at age 6 years.

twelfth-year m. The permanent second molar tooth, which usually erupts at age 12 years.

molarity *(mō-lăr'-ĭ-tē).* The number of moles of a solute per liter of solution.

molding *(mōld'ĭng).* 1. Creating a shape or fashioning of an object. 2. Shaping of the fetal head to adjust to the shape and size of the birth canal.

mole *(mōl).* 1. Short for molecular weight, it is a unit of weight defined as the number of atoms in exactly 12 grams of carbon-12. 2. Any pigmented fleshy growth on the skin.

molecule *(mŏl'-ĕ-kūl).* The smallest quantity into which a substance may be divided without losing its characteristics.

molluscum contagiosum *(mō-lŭs'-kŭm).* A contagious skin disease characterized by soft, rounded tumors caused by a poxvirus. Seen mainly in children; in adolescents and adults it is often sexually transmitted. It is benign and usually disappears without treatment.

molybdenum *(mĕ-lĭb'-dĕ-nĕm).* An essential trace element in human nutrition.

mometasone *(mō-mĕt'-ĕ-sōn).* *See* Appendix, Common Prescription and OTC Drugs: By Generic Name.

monarticular *(mŏn-ăr-tĭk'-ū-lăr).* Referring to or affecting one joint.

mongolian spot *(mŏn-gō'-lē-ăn).* Blue-gray spots on the skin of the thighs, lower back, and sometimes the shoulders that appear on about 80 percent of nonwhite and 10 percent of white infants. These spots gradually fade.

moniliasis *(mŏn-ĭ-lĭ'-ĕ-sĭs).* Candidiasis.

Monistat *(mŏn'-ĭ-stăt).* Trade name for preparations of miconazole, used to treat fungal infections, including candidiasis.

monitor *(mŏn'-ĭ-tŏr).* A device that observes or records physiological signs.

ambulatory ECG m. A portable continuous electrocardiograph recorder used to detect heart rhythm disruptions.

fetal m. Device that detects and displays fetal heartbeat.

Holter m. A type of ambulatory ECG monitor.

monoamine oxidase inhibitor (MAOI) *(mŏn-ō-ĕ-mēn').* One of a class of medications used to treat depression.

monoclonal antibodies *(mŏn-ō-klōn'-ĕl).* Antibodies that are all derived from a single clone of cells, thus they all contain similar molecules. They have demonstrated some ability to increase resistance to cancer growth.

monocyte *(mŏn'-ō-sīt).* A type of white blood cell that forms in bone marrow. They typically increase in number during certain infections.

monocytosis *(mŏn-ō-sī-tō'-sĭs).* An increase in the proportion of monocytes in the blood.

monomania *(mŏn-ō-mā'-nē-ă).* A type of mental disorder characterized by a preoccupation with a single idea or topic.

mononucleosis *(mŏn-ō-nū-klē-ō'-sĭs).* Condition characterized by an excessive number of monocytes, usually abnormal types.

cytomegalovirus m. A type of the disease resembling infectious mononucleosis, but it does not include pharyngitis. It may occur sporadically or after receiving multiple blood transfusions.

infectious m. A common, acute, infectious disease caused by the Epstein-Barr virus, characterized by fever, abnormal lymphocytes, enlarged lymph nodes and spleen, membranous pharyngitis, and abnormal liver function.

monooctanoin *(mŏn-ō-ŏk-tē-nō'-ĭn).* A substance used to dissolve cholesterol stones in the bile ducts. It is administered as a continuous perfusion via catheter.

monoplegia *(mŏn-ō-plē'jă).* Paralysis of a single limb.

monorchism *(mŏn'-ŏr-kĭz-ĕm).* Condition in which there is only one testis in the scrotum.

monosodium glutamate (MSG) *(mŏn-ō-sō'-dē-um).* A sodium salt of the amino acid glutamic acid, it is used to enhance the flavor of certain foods. It is made by fermenting corn, potatoes, and rice.

monozygotic *(mŏn-ō-zī-gŏt'-ĭk).* Originating from a single fertilized egg.

mons pubis *(mŏns pū'-bĭs).* The rounded fleshy prominence over the symphysis pubis.

montelukast *(mŏn-tĕ-loo'-kăst).* *See* Appendix, Common Prescription and OTC Drugs: By Generic Name.

morbid *(mŏr'-bĭd).* Referring to, affected with, or causing disease.

morbidity rate. The number of cases of a certain disease that occur during a specified period of time per unit of population (usually 1,000 or 10,000 or 100,000).

morcellation *(mŏr-sĕl-ā'-shĕn).* The breaking of solid tissue into pieces so it can be removed.

moribund *(mŏr'-ĭ-bĕnd).* Dying; in a dying condition.

morning after pill. A form of contraception used after (up to three days after) rather than before sexual intercourse. The pill interferes with pregnancy by blocking implantation of a fertilized egg in the uterus.

morning sickness. The nausea and vomiting that affects some women (usually in the morning) during the first few months of pregnancy. Other symptoms may include headache, dizziness, and fatigue.

morphine *(mŏr'-fēn).* The primary and most active ingredient in opium, it is a powerful painkiller with some stimulant action used to treat severe pain. Abuse of morphine can lead to dependence.

morphology *(mŏr-fŏl'-ĕ-jē).* The study of the structure of organisms.

mortality *(mŏr-tăl'-ĭ-tē).* The quality of being mortal.

mortality rate. *See* **death rate.**

Morton neuroma *(mŏr'-tĕn).* A swollen inflamed nerve in the ball of the foot, caused by chronic pressure against a branch of the plantar nerve. This condition is common in women who wear high-heeled and/or narrow shoes.

mother tincture. Mixture of medicinal extracts, alcohol and distilled water that forms the basis of homeopathic remedies.

motility *(mō-tĭl'-ĭ-tē).* Spontaneous movement, or the ability to move spontaneously.

motion sickness. Condition marked by nausea, vomiting, and dizziness or vertigo caused by rhythmic or irregular movements.

Motrin *(mō'-trĭn). See* Appendix, Common Prescription and OTC Drugs: By Trade Name.

mountain sickness *(mŏwn'-tĭn).* A type of high-altitude sickness caused by exposure to altitude high enough to cause hypoxia (inadequate oxygen to the lungs and blood).

acute ms. Type that appears a few hours after one is exposed to high altitude, characterized by fatigue, dizziness, breathlessness, headache, nausea, vomiting, insomnia, and impaired mental capacity.

chronic ms. Type characterized by loss of hypoxia in an individual who previously could tolerate high altitudes.

subacute ms. Type that is milder than the chronic form and similar to the acute form except for being persistent and likely to resolve once the individual descends to a lower altitude.

mouth *(mouth).* 1. The opening of any cavity. 2. The cavity within the cheeks, containing the teeth and tongue.

mouth-to-mouth breathing. A form of artificial respiration in which the first-aider places his/her mouth against the mouth of the patient and rhythmically forces air into the patient's lungs, about 20 times per minute, until the patient resumes breathing on his/her own.

moxa *(mŏk'săh).* The dried leaves of *Artemisia vulgaris,* burned on or near acupoints in moxibustion.

moxibustion *(mŏks-ĭ-bŭs'-chŭn).* Stimulation of an acupoint by burning a cone or cylinder of moxa placed at or near the point.

MRSA. Acronym for methicillin-resistant staphylococcus aureus. It first appeared among persons in health-care facilities, including nursing homes, and continues to be a concern in these settings. It can cause deadly infections and be difficult to treat because few antibiotics are effective against it.

mucin *(mū'sĭn).* A protein-containing substance found in mucus and in saliva, bile, skin, connective tissues, tendons, and cartilage.

mucocele *(mū'-kō-sēl).* Enlargement of the lacrimal sac.

mucocutaneous junction *(mū-kō-kū-tā'-nē-ŭs).* The site where the skin meets mucous membrane, as where the lips meet the mucous membrane of the mouth.

mucopolysaccharides *(mū-kō-pŏl-ē-săk'ĕ-rīdz).* Polysaccharides that form chemical bonds with water.

mucopolysaccharidosis *(mū-kō-pŏl-ē-săk-ĕ-rī-dō'-sĭs).* A group of inherited disorders in which there is a deficiency of enzymes that are necessary for the breakdown of mucopolysaccharides. Symptoms include stiff joints, corneal clouding, hernia, mental deficiency, enlargement of the liver and spleen, and skeletal impairment.

mucopurulent *(mū-kō-pŭr'-ū-lĕnt).* Consisting of mucus and pus.

mucosa *(mū-kō'-să).* Mucous membrane.

mucosectomy *(mū-kō-sĕk'-tĕ-mē).* Surgical removal of the mucous membrane while leaving the submucosal tissues and muscle wall intact. This procedure is sometimes used to retain the rectum rather than remove it completely.

mucous *(mū'-kŭs).* Referring to mucus.

mucous membrane. Membrane that is composed of cells that secrete mucus and which lines passages and cavities that communicate with the air.

mucus *(mū'-kŭs).* A thick liquid secreted by mucous membranes and glands.

muira-puama. *Ptychopetalum olacoides,* also known as potency wood, is derived from the roots and bark of the muira-puama tree. It has been used as a sexual stimulant for centuries in the Amazon and, more recently, in Europe and the United States.

multipara *(mŭl-tip'-ĕ-rĕ).* A woman who has borne more than one viable fetus.

multiple myeloma. *See* **myeloma.**

multiple organ failure. A condition in which two or more organ systems fail to function effectively.

multiple personality disorder. A mental condition in which an individual develops two or more personalities, each of which becomes dominant and controls behavior from time to time while excluding the other personalities.

multiple sclerosis. *See* **sclerosis.**

mummification *(mŭm-ĭ-fĭ-kā'-shĕn).* Conversion into a state resembling a mummy, which is seen in dry gangrene or the drying up of a dead fetus.

mumps *(mŭmps).* An acute infectious disease caused by a paramyxovirus and usually seen in children younger than fifteen years. Characterized by painful swelling of one or both parotid glands.

Munchausen syndrome *(moon'-chou-zĕnz).* A disorder in which individuals routinely seek treatment at hospitals for an apparent acute physical illness and give a logical explanation for their symptoms, but no organic disease exists.

murmur *(mŭr'-mĕr).* A blowing or rasping sound heard through a stethoscope. It is the result of vibrations produced by movement of the blood within the heart and nearby large blood vessels.

muscle *(mŭs'-ĕl).* A type of tissue composed of contractile fibers or cells that cause movement of an organ or part of the body.

muscular dystrophy. A group of degenerative muscle diseases characterized by a wasting away and atrophy of muscles.

musculoskeletal system *(mŭs-kū-lō-skĕl'-ĕ-tĕl).* Body system that consists of the skeleton and the muscles.

mutagenesis *(mū-tĕ-jĕn'-ĕ-sĭs).* The production of change or the induction of genetic mutation.

Mutamycin *(mū-tĕ-mī'-sĭn).* Trade name for a preparation of mitomycin.

mutant gene *(mū'-tĕnt).* A gene that has undergone genetic mutation.

mutation *(mū-tā'-shĕn).* In genetics, a permanent change in the genetic material, usually occurring in a single gene.

myalgia *(mī-ăl'jē-ă).* Pain in one or more muscles.

myasthenia *(mī-ĕs-thē'-nē-ă).* Muscle weakness.

myasthenia gravis. A disease characterized by significant muscle weakness and progressive fatigue.

myatonia *(mī-ă-tō'-nē-ă).* Deficiency or loss of muscular tone.

Mycobacteria *(mī-kō-băk-tĕr'-ē-ă).* A genus of organisms that belong to the Mycobacteriaceae family and contains many species that cause tuberculosis.

mycology *(mī-kŏl'-ō-jē).* The science and study of fungi.

Mycoplasma *(mī'-kō-plăz-mă).* A genus of bacteria of the family *Mycoplasmataceae* for which cholesterol or other sterols are required for growth. The organisms can cause various infections in humans and animals.

M. hominis. Species that inhabit the vagina and cervix, causing infections in the reproductive tracts; also associated with respiratory disease and pharyngitis.

M. pneumoniae. A species that can cause mild respiratory tract disease and a type of primary atypical pneumonia.

mycosis *(mī-kō'-sĭs).* Any disease caused by a fungus.

mydriasis *(mĭ-drī'-ĕ-sĭs).* Abnormal dilation of the pupil.

mydriatic *(mĭd-rē-ăt'-ĭk).* 1. Causing papillary dilatation. 2. Any drug that dilates the pupil.

myectomy *(mī-ĕk'-tō-mē).* Surgical removal of a portion of a muscle.

myelemia *(mī-ĕl-ē'-mē-ă).* An abnormal number of marrow cells in the blood.

myelin *(mī'-lĭn).* A fatlike substance, composed of protein and lipids, that forms a covering around the axons of certain nerves.

myelitis *(mī-ĕ-lī'-tĭs).* 1. Inflammation of bone marrow. 2. Inflammation of the spinal cord.

myeloblast *(mī'-ē-lō-blăst).* An immature cell found in the bone marrow.

myelocele *(mī'-ĕ-lō-sēl).* Protrusion of the substance of the spinal cord through a defect in the vertebral arch.

myelocyte *(mī'-ĕ-lō-sīt).* A large cell in red bone marrow from which white blood cells are derived.

myelography *(mī-ĕ-lŏg'-ră-fē).* Radiography of the spinal cord using a radiopaque medium injected into the vertebral space.

myeloma *(mī-ĕ-lō'-mă).* A tumor composed of cells of those typically seen in bone marrow.

myelomalacia *(mī-ĕ-lō-mĕ-lā'-shă).* Morbid softening of the spinal cord.

myeloradiculitis *(mī-ĕ-lō-rē-dĭk-ū-lī'-tĭs).* Inflammation of the spinal cord and the posterior nerve roots.

myelosuppression *(mī-lō-sū-prĕsh'-ĕn).* Decrease in blood cell production by the bone marrow.

myenteric *(mī-ĕn-tĕr'-ĭk).* Referring to the muscles of the intestines.

myesthesia *(mī-ĕs-thē'-zhă).* Consciousness of muscle contractions.

Mylanta *(mī-lăn'-tă).* Trade name of an over-the-counter antacid medication.

Mylicon *(mī'-lī-kŏn).* Trade name for simethicone, a drug used to treat excess gas in the intestinal tract.

myoblast *(mī'-ō-blăst).* An embryonic cell that develops into a muscle fiber cell.

myoblastoma *(mī-ō-blăs-tō'-mă).* A benign, tumorlike lesion of soft tissue, often consisting of myoblasts.

myocardial *(mī-ō-kăr'-dē-ăl).* Referring to the myocardium.

myocardial infarction. A condition caused by blockage of one or more of the coronary arteries. Symptoms include squeezing pain or prolonged heavy pressure in the chest behind the sternum; pain may spread to the shoulders, neck, arm, back, and jaw.

myocardial ischemia. Deficiency of blood supply to the heart due to constriction or obstruction of the coronary arteries.

myocarditis *(mī-ō-kăr-dī'-tĭs).* Inflammation of the muscular walls of the heart.

myocardium *(mī-ō-kăr'-dē-ŭm).* The middle layer of the walls of the heart, composed of cardiac muscle.

myoclonus *(mī-ŏk'-lō-nŭs).* Twitching or clonic spasm of one or more muscles.

cortical m. Myoclonus caused by an electrical charge to the cerebral cortex.

epileptic m. Myoclonus that occurs as part of an epileptic seizure.

essential m. Myoclonus of unknown cause that may involve a single or several muscles.

intention m. Myoclonus that occurs when an individual initiates voluntary muscle activity.

myoepithelioma *(mī-ō-ĕp-ĭ-thē-lē-ō'-mă).* A benign tumor composed primarily of myoepithelial cells.

myoepithelium *(mī-ō-ĕp-ĭ-thē'-lē-ĕm).* A specialized type of tissue that contains epithelial cells that have the ability to contract.

myofacial pain *(mī-ō-fāsh'-ĕl).* Pain in the facial area, including orofacial and craniofacial pain; associated conditions include inflammatory and neoplastic disorders, as well as nerve syndrome that involve the trigeminal, facial, and glossopharyngeal nerves.

myofascial release. A bodywork technique in which practitioners use their fingers, palms, elbows, and forearms to apply long, sustained pressure designed to stretch the fascia to relieve tension and constriction and thus improve mobility and function.

myofascitis *(mī-ō-fă-sī'-tĭs).* Inflammation of a muscle and its fascia.

myofibroma *(mī-ō-fī-brō'-mă).* A tumor that contains fibrous and muscular tissue.

myofibrosis *(mī-ō-fī-brō'-sĭs).* Replacement of muscular tissue with fibrous tissue.

myogram *(mī'-ō-grăm).* The recording produced by a myograph.

myograph *(mī'-ō-grăf).* A device that records the effects of a muscular contraction.

myolipoma *(mī-ō-lī-pō'-mă).* A muscle tissue tumor that contains fatty substances.

myoma *(mī-ō'-mă).* A benign tumor composed of muscular elements.

myomalacia *(mī-ō-mă-lā'-shă).* Softening of muscular tissue.

myomectomy *(mī-ō-mĕk'-tē-mē).* Surgical removal of a myoma.

myometritis *(mī-ō-mē-trī'-tĭs).* Inflammation of the myometrium of the uterus.

myometrium *(mī-ō-mē'-trē-ŭm).* The muscular wall of the uterus.

myoneural junction *(mī-ō-nū'-rĕl).* The point at which nerves enter a muscle.

myopathy *(mī-ŏp'-ă-thē).* Any disease or abnormal condition of striated muscle.

distal m. A form of muscular dystrophy that appears in two types: one appears in infancy but does not progress past adolescence; the other first appears in adulthood.

myopia *(mī-ō'-pē-ă).* A defect in vision in which parallel light rays that enter the eye focus in front of the retina, due to the eyeball being too long from front to back. Also called *nearsightedness.*

myosarcoma *(mī-ō-săhr-kō'-mă).* A malignant tumor derived from muscle tissue.

myositis *(mī-ō-sī'-tĭs).* Inflammation of a voluntary muscle.

myotherapy *(mī-ō-thĕr'-ĕ-pē).* A muscle therapy in which practitioners apply direct pressure to trigger points (painful areas where muscles have formed a knot) using their fingers, knuckles, and elbows. This therapy is painful but relief comes when the knot is eliminated.

myotomy *(mī-ŏt'-ĕ-mē).* The cutting or dissection of a muscle or muscular tissue.

myotonia *(mī-ō-tō'-nē-ă).* Dystonia involving increased muscular irritability and contractility with a reduced ability to relax.

myringectomy *(mĭr-ĭn-jĕk'-tĕ-mē).* Surgical removal of the eardrum; also called *tympanectomy.*

myringitis *(mĭr-ĭn-jī'-tĭs).* Inflammation of the eardrum.

myringoplasty *(mĭ-rĭng'-gō-plăs-tē).* Surgical procedure to restore the integrity of a perforated eardrum by grafting.

myringotomy *(mĭr-ĭng-gŏt'-ō-mē).* Creation of a hole in the eardrum.

myrrh *(mŭr).* A gummy substance obtained from species of *Commiphora.* It is an ingredient in some topical medications to treat canker sores and in some mouthwash.

myxedema *(mĭks-ĕ-dē'-mĕ).* A condition caused by abnormally reduced function of the thyroid gland, characterized by anemia, large tongue, slow speech, puffy hands and face, coarse skin, hair loss, drowsiness, and sensitivity to cold.

myxochondroma *(mĭks-ō-kŏn-drō'-mă).* A benign tumor that is composed of myxomatous and chondromatous elements.

myxoma *(mĭk-sō'-mă).* A tumor that is composed of mucous connective tissue. It is usually soft, gray, and translucent.

myxosarcoma *(mĭk-sō-săr-kō'-mă).* A tumor that is partly myxomatous, partly sarcomatous.

N

Nabothian follicle *(nē-bō'-thē-ăn).* Cystlike formations caused by blockage of the lumina of glands in the mucosa of the uterine cervix.

nabumetone *(nĕ-bū'-mĕ-tŏn).* Generic name for a nonsteroidal anti-inflammatory drug used to treat osteoarthritis and rheumatoid arthritis; trade name: Relafen.

nail *(nāl).* A horny cell structure, composed of keratin, of the outer layer of skin that forms flat plates on the ends of the phalanges.

nail bed. The part of the finger or toe that is covered by the nail.

nailing. Fixing or fastening a fractured bone with a nail.

naloxone *(năl-ŏk'-sŏn).* A drug related to oxymorphone, used to treat opioid toxicity, to reverse opioid-induced respiratory depression, and to treat hypotension associated with septic shock.

nape *(nāp).* The back of the neck.

naphthalene *(năf'-thĕ-lēn).* A crystalline hydrocarbon derived from coal tar oil and used as a fungicide, moth repellent, and preservative. Toxic if ingested, inhaled, or absorbed through the skin.

Naprosyn *(năp'-rō-sĭn). See* Appendix, Common Prescription and OTC Drugs: By Trade Name.

naproxen *(nă-prŏk'-sĕn). See* Appendix, Common Prescription and OTC Drugs: By Generic Name.

narcissism *(năhr'-sĭ-sĭz-ĕm).* Self-love; primarily interested in oneself.

narcolepsy *(năhr'-kō-lĕp-sē).* Recurring, uncontrollable episodes in which an individual falls asleep for brief periods but is easily awakened.

narcosis *(năr-kō'-sĭs).* An unconscious state caused by use of narcotics.

narcotic *(năhr-kŏt'-ĭk).* A drug that in moderate amounts depresses the central nervous system, relieves pain, and induces sleep; in larger amounts, it produces unconsciousness, coma, and possibly death.

naris (*pl.* **nares)** *(nā'-rĭs).* The nostril.

nasal *(nā'-zĕl).* Referring to the nose.

nasal cavity. The space between the floor of the cranium and the roof of the mouth.

nasal septum. A wall or septum positioned between the two nasal cavities.

nascent *(nās'-ĕnt).* Beginning; just born.

nasociliary nerve *(nā-zō-sĭl'-ē-ăr-ē).* The nerve that supplies the nose, eyebrows, and eyes.

nasogastric tube *(nā-zō-găs'-trĭk).* A tube that is inserted through a nostril and extended into the stomach for the purpose of giving liquids to the patient or removing gas or liquids from the stomach.

nasolabial *(nā-zō-lā'-bē-ĕl).* Referring to the nose and lip.

Nasonex *(nā'-zō-nĕks). See* Appendix, Common Prescription and OTC Drugs: By Trade Name.

nasopharyngitis *(nā-zō-făr-ĭn-jī'-tĭs).* Inflammation of the nasopharynx.

nasopharynx *(nā-zō-făr'-ĭnks).* The part of the pharynx that is located above the soft palate.

natality *(nā-tăl'-ĭ-tē).* Birth rate.

nates *(nā'-tēz).* The buttocks.

natriuresis *(nā-trē-ū-rē'-sĭs).* The excretion of excessive amounts of sodium in the urine.

natrum muriaticum *(nā'-trŭm mŭr-ē-ăt'-ĭ-kŭm).* A common, powerful homeopathic remedy made from common salt and used to treat headache related to eyestrain, head pain accompanied by chills and fever, and emotionally sensitive individuals.

natural childbirth. *See* **childbirth.**

natural selection theory. A concept of evolution proposed by Charles Darwin which states that chances for survival and reproduction of organisms are most likely to occur among those that are best adapted to their environment.

naturopath *(nā'-chĕr-ō-păth).* A practitioner of naturopathy.

naturopathy *(nā-chĕr-ŏp'-ĕ-thē).* A system of health care in which treatments consist not of drugs but of other therapies, such as nutrition, hydrotherapy, massage, herbs, heat, and light.

nausea *(naw'-zē-ă).* An unpleasant sensation in the abdomen that often precedes vomiting. It is a symptom of many conditions, including motion sickness, morning sickness, and some central nervous system diseases, and is often present in influenza and gallbladder disturbances.

navel *(nā'-vĕl).* The depression in the middle of the abdomen where the umbilical cord was attached to the fetus; also called *umbilicus.*

navicular bones *(nĕ-vĭk'-ū-lăr).* Scaphoid bones in the wrist and ankle.

nearsightedness *(nēr-sīt'-ĕd-nĕs).* Myopia; the ability to see clearly only those objects that are close to the eyes.

nebulizer *(nĕb'-ū-lī-zĕr).* A device that produces a fine aerosol spray or mist. Often used by individuals who have asthma or other respiratory conditions that require medication dispensed in this manner.

neck *(nĕk).* The part of the body that lies between the head and shoulders.

necrobiosis *(nĕk-rō-bī-ō'-sĭs).* Gradual degeneration and swelling of the collagen.

necrology *(nĕk-rŏl'-ĕ-jē).* The study of mortality statistics.

necromania *(nĕk-rō-mā'-nē-ă).* An abnormal preoccupation with death or dead bodies.

necropsy *(nĕk'-rŏp-sē).* Examination of a body after death to determine the cause of death or to study disease conditions.

necrosis *(nĕ-krō'-sĭs).* Death of areas of tissue or bone surrounded by healthy parts. It can be caused by insufficient blood supply, trauma, chemical substances, and other causes.

necrospermia *(nĕk-rō-spĕr'-mē-ă).* A condition in which the spermatozoa of the semen are dead or motionless.

necrotizing *(nĕk'-rō-tīz-ĭng).* Causing necrosis.

needle *(nē'-dĕl).* A pointed instrument used to stitch or puncture the skin or other tissue.

aspirating n. A hollow needle usually fitted to a syringe and used to withdraw fluids from a body cavity.

cataract n. A needle used to extract a cataract.

hypodermic n. A slender, hollow needle used to inject drugs under the skin.

needlestick injury. A penetrating puncture wound from a needle or other sharp object that may result in exposure to bodily fluids. The main concern of such an injury is that the bodily fluid may carry infectious disease, such as human immunodeficiency virus, hepatitis B, or hepatitis C.

neem *(nēm)*. An evergreen tree native to India and Sri Lanka, whose parts have antifungal, antibacterial, antiviral, and antimalarial properties.

Neisseria *(nī-sē'-rē-ă)*. A genus of bacteria of the Neisseriaceae family, consisting of gram-negative, aerobic, and anaerobic organisms. They are normally present in the oropharynx, nasopharynx, and genitourinary tract.

N. flavescens. A species sometimes found in people who have meningitis and septicemia.

N. meningitides. A common cause of meningitis, it can also cause bacterial pneumonia.

nematode *(nēm'-ĕ-tōd)*. Any member of the class Nematoda, which are roundworms. Many of the species in this class are parasites.

Nembutal *(nĕm'-bū-tăl)*. Trade name for preparations of pentobarbital sodium, a sedative.

neomycin *(nē-ō-mī'-sĭn)*. Generic name for an antibiotic complex derived from *Streptomuces fradiae,* it is effective against many aerobic gram-negative bacteria and some gram-positive bacteria.

neonatal *(nē-ō-nā'-tăl)*. Referring to the first four weeks after birth.

neonatology *(nē-ō-nā-tŏl'-ĕ-jē)*. The science of diagnosis and treatment of disorders that affect newborns.

neoplasia *(nē-ō-plā'-zhĕ)*. The formation or development of a neoplasm.

neoplasm *(nē'-ō-plăzm)*. A new and abnormal formation of tissue, such as a tumor, in which the growth is uncontrolled and progressive.

benign n. A neoplasm whose growth is not infiltrating tissue or metastasizing.

malignant n. Growth that infiltrates other tissue, metastasizes, and may recur after surgical removal.

Neosporin *(nē-ō-spŏ'-rĭn)*. Trade name for a combination preparation of neomycin sulfate, polymyxin B sulfate, and (depending on the preparation) either gramicidin or bacitracin zinc; used as a topical antibiotic.

neostigmine *(nē-ō-stĭg'-mēn)*. Generic name of a drug used to improve transmission of nerve signals and muscle tone, especially in people with myasthenia gravis; also used for glaucoma.

nephrectomy *(nĕ-frĕk'-tĕ-mē)*. Surgical procedure to remove a kidney.

abdominal n. Removal of a kidney through an incision in the abdominal wall.

laparoscopic n. A minimally

invasive type of nephrectomy which uses laparoscopic techniques.
lumbar n. Removal of a kidney through an incision in the lumbar area.
radical n. Removal of a kidney with its fascia, plus the adjacent adrenal gland and all lymph nodes in the area. This approach is usually done in cases of renal cancer.

nephritis *(nĕ-frī'-tĭs).* Inflammation of the kidney.

nephrolithiasis *(nĕf-rō-lĭth-ī'-ă-sĭs).* The presence of stones in the kidney.

nephrolithotomy *(nĕf-rō-lĭth-ŏt'-ĕ-mē).* Surgical incision to remove stones from a kidney.

nephrology *(nĕ-frŏl'-ĕ-jē).* Science of the structure and function of the kidney.

nephroma *(nĕ-frō'-mă).* Tumor of the kidney.

nephromegaly *(nĕf-rō-mĕg'-ĕ-lē).* Enlargement of the kidney.

nephons *(nĕf'-rŏnz).* The structural and functional elements of the kidney, which number about one million in each kidney.

nephropathy *(nĕ-frŏp'-ĕ-thē).* Any disease of the kidneys.
diabetic n. Nephropathy that often occurs during the latter stages of diabetes mellitus, it involves progressive renal failure leading to end-stage renal disease.
obstructive n. Nephropathy caused by a blockage of the urinary tract.

nephropexy *(nĕf-rō-pĕk'-sē).* Surgical procedure to attach or fixate a floating kidney.

nephroptosis *(nĕf-rŏp-tō'-sĭs).* Prolapse of the kidney.

nephrosclerosis *(nĕf-rō-sklĕ-rō'-sĭs).* Hardening of the kidney arteries.

nephrosis *(nĕf-rō'-sĭs).* Condition in which the kidneys degenerate, especially the renal tubules, without the presence of inflammation.

nephrostomy *(nĕ-frŏs'-tĕ-mē).* Surgical creation of an artificial fistula into the renal pelvis.

nephrotoxin *(nĕf-rō-tŏk'-sĭn).* A poison that can damage or destroy kidney cells.

nerve *(nĕrv).* A cordlike structure composed of fibers that carries impulses between a part of the central nervous system and another area of the body.

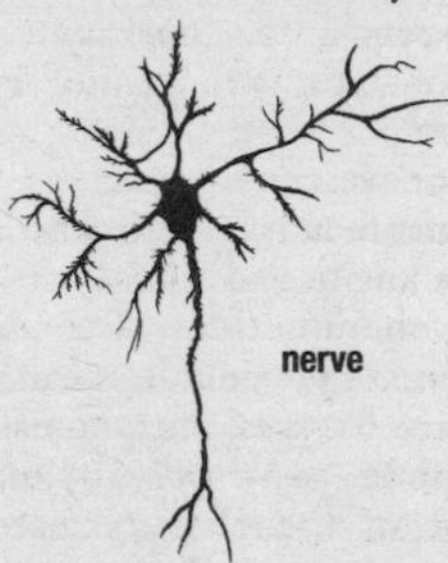

afferent n. Any nerve that transmits impulses from the periphery to the central nervous system.
autonomic n. Nerve of the autonomic nervous system.
cranial n. One of twelve pairs of nerves that originate in the

brain and exit through the foramina of the cranium.

efferent n. Nerve that transmits impulses from a nerve center to the periphery.

facial n. The seventh cranial nerve.

motor n. Nerve that transmits motor impulses.

parasympathetic n. Nerve of the parasympathetic segment of the autonomic nervous system.

sensory n. Nerve that transmits afferent impulses from sensory receptor sites to the brain or spinal cord.

splanchnic n. A nerve that supplies the visceral organs.

trigeminal n. The fifth cranial nerve and the most important sensory and motor nerve of the oral region.

nerve block. The use of localized anesthesia, such as injecting a painkiller near a nerve, to prevent nerve impulses from reaching the brain, and resulting in a temporary loss of sensation.

nerve ending. The point at which a nerve enters the structure it supplies and transmits or receives impulses. Your skin, for example, has millions of nerve endings.

nervousness *(nĕr'-vĕs-nĕs).* Excessive excitability and irritability.

nettle *(nĕt'-ĕl).* Any plant of the genus *Urtica.* The species *U. dioica,* or stinging nettle, is used to treat urinary problems associated with benign prostatic hyperplasia, urinary tract infections, kidney and bladder stones, and hay fever.

network *(nĕt'wŏrk).* A meshlike structure composed of interlocking fibers, tubules, or strands.

neural *(nūr'-ĕl).* Referring to one or more nerves.

neuralgia *(nū-răl'-jă).* Pain that extends along the length of one or more nerves. The type of neuralgia is determined by the part of the body affected or the cause.

cranial n. Neuralgia along a cranial nerve.

idiopathic n. Neuralgia of unknown cause and not accompanied by any structural changes.

migrainous n. Cluster headache.

Morton's n. A form of foot pain caused by pressure against a branch of the plantar nerve; *see also* **Morton neuroma.**

trigeminal n. Excruciating episodic pain in the area supplied by the trigeminal nerve—face, teeth, mouth, and nasal cavity.

neurectomy *(nū-rĕk'-tĕ-mē).* Surgical removal of part of a nerve.

neurilemma *(noor-ĭ-lĕ'-mă).* The thin membrane that wraps the myelin layers of certain fibers, especially the peripheral nerves or some unmyelinated nerve fibers. Also called *Schwann's membrane.*

neurilemmoma *(nū-ĭ-lĕm-ō'-mă).* A

tumor of the neurilemma, most of which are benign.

neuritis *(nū-rī'-tĭs)*. Inflammation of a nerve, characterized by pain and tenderness, loss of reflexes, wasting, and paralysis.

neuroblast *(nū'-rō-blăst)*. Any embryonic cell that develops into a nerve cell.

neuroblastoma *(nŭ-rō-blăs-tō'-mă)*. A sarcoma that consists of malignant neuroblasts; it most often affects infants and children up to 10 years of age.

neurochemistry *(nū-rō-kĕm'-ĭs-trē)*. The branch of neurology that studies the chemistry of the nervous system.

neurodermatitis *(nū-rō-dĕr-mĕ-tī'-tĭs)*. Condition characterized by scaly patches of skin on the head, lower legs, wrists, or forearms, caused by a localized itch that becomes severely irritated when scratched.

neurofibroma *(nū-rō-fī-brō'-mă)*. A tumor, usually benign, of the peripheral nerves, caused by abnormal growth of Schwann cells.

neurofibromatosis *(nū-rō-fī-brō-mă-tō'-sĭs)*. A familial condition characterized by developmental changes in the nervous system, bones, muscles, and skin and manifested by soft, pigmented tumors (neurofibromas) over the entire body.

neurogenic *(nū-rō-jĕn'-ĭk)*. 1. Forming nervous tissue. 2. Originating in the nervous system.

neurogenic bladder. *See* **bladder.**

neurohormone *(nū-rō-hŏr'-mōn)*. A hormone secreted by a specialized neuron into the bloodstream, cerebrospinal fluid, or the intercellular spaces of the nervous system.

neuroleptanesthesia *(nū-rō-lĕp-tăn-ĕl-jē'-zhă)*. A state of altered awareness and analgesia caused by the administration of a combination of an opioid analgesic and a neuroleptic agent.

neuroleptic *(nū-rō-lĕp'-tĭk)*. Agents used to suppress emotions and initiative, as part of the effects of anesthesia or analgesia.

Neuro-Linguistic Programming (NLP). A study of the way people receive and process information, how they act upon it, and how they relate to the world. NLP can be used to learn new behaviors and unlearn undesirable habits.

neurologist *(nū-rŏl'-ō-jĭst)*. A physician whose practice involves diagnosis and treatment of conditions affecting the nervous system.

neuroma *(nū-rō'-mă)*. A tumor that arises from a nerve or that is composed largely of nerve cells and nerve fibers.

neuromuscular junction *(nū-rō-mŭs'-kū-lăr)*. Point at which a nerve ends in a muscle and transmits its impulse to the muscle.

neuromylitis *(nū-rō-mī-lī'-tĭs)*. Inflammation of nerves and the spinal cord.

neuron (also neurone) *(nū'-rŏn)*. A nerve cell, which consists of a cell body, an axon, and one or more dendrites. *See* illustration at nerve cell.

neuronitis *(nūr-ō-nī'-tĭs).* Inflammation of one or more neurons.

neuronopathy *(nūr-ŏn-ŏp'-e-thē).* Polyneuropathy involving destruction of the cell bodies of neurons.

neurooncology *(nū-rō-ŏn-kŏl'-ō-jē).* A specialized field of medicine that deals with tumors of the nervous system.

neuropathy *(nū-rŏp'-ĕ-thē).* A dysfunction of pathological change in the peripheral nervous system. Neuropathies that affect a specific nerve are often named for the nerve involved.

alcoholic n. Neuropathy caused by a thiamine deficiency in chronic alcoholism.

autonomic n. Any neuropathy of the autonomic nervous system, which causes symptoms such as sexual dysfunctions, orthostatic hypotension, and bladder dysfunction.

diabetic n. Any of several polyneuropathies associated with diabetes mellitus, the most common of which affects the nerves of the lower limbs.

hepatic n. Neuropathy caused by liver disease.

ischemic n. An injury to a peripheral nerve due to reduced blood supply, the most common of which is diabetic neuropathy.

sensorimotor n. Neuropathy or polyneuropathy that involves both sensory and motor nerves.

neuropeptide *(nūr-ō-pĕp'tīd).* Any of several types of molecules present in brain tissue, composed of amino acids and including endorphins, enkephalins, and others.

neurophysiology *(nū-rō-fĭz-ē-ŏl'-ō-jē).* The physiology of the nervous system.

neuropsychiatrist *(nū-rō-sī-kī'-ă-trĭst).* A physician who specializes in neuropsychiatry.

neuroretinitis *(nū-rō-rĕt-ĭ-nī'-tĭs).* Inflammation of the optic nerve and the retina.

neuroscience *(nū'-rō-sī-ĕns).* Any one of the branches of science that deals with the embryology, physiology, biochemistry, anatomy, pharmacology, and so on, of the nervous system.

neurosis *(nū-rō'-sĭs).* Generally, a term referring to mental disorders in which the symptoms are distressing to the individual, reality is intact, behavior does not violate gross social norms, and there is no apparent organic cause.

neurosurgeon *(nūr'-ō-sŭr-jĭn).* A physician who specializes in neurosurgery.

neurosurgery *(nūr'-ō-sĕr-jĕr-ē).* Surgery of the nervous system.

neurosyphilis *(nū-rō-sĭf'-ĭ-lĭs).* Syphilis that has advanced to the central nervous system, which may or may not cause symptoms.

neurotic *(nū-rŏt'-ĭk).* Referring to or characterized by neurosis, or an individual who is affected by neurosis.

Neurotin *(nū'-rō-tĭn).* *See* Appendix, Common Prescription and OTC Drugs: By Trade Name.

neurotoxin *(nū'-rō-tŏk-sĭn).* A substance that is poisonous to or destroys nerve tissue.

neurotransmitter *(nū-rō-trăns'-mĭt-ĕr).* Any of a group of substances that is released upon stimulation from a presynaptic neuron of the central or peripheral nervous system and travels across the synapse to a target cell.

neurotrauma *(nū-rō-trăw'-mă).* Mechanical injury of a nerve.

neuter *(nū'-tĕr).* To castrate an animal; *see also* **spay.**

neutrino *(nū-trē'-nō).* An elementary particle that has no electrical charge and no mass; it rarely reacts with matter.

neutron *(nū'-trŏn).* An electrically neutral or uncharged particle of matter that exists along with protons in the atoms of all elements except the isotope of hydrogen.

neutropenia *(nū-trō-pē'-nē-ă).* A decrease in the number of neutrophils in the blood.

chronic n. A type of neutropenia that keeps recurring and is characterized by malaise, fever, and various infections.

drug-induced n. Neutropenia caused by medications.

neonatal n. Neutropenia in newborns caused by in utero incompatibility between the infant's immunoglobulin G antigens and those of the mother's blood. Symptoms may include fever, pneumonia, septicemia, and other infections. May be fatal.

neutrophil *(nū'-trō-fĭl).* 1. A mature granular white blood cell. 2. A white blood cell that stains easily with neutral dyes.

neutrophilia *(nū-trō-fĭl'-ē-ă).* An increase in the number of neutrophils in the blood; the most common form of leukocytosis.

nevus *(nē'-vŭs).* Any congenital lesion of the skin; *see also* **birthmark.**

newborn *(nū'-bŏrn).* Recently born infant.

Nexium *(nĕk'-sē-ŭm).* *See* Appendix, Common Prescription and OTC Drugs: By Trade Name.

niacin *(nī'-ă-sĭn).* A B complex vitamin; also known as *nicotinic acid.*

niacinamide *(nī-ĕ-sĭn'-ĕ-mīd).* A B complex vitamin used in the prevention and treatment of pellagra; also known as *nicotinamide.*

niche *(nĭch).* A defect or imperfection in an otherwise even surface; especially refers to a depression in an organ wall as seen on a radiograph or visible to the unaided eye.

nicotine *(nĭk'-ō-tēn).* A poisonous, colorless, soluble alkaloid derived from tobacco or produced synthetically.

nicotinic acid *(nĭk-ō-tĭn'-ĭk).* Niacin.

nictation *(nĭk-tā'-shŭn).* Winking.

nidus *(nī'dĕs)* (*pl.* nidi). The originating point or focus of a morbid process.

Niemann-Pick disease *(nē'-măhn pĭk).* A hereditary disease characterized by an enlarged liver and spleen, anemia, disease of the lymph nodes (lymphadenopathy), and progressive physical and mental deterioration.

nightmare *(nīt'-măr)*. A very disturbing dream accompanied by great fear or anxiety.

night sweats. Profuse sweating that occurs during sleep; it is often an early sign of disease, such as acquired immune deficiency syndrome or tuberculosis, especially if accompanied by intermittent fever.

night vision. Visual perception under conditions of reduced light, especially the darkness of night.

NIH. Acronym for the National Institutes of Health.

nihilism *(nī'-ĭl-ĭz-ĕm)*. Skepticism regarding traditional values and beliefs, or complete rejection of them.

nipple *(nĭp'-ĕl)*. The conical protuberance in each breast consisting of erectile tissue, from which the milk-producing ducts release milk in female mammals.

nit *(nĭt)*. The egg of a louse or any other parasitic insect.

nitric acid *(nī'-trĭk)*. An extremely corrosive acid that is highly toxic when inhaled and corrosive to skin and mucous membranes.

nitric oxide. A toxic compound that also has important roles in the body, such as controlling blood flow to tissues, relaxing blood vessels, killing parasitic organisms, and stimulating the production of new mitochondria.

nitrite *(nī'-trīte)*. Any salt or ester of nitrous acid. Nitrites can be used to dilate blood vessels, reduce blood pressure, and stop spasms. Amyl nitrite, for example, is used as a coronary vasodilator in the treatment of angina pectoris.

Nitro-Dur *(nī'-trō-dĕr)*. Trade name for a preparation of nitroglycerin, used to treat angina pectoris.

nitrogen *(nī'-trō-jĕn)*. A colorless, gaseous element found in the air. It is also found in all living cells and is an important component of protein and nucleic acids.

nitrogen mustard. A term for several therapeutic mustard compounds that have the ability to disturb cell growth and to destroy lymphoid tissue in Hodgkin's disease.

nitroglycerin *(nī-trō-glĭs'-ĕr-ĭn)*. 1. A colorless to yellow liquid formed by the action of nitric acid and sulfuric acids on glycerin. 2. A drug preparation of diluted nitroglycerin, used to prevent and treat angina pectoris, congestive heart failure, and high blood pressure.

Nitrostat *(nī'-trō-stăt)*. Trade name for a preparation of nitroglycerin, used to treat angina pectoris.

nitrous oxide *(nī'-trĕs)*. A colorless, odorless gas that has weak anesthetic qualities. Popularly known as *laughing gas*.

nociceptor *(nō-sī-sĕp'-tĕr)*. A receptor for pain that is caused by injury to tissues. Most nociceptors are in the skin or the walls of the viscera.

noctambulation *(nŏk-tăm-bū-lā'-shĕn).* Somnambulism; sleepwalking.

nocturia *(nŏk-tū'-rē-ă).* Urinary frequency that occurs at night. Also called *nycturia.*

nocturnal *(nŏk-tŭr'-nĕl).* Referring to, occurring at, or active at night.

node *(nōd).* A small mass of tissue in the form of a knot, swelling, or protuberance.

Bouchard's n. In osteoarthritis, a bony enlargement of the proximal interphalangeal joints of the fingers.

Heberden's n. In osteoarthritis, nodes on the terminal phalangeal joints.

lymph n. Mass of lymphoid tissue that develops along lymphatic vessels.

Meynet's n. In rheumatism, nodes in capsules of joints and in tendons, especially in children.

Osler's n. Tender nodes, usually seen in fingers and toes, that occur with subacute bacterial endocarditis.

nodule *(nŏd'-ūl).* A small node.

noesis *(nō-ē'-sĭs).* The act of thinking.

Nolvadex *(nŏl'-vă-dĕks).* Trade name for a preparation of tamoxifen citrate.

nomenclature *(nō'-mĕn-klā-chĕr).* A classified system of scientific or technical terms.

non compos mentis *(nŏn-kŏm'-pōs mĕn'-tĭs).* A legal term that means not of sound mind or not legally responsible.

nonconductor *(nŏn-kĕn-dŭk'-tĕr).* Any substance that does not readily transmit light, heat, or electricity.

noninvasive procedure *(nŏn-ĭn-vā'-sĭv).* A medical procedure that does not require puncturing or entering the body.

nonproprietary *(nŏn-prō-prī'-ĕ-tăr-ē).* Not protected against free competition by patent, trademark, or other means.

nonproprietary name. The name of a drug other than its trademarked or proprietary name. In some cases, the nonproprietary name and generic name of a drug are the same.

nonspecific *(nŏn-spĕ-sĭf'-ĭk).* Term to refer to the cause of disease when the exact causative organism or agent has not been identified.

nonunion of fracture. *See* **fracture.**

nonviable *(nŏn-vī'-ă-bĕl).* Not capable of living.

noradrenaline *(nŏr-ă-drĕn'-ĕ-lĭn).* *See* **norepinephrine.**

norepinephrine *(nŏr-ĕp-ĭ-nĕf'-rĭn).* One of the naturally occurring catecholamines, it is released by specific nerves and brain neurons, as well as secreted by the adrenal glands. It is both a neurohormone (and acts as a vasoconstrictor) and a neurotransmitter.

norgestimate/ethinyl estradiol *(nŏr-jĕs'-tĭ-māt).* *See* Appendix, Common Prescription and OTC Drugs: By Generic Name.

norm *(nŏrm).* An ideal or fixed standard for a specific group.

normoblast *(nŏr'-mō-blăst).* 1. A term often used as a synonym of erythroblast. 2. A nucleated red blood cell.

Normodyne *(nŏr'-mō-dīn).* Trade name for a preparation of labetalol hydrochloride, used to treat high blood pressure.

normotensive *(nŏr-mō-tĕn'-sĭv).* Characterized by normal tension, tone, or pressure, especially when referring to normal blood pressure.

normovolemia *(nŏr-mō-vō-lē'-mē-ă).* Normal blood volume.

Norvasc *(nŏr'-văsk). See* Appendix, Common Prescription and OTC Drugs: By Trade Name.

nose *(nōs).* The organ of smell and the entrance that filters, warms, and moistens air that is inhaled and flows to the lungs.

nosocomial *(nōs-ō-kō'-mē-ăl).* Referring to a hospital. Often used to refer to an infection that was not present or incubating before admittance to the hospital.

nosology *(nō-sŏl'-ĕ-jē).* The science of the classification of diseases.

nosophobia *(nō-sō-fō'-bē-ă).* An irrational fear of illness or of a particular disease.

nostril *(nŏs'-trĭl). See* **naris.**

nostrum *(nŏs'-trĕm).* A secret or quack remedy.

Novaldex *(nō-văl'-dĕks).* Trade name for tamoxifen, used to treat osteoporosis.

Novocain *(nō'-vĕ-kān).* Trade name for preparations of procaine hydrochloride, used in peripheral nerve blocks, for spinal anesthesia, and for temporary anesthesia, especially in dentistry.

noxious *(nŏk'-shĕs).* Harmful; damaging to tissues.

nuclear *(noo'-klē-ĕr).* Referring to a nucleus.

nucleated *(noo'-klē-ā-tĕd).* Having a nucleus.

nucleic acid *(noo-klē'-ĭk).* High molecular weight substances present in the cells of all living organisms. There are two types: deoxyribonucleic acid (DNA) and ribonucleic acid (RNA).

nucleolus *(noo-klē'-ĕ-lĕs).* A rounded body present in the nucleus of most cells that is where ribosomal RNA is synthesized.

nucleus *(noo'-klē-ŭs).* 1. The core or a body or object. 2. In a cell, the body within it that is crucial for the growth, metabolism, reproduction, and transmission of characteristics of a cell.

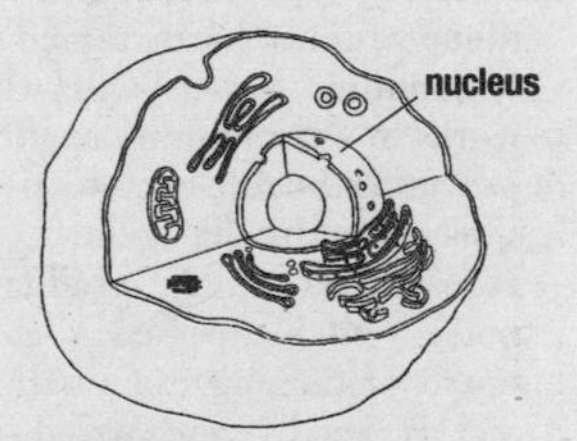

nulliparous *(nŭl-lĭp'-ăr-ŭs).* Never having borne a child.

numbness *(nŭm'-nĕs).* Lack of sensation in a body part.

nummular *(nŭm'-ū-lăr).* Coin-shaped.

nurse *(nŭrs).* An individual who has received specialized training in the scientific basis of nursing and who provides health-care services.

general duty n. A registered nurse who has not received training beyond basic nursing and who provides general nursing care.

head n. Nurse who is in charge of a group of patients and a nursing staff required to care for those patients.

licensed practical n. A graduate of a school of practical nursing and who has passed a state board of nursing examination and is legally authorized to practice as a licensed practical nurse.

practical n. An individual who has practical nursing experience but who has not graduated from a nursing school.

private duty n. Nurse who cares for an individual patient, usually on a fee-for-service basis.

public health n. A registered nurse who has been trained in community health and who works in a community agency, providing health care to persons served by the agency.

registered n. A graduate nurse who has passed a state board of nursing examination and is legally authorized to practice and use the designation RN.

scrub n. A nurse who directly assists surgeons in the operating room.

wet n. A woman who breast feeds the infant of another woman.

nutrient *(noo'-trē-ĕnt).* Any substance that provides the body with elements necessary for metabolism.

nutrition *(noo-trĭ'-shĕn).* The ingestion and metabolism of nutrients by an organism for the purpose of maintaining life and promoting growth.

nux vomica *(nŭks).* The dried ripe seed of *Strychnos nux-vomica* that is used as a central nervous system stimulant and a tonic. It is a common remedy in homeopathy.

nyctalgia *(nĭk-tăl'-jē).* Pain that occurs during sleep only.

nyctalopia *(nĭk-tĕ-lō'-pē-ă).* Night blindness; an inability to see at night or in dim light.

nymphomania *(nĭm-fō-mā'-nē-ă).* An abnormally excessive sexual desire in females.

nystagmus *(nĭs-tăg'-mŭs).* Involuntary, rapid, rhythmic movement of the eyeball in any direction.

nystatin *(nī-stăt'-ĭn).* An antifungal agent that is effective against *Candida albicans* and other *Candida* species, and is used in the treatment of vaginal, intestinal, and skin infections.

O

oat cell cancer. A type of lung cancer that spreads rapidly through the lymphatic channels and is usually inoperable.

obduction *(ŏb-dŭk'-shĕn).* An inspection of a dead body to determine the cause of death; *see also* **autopsy**.

obesity *(ō-bē'-sĭ-tē).* An excessive amount of body weight resulting from a large accumulation of fat in the body.
adult-onset o. Obesity that begins in adulthood and is characterized by an increase in the size but not in the number of fat cells.
endogenous o. Obesity caused by metabolic dysfunction or genetic defects that have an impact on the synthesis of enzymes involved in metabolism.
exogenous o. Obesity caused by overeating.
lifelong o. Obesity that begins in childhood and is characterized by an increase both in the number and size of fat cells.
morbid o. A condition in which an individual weighs two or more times his or her ideal weight.

obfuscation *(ŏb-fŭs-kā'-shŭn).* The process of becoming confused.

objective *(ŏb-jĕk'-tĭv).* 1. Perceptible to other people; *see also* **objective symptom.** 2. The lens or lenses in a microscope closest to the object being examined.

objective sign. In physical diagnosis, a sign that can be observed by the physician or other diagnostician. This is in contrast to a subjective sign, which only the patient experiences.

objective symptom. A symptom that is obvious to the senses of the physician or other diagnostician.

obliteration *(ŏb-lĭt-ĕr-ā'-shŭn).* Complete removal of a part by disease, surgical procedure, irradiation, degeneration, or other means.

obsession *(ŏb-sĕsh'-shŭn).* A recurrent, persistent, and uncontrollable impulse, thought, idea, or image that comes to mind despite attempts to ignore or suppress it.

obsessive-compulsive disorder. A mental condition in which an individual has an uncontrollable desire to perform certain rituals repetitively as a means to relieve anxiety.

obsolescence *(ŏb-sō-lĕs'-ĕns).* The end or the beginning of the end of a physiologic process; the condition of becoming useless.

obstetrician *(ŏb-stĕ-trĭ'-shĕn).* A physician who practices obstetrics.

obstetrics *(ŏb-stĕt'-rĭks).* The branch of surgery that deals with pregnancy, labor, and the puerperium.

obstipation *(ŏb-stĭ-pā'-shŭn).* Complete inability to move the bowels.

obstruction *(ŏb-strŭk'-shŭn).* The act of blocking or the state of being blocked.
bladder outlet o. A blockage in the outflow of urine from the bladder, which can be caused by benign prostatic hyperplasia, prostate cancer, and various other conditions.
closed-loop o. Blockage of the intestines caused by the closing of both ends of a bowel segment.

intestinal o. Any blockage to the passage of intestinal contents.

obtund *(ŏb-tŭnd').* To make a sensation dull or less acute.

obturator *(ŏb'-tĕ-rā-tŏr).* Any structure, natural or artificial, that closes an opening.

occipital *(ŏk-sĭp'-ĭ-tĕl).* Referring to the occiput.

occiput *(ŏk'-sĭ-pŭt).* The lower portion of the back of the head.

occlusion *(ō-klū'-zhĕn).* 1. An obstruction. 2. The trapping of a liquid or gas within the cavities in a solid or on its surface.

abnormal o. Malocclusion.

coronary o. Complete blockage of an artery of the heart, usually due to progressive atherosclerosis.

lateral o. Occlusion of the teeth when the lower jaw is moved to either side of the center position.

normal o. Contact of the upper and lower teeth in the center position.

occult *(ŏ-kŭlt').* Hidden from view or difficult to understand.

occult blood. Blood that is present in such minute quantities that it can be seen only using a microscope or by chemical means.

occult blood test. A chemical test or microscopic examination for blood in a specimen, especially stool.

occupational disease *(ŏk-ū-pā'-shŭn-ăl).* A disease associated with the type of work an individual performs, such as lung inflammation among miners.

occupational therapy. A therapeutic approach in which individuals perform work, self-care, and play activities designed to increase independent function, improve development, and prevent disability.

ochrometer *(ō-krŏm'-ĕ-tĕr).* A device that measures the capillary blood pressure by recording the pressure required to compress a finger until the skin becomes blanched.

ocular *(ŏk'-ū-lăr).* Referring to or affecting the eye.

oculomotor *(ŏk-ū-lō-mō'-tŏr).* Referring to or affecting movements of the eye.

oculist *(ŏk'-ū-lĭst).* Ophthalmologist.

OD. Acronym for Doctor of Optometry; also means overdose.

Oddi's sphincter *(ōd'-ēz).* A sphincter at the opening of the common bile duct into the duodenum.

oddities *(ōd-ī'-tĭs).* Inflammation of the sphincter of Oddi.

odontectomy *(ō-dŏn-tĕk'-tĕ-mē).* Removal of a tooth.

odontology *(ō-dŏn-tŏl'-ĕ-jē).* The science of dentistry.

odontoma *(ō-dŏn-tō'-mă).* Any tumor of a tooth or that develops in dental tissue.

odoriferous *(ō-dĕr-ĭf'-ĕr-ĕs).* Emitting a fragrance or odor.

odorimetry *(ō-dĕr-ĭm'-ĕ-trē).* Measurement of the ability of a substance to stimulate olfactory sensations.

Oedipus complex *(ĕd'-ĭ-pŭs).* In adulthood, the abnormal intense love of a child for the parent of the opposite sex, usually love of a son for his mother.

off-label use. A practice by which physicians prescribe drugs for uses other than their intended indications. This is allowed per regulations of the Food and Drug Administration.

ointment *(oint'-mĕnt).* A semisolid preparation, usually containing a medicinal substance, used as a topical treatment for various skin and mucous membrane conditions.

olanzapine *(ō-lăn'-zĕ-pēn). See* Appendix, Common Prescription and OTC Drugs: By Generic Name.

olecranon *(ō-lĕk'-rĕ-nŏn).* The bony projection of the ulna at the elbow.

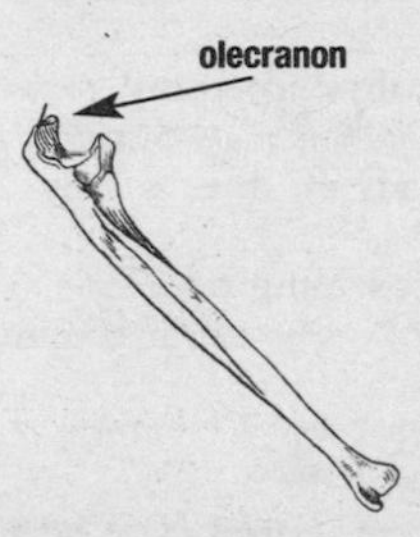

olfaction *(ŏl-făk'-shĕn).* The sense of smell.

olfactory *(ŏl-făk'-tĕ-rē).* Referring to the sense of smell.

oligemia *(ŏl-ĭ-jē'-mē-ă). See* **hypovolemia.**

oligodendroglioma *(ŏl-ĭ-gō-dĕn-drō-glī-ō'-mă).* A neoplasm, usually benign, composed of oligodendrocytes. Most occur in adults in the white matter of the brain.

oligomenorrhea *(ŏl-ĭ-gō-mĕn-ō-rē'-ă).* Infrequent menstruation, occurring at intervals of 35 days to 6 months.

oligospermia *(ŏl-ĭ-gō-spĕr'-mē-ă).* A decreased number of spermatozoa in the semen.

oliguria *(ŏl-ĭ-gū'-rē-ă).* Reduced production and excretion of urine as compared with the amount of fluid taken in; usually defined as less than 500 mL per 24 hours.

omega-3 fatty acids. A class of fatty acids present in cold-water fish (e.g., salmon, tuna) that helps lower levels of cholesterol and low-density lipoprotein in the blood.

omega-6 fatty acids. A class of fatty acids present mainly in vegetable and seed oils. Diets high in these fatty acids have been linked to cancer.

omentectomy *(ō-mĕn-tĕk'-tĕ-mē).* Surgical removal of all or part of the omentum.

omentopexy *(ō-mĕn'-tō-pĕk-sē).* A surgical procedure in which the omentum is fastened to another tissue.

omentum *(ō-mĕn'-tŭm).* A double fold of peritoneum attached to the stomach and connecting it with other abdominal large organs.

omeprazole *(ō-mĕp'-rā-zōl). See* Appendix, Common Prescription and OTC Drugs: By Generic Name.

omnivorous *(ŏm-nĭv'-ĕ-rĕs).* Eating both plant and animal foods.

omphalectomy *(ŏm-fĕ-lĕk'-tĕ-mē).*

Surgical removal of the umbilicus.

omphalitis *(ŏm-fĕ-lī'-tĭs).* Inflammation of the umbilicus.

omphalocele *(ŏm'-fĕ-lō-sēl).* At birth, the protrusion of part of the intestine through a defect in the abdominal wall.

omphalotomy *(ŏm-fĕ-lŏt'-ĕ-mē).* Cutting of the umbilical cord.

oncogene *(ŏng'-kō-jēn).* A gene capable of causing the initial and progressive conversion of normal cells into cancer cells.

oncogenic *(ŏng-kō-jĕn'-ĭk).* Giving rise to benign or malignant tumors or causing tumor formation.

oncology *(ŏng-kŏl'-ĕ-jē).* The study of tumors.

onlay *(ŏn'-lā).* A graft applied to the surface of an organ or structure.

onychia *(ō-nĭk'-ē-ă).* Inflammation of the matrix of the nail, which causes the nail to fall off.

onychomalacia *(ŏn-ĭ-kō-mĕ-lā'-shă).* Softening of the nails.

onychomycosis *(ŏn-ĭ-kō-mī-kō'-sĭs).* Infection of the nails caused by a combination of fungi and bacteria, especially species of *Candida.* Also called *tinea unguium.*

onychopathic *(ŏn-ĭ-kō-păth'-ĭk).* Referring to any disease of the nails.

onychophagia *(ŏn-ĭ-kō-fā'-jă).* The habit of biting the nails.

onychotomy *(ŏn-ĭ-kŏt'-ĕ-mē).* Incision of a nail.

ooblast *(ō'-ō-blăst).* A primitive cell from which an ovum (oocyte) develops.

oocyte *(ō'-ō-sīt).* A developing egg cell, which goes through two stages, primary and secondary, which is followed by ovulation. If the secondary oocyte is fertilized, it divides into an ootid.

oogenesis *(ō-ō-jĕn'-ĕ-sĭs).* The process by which female gametes (oocytes) are formed.

oophorectomy *(ō-ŏf-ĕ-rĕk'-tĕ-mē).* Surgical removal of one or both ovaries. Also called *ovariectomy.*

oophoritis *(ō-ŏf-ĕ-rī'-tĭs).* Inflammation of an ovary.

oophorocystectomy *(ō-ŏf-ĕ-rō-sĭs-tĕk'-tĕ-mē).* Surgical removal of an ovarian cyst.

oophorocystosis *(ō-ŏf-ĕ-rō-sĭs-tō'-sĭs).* The development of ovarian cysts.

oophorohysterectomy *(ō-ŏf-ĕ-rō-hĭs-tĕr-ĕk'-tĕ-mē).* Surgical removal of the uterus and ovaries.

oophorosalpingitis *(ō-ŏf-ĕ-rō-săl-pĭn-jī'-tĭs).* *See* **salpingooophoritis.**

ootid *(ō'-ō-tĭd).* A mature ovum (oocyte).

opacity *(ō-păs'-ĭ-tē).* The state of being opaque.

opaque *(ō-păk').* Impenetrable by visible light rays, x-rays, or other electromagnetic radiation.

open heart surgery. Surgery that involves opening the chest and performing procedures on the heart.

open label trial. A research study in which both the researchers and the participants know the identity of the treatment being used.

open reduction. In orthopedics, a procedure in which the broken segments of a fractured bone are realigned after incision of the skin and tissues around the break.

operable *(ŏp'-ĕr-ă-bĕl).* Something that can be operated upon with a reasonable degree of safety and expectation of improvement.

operation *(ŏp-ĕr-ā'-shŭn).* A surgical procedure.

cosmetic o. Procedure intended to correct or eliminate a deformity in an aesthetically pleasing manner.

exploratory o. Surgical incision made for the purpose of investigating the cause of unexplained symptoms or to aid a diagnosis.

operative risk *(ŏp'-ĕr-ă-tĭv).* A determination of how well or how poorly a patient will withstand surgery.

ophthalmectomy *(ŏf-thĕl-mĕk'-tĕ-mē).* Surgical removal of an eye.

ophthalmia *(ŏf-thăl'-mē-ă).* Severe inflammation of the eye or of structures within the eye.

catarrhal o. A severe form of simple conjunctivitis.

purulent o. Ophthalmia characterized by discharge of pus; commonly associated with gonorrhea.

ophthalmologist *(ŏf-thĕl-mŏl'-ĕ-jĭst).* A physician who specializes in the diagnosis and treatment (medical and surgical) of eye defects and diseases.

ophthalmomalacia *(ŏf-thăl-mō-mĕ-lā'-shĕ).* Abnormal softness of the eye.

ophthalmometer *(ŏf-thăl-mŏm'-ĕ-tĕr).* A device that measures the curvature of the cornea. Also called a *keratometer.*

ophthalmomyotomy *(ŏf-thăl-mō-mī-ŏt'-ĕ-mē).* Surgical procedure in which the muscles of the eye are divided.

opthalmoplegia *(ŏf-thăl-mō-plē'-jĕ).* Paralysis of the eye muscles.

ophthalmorrhagia *(ŏf-thăl-mō-rā'-jē-ĕ).* Bleeding from the eye.

ophthalmoscope *(ŏf-thăl'-mĕ-skōp).* A device composed of a mirror and lenses used to examine the interior of the eye.

ophthalmotomy *(ŏf-thĕl-mŏt'-ĕ-mē).* The surgical procedure that involves cutting the eyeball.

opiate *(ō'-pē-ăt).* Any drug that contains or is derived from opium.

opisthotonos *(ō-pĭs-thŏt'-ĕ-nĕs).* A type of spasm in which the head and heels are bent backward and the body bows forward.

opium *(ō'-pē-ŭm).* The substance obtained by air drying the juice from unripe capsules of *Papaver somniferum* or *P. album.* Some of opium's derivatives include codeine, morphine, and papaverine, which have narcotic and analgesic effects.

opportunistic infection *(ŏp-ĕr-tū-nĭs'-tĭk).* A microorganism that ordinarily does not cause disease but will, under certain conditions, such as treatment with some drugs or presence of a compromised immune system.

opsonin *(ŏp'-sĕ-nĭn).* Substance in blood serum that binds to antigens and causes them to be digested by macrophages and neutrophils, a process called phagocytosis.

opsonization *(ŏp-sĕ-nĭ-zā'-shĕn).* The action of opsonins to cause phagocytosis.

optical *(ŏp'-tĭ-kăl).* Referring to vision, the eye, or optics.

optician *(ŏp-tĭsh'-ĕn).* An individual specially trained in fulfilling ophthalmic prescriptions (e.g., eyeglasses, contact lenses) and products.

optic nerve *(ŏp'-tĭk).* The nerve that carries impulses for the sense of light to the eye.

optometrist *(ŏp-tŏm'-ĕ-trĭst).* An individual trained and then licensed by the state to diagnose, manage, and treat conditions affecting human vision.

optometry *(ŏp-tŏm'-ĕ-trē).* The professional service consisting of examination of the eyes to determine visual abilities, diagnosis of eye diseases and conditions that affect vision, and provision of treatment.

OR. Acronym for operating room.

oral *(ŏr'-ĕl).* Referring to the mouth.

oral aversion. Refusal or reluctance to eat. An example is an infant's refusal to breastfeed.

oral contraceptive. A form of birth control taken by mouth that blocks ovulation and prevents pregnancy. Usually refers to the contraceptive pill or "the pill."

orbicular muscles *(ŏr-bĭk'-ū-lăr).* The sphincter muscle that surrounds the eyelids and is responsible for closing the eyelids, wrinkling the forehead, and compressing the lacrimal (tear) sac.

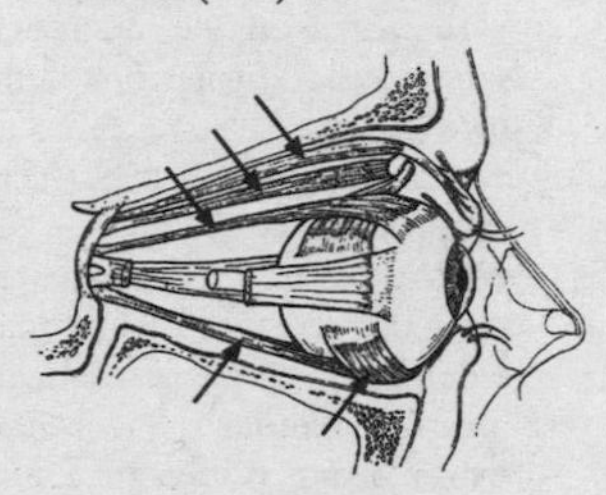

orbicular muscles

orbit *(ŏr'-bĭt).* The bony cavity of the skull that contains and protects the eyeball.

orchialgia *(ŏr-kē-ăl'-jē).* A painful testis.

orchiectomy *(ŏr-kē-ĕk'-tĕ-mē).* Surgical removal of a testicle.

orchioncus *(ŏr-kē-ŏng'-kĕs).* A testicular tumor.

orchiopexy *(ŏr'-kē-ō-pĕk-sē).* Surgical fixation in the scrotum of an undescended testis.

orchitis *(ŏr-kī'-tĭs).* Inflammation of a testis, characterized by swelling, pain, and a feeling of heaviness. The cause may be unknown or associated with mumps, gonorrhea, syphilis, or tuberculosis.

orderly *(ŏr'-dĕr-lē).* A hospital attendant who works under the supervision of a nurse.

organ *(ŏr'-găn).* A somewhat independent part of the body that performs a specific function or functions and is composed of various tissues.
end o. The specialized termination of a sensory nerve fiber that acts as a receptor.
gustatory o. Organ that regulates taste; a taste bud.
sense o. A structure that has specialized sensory nerve endings that can react to a stimulus by giving rise to nerve signals that travel through afferent nerves to the central nervous system.
target o. Organ upon which a hormone or chemical acts.
vestigial o. Organ that is underdeveloped in people but is completely functional in some animals.

organelle *(ŏr-gĕ-nĕl').* Any one of the self-contained structures present in all eukaryotic cells (those that have a true nucleus).

organ failure. The shutdown or failure to function by an essential system in the body. *Multiple organ failure* is cessation of function by more than one system.

organic brain syndrome. Psychiatric or neurological symptoms, such as problems with memory, concentration, anxiety, depression, and attention, that arise from damage to or disease in the brain.

organism *(ŏr'-găn-ĭzm).* Any individual living thing, animal or plant.

organomegaly *(ŏr-gĕ-nō-mĕg'-ă-lē).* *See* **visceromegaly.**

organotherapy *(ŏr-gĕ-nō-thĕr'-ă-pē).* The treatment of disease by administering extracts or portions of animal endocrine organs.

orgasm *(ŏr'-găz-ĕm).* The culmination of sexual arousal.

oriented *(ŏr'-ē-ĕn-tĕd).* Awareness of one's surroundings, especially regarding place, people and time.

orifice *(ŏr'-ĭ-fĭs).* The entrance or departure point of a body cavity.

origin *(ŏr'-ĭ-jĭn).* The beginning or source of something.

ornithine *(ŏr'-nĭ-thĕn).* An amino acid, not found in proteins, that is formed in the liver when the enzyme arginase acts upon the amino acid arginine.

ornithosis *(ŏr-nĭ-thō'-sĭs).* An acute infectious disease that affects birds and domesticated fowl, which can be transmitted to humans.

oropharynx *(ŏr-ō-făr'-ĭnks).* The mid portion of the pharynx located between the soft palate and the upper part of the epiglottis.

orphan disease. Based on US criteria, a disease that affects fewer than 200,000 people. There are more than 5,000 orphan diseases. The term also refers to a common disease that is largely ignored because it is far more common in developing countries than in the developed world (e.g., tuberculosis, malaria).

orphan drug. A drug developed to treat an orphan disease.

Ortho Cyclen *(ŏr'-thō sī'-klĕn).* Trade name for combination preparations of norgestimate and ethinyl estradiol, oral contraceptives.

orthodontics *(ŏr-thō-dŏn'-tĭks).* The branch of dentistry involved in the supervision, guidance, and correction of the developing and mature teeth and associated structures.

orthodontist *(ŏr-thō-dŏn'-tĭst).* A dentist who specializes in orthodontics.

orthomolecular medicine *(ŏr-thō-mō-lĕk'-ū-lĕr).* A branch of nutrition science in which illness and disease are treated with large doses of nutrients.

Ortho Novum *(ŏr'-thō nō'-vŭm).* Trade name for a hormone treatment medication that contains estrogen and progestogen.

orthopedics *(ŏr-thō-pē'-dĭks).* Branch of surgery that focuses on the preservation and restoration of the skeletal system and its functions.

orthopedist *(ŏr-thō-pē'-dĭst).* An orthopedic surgeon.

orthopnea *(ŏr-thŏp-nē'-ă).* Breathlessness or shortness of breath that is relieved when the affected individual gets into an upright position.

orthopsychiatry *(ŏr-thō-sī-kī'-ă-trē).* An interdisciplinary field that combines psychiatry with the concepts associated with psychology, social work, sociology, and other areas concerned with maintaining or restoring mental health with a focus on prevention.

orthoptics *(ŏr-thŏp'-tĭks).* Eye exercises designed to correct faulty eye coordination that affects binocular vision.

orthosis *(ŏr-thō'-sĭs).* An orthopedic device used to support, align, prevent, or correct deformities or to improve the function of moveable body parts.

cervical o. A rigid plastic device used to treat injuries to the cervical spine. It goes around the neck and supports the chin and the back of the head.

halo o. A device consisting of a metal or plastic halo attached to the upper skull by tongs or pins and connected with vertical bars to a rigid jacket on the chest.

spinal o. A corset-type device that surrounds part or all of the trunk to support or align the spinal column or prevent movement following trauma.

orthostatic *(ŏr-thō-stăt'-ĭk).* Referring to an upright position.

orthotics *(ŏr-thŏt'-ĭks).* The science that relates to orthoses and their use.

Ortho-Tri-Cyclen *(ŏr-thō-trī-sī'-klēn).* *See* Appendix, Common Prescription and OTC Drugs: By Trade Name.

os *(ŏs).* 1. Any opening of the body. 2. Bone.

os calcis. The heel bone.

os coxae. The hip bone.
os pubis. The pubic bone.
oscillation *(ŏs-ĭ-lā'-shŭn).* Backward and forward motion, resembling a pendulum.
oscitation *(ŏs-sĭ-tā'-shŭn).* Yawning.
OSHA. Occupational Safety and Health Administration. An agency of the US government responsible for ensuring that US workers have safe, healthy work environments.
osmesis *(ŏz-mē'-sĭs).* The act of smelling.
osmidrosis *(ŏz-mĭ-drō'-sĭs).* Apocrine sweat that has a foul smell due to its bacterial decomposition.
osmolarity *(ŏz-mō-lăr'-ĭ-tē).* Concentration of osmotically active particles in solution.
osmology *(ŏz-mŏl'-ō-jē).* The study of odors.
osmosis *(ŏz-mō'-sĭs).* The movement of solvent through a semipermeable membrane that separates solutions of different concentrations.
osmotic pressure. Pressure that develops when two solutions of different concentrations are separated by a semipermeable membrane.
osphytis *(ŏs-fē-ī'-tĭs).* Inflammation of the lumbar area..
ossein *(ŏs'-ē-ĭn).*The collagen present in bone.
osseous *(ŏs'-ē-ŭs).* Bonelike; referring to bone.
ossicle *(ŏs'-ĭ-kĕl).* A small bone.
ossiculectomy *(ŏs-ĭ-kū-lĕk'-tĕ-mē).* Surgical removal of one or more ossicles of the ear.
ossification *(ŏs-ĭ-fĭ-kā'-shŭn).* The formation of bone or a bony substance.
osteal *(ŏs'-tē-ăl).* Bony.
ostealgia *(ŏs-tē-ăl'-jē).* Pain in bone.
ostectomy *(ŏs-tĕk'-tĕ-mē).* Surgical excision of a bone or part of a bone.
osteitis *(ŏs-tē-ī'-tĭs).* Inflammation of bone, characterized by enlargement of the bone, tenderness, and aching pain.
osteoarthritis *(ŏs-tē-ō-ăr-thrī'-tĭs).* A noninflammatory degenerative joint disease characterized by pain, usually after prolonged activity; stiffness, especially in the morning or when inactive.
osteoarthropathy *(ŏs-tē-ō-ăhr-thrŏp'-ĕ-thē).* Any disease that affects the joints and bones.
osteoblast *(ŏs'-tē-ō-blăst).* A cell that arises from a fibroblast and eventually is associated with bone production.
osteoblastoma *(ŏs-tē-ō-blăs-tō'-mă).* A benign, painful bone tumor characterized by the formation of osteoid tissue and primitive bone.
osteocalcin *(ŏs-tē-ō-kăl'-sĭn).* The most abundant noncollagen protein present in bone. Increased serum concentrations indicate increased bone turnover in some diseases.
osteochondritis *(ŏs-tē-ō-kŏn-drī'-tĭs).* Inflammation of bone and cartilage.
osteochondrodystrophy *(ŏs-tē-ō-kŏn-drō-dĭs'-trĕ-fē).* A disorder of

skeletal growth that results from malformation of bone and cartilage; also known as *Morquio's syndrome.*

osteochondroma *(ŏs-tē-ō-kŏn-drō'-mă).* A benign tumor that consists of adult bone and cartilage.

osteochondrosis *(ŏs-tē-ō-kŏn-drō'-sĭs).* A disease of the growth or ossification centers in children that begins with degeneration or necrosis of bone, followed by slow healing and repair.

osteoclasia *(ŏs-tē-ō-klā'-zhă).* The absorption and destruction of bone tissue.

osteoclast *(ŏs'-tē-ō-klăst).* A large multinuclear cell associated with the absorption and removal of bone. When in the presence of parathyroid hormone, these cells cause increased bone resorption.

osteodynia *(ŏs-tē-ō-dĭn'-ē-ă).* Pain in a bone.

osteodystrophy *(ŏs-tē-ō-dĭs'-trĕ-fē).* A defect in bone formation.

osteoectasia *(ŏs-tē-ō-ĕk-tā'-zhă).* Bowing of the bones.

osteofibroma *(ŏs-tē-ō-fī-brō'-mă).* A benign tumor that contains both osseous and fibrous elements.

osteogenesis *(ŏs-tē-ō-jĕn'-ĕ-sĭs).* The formation of bone.

osteogenesis imperfecta. An inherited collagen disorder characterized by brittle, easily fractured bones, early deafness, translucent skin, and joint instability.

osteology *(ŏs-tē-ŏl'-ĕ-jē).* The scientific study of bone.

osteolysis *(ŏs-tē-ŏl'-ĭ-sĭs).* The removal or loss of calcium from bone.

osteoma *(ŏs-tē-ō'-mă).* A benign, slow-growing tumor that usually arises in membrane bones, especially the skull and facial bones.

osteomalacia *(ŏs-tē-ō-mĕ-lā'-shă).* A disease characterized by increasing softening of the bones. It is the adult equivalent of rickets in children.

hepatic o. Osteomalacia as a complication of cholestatic liver disease, which can result in severe bone pain and multiple fractures.

puerperal o. Form of osteomalacia that results from depletion of calcium and phosphorus due to repeated pregnancies and lactation.

senile o. Form of osteomalacia in the elderly caused by a vitamin D deficiency.

osteomyelitis *(ŏs-tē-ō-mī-lī'-tĭs).* Inflammation of bone caused by an infection, usually a fever-causing organism.

osteonecrosis *(ŏs-tē-ō-nĕ-krō'-sĭs).* Death of bone cells due to blockage of their blood supply.

osteopathy *(ŏs-tē-ŏp'-ĕ-thē).* A system of medicine based on the concept that the body can heal itself and rectify itself against toxic elements when it has favorable environmental circumstances and adequate nourishment.

osteopetrosis *(ŏs-tē-ō-pĕ-trō'-sĭs).* A hereditary condition characterized by excessive calcification of bone, leading to spontaneous fractures, compression of the skull, and anemia.

osteoplasty *(ŏs'-tē-ō-plăs-tē).* Plastic surgery of the bones.

osteoporosis *(ŏs-tē-ō-pĕ-rō'-sĭs).* Decrease in bone mass, resulting in fractures after minimal trauma.

postmenopausal o. Osteoporosis that occurs within 3 to 20 years after menopause and is manifested primarily by vertebral and hip fractures, Colles' fracture, and increased tooth loss.

post-traumatic o. Bone loss that occurs following an injury to a nerve.

senile o. Osteoporosis that occurs in men and women older than 70, manifested primarily by vertebral and hip fractures.

osteosarcoma *(ŏs-tē-ō-săr-kō'-mă).* Malignant neoplasm of bone.

osteosclerosis *(ŏs-tē-ō-sklĕ-rō'-sĭs).* Hardening or abnormal density of bone.

osteosteatoma *(ŏs-tē-ō-stē-ă-tō'-mă).* A benign fatty tumor consisting of bony elements.

osteotomy *(ŏs-tē-ŏt'-ĕ-mē).* The surgical procedure for cutting through a bone.

ostium *(ŏs'-tē-ŭm).* An opening, orifice, or aperature.

otalgia *(ō-tăl'-jē-ă).* Ear pain.

OTC. Acronym for over-the-counter.

otitis media *(ō-tī'-tĭs mē'-dē-ă).* Inflammation of the middle ear, accompanied by pain, fever, vertigo, tinnitus, and hearing loss.

otoencephalitis *(ō-tō-ĕn-sĕf-ĕ-lī'-tĭs).* Inflammation of the brain due to infiltration from an inflamed middle ear.

otolaryngology *(ō-tō-lăr-ĭng-gŏl'-ĕ-jē).* The branch of medicine concerned with the diagnosis and treatment (including surgical) of conditions affecting the head and neck.

otoliths *(ō'-tō-lĭths).* Minute particles, composed mainly of calcium carbonate, present in the inner ear.

otology *(ō-tŏl'-ĕ-jē).* The branch of medicine concerned with the anatomy, physiology, and pathology of the ear, as well as the diagnosis and treatment of conditions of the ear.

otomycosis *(ō-tō-mī-kō'-sĭs).* A fungal infection of the external ear, usually caused by a species of *Aspergillus.* It is characterized by itching, inflammation, and drainage from the ear.

otoplasty *(ō'-tō-plăs-tē).* Plastic surgery of the ear, performed to correct defects.

otorrhea *(ō-tō-rē'-ă).* Discharge from the ear, usually consisting of pus.

otosclerosis *(ō-tŏ-sklĕ-rō'-sĭs).* Condition in which spongy bone develops in the labyrinth of the ear, resulting in chronic, progressive deafness.

otoscope *(ŏ'-tō-skōp).* Device used to examine the ear.

ounce *(owns).* A measure of weight and liquid volume equal to 1/16 of a pound or 28.35 grams.

outpatient *(owt'-pā-shĕnt).* An individual who receives treatment at a hospital or clinic but is not hospitalized.

output *(owt'-pŭt)*. The total of anything produced by any functioning system in the body.

cardiac o. Volume of blood pumped into the arteries per unit of time, usually per minute.

energy o. Work expended by the body per unit of time.

urinary o. The amount of urine excreted by the kidneys.

ovarian *(ō-văr'-ē-ăn)*. Referring to an ovary or ovaries.

ovary *(ō'-vĕ-rē)*. The female gonad, one of the two sexual glands in which the eggs (ova) are formed.

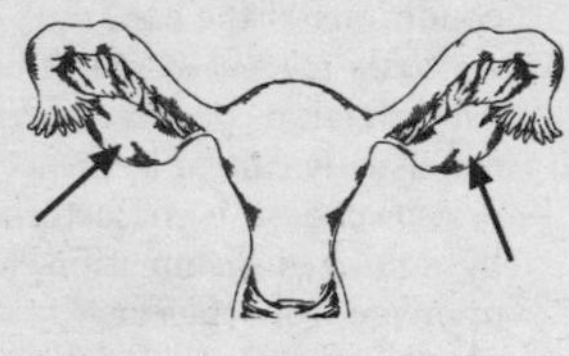

ovaries

overactive bladder. A condition characterized by a sudden, involuntary contraction of the bladder wall, causing an immediate need to urinate. It affects about 9 percent of adults, mostly older adults, also called *urge incontinence.*

overbite *(ō'-vĕr-bīt)*. Vertical extension of the upper front teeth over the lower front teeth when the jaws are closed.

overextension *(ō-vĕr-ĕk-stĕn'-shĕn)*. Extension beyond the normal limit, usually referring to a limb.

overlay *(ō'-vĕr-lā)*. An addition superimposed upon an already existing object.

overreaction *(ō-vĕr-rē-ăk'-shŭn)*. Abnormally excessive response to a stimulus.

overriding *(ō-vĕr-rīd'-ĭng)*. When part of a fractured bone slips past the other part.

over-the-counter drug. A drug for which no prescription is required.

oviduct *(ō'-vĭ-dŭkt)*. One of two muscular tubes that extend laterally from the uterus. It carries the ovum and provides a channel through which sperm travels from the uterus toward the ovary.

oviparous *(ō-vĭp'-ĕr-ŭs)*. Producing eggs that are hatched outside the body.

ovotestis *(ō-vō-tĕs'-tĭs)*. A gonad that contains both ovarian and testicular tissue.

ovoviviparous *(ō-vō-vī-vĭp'-ĕ-rĕs)*. Bearing living young that hatch from eggs inside the body of the female organism.

ovular *(ŏv'-ū-lăr)*. Referring to an ovule.

ovulation *(ŏv-ū-lā'-shŭn)*. The periodic discharge of an oocyte from an ovarian follicle.

ovule *(ō'-vŭl)*. The ovum within the ovarian follicle.

ovum *(ō'-vŭm)*. The reproductive cell of a female which, when fertilized, eventually develops into a new member of the same species, also called *egg.*

oxacillin *(ŏks-ĕ-sĭl'-ĭn)*. A semisynthetic penicillin used to treat infections caused by penicillin-resistant staphylococci.

oxaluria *(ŏk-sē-lū'-rē-ă).* Excessive excretion of oxalates in the urine, which may lead to development of urinary stones, or calculi.

oxidase *(ŏk'-sĭ-dās).* A class of enzymes that catalyzes oxidation reactions.

oxidation *(ŏk-sĭ-dā'-shŭn).* The process by which a substance combines with oxygen.

oxycodone *(ŏk-sē-kō'-dŏn). See* Appendix, Common Prescription and OTC Drugs: By Generic Name.

OxyContin *(ŏk-sē-kŏn'-tĭn). See* Appendix, Common Prescription and OTC Drugs: By Trade Name.

oxygen *(ŏk'-sĭ-jĕn).* A colorless, tasteless, odorless, gaseous element present in the air and necessary for respiration for most animal and plant life.

oxygenase *(ŏk'-sĭ-jĕn-ās).* An enzyme that helps an organism use atmospheric oxygen in respiration.

oxygenation *(ŏk-sĭ-jĕn-ā'-shŭn).* The process of combining with oxygen.

oxyhemoglobin *(ŏk-sē-hē-mō-glō'-bĭn).* The combined form of hemoglobin and oxygen, which is formed when hemoglobin is exposed to alveolar gas in the lungs.

oxytetracycline *(ŏk-sē-tĕt-ră-sī'-klēn).* A broad-spectrum synthetic antibiotic effective against both gram-positive and gram-negative organisms.

oxytocic *(ŏk-sĭ-tō'-sĭk).* Referring to or characterized by rapid labor.

oxytocin *(ŏk-sĭ-tō'-sĭn).* A substance secreted by the hypothalamus, it promotes uterine contractions and milk ejection.

ozone *(ō'-zōn).* A bluish gas or liquid that is formed when oxygen is exposed to a silent discharge of electricity. It is toxic and irritates the respiratory system, and is an air pollutant.

ozone therapy. The introduction of ozone into the body for medicinal purposes. Ozone increases the level of oxygen in the body, which reportedly helps the immune system and facilitates cell functioning.

ozostomia *(ō-zō-stō'-mē-ă).* Bad breath.

P

pacemaker *(pās'-mā-kĕr).* An object or substance that has an impact on the rate at which a certain event occurs, especially when referring to the heart.

artificial p, artificial cardiac p. A device that uses electrical impulses to regulate or reproduce the rhythms of the heart.

DDD p. An artificial cardiac pacemaker that can sense and pace both the atria and ventricles.

DDDR p. A universal pacemaker that can respond to a patient's respiratory rate.

external p. An artificial cardiac pacemaker positioned outside the body with a wire or an electrode connecting with the heart.

internal p. One that is completely implanted into the subcutaneous tissue of the chest.
natural p. The sinus node, one of the main elements in the cardiac conduction system, is the natural pacemaker of the heart.
universal p. A term that describes a DDD pacemaker, which can be programmed to operate in one of several possible modes.

pachydermoperiostosis *(păk-ē-dĕr-mō-pĕr-ē-ŏs-tō'-sĭs).* A condition, believed to be inherited, characterized by thickened skin of the head and distal extremities; deep folds of the skin of the forehead, scalp, and cheeks; seborrhea; profuse sweating; clubbed fingers and toes; and enlarged hands and feet.

pachymeningitis *(păk-ē-mĕn-ĭn-jī'-tĭs).* Inflammation of the dura mater, accompanied by symptoms similar to those of meningitis.

pack *(păk).* Treatment that consists of wrapping a patient or a body part (e.g., leg, arm) in blankets, sheets, or towels, either hot or cold, wet or dry.

packed cell transfusion. A procedure in which white and red blood cells are transfused, but not serum.

Paget's disease *(păj'-ĕts).* Skeletal disease of the elderly characterized by chronic inflammation of the bones, which results in thick, soft bones and bowing of long bones.

pain *(pān).* The sensation of distress, discomfort, or agony resulting from the stimulation of specialized nerve endings and associated with actual or potential tissue damage.
baseline p. The average intensity of pain a person experiences for 12 or more hours during a 24-hour period.
breakthrough p. A transient increase in pain intensity from baseline of no greater than moderate.
false p. Pains that resemble labor pains, but which are not accompanied by dilatation of the cervix.
labor p. Rhythmic pains that increase in severity and frequency as delivery time approaches, and which are caused by contractions of the uterus.
referred p. Pain felt in a part of the body other than where it is actually produced.

pain management. The process of providing medical care that reduces or eliminates pain. Treatment can take many forms, from use of OTC medications such as aspirin or ibuprofen, to opiates for more severe pain; use of nerve blocks and surgical procedures; and complementary therapies such as physical therapy, acupuncture, massage, and chiropractic.

pain threshold. The point at which a person feels pain. A person with a low threshold feels pain sooner than one who has a high threshold.

palate *(păl'-ăt)*. The partition that separates the nasal and oral cavities.

cleft p. Congenital fissure of the soft palate or both the soft and hard palates.

hard p. The front part of the palate.

soft p. The fleshy part of the roof of the mouth.

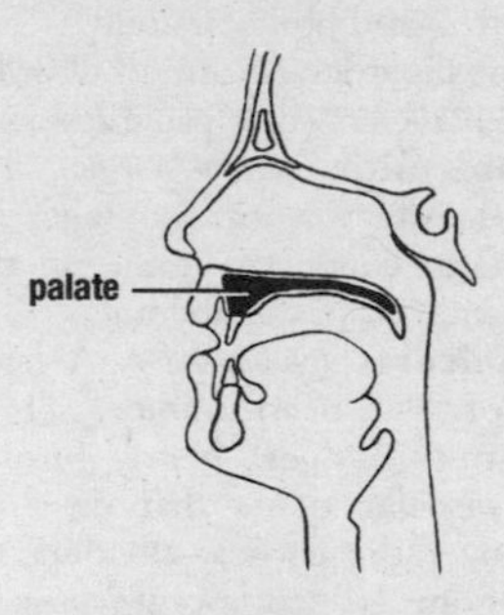

palatoplasty *(păl'-ă-tō-plăs-tē)*. Plastic reconstruction of the palate, including surgery to correct a cleft palate.

palinopsia *(păl-ĭ-nŏp'-sē-ă)*. Persistence of a visual image after the object is no longer physically visible.

pallesthesia *(păl-ĕs-thē'-zhē-ă)*. The ability to feel physical vibrations on or near the body.

palliative *(păl'-ē-ā-tĭv)*. Something that provides relief from pain or discomfort but not a cure.

pallidectomy *(păl-ĭ-dĕk'-tō-mē)*. Surgical removal of the globus pallidus, which is done to relieve the palsy of Parkinson's disease.

pallidum *(păl'-ĭ-dĕm)*. The globus pallidus, an area of the brain that contains important nerve centers.

pallor *(păl'-ŏr)*. Paleness.

palm *(păhm)*. The inner surface of the hand.

palmature *(păhl'-mă-tūr)*. Webbed fingers.

palpable *(păl'-pĕ-bĕl)*. Perceptible by touch.

palpation *(păl-pā'-shŭn)*. The act of feeling with the hand or fingers to the surface of the body.

palpebral *(păl'-pĕ-brăl)*. Referring to the eyelid.

palpebrate *(păl'-pĕ-brāt)*. To wink.

palpitation *(păl-pĭ-tā'-shŭn)*. A sensation of a rapid or irregular heartbeat.

palsy *(păwl'-zē)*. Paralysis.

Bell's p. Facial paralysis on one side of the face, which comes on suddenly, and is caused by a lesion of the facial nerve.

birth p. Paralysis caused by injury received at birth.

bulbar p. Palsy caused by degeneration of nuclear cells of the lower cranial nerves.

cerebral p. A persistent, nonprogressive motor disorder that appears in young children and results from brain damage during pregnancy or birth. Characteristics include delayed or abnormal motor development, and often by mental retardation, seizures, or ataxia.

shaking p. Progressive muscular weakness and tremor with impaired voluntary motion.

panacea *(păn-ĕ-sē'-ă).* A universal remedy.

panarthritis *(păn-ăhr-thrī'-tĭs).* Inflammation of all the joints or all of the structures of a joint.

pancarditis *(păn-kăhr-dī'-tĭs).* Inflammation of the heart, including the endocardium, myocardium, and pericardium.

pancolitis *(păn-kō-lī'-tĭs).* Inflammation of the entire colon.

pancreas *(păn'-krē-ăs).* An elongated gland located behind the stomach and between the spleen and duodenum. It secretes pancreatic juice, which plays a role in digestion; and insulin and glucagons, which have a role in carbohydrate metabolism.

pancreatectomy *(păn-krē-ăt-ĕk'-tō-mē).* Surgical procedure to remove all or part of the pancreas.

pancreatic duct *(păn-krē-ăt'-ĭk).* Channel that carries pancreatic juice to the duodenum.

pancreatitis *(păn-krē-ă-tī'-tĭs).* Inflammation of the pancreas.

pancreatoduodenectomy *(păn-krē-ă-tō-dū-ō-dē-nĕk'-tō-mē).* Surgical excision of the head of the pancreas and the adjacent portion of the duodenum.

pancreatolithiasis *(păn-krē-ă-tō-lĭ-thī'-ă-sĭs).* Presence of calculi in the ducts of the pancreas.

pancytopenia *(păn-sī-tō-pē'-nē-ă).* A deficiency of all the cellular components of the blood.

pandemic *(păn-dĕm'-ĭk).* A widely distributed disease.

panhysterectomy *(păn-hĭs-tĕr-ĕk'-tĕ-mē).* A total hysterectomy, in which the uterus and cervix are completely removed.

panic *(păn'-ĭk).* Acute anxiety, fright, or terror, that usually comes on suddenly.

panic attack. An acute episode of intense anxiety, which may be characterized by sweating, vertigo, palpitations, chest pain, shortness of breath, nausea, dread, feelings of losing control, and blurry vision.

panic disorder. An anxiety disorder characterized by panic attacks.

panniculitis *(păn-ĭk-ū-lī'-tĭs).* Inflammation of the layer of fatty connective tissue in the anterior of the abdomen.

panniculus *(păn-ĭk'-ū-lŭs).* A layer of tissue or membrane.

pannus *(păn'-nĕs).* Newly formed vascular tissue that develops over the cornea, clouding vision. It may occur in acne rosacea, eczema, trachoma, and granular conjunctivitis.

panophthalmitis *(păn-ŏf-thĕl-mī'-tĭs).* Inflammation of all tissues or structures of the eye.

pansinusitis *(păn-sī-nŭs-ī'-tĭs).* Inflammation of all the paranasal sinuses on one side.

pantoprazole *(păn-tō'-prĕ-zŏl).* *See* Appendix, Common Prescription and OTC Drugs: By Generic Name.

pantothenic acid *(păn-tō-thĕn'-ĭk).* Also known as vitamin B5, this vitamin helps convert proteins, carbohydrates and fats into energy and has the ability to relieve depression and anxiety. It is one of the B-complex family of vitamins and is found in a wide variety of foods.

Papanicolaou test *(pă-pĕ-nĭ-kō-lă'-ū).* Commonly known as the Pap smear test, it is used for early detection of cancer cells, especially in the cervix and vagina.

papaverine *(pă-păv'-ĕr-ēn).* A salt of an opium alkaloid, it is used as a smooth muscle relaxant and vasodilator, especially in the treatment of cerebral and peripheral ischemia; also used in the treatment of erectile dysfunction.

papilla *(pă-pĭl'-ă).* A small, nipple-like projection or elevation.

papilledema *(păp-ĭl-ĕ-dē'-mă).* Swelling of the optic disk (papilla), which may be caused by an increase in intracranial pressure, malignant hypertension, or other factors.

papillitis *(păp-ĭ-lī'-tĭs).* Inflammation of a papilla.

papilloma *(păp-ĭ-lō'-mă).* A benign neoplasm that produces fingerlike projections from its surface.

intraductal p. A tumor in a milk-producing duct in the breast. The solitary type, found in just one duct, is usually benign; the multiple type, found in several ducts, is often premalignant.

inverted p. Papilloma that usually develops in the nasal cavity, urinary bladder, or oral soft tissues in middle-aged males.

squamous p. A papilloma composed of squamous epithelium.

papillomatosis *(păp-ĭ-lō-mă-tō'-sĭs).* The development of many papillomas.

papillotomy *(păp-ĭ-lŏt'-ĕ-mē).* Surgical incision of a papilla.

papule *(păp'-ūl).* A small, superficial, solid elevation of the skin less than 1 cm (or less than 0.5 cm according to some experts) in diameter.

para-aminobenzoic acid (PABA) *(păr-ă-ă-mē-nō-bĕn-zō'-ĭk).* A substance that is part of the vitamin B complex and required for the synthesis of folic acid. It is also used as a topical sunscreen because of its ability to absorb ultraviolet light.

paracentesis *(păr-ă-sĕn-tē'-sĭs).* Surgical puncture of a cavity with a needle or other hollow instrument to aspirate fluid for diagnostic or therapeutic purposes.

paraffin dip *(păr'-ă-fĭn).* A treatment for the pain and stiffness associated with joint and muscle conditions, such as arthritis. Melted mineral wax derived from petroleum is applied to the affected area, allowed to harden, then peeled off. This process is repeated numerous times.

paraganglia *(păr-ă-găng'-lē-ă).* A collection of chromaffin cells, usually seen near the sympathetic ganglia and the aorta and its branches. Most paraganglia secrete epinephrine and norepinephrine.

paraganglioma *(păr-ă-găng-glē-ō'-mă).* A tumor of the tissue that makes up the paraganglia.

paragraphia *(păr-ă-grăf'-ē-ă).* Condition in which an individual makes errors in spelling or writes one word in place of another.

parahormone *(păr-ă-hŏr'-mōn).* A substance not conventionally accepted as a true hormone but which has hormonelike qualities.

parakinesia *(păr-ă-kĭ-nē'-zhă).* Condition characterized by abnormal motor functioning resulting in distorted movements.

paraldehyde *(păr-ăl'-dĕ-hīd).* A substance that has rapid-acting hypnotic and sedative properties. It is used to treat insomnia, agitation, delirium, and convulsions.

parallax *(păr'-ĕ-lăks).* The apparent displacement of an object because of a change in the observer's position.

paralogia *(păr-ă-lō'-jē).* Dysfunction in an individual's ability to reason, characterized by delusion or illogical speech.

paralysis *(pă-răl'-ĭ-sĭs).* Loss or impairment of motor function and sensory function, partly due to a lesion of a neural or muscular region.

compression p. Paralysis caused by pressure on a nerve; also called *pressure p.*

facial p. Paralysis or weakening of the facial nerve, as seen in Bell's palsy or Millard-Gubler syndrome.

functional p. A temporary paralysis that does not seem to be caused by a nerve lesion.

ischemic p. Localized paralysis caused by an impairment of circulation, related to trauma, an embolism, or thrombosis.

posthemiplegic p. Residual muscle weakness after a stroke.

spinal p. Paralysis caused by a lesion of the spine.

paramedical *(păr-ă-mĕd'-ĭ-kĕl).* Having an association with the science or practice of medicine. Paramedical occupations include speech therapy, occupational therapy, pharmacists, and physical therapists, among others.

parameter *(pă-răm'-ĕ-tĕr).* A constant in a mathematical expression that has a fixed value in one situation but different values in other situations.

parametritis *(păr-ă-mĕ-trī'-tĭs).* Inflammation of the parametrium.

parametrium *(păr-ă-mē'-trē-ŭm).* Loose connective tissue around the uterus.

paranasal sinuses *(păr-ă-nā'-săl).* *See* **sinuses.**

paranoia *(păr-ĕ-noy'-ă).* A disorder characterized by persistent delusions of persecution and/or delusions of grandeur.

paranormal *(păr-ă-nŏr'-măl).* Beyond the natural or normal state; often used to refer to phenomena such as extrasensory perception.

paraphilia *(păr-ĕ-fĭl'-ē-ă).* A psychosexual disorder characterized by persistent, intense sexual urges, sexual fantasies, or behavior that involves use of a nonhuman object, the humiliation or suffering of oneself or one's partner, children, or other nonconsenting individuals.

paraphimosis *(păr-ĕ-fī-mō'-sĭs).* Retraction of the foreskin, causing a painful swelling of the glans.

paraplegia *(păr-ă-plē'-jē-ă).* Paralysis of the lower region of the body, including both legs.

paraprofessional *(păr-ă-prō-fĕsh'-ĕn-ăl).* An individual who is specially trained in a specific field or occupation to assist a professional.

parapsychology *(păr-ă-sī-kŏl'-ĕ-jē).* The study of psychical effects and experiences that appear to be outside the scope of physical law, such as clairvoyance and telepathy.

parasite *(păr'-ă-sīt).* An animal or plant that lives upon or within another living organism at the latter's expense and obtains benefits from this relationship.

parasitology *(păr-ă-sī-tŏl'-ĕ-jē).* The study of parasites and parasitism.

parasympathetic *(păr-ă-sĭm-pă-thĕt'-tĭk).* Referring to the craniosacral division of the autonomic nervous system.

parasympathetic nervous system. The craniosacral division of the autonomic nervous system, responsible for many functions, including constriction of the pupil, contraction of smooth muscles in the alimentary canal, slowing of heart rate, and increased secretion by glands.

parasympathomimetic *(păr-ă-sĭm-păth-ō-mĭ-mĕt'-ĭk).* Producing effects that are similar to those that result from stimulating the parasympathetic nervous system.

parathormone *(păr-ă-thŏr'-mōn).* A hormone secreted by the parathyroid glands; it regulates calcium and phosphorus metabolism.

parathyroid glands *(păr-ă-thī'-royd).* Four small endocrine glands located on the back of and at the lower edge of the thyroid gland. They secrete parathormone, which regulates calcium and phosphorus metabolism.

parathyroidectomy *(păr-ă-thī-royd-ĕk'-tĕ-mē).* Surgical removal of one or more of the parathyroid glands.

paratrichosis *(păr-ă-trī-kō'-sĭs).* An abnormal amount of hair or having hair in an unusual location.

paratyphoid fever *(păr-ă-tī'-foyd).* An infectious fever that resembles typhoid.

paravertebral *(păr-ă-vĕr'-tĕ-brăl).* Located alongside or near the vertebral column.

parenchymal tissue *(pă-rĕng'-kĭ-măl).* The part of an organ that is responsible for its main function; for example, the air cells in the lungs.

parenteral *(pă-rĕn'-tĕr-ĕl).* Administered through means other than the alimentary canal, such as by injection: subcutaneous, intraorbital, intraspinal, intrasternal, intravenous, intramuscular, or intracapsular.

paresis *(pă-rē'-sĭs).* Slight or incomplete paralysis.

paresthesia *(păr-ĕs-thē'-zhĕ).* An abnormal sensation of touch, such as burning or prickling, usually occurring without any external stimulation.

paries *(păr'-ĕ-ēz)* (*pl.* parietes). Wall; a term used to refer to the structure that makes up one side of an organ or body cavity.

parietal *(pă-rī'-ĕ-tĕl).* Referring to the walls of a cavity.

parietal bones. The primary bones of the skull, which make up the top of the head.

parkinsonism *(păhr'-kĭn-sĕn-ĭz-ĕm).* A group of neurological disorders characterized by tremor, muscular rigidity, and abnormal gait.

parodontitis *(păr-ō-dŏn-tī-tĭs).* Inflammation of the tissues around a tooth.

paronychia *(păr-ō-nĭk'ē-ĕ).* Acute or chronic infection of the marginal structures of a nail.

parosmia *(păr-ŏz'-mē-ĕ).* Any disorder or disturbance of the sense of smell. Pleasant odors are considered offensive and unpleasant ones are accepted.

parotidectomy *(pă-rŏt-ĭ-dĕk'-tō-mē).* Surgical removal of the parotid gland.

parotid gland *(pă-rŏt'-ĭd).* One of the salivary glands of the mouth, it secretes saliva, which helps to lubricate food and makes it easier to chew and swallow.

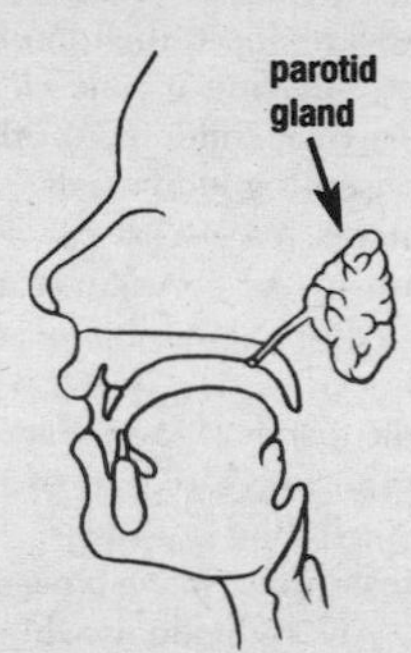

parotitis *(pă-rō-tī'-tĭs).* Inflammation of the parotid gland.

parous *(pă'-rŭs).* Having borne at least one child.

paroxetine *(pă-rŏk'-sĕ-tĕn).* *See* Appendix, Common Prescription and OTC Drugs: By Generic Name.

paroxysm *(păr'-ŏk-sĭz-ĕm).* A sudden recurrence or intensification of symptoms.

parrot fever *(păr'-ŏt).* *See* **psittacosis.**

pars *(păhrz).* Term for a particular division or portion of a larger organ, structure, or area of the body.

partial mastectomy. *See* **mastectomy.**

particle *(păhr'-tĭ-kĕl).* A tiny mass of material.

particulate *(păhr-tĭk'-ū-lĕt).* Composed of separate particles.

parturition *(păhr-tū-rĭ'-shĕn).* Childbirth.

paraumbilical *(păr-ĕm-bĭl'-ĭ-kĕl).* Alongside the umbilicus.

parvovirus *(păhr'-vō-vī-rŭs).* A genus of viruses that infect mammals and birds. In humans they can cause acute arthritis, spontaneous abortion, fetal death, and other conditions.

passage *(păs'-ĕj).* A channel.

passive *(păs'-ĭv).* Not active or spontaneous; not produced by active efforts.

passive-aggressive. Referring to behavior in which an individual expresses aggression in passive ways; for example, through stubbornness, procrastination, sullenness, or intentional inefficiency.

passive smoking. Situation in which individuals who do not smoke are exposed to residual smoke from others around them who are smoking; also called *secondhand smoke.*

pasteurization *(păs-chĕr-ĭ-zā'-shŭn).* A method of heating milk or other liquids to destroy microorganisms that can cause spoilage, while maintaining the flavor of the liquid.

past-pointing. A neurological test of balance in which an individual is rotated in a revolving chair and then asked to touch the tip of his/her nose with the eyes closed.

patch test. A skin test used to diagnose allergies. Small pieces of gauze or paper impregnated with suspected allergens are applied to the skin for a specific amount of time, and a response (swelling, redness) indicates an allergic reaction.

patella *(pă-tĕl'-ă).* A triangular bone located at the front of the knee; also called the *kneecap.*

patent *(pă'-tĕnt).* 1. Open, unobstructed. 2. Evident, apparent.

patent medicine. A drug that is protected by patent and sold without a physician's prescription.

paternity test *(pă-tĕr'-nĭ-tē).* Test to determine whether it is possible for an individual to have fathered a specific child.

pathogen *(păth'-ŏ-jĕn).* A microorganism or substance that is capable of producing a disease.

pathogenesis *(păth-ō-jĕn'-ĕ-sĭs).* Origin and development of a disease.

pathogenic *(păth-ō-jĕn'-ĭk).* Giving origin to disease.

pathological *(păth-ō-lŏj'-ĭ-kăl).* Diseased, or due to disease.

pathologist *(păth-ŏl'-ō-jĭst).* An individual who specializes in the diagnosis of abnormal changes in tissues removed surgically and postmortem.

pathology *(păth-ŏl'-ō-jē).* Study of the nature and cause of disease.

pathway *(păth'-wā).* 1. A path or course that is followed to reach a specific end. 2. The structures through which a nerve impulse travels between groups of nerve cells or between the central nervous system and a muscle or organ.

pathopsychology *(păth-ō-sī-kŏl'-ĕ-jē).* The psychology of mental disease.

patient *(pā'-shĕnt).* An individual who will or who is receiving treatment for a disease or other medical condition.

patient controlled analgesia (PCA). A method of administering medication in which the patient can control the rate of drug delivered for the control of pain.

pau d'arco *(paw dē-ăr'-kō).* A tree that grows in Brazil (*Tabebuia heptaphylla*) whose inner bark is used by herbalists to treat ulcers, diabetes, leukemia, bronchitis, hemorrhage, gastritis, and other conditions. The herb contains cancer-fighting substances.

Paxil *(păk'-sĭl).* *See* Appendix, Common Prescription and OTC Drugs: By Trade Name.

peau d'orange *(pō dō-răhj').* A condition in which the skin is dimpled, resembling that of an orange.

pectin *(pĕk'-tĭn).* The purified carbohydrate derived from the peel of citrus fruits or apple pulp.

pectinase *(pĕk'-tĭ-nās).* An enzyme that digests pectin.

pectoral muscles *(pĕk'-tō-răl).* Muscles located in the upper chest region; they bring the arm close to the body and move the shoulders forward.

pectus *(pĕk'-tŭs).* The breast, thorax, or chest.

pediatrician *(pē-dē-ă-trĭsh'-ŭn).* An individual who specializes in the treatment of disease in children.

pediatrics *(pē-dē-ăt'-rĭks).* Medical science that relates to the care and treatment of children and their diseases.

pedicle *(pĕd'-ĭ-kĕl).* A stalk or stem-like structure that attaches a new growth, as a stalk that attaches a tumor to normal tissue. Also called a *pediculus.*

pedicular *(pĕ-dĭk'-ū-lĕr).* Referring to or caused by lice.

pediculosis *(pĕ-dĭk-ū-lō'-sĭs).* An infestation with lice of the family *Pediculidae.*

capitis p. Lice infestation of the hair on the head.

corporis p. Infestation of the body by lice.

palpebrarum p. Infestation of the eyelashes by lice.

pubis p. Infestation of the pubic area by lice.

pedophilia *(pĕ-dō-fĭl'-ē-ă).* A paraphilia in which an adult has persistent, intense sexual urges or sexually stimulating fantasies of engaging in sexual activity with a prepubertal child.

peduncle *(pĕ-dŭng'-kĕl).* The stalk or stem by which a nonsessile tumor (one not attached by a base) is attached to normal tissue.

peeling *(pēl'-ĭng).* Shedding of the surface layer of skin.

pelage *(pĕl'-ĭj).* The collective hair of the body.

pellagra *(pĕl-ă'-gră).* A syndrome caused by a niacin deficiency or failure to transform tryptophan to niacin, and characterized by inflamed mucous membranes, diarrhea, dermatitis, and mental disorders, such as depression, anxiety, and confusion.

pellicle *(pĕl'-ĭ-kĕl).* A thin piece of cuticle or skin, or a thin film on the surface of a liquid.

pelotherapy *(pĕl-ō-thĕr'-ă-pē).* The therapeutic use of mud or earth.

pelvic *(pĕl'-vĭk).* Referring to the pelvis.

pelvic inflammatory disease (PID). An infection that spreads from the vagina or cervix to the uterus, fallopian tubes, and broad ligaments and characterized by vaginal discharge and often anorectal pain and abdominal pain as well.

pelvis *(pĕl'-vĭs).* The region of the body bounded by the two hip bones and the sacrum and coccyx.

contracted p. A pelvis in which there is a reduction of 1.5 to 2 cm in any important diameter.

frozen p. Fixation of organs in the pelvic area due to infection or neoplastic growth.

infantile p. An adult pelvis that retains infantile characteristics.

pelvisection *(pĕl-vĭ-sĕk'-shŭn).* Surgical cutting of the pelvis bones, such as is done during a symphysiotomy.

pemphigus *(pĕm'-fĭ-gŭs).* Any of one of several chronic, recurrent, sometimes fatal skin diseases characterized by successive episodes of bullae that may be accompanied by itching and burning.

pendulous *(pĕn'-dū-lĕs).* Hanging loosely.

penetrating *(pĕn'-ĕ-trāt-ĭng).* Entering deeply; piercing.

penicillin *(pĕn-ĭ-sĭl'-ĭn).* Any of a large number of natural or semisynthetic antibacterial antibiotics that have been derived from strains of the fungus *Pencillium* and other fungi grown in special culture media and are effective against many different bacteria.

penicillinase *(pĕn-ĭ-sĭl'-ĭ-nās).* A bacterial enzyme that inactivates many but not all penicillins.

penile prosthesis *(pē'-nīl).* A device, usually silicone or plastic, implanted in the penis that assists in making it become erect. It is used in men who are impotent due to organic causes, such as diabetes, trauma, or surgery.

penis *(pē'-nĭs).* The male organ of copulation and urinary excretion.

penis envy. In psychotherapy, a term that refers to the envy of a female child for the organ she does not have.

penrose drain *(pĕn'-rōs).* *See* **drain.**

pentobarbital *(fĕ-nō-băr'-be-tăl).* A barbiturate used as a sedative and anticonvulsant drug.

Pepcid *(pĕp'-sĭd).* Trade name of a medication used to treat excess stomach acid, which relieves heartburn and indigestion and helps in the healing of ulcers.

peppermint *(pĕp'-ĕr-mĭnt).* The *Mentha piperita* plant, the dried leaves and flowers and the oil of which are taken to relieve gastric conditions, such as bloating and intestinal colic.

pepsin *(pĕp'-sĭn).* The primary enzyme of gastric juice, it converts proteins into polypeptides.

peptic *(pĕp'-tĭk).* Referring to pepsin or to the digestive process.

peptic ulcer. *See* **ulcer.**

peptone *(pĕp'-tōn).* A type of protein that is soluble in water.

percentile *(pĕr-sĕn'-tīl).* One of 100 equal divisions of a series of items or data. For example: if a test score is greater than 90% of all other scores in a series, that result is above the 90th percentile of the range of scores.

perception *(pĕr-sĕp'-shŭn).* The conscious mental awareness and registration of sensory stimuli.

Percocet *(pĕr'-kō-sĕt). See* Appendix, Common Prescription and OTC Drugs: By Trade Name.

Percodan *(pĕr'-kō-dăn).* Trade name for a combination preparation of oxycodone hydrochloride, oxycodone terephthalate, and aspirin, used to treat moderate to severe pain.

percussion *(pĕr-kŭsh'-ŭn).* Use of the fingertips to lightly but sharply tap the body to identify the position, size, and consistency of internal structures or the presence of fluid in a cavity.

percutaneous *(pĕr-kū-tā'-nē-ŭs).* Through the skin, as an injection or the removal of tissue for biopsy.

percutaneous transluminal coronary angioplasty (PTCA). X-ray of the coronary arteries of the heart after passing a catheter into the heart and dye is injected to outline the arteries.

perennial *(pĕ-rĕn'-ē-āl).* Lasting throughout the year or for more than one year.

perforate *(pĕr'-fĕ-rāt).* To puncture or to make holes.

perforated ulcer. *See* **ulcer.**

perfusion *(pĕr-fū'-shŭn).* The passage of a fluid through holes or spaces or through the vessels of an organ.

perianal *(pĕr-ē-ā'-năl).* Located near or around the anus.

periarteritis *(pĕr-ē-ăhr-tĕr-ī'-tĭs).* Inflammation of the external layers of an artery and the tissues around the artery.

periarthritis *(pĕr-ē-ăhr-thrī'-tĭs).* Inflammation of the tissues around a joint.

pericardiectomy *(pĕr-ĭ-kăhr-dē-ĕk'-tĕ-mē).* Surgical removal of the pericardium.

pericardiocentesis *(pĕr-ĭ-kăhr-dē-ō-sĕn-tē'-sĭs).* Surgical puncture of the pericardial cavity to aspirate fluid.

pericardiometry *(pĕr-ĭ-kăhr-dē-ŏt'-ĕ-mē).* Surgical incision of the pericardium.

pericarditis *(pĕr-ĭ-kăhr-dī'-tĭs).* Inflammation of the pericardium.

bacterial p. Percarditis caused by a bacterial infection, especially by staphylococci or gram-negative bacilli.

chronic constrictive p. Form in which a thickened pericardium hinders diastolic filling and cardiac output.

purulent p. Form characterized by the presence of pus, usually caused by a bacterial infection and less often by fungal or viral infection.

traumatic p. Pericarditis caused by an injury to the pericardium, such as a puncture or bullet wound or radiation.

pericardium *(pěr-ĭ-kăhr'-dē-ŭm).* The membranous fibroserous sac that surrounds the heart and the roots of the great blood vessels.

pericholecystitis *(pěr-ĭ-kō-lě-sĭs-tī'-tĭs).* Inflammation of the tissues around the gallbladder.

perichondritis *(pěr-ĭ-kŏn-drī'-tĭs).* Inflammation of the perichondrium.

perichondrium *(pěr-ĭ-kŏn'-drē-ŭm).* The layer of thick, fibrous connective tissue around the surface of all cartilage except the articular cartilage of synovial joints.

periduodenitis *(pěr-ĭ-dū-ō-dě-nī'-tĭs).* Inflammation around the duodenum, often causing adhesions attaching it to the peritoneum.

perimetritis *(pěr-ĭ-mē-trī'-tĭs).* Inflammation of the peritoneal covering of the uterus.

perinatal *(pěr-ĭ-nā'-tăl).* Referring to or occurring in the period shortly before and after birth.

perinatology *(pěr-ĭ-nā-tŏl'-ĕ-jē).* The branch of medicine that deals with the fetus and the infant during the perinatal period.

perineotomy *(pěr-ĭ-nē-ŏt'-ō-mē).* Surgical incision into the perineum.

perineum *(pěr-ĭ-ēn'-ĕm).* The external region that lies between the vulva and the anus in females or between the scrotum and the anus in males.

perineuritis *(pěr-ĭ-nū-rī'-tĭs).* Inflammation of the sheath that covers nerve fibers.

perinuclear *(pěr-ĭ-nū'-klē-ăr).* Around a nucleus.

period *(pēr'-ē-ĕd).* 1. The interval of time between two successive occurrences. 2. A term commonly used to refer to menstruation.

gestational p. The duration of pregnancy, from conception to delivery of the fetus.

incubation p. The period of time needed for development.

latent p. A seemingly inactive period; the time between stimulation and the resulting response.

periodicity *(pěr-ē-ō-dĭs'-ĭ-tē).* Recurrence at regular intervals of time.

periodontal disease. A bacterial infection that damages the fibers and bone that support and hold the teeth in the mouth. The main cause of periodontal disease is a sticky, colorless film called bacterial plaque that constantly forms on teeth.

periodontics *(pěr-ē-ō-don'-tiks).* The branch of dentistry that deals with the study and treatment of diseases of the periodontium.

periodontist *(pĕr-ē-ō-dŏn'-tĭst)*. A dentist who specializes in periodontics.

perioperative *(pĕr-ē-ŏp'-ĕr-ĕ-tĭv)*. Referring to the period extending from the time of hospitalization for surgery to discharge from hospital.

periosteum *(pĕr-ē-ŏs'-tē-ĕm)*. The specialized connective tissue in two layers that covers all the bones of the body.

periostitis *(pĕr-ē-ŏs-tī'-tĭs)*. Inflammation of the periosteum.

peripheral *(pĕr-ĭf'-ĕr-ăl)*. Referring to or located at the periphery.

perisplenitis *(pĕr-ĭ-splē-nī'-tĭs)*. Inflammation of the peritoneal covering of the spleen and the surrounding structures.

perispondylitis *(pĕr-ĭ-spŏn-dĭl-ī'-tĭs)*. Inflammation of the structures around a vertebra.

peristalsis *(pĕr-ĭ-stăl'-sĭs)*. The progressive wave of contractions by which the alimentary canal and other tubular organs move along their contents.

peritendinitis *(pĕr-ĭ-tĕn-dĭ-nī'-tĭs)*. Inflammation of the sheath of tendons.

peritoneal *(pĕr-ĭ-tō-nē'-ăl)*. Referring to the peritoneum.

peritoneum *(pĕr-ĭ-tō-nē-ŭm)*. The serous membrane that lines the abdominal cavity and covers the viscera.

peritonitis *(pĕr-ĭ-tō-nī'-tĭs)*. Inflammation of the peritoneum.

adhesive p. Form of peritonitis characterized by adhesions between adjacent serous surfaces.

bacterial p. Peritonitis caused by a bacterial infection, typically with *Staphylococcus, Pseudomonas,* or *Mycobacterium.*

general p. Inflammation of the greater part of the peritoneum.

localized p. Type limited to a portion of the peritoneum.

puerperal p. Peritonitis that develops after childbirth.

purulent p. Peritonitis accompanied by the formation of pus.

septic p. Type due to a pus-producing microorganism.

peritonsillar *(pĕr-ĭ-tŏn'-sĭ-lăr)*. Located around a tonsil.

perle *(pĕrl)*. A soft capsule that contains medication.

perleche *(pĕr-lĕsh')*. Disorder characterized by fissures at the corners of the mouth that may spread to the lips and cheeks.

permanent makeup. A form of tattooing (also known as *micropigmentation*) that provides individuals with permanent cosmetic enhancement; for example, tattooing eyebrows for someone who has little or no eyebrows; tattooing the lips to provide permanent color.

permeability *(pĕr-mē-ă-bĭl'-ĭ-tē)*. The state of being permeable.

permeable *(pĕr'-mē-ă-bĕl)*. Allowing passage of a substance.

permeate *(pĕr'-mē-āt)*. To penetrate or pass through, as passing through a filter.

permethrin *(pĕr-mĕth'-rĭn)*. An insecticide applied topically to treat infestations by mites, lice, and ticks; also applied to objects such as bedding and furniture to help eliminate these pests. Also an ingredient in some flea collars for cats and dogs.

pernicious anemia. *See* **anemia.**

peroral *(pĕr-ŏr'-ĕl).* Done through or given through the mouth.

per os *(pĕr ŏs).* By mouth.

peroxide *(pĕr-ŏk'-sīd).* A compound that contains more oxygen than the other oxides of the element in question.

perseveration *(pĕr-sĕv-ĕr-ā'-shŭn).* Persistent repetition of meaningless words or phrases or repetition of answers that do not relate to questions asked. Usually associated with brain lesions or schizophrenia.

personality *(pĕr-sŏ-năl'-ĭ-tē).* The characteristic way an individual acts, feels, and thinks, including attitudes, styles, and values.

personality disorder. A condition characterized by chronic use of inappropriate coping mechanisms and by behavior that differs significantly from the expectations of the person's culture. Some such disorders include paranoid, schizoid, antisocial, and obsessive/compulsive personality disorder.

perspiration *(pĕr-spĭ-rā'-shŭn).* Sweating.

perspire *(pĕr-spīr').* To excrete fluid through the pores of the skin.

Perthes' disease *(pĕr'-tēz).* A disease characterized by changes in the bone at the head of the femur, resulting in deformity.

perturbation *(pĕr-tĕr-bā'-shŭn).* State of being agitated, uneasy, or greatly disturbed.

pertussis *(pĕr-tŭs'-ĭs).* An acute, contagious infection of the respiratory tract, usually occurring in young children; also called *whooping cough.*

perversion *(pĕr-vĕr'-zhŭn).* Deviation from the normal course, which may involve an individual's actions, intellect, emotions, or reactions.

pes *(pĕs).* The foot or a footlike structure.

pes cavus. An abnormal concavity or hollowness of the sole of the foot.

pessary *(pĕs'-ĕ-rē).* A device placed in the vagina to support the uterus or rectum or to act as a contraceptive.

pesticide *(pĕs'-tĭ-sīd).* A poison, such as an insecticide, herbicide, fungicide, or rodenticide, used to destroy pests.

petechia *(pĕ-tē'-kē-ā).* Pinpoint, purplish, nonraised spot on the skin caused by submucous or intradermal bleeding.

petit mal. *See* **epilepsy.**

petri dish *(pĕ'-trē).* A round, shallow, transparent glass or plastic dish used for the culture of microorganisms and for tissue cell cultures.

petrissage *(pă'-trĭ-săhzh).* A massage technique in which the muscles are pressed and kneaded.

pet therapy. The practice of bringing together companion animals with humans for therapeutic reasons. Animals have been shown to help reduce or eliminate depression, stress, loneliness, and destructive behavior, and to lower blood pressure.

Peyer's patches *(pī'-ĕrz)*. An accumulation of lymph nodules located mainly in the ileum near its junction with the colon.

Peyronie's disease *(pā-rō-nēz')*. Hardening of the corpora cavernosa of the penis, which causes curvature of the penis, especially when it is erect.

pH. Abbreviation for *potential of hydrogen.* It is the degree of acidity or alkalinity of a substance. The neutral point is pH 7; numbers less than 7 reflect increasing acidity while numbers greater than 7 indicate increasing alkalinity.

phakitis *(fă-kī'-tĭs)*. Inflammation of the crystalline lens.

phacolysis *(fă-kō-lī'-sĭs)*. Dissection and removal of the crystalline lens of the eye, usually as a treatment for cataracts.

phagocyte *(făg'-ō-sīt)*. A cell that has the ability to consume and destroy particulate substances, such as bacteria, cells, cell debris, protozoa, and dust particles.

phagocytosis *(făg-ō-sī-tō'-sĭs)*. The ingestion and digestion of bacteria and other microorganisms by phagocytes.

phalanges *(fĕ-lăn'-jēz)*. *See* **phalanx.**

phalanx *(făl'-ănks)*. Any of the bones of the fingers or toes.

phallic *(făl'-ĭk)*. Pertaining to the penis.

phallus *(făl'-ŭs)*. The penis.

phantom limb *(făn'-tŭm)*. The sensual perception, including pain, that a limb still exists after it has been amputated.

pharmaceutical *(făhr-mĕ-sū'-tĭ-kĕl)*. Referring to pharmacy or to drugs.

pharmacist *(făhr'-mĕ-sĭst)*. An individual who is licensed to prepare and dispense drugs and to make up prescriptions.

pharmacodynamics *(făhr-mĕ-kō-dī-năm'-ĭks)*. The study of the biochemical and physiological effects of drugs and their actions on living organisms.

pharmacognosy *(făhr-mĕ-kŏg'-nĕ-sē)*. The branch of pharmacology that deals with the biological, biochemical, and economic factors of natural drugs and their constituents.

pharmacokinetics *(făhr-mĕ-kō-kĭ-nĕt'-ĭks)*. Study of the metabolism, action, and fate of drugs in the body over time.

pharmacology *(făhr-mĕ-kŏl'-ĕ-jē)*. The science that deals with the origin, nature, chemistry, impact, and uses of drugs.

pharmacopeia *(făr-mă-kō'-pē-ă)*. Authorized treatise on drugs and their preparations.

pharmacy. A location at which prescription drugs are sold. It must be supervised by a licensed pharmacist at all times of operation.

compounding p. A facility that both makes and sells prescription medications, and which can provide specially formulated drugs, such as a liquid form of a drug usually only available in capsule form.

pharma food. A food product that contains a pharmacological additive designed to improve health, such as a cholesterol-lowering item.

pharyngeal *(făr-ĭn-jē'-ĕl).* Referring to the pharynx.

pharyngitis *(făr-ĭn-jī'-tĭs).* Inflammation of the pharynx.

pharyngotomy *(făr-ĭng-gŏt'-ĕ-mē).* Surgical incision of the pharynx.

pharynx *(făr'-ĭnks).* The passageway for air that runs from the nasal cavity to the larynx, and for food from the mouth to the esophagus. It consists of three sections: nasopharynx, oropharynx, and laryngopharynx. Also called the *throat.*

phenelzine sulfate *(fĕn'-ĕl-zēn).* A monoamine oxidase inhibitor drug used as an antidepressant and to help prevent migraine.

Phenergan *(fĕn'-ĕr-gĕn).* Trade name for preparations of promethazine hydrochloride, used to treat allergic conditions, to manage nausea and vomiting, and as an ingredient in cold remedies.

phenmetrazine *(fĕn-mĕt'-ră-zēn).* A central nervous system stimulant used to suppress appetite.

phenobarbital *(fē-nō-băhr'-bĭ-tăl).* A long-acting barbiturate used as a sedative, anticonvulsant, and hypnotic.

phenol *(fē'-nŏl).* An extremely poisonous, colorless to light pink compound derived from coal tar, used as a pharmaceutical preservative.

phenomenon *(fĕ-nŏm'-ĕ-nŏn).* Any observable event, occurrence, or fact.

phenotype *(fē'-nō-tīp).* The complete physical, biochemical, and physiological makeup of an individual.

phenylalanine *(fēn-ĕl-ăl'-ĕ-nēn).* An essential amino acid which is used for protein synthesis.

phenylbutazone *(fēn-ĕl-bū'-tĕ-zōn).* A nonsteroidal anti-inflammatory drug used to treat severe rheumatoid disorders that do not respond to less toxic substances.

phenylketonuria *(fēn-ĕl-kē-tō-nū'-rē-ă).* A hereditary disease caused by the body's failure to oxidize phenylalanine (an amino acid) to tyrosine because of a defective enzyme.

phenylpropanolamine *(fĕn-ĭl-prō-pă-nŏl'-ă-mēn).* A drug used as a bronchodilator and vasoconstrictor.

phenytoin *(fĕn'-ĭ-tō-ĭn).* An anticonvulsant drug; trade name: Dilantin.

pheochromocytoma *(fē-ō-krō-mō-sī-tō'-mĕ).* A usually benign, vascular tumor of the adrenal gland typically associated with persistent or intermittent elevated blood pressure.

pheromone *(fĕr'-ō-mōn).* A substance secreted to the outside of the body and perceived, by smell, by another individual of the same species, who responds in a specific manner.

philtrum *(fĭl'-trŭm).* The vertical

groove in the middle region immediately above the upper lip.

phimosis *(fī-mō'-sĭs).* A condition in which the foreskin cannot be retracted back over the glans. In such cases, circumcision is usually necessary.

pHisoHex *(fī'-sō-hĕks).* The trade name for an emulsion that contains hexachlorophene; used as a topical antibacterial.

phlebectomy *(flē-bĕk'-tō-mē).* Surgical excision of a vein or part of a vein.

phlebitis *(flē-bī'-tĭs).* Inflammation of a vein.

adhesive p. Type in which the vein tends to become obliterated.

obliterative p. Phlebitis that completely blocks a vein.

puerperal p. Septic inflammation of uterine or other veins after childbirth.

sinus p. Inflammation of a cerebral sinus.

phlebolith *(flĕb-ō-lĭth).* A stone (calculus) in a vein; also called a *vein stone.*

phleboplasty *(flĕb'-ō-plăs-tē).* Plastic operation to repair a vein.

phlebothrombosis *(flĕb-ō-thrŏm-bō'-sĭs).* Presence of a clot in a vein, not associated with inflammation of the vein wall.

phlebotomy *(flē-bŏt'-ĕ-mē).* 1. Incision of a vein to release blood. 2. Needle puncture of a vein for drawing a blood sample.

phlegm *(flĕm).* Abnormally thick mucus secreted by the mucosa of the lungs or bronchial tubes during an infection.

phlegmon *(flĕg'-mŏn).* A spreading inflammatory reaction to infection with streptococci, which forms a lesion that may infiltrate the deep subcutaneous tissues and muscles, creating pockets of pus.

phobia *(fō'-bē-ă).* An irrational, persistent, intense fear of a specific object, activity, or situation, a fear that the individual recognizes as being unreasonable.

phonetics *(fōn'-ĭks).* The science of vocal sounds.

phosphatase, acid. An enzyme that works under acid conditions and is produced in the liver, bone marrow, spleen, and the prostate gland. Abnormally high serum levels may indicate prostate disease.

phosphate *(fŏs'-fāt).* Any salt of phosphoric acid or its anions. Calcium phosphate makes bones and teeth hard.

phosphatidylserine *(fŏs-fū-tĭd-ĕl-sĕr'-ĕn).* A naturally occurring phospholipid nutrient that plays a role in the functioning of nearly every cell in the body, but which is most prevalent in the brain, where it performs many tasks among nerve cells.

phosphaturia *(fŏs-fă-tū'-rē-ă).* Excretion of phosphates in the urine.

phosphene *(fŏs'-fēn).* An objective visual sensation that an individual sees with eyes closed and in the absence of visual light.

phospholipids *(fŏs-fō-lĭp'-ĭdz).*

Lipids that contain phosphorus. They are the major form of lipid in all cell membranes.

phosphorus *(fŏs'-fŏr-ŭs).* An essential element in the diet and a major component of bone and all bodily tissues. It is involved in nearly all metabolic processes.

photoablation *(fōt-tō-ăb-lā'-shŭn).* Volatilization of tissue using ultraviolet rays emitted by a laser.

photoaging *(fō-tō-āj'-ĭng).* Premature aging of the skin due to prolonged exposure to ultraviolet rays, such as in sunlight.

photochemotherapy *(fō-tō-kē-mō-thĕr'-ĕ-pē).* Treatment with drugs that react to ultraviolet radiation or sunlight.

photodermatitis *(fō-tō-dĕr-mă-tī'-tĭs).* An abnormal skin response to exposure to ultraviolet light, especially sunlight. Characteristics include swelling, redness, blisters, itchy bumps, fever, chills, and headache.

photodynamic therapy *(fō-tō-dī-năm'-ĭk).* A type of cancer treatment that uses a photosensitizing substance, given intravenously, that concentrates selectively in cancer cells, followed by exposure of the tumor to a red laser light to destroy as much of the tumor as possible.

photophobia *(fō-tō-fō'-bē-ă).* Painful oversensitivity to light.

photopsia *(fō-tŏp'-sē-ă).* Subjective sensation of flashes or sparks of light sometimes associated with eye or brain diseases.

photorefractive keratectomy (PRK) *(fō-tō-rē-frăk'-tĭv kĕr-ă-tĕk'-ĕ-mē).* A kind of laser eye surgery designed to change the shape of the cornea to reduce or eliminate the need for glasses or contact lenses.

photoreceptor cells *(fō-tō-rē-sĕp'-tŏr).* Sensory nerve cells that are capable of being stimulated by light. In humans, these cells are called rods and cones.

photoretinitis *(fō-tō-rĕt-ĭ-nī'-tĭs).* Inflammation of the retina due to exposure to intense light.

photosensitivity *(fō-tō-sĕn-sĭ-tĭv'-ĭ-tē).* The ability of a cell, organ, or organism to react to light.

photosynthesis *(fō-tō-sĭn'-thĕ-sĭs).* The chemical process by which plants make carbohydrates by combining carbon dioxide from air and water from the soil and light energy in the presence of chlorophyll.

phototherapy *(fō-tō-thĕr'-ă-pē).* The therapeutic use of sunlight or artificial light.

phrenology *(frĕ-nŏl'-ō-jē).* The imagined ability to predict mental abilities by studying the shape, bumps, and other features on the skull.

physical examination *(fĭz'-ĭ-kăl).* Examination of the body using various techniques, including palpation, auscultation, percussion, and inspection.

physical sign. An abnormal finding observed during the physical examination of a patient.

physical therapy. Rehabilitation that focuses on restoring function and preventing disability associated with disease, trauma, or loss of a body part.

physician *(fĭ-zĭsh'-ŭn).* An individual who has graduated from an authorized college of medicine or osteopathy and is licensed by the appropriate board.

attending p. A physician who attends a hospital at stated times to visit patients and give directions concerning treatment.

family p. A physician who plans and provides comprehensive primary health care for family members of all ages.

physician's assistant (PA). An individual who has been trained in an accredited program and is certified to perform certain of a physician's duties, including history taking, physical examination, diagnostic tests, some minor surgical procedures, and other tasks, under the supervision of a licensed physician.

physics *(fĭz'-ĭks).* The science of the laws, forces, and general properties of nature.

physiognosis *(fĭz-ē-ŏg-nō'-sĭs).* Diagnosis determined from a person's facial expression and appearance.

physiology *(fĭz-ē-ŏl'-ō-jē).* The science that treats the functions of a living organism and its parts.

physiotherapy *(fĭz-ē-ō-thĕr'-ă-pē).* Physical therapy.

physostigmine salicylate *(fī-zō-stĭg'-mēn).* A substance obtained from the dried ripe seed (Calabar bean); used topically to decrease intraocular pressure in glaucoma and internally to reverse central nervous system effects caused by an overdose of anticholinergic drugs.

phytochemical *(fī-tō-kĕm'-ĭ-kăl).* A general term for one of many health-protecting compounds found as components of plants. The terms "phytochemical" and "phytonutrient" are often used interchangeably.

phytoestrogen *(fī-tō-ĕs'-trĕ-jĕn).* Any of a group of weakly estrogenlike, nonsteroidal compounds found in many plants. Eating foods containing these compounds appears to have a preventive effect on cardiovascular disease and osteoporosis.

phytonutrient *(fī-tō-nū'-trē-ĕnt). See* **phytochemical.**

phytotoxin *(fī-tō-tŏk'-sĭn).* Any toxic substance of plant origin.

pia arachnoid *(pē'-ă ĕ-răk'-noid).* The pia mater and the arachnoid together as one functional unit.

pia mater *(pē'-ă mā'-tĕr).* The innermost of the three membranes (meninges) that cover the brain and spinal cord.

pica *(pē'-kă).* Compulsive eating of nonnutritive substances, such as dirt, gravel, clay, hair, or ice.

Pick's disease *(pĭks).* A type of progressive degenerative disease of the brain characterized by irreversible loss of memory, disordered emotions, apathy, speech disturbances, and disorientation.

PID. Acronym for pelvic inflammatory disease.

piebaldism *(pī'-băwld-ĭz-ĕm).* A congenital skin disorder characterized by patchy areas of depigmentation or hypopigmentation due to the absence of functioning melanocytes and melanin.

pigeon-breast *(pĭ'-jĭn-brĕst).* A deformity of the breastbone in which the sternum projects forward, resembling the breast of a pigeon.

pigeon-toed. Walking with the feet pointed inward.

pigment *(pĭg'-mĕnt).* Any organic coloring substance in the body.
endogenous p. Pigment that is produced within the body, such as melanin.
exogenous p. Pigment produced outside the human body.
skin p. Melanin, melanoid, and carotene.

pilar *(pī'-lăr).* Referring to the hair.

Pilates method *(pĭ-lŏ'-tāz).* A type of movement program consisting of more than 500 unique controlled exercises designed to strengthen muscles, release body tension, improve alignment and balance, and increase joint flexibility.

pile *(pīl).* A hemorrhoid.

pill *(pil).* Tablet.

Pill, the. A slang term for any one of various birth control medications.

pilocarpine *(pī-lō-kăhr'-pēn).* An alkaloid obtained from plants of the genus *Pilocarpus* and used to reduce intraocular pressure in individuals who have glaucoma.

pilonidal *(pī-lō-nī'-dăl).* Referring to or characterized by a tuft of hairs.

pilosis *(pī-lō'-sĭs).* Excessive formation of hair.

pimelorrhea *(pĭm-ĕl-ŏr-ē'-ă).* Elimination of fat in loose stools.

pimple *(pĭm'-pĕl).* A papule or pustule, usually seen on the face, neck, or upper body, and usually associated with acne vulgaris.

pinealectomy *(pĭn-ē-ăl-ĕk'-tō-mē).* Surgical removal of the pineal gland.

pineal gland. A glandlike structure in the brain that is believed to secrete melatonin, and thus may be part of the body's sleep-regulation system.

pinealoma *(pĭn-ē-ă-lō'-mă).* A tumor of the pineal gland, often associated with precocious puberty.

pinkeye *(pĭnk'-ī).* Conjunctivitis, which may be viral or bacterial in origin. Viral pinkeye is highly contagious and characterized by a watery discharge and coldlike symptoms. Symptoms of bacterial pinkeye include eye pain, swelling, redness, and some discharge.

pinna *(pĭn'-ă).* The outer ear, which collects and directs sound waves into the middle and inner ear.

pint *(pīnt).* In the United States, a unit of liquid measure equal to 16 ounces or to 0.473 liter.

pinworm *(pĭn'-wĕrm).* A parasitic nematode, *Enterobius vermicu-*

laris, that causes infection of the intestines and rectum.

pioglitazone *(pī-ō-glĭt'-ĕ-zōn).* An antidiabetic substance that decreases insulin resistance in the peripheral tissues and liver, used to treat type II diabetes.

piperazine *(pī-pĕr'-ă-zēn).* A chemical used to eliminate certain worm infestations, including pinworms and roundworms.

pipette *(pī-pĕt').* Narrow glass tube used to transfer and measure small amounts of liquid.

pit *(pĭt).* A tiny hollow or depression.

pitting edema. Edema, usually affecting the extremities, that when pressed firmly with a finger maintains the depression made by the finger.

pituitary gland *(pĭ-tū'-ĭ-tăr-ē).* An endocrine gland, located at the base of the brain, that secretes various hormones that regulate many bodily functions, including growth, reproduction, and metabolism.

pityriasis rosea *(pĭt-ĭ-rī'-ă-sĭs).* Any one of various skin diseases characterized by branny scales.

placebo *(plă-sē'-bō).* 1. An inactive substance given to a patient solely for its psychophysiological effects. 2. An inactive substance administered to a control group in a controlled clinical trial in order that any effects of the experimental treatment can be determined.

placebo effect. A phenomenon in which a placebo may improve a patient's condition just because the individual expects it will be helpful. The greater the expectation, the more likely the individual will experience a benefit. Also called a *placebo response.*

placenta *(plă-cĕn'-tă).* The oval spongy structure in the uterus that supplies nourishment to a developing fetus. It is expelled from the uterus after the fetus is delivered, at which time it is known as the *afterbirth.*

abruption p. Premature of the placenta from the uterus.

previa p. A placenta that develops in the lower part of the uterus, commonly causing painless bleeding during the last trimester.

retained p. A placenta that is not expelled after childbirth.

placentography *(plă-sĕn-tŏg'-ră-fē).* X-ray visualization of the placenta after injecting a contrast medium.

plague *(plāg).* An infectious disease caused by a bacterium, *Yersinia pestis,* which mainly infects rodents. Fleas bite infected rodents and carry the bacteria, biting people and transmitting the disease to them.

plane *(plān).* A surface on which a straight line connecting any two points will lie completely in the surface.

planing, skin. A cosmetic procedure in which disfigured skin is abraded to stimulate new skin growth with minimal scarring.

plantar fasciitis *(plăn'-tĕr).* Inflammation of the plantar fascia (fasciitis), the tissue in the foot that runs from the heel to the front of the foot. This condition is often associated with a bony spur that projects from under the heel and makes walking painful.

plantar wart. *See* **wart.**

plaque *(plăk).* A patch or flat area on the skin.

dental p. A thin film of food debris, mucin, and dead epithelial cells that are deposited on the teeth.

fibrous p. A lesion associated with atherosclerosis and located within an artery.

senile p. Microscopic masses seen in small amounts in the cerebral cortex of elderly people and in greater amounts in people with Alzheimer's disease.

plasma *(plăz'-mĕ).* The liquid part of blood and lymph in which the particulate components are suspended. *See also* **serum.**

citrated p. Plasma that has been treated with sodium citrate, which prevents clotting.

fresh frozen p. Plasma that has been separated from whole blood and frozen within eight hours. It contains all coagulation factors.

normal human p. Sterile plasma obtained by collecting approximately equal amounts of plasma of citrated whole blood from eight or more adults and then used as a blood volume replenisher.

pooled p. Plasma collected from several donors.

plasma cell. A type of white blood cell similar to a lymphocyte, found in bone marrow.

plasma protein. The type of protein found in plasma, usually albumin and globulin.

plaster cast *(plăs'-tĕr).* A hard appliance made from a gypsum material that hardens when mixed with water, and used to immobilize or make impressions of body parts.

plastic surgery. Surgery that alters the shape of one or more body parts or restores lost tissues. Also called *cosmetic surgery.*

plastron *(plăs'-trŏn).* The breastbone and the rib cartilages that are attached to it.

platelet, blood *(plāt'-lĕt).* A disk-shaped structure found in the blood of all mammals. It plays a major role in coagulation.

plating *(plāt'-ĭng).* The process of preparing a bacterial culture on a plate of medium in a Petri dish.

Plavix *(plă'-vĭks). See* Appendix, Common Prescription and OTC Drugs: By Trade Name.

plethoric *(plĕ-thŏr'-ĭk).* Having an excess amount of blood in the body or a body part, giving one a very red complexion.

pleura *(ploor'-ĕ).* The serous membrane that encloses the lungs and lines the thoracic cavity. It is moistened with a secretion that aids in the movement of the lungs in the chest.

pleurectomy *(plū-rĕk'-tĕ-mē).* Surgical removal of a portion of the pleura.

pleurisy *(ploor'-ĭ-sē).* Inflammation of the pleura, characterized by localized chest pain and dry cough.

diaphragmatic p. A type of pleurisy limited to parts near the diaphragm.

diffuse p. Pleurisy in which the entire surface of the pleura is inflamed.

fibrinous p. Pleurisy in which fibrinous adhesions form.

hemorrhagic p. Pleurisy with effusion that contains blood.

indurative p. Dry pleurisy characterized by a thickening and hardening of the pleura.

pleurodynia *(ploor-ō-dĭn'-ē-ă).* Pain in the pleural cavity.

plexus *(plĕk'-sĕs).* A network of lymphatic vessels, nerves, or veins.

plica *(plī'-kă).* A fold or ridge, as of a membrane.

plicate *(plī'-kāt).* Folded or plaited.

pneumarthrography *(noo-măhr-thrŏg'-rĕ-fē).* X-ray of a joint after it has been injected with gas or air as a contrast medium.

pneumaturia *(noo-mă-tū'-rē-ă).* Excretion of gas in the urine.

pneumococcemia *(noo-mō-kŏk-sē'-mē-ă).* The presence of pneumococci in the blood.

Pneumococcus *(noo-mō-kŏk'-ĕs).* An individual organism of the species *Streptococcus pneumoniae.*

pneumocystis carinii pneumonia *(noo-mō-sĭs'-tĭs).* The organism that causes interstitial plasma cell pneumonia, which can also cause extrapulmonary disease in individuals who are immunocompromised.

pneumonectomy *(noo-mō-nĕk'-tĕ-mē).* Surgical removal of lung tissue, especially the entire lung.

pneumonia *(noo-mō'-nē-ă).* Inflammation of the lungs accompanied by exudates that fill the air spaces, causing breathing difficulties.

pneumonitis *(noo-mō-nī'-tĭs).* Inflammation of the lungs. *See* **pneumonia.**

pneumoperitoneum *(noo-mō-pĕr-ĭ-tō-nē'-ŭm).* The presence of gas or air in the peritoneal cavity.

pneumoperitonitis *(noo-mō-pĕr-ĭ-tō-nī'-tĭs).* Peritonitis accompanied by an accumulation of air or gas in the peritoneal cavity.

pneumothorax *(noo-mō-thŏr'-ăks).* The accumulation of gas or air in the pleural space.

closed p. Pnemothorax in which pulmonary air leaks into the pleural cavity through a wound in a lung.

spontaneous p., primary. A pneumothorax without an obvious external cause.

spontaneous p., secondary. Pneumothorax with a known cause.

traumatic p. Pneumothorax that results from trauma to the chest, such as a fractured rib or a penetrating wound.

podiatrist *(pō-dī'-ĕ-trĭst).* An individual who specializes in podiatry, the study and care of the foot, including diagnosis and medical and surgical treatment.

podiatry *(pō-dī'-ĕ-trē).* A special field of medicine that deals with the study and care of the foot.

poikilocyte *(poi'-kĭ-lō-sīt).* An abnormally shaped red blood cell, such as a sickle cell or target cell.

poikilocytosis *(poi-kĭ-lō-sī-tō'-sĭs).* The presence of poikilocytes in the blood.

point of focus. The spot within the eyeball where visual information is most sharply focused by the cornea and lens. Optimal vision occurs when the point of focus is directly on the retina.

poison *(poi'zĕn).* Any substance that, when relatively small amounts are ingested, inhaled, absorbed, applied, injected, or developed with the body, causes damage to the function or structure of the body and results in various symptoms, illness, or death.

poison ivy. A climbing vine, *Rhus toxicodendron,* which contains urushiol, an irritating resin that causes a severe type of dermatitis upon contact with the skin.

poison oak. A climbing vine, *Rhus radicans* or *R. diversiloba,* that is similar to poison ivy and causes the same symptoms upon contact with the skin.

poison sumac. A shrublike plant, *Toxicodendron vernix,* that contains urushiol. It is similar to poison ivy and poison oak and causes the same symptoms upon contact with the skin.

polarity *(pō-lăr'-ĭ-tē).* 1. A state of having poles. 2. Exhibiting opposite effects at the two extremities in physical therapy.

polarity therapy. A healing technique based primarily on ayurvedic philosophy, in which practitioners use their hands to establish a connection between themselves and their client and help rebalance the body's energy flow.

polio *(pō'-lē-ō). See* **poliomyelitis.**

polioencephalitis *(pō-lē-ō-ĕn-cĕf-ă-lī'-tĭs).* Condition characterized by inflammatory lesions of the gray matter of the brain.

poliomyelitis *(pō-lē-ō-mī-lī'-tĭs).* Inflammation of the gray matter of the spinal cord, characterized by fever, headache, vomiting, sore throat, and stiffness of the neck and back.

bulbar p. A serious form of poliomyelitis in which there may be respiratory and circulatory distress, as well as problems with swallowing.

cerebral p. Poliomyelitis that has reached the brain.

pollen *(pŏl'-ĕn).* Microspores (male fertilizing substances) of flowering plants which in many cases are allergens.

pollen count. The number of pollen grains that land on a given area during a specified time. A pollen count of 50 or less is usually considered low; counts of 1,000 or more are considered to be very high.

pollenosis *(pŏl-ĕ-nō'-sĭs).* Hay fever.

pollex *(pŏl'-ĕks).* The thumb.

pollicization *(pŏl-ĭs-ĭ-zā'-shĕn).* Replacement or rehabilitation of a thumb, especially surgical construction of a thumb using the great toe or index finger.

polyarteritis *(pŏl-ē-är-tĕr-ī'-tĭs)*. A disease that affects the medium and small arteries and is characterized by inflammation and necrosis of the blood vessel lining and walls, resulting in reduced blood flow.

polyarthritis *(pŏl-ē-är-thrī'-tĭs)*. Inflammation of more than one joint.

polychondritis *(pŏl-ē-kŏn-drī'-tĭs)*. Inflammation of several cartilages of the body.

polycythemia *(pŏl-ē-sī-thē'-mē-ă)*. An increase in the total red cell mass of the blood.

polydactyly *(pŏl-ē-dăk'-tĕ-lē)*. An abnormality characterized by the presence of more than the usual number of fingers or toes.

polydipsia *(pŏl-ē-dĭp'-sē-ă)*. Chronic excessive thirst and intake of fluids.

polyethylene *(pŏl-lē-ĕth'-ĕ-lēn)*. A synthetic plastic that is frequently used in reparative surgery.

polymorphic *(pŏl-ē-mŏr'-fĭk)*. Occurring or appearing in many or different forms or stages of development.

polymorphonuclear *(pŏl-ē-mŏr-fō-noo'-klē-ĕr)*. Having a nucleus that is deeply lobed or is very divided so that it appears to be multiple.

polymyositis *(pŏl-ē-mī-ō-sī'-tĭs)*. A chronic, progressive inflammatory disease that affects the skeletal muscles, characterized by weakness of the neck, pharynx, and limbs, and usually accompanied by tenderness and pain.

polymyxin B *(pŏl-ē-mĭk'-sĭn)*. An antibiotic derived from strains of the bacterium *Bacillus polymyxa*, and used to treat various urinary tract, ophthalmic, skin, and systemic infections caused by gram-negative bacteria.

polyneuritis *(pŏl-ē-noo-rī'-tĭs)*. Inflammation of several peripheral nerves simultaneously.

polyp *(pŏl'-ĭp)*. A protruding growth from a mucous membrane.

adenomatous p. A benign neoplastic growth in the sigmoid colon, rectum, or stomach, seen in up to 50% of people older than sixty.

cervical p. A common, relatively harmless tumor of the uterine cervix, many of which cause irregular bleeding.

fibrinous p. An intrauterine polyp composed of fibrin from retained blood; it may develop from an ovum or from a thrombus at the placental site.

nasal p. Accumulation of edema fluid in the mucosa of the nose.

polypectomy *(pŏl-ĭ-pĕk'-tĕ-mē)*. Surgical removal of a polyp.

polyphagia *(pŏl-ē-fā'-jă)*. Excessive eating.

polyposis *(pŏl-ĭ-pō'-sĭs)*. The development of multiple polyps in an organ or tissue.

polypropylene *(pŏl-ē-prō'-pĕ-lēn).* A commonly used synthetic polymer used in the manufacture of nonabsorbable sutures, surgical casts, and semipermeable membranes.

polysaccharide *(pŏl-ē-săk'-ĕ-rīd).* A type of complex carbohydrate that is usually insoluble in water but when soluble, forms colloidal solutions. They include two groups: starch and cellulose.

polyserositis *(pŏl-ē-sĕr-ō-sī'-tĭs).* General inflammation of serous membranes accompanied by serous effusion.

polytendinitis *(pŏl-ē-tĕn-dĭ-nī'-tĭs).* Inflammation of several tendons.

polyunsaturated *(pŏl-ē-ŭn-săch-ĕr-āt'-ĕd).* A chemical compound that contains two or more double or triple bonds.

polyuria *(pŏl-ē-ū'-rē-ă).* Excessive secretion and elimination of urine, often seen in diabetes.

pons *(pŏnz).* 1. Any piece of tissue that connects two parts of an organ. 2. The part of the brain that lies between the medulla oblongata and the mesencephalon.

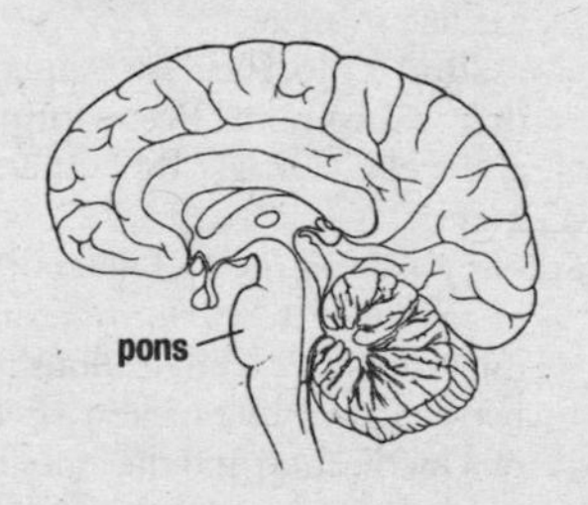

popiliteal *(pŏp-lĭt'-ē-ăl).* Referring to the posterior area of the knee.

pore *(pŏr).* A small opening.

porencephaly *(pŏr-ĕn-sĕf'-ĕ-lē).* The presence of one or more cavities in the brain, most often seen in a fetus or during early infancy.

porosis *(pō-rō'-sĭs).* Formation of callus during the repair of fractured bones.

porphyria *(pŏr-fĭ'-rē-ă).* A group of conditions that result from a dysfunction in porphyrin metabolism, which leads to increased formation and excretion of porphyrin or its precursors.

porphyrin *(pŏr'-fĭ-rĭn).* Any of a group of nitrogen-containing organic compounds that are the basis of the respiratory pigments in animals and plants, and which are obtained from hemoglobin and chlorophyll.

portacaval shunt *(pŏr-tĕ-kā'-vĕl).* A shunt (passage) that connects the portal vein and the vena cava.

portal hypertension. *See* **hypertension.**

portal system. The system of veins that runs from the intestines into the liver.

port wine stain. A mark on the skin that is ruby red, resembling port wine, and caused by an abnormal aggregation of capillaries.

positron emission tomography (PET) *(pŏz'-ĭ-trŏn).* An imaging technique used to measure blood flow and metabolism of the body's tissues, including the heart and brain.

posterior *(pŏs-tĕr'-ē-ŏr).* Located in back of, or in the back part of, a structure.

posthumous *(pōs'-tū-mĕs).* Occurring after death.

posthypnotic *(pōst-hĭp-nŏt'-ĭk).* Following a hypnotic state.

posthypnotic suggestion. A suggestion given by the hypnotizer during the hypnotic session that the individual will carry out after the session has ended.

postictal *(pōst-ĭk'-tĕl).* Occurring after a seizure or sudden attack.

postmenopausal *(pōst-mĕn-ō-păw'-zĕl).* The time immediately following cessation of a menstrual period.

postmortem *(pōst-mŏr'-tĕm).* After death.

postnasal drip *(pōst-nā'-zĕl).* A discharge of mucus or pus that flows down the back of the nose into the throat.

postnatal *(pōst-nā'-tĕl).* Referring to the period after birth.

postoperative *(pōst-ŏp'-ĕr-ă-tĭv).* Occurring after a surgical procedure.

postpartum *(pōst-păhr'-tŭm).* Occurring after childbirth, with reference to the mother.

postpartum depression. A depressed state of mind experienced by some birth mothers following delivery. It typically lasts several weeks or months and clears up spontaneously.

postpolio syndrome (PPS). A constellation of symptoms and signs that appear from 15 to 40 years after recovery from poliomyelitis, and characterized by newly progressive fatigue, pain, breathing difficulties, swallowing problems, muscle twitching, and weakness, and in some cases, muscle wasting.

postprandial *(pōst-prăn'-dē-ăl).* Occurring after a meal.

post-traumatic *(pōst-traw-măt'-ĭk).* Occurring as a result of or after injury or trauma.

post-traumatic stress disorder. An anxiety disorder caused by exposure to an extremely traumatic event and characterized by reexperiencing the event in flashbacks or nightmares, sleeping problems, difficulty concentrating, and hyperalertness.

posture *(pŏs'-tŭr).* Position or attitude of the body.

potassium *(pō-tăs'-ē-ŭm).* The main cation in intracellular fluid and an important element in cell maintenance. A proper balance of potassium and other elements is essential for muscle health, especially cardiac muscle.

potassium chloride. *See* Appendix, Common Prescription and OTC Drugs: By Generic Name.

potency *(pō'-tĕn-sē).* 1. The ability of the male to perform sexual intercourse. 2. The relationship between the therapeutic effect of a medication and the amount needed to achieve that effect.

potentiation *(pō-tĕn-shē-ā'-shĕn).* The increasing of potency, especially the synergistic action of two drugs, leading to an effect that is greater than the sum of each drug used alone.

poultice *(pōl'-tĭs).* A soft, moist substance with the consistency of cooked oatmeal, spread between layers of cloth and applied hot to a given area to provide moist heat or relieve irritation.

powder *(pou'dĕr).* An aggregation of fine particles of one or more substances.

pox *(poks).* Any disease characterized by eruptions or pustulars, especially one caused by a virus; for example, chicken pox, cowpox.

prana *(prăh'-nă).* In ayurvedic medicine, the term for the vital life energy, akin to qi, that flows throughout the body.

prandial *(prăn'-dē-ĕl).* Referring to a meal, especially dinner.

pranayama *(pră-nĕ-yăh'-mă).* In ayurveda, a method of breath control, used to control the energy within the body and the mind, and acting as a vitalizing and regenerating force.

Pravachol *(prăv'-ă-kōl). See* Appendix, Common Prescription and OTC Drugs: By Trade Name.

pravastatin *(prăv'-ĕ-stăt'-ĭn). See* Appendix, Common Prescription and OTC Drugs: By Generic Name.

precancerous *(prē-kăn'-sĕr-ŭs).* Referring to a pathologic process that tends to become cancerous (malignant).

precipitate *(prē-sĭp'-ĭ-tāt).* To cause a substance in solution to settle down in solid particles.

precipitation *(prē-sĭp-ĭ-tā'-shŭn).* The process of precipitating.

preclinical *(prē-klĭn'-ĭ-kĕl).* Referring to a disease before it becomes clinically apparent.

preclinical study. A study in which a drug, procedure, or other medical treatment is performed on animals to test for safety. Such studies must be done before clinical trials can begin.

precocious *(prē-kō'-shĕs).* Mental or physical development that occurs earlier than would be expected for a given age.

precoital *(prē-kō'-ĭ-tăl).* Prior to sexual intercourse.

preconscious *(prē-kŏn'-shŭs).* Not present in consciousness, but able to be recalled when desired.

precordial *(prē-kŏr'-dē-ăl).* Referring to the precordium.

precordium *(prē-kŏr'-dē-ŭm).* The area on the front surface of the body overlying the heart and the lower part of the thorax.

precursor *(prē-kĕr'-sŏr).* A substance that precedes another substance. The latter substance is usually more active or more mature than the one from which it was formed.

predisposition *(prē-dĭs-pō-zĭsh'-ŭn).* The potential to develop a certain disease or condition under certain conditions.

prednisone *(prĕd'-nĭ-sōn). See* Appendix, Common Prescription and OTC Drugs: By Generic Name.

prednisolone *(prĕd-nĭs'-ĕ-lōn).* A synthetic glucocorticoid derived from cortisol, used as replacement therapy for adrenocortical insufficiency and as an anti-inflammatory and immunosuppressant.

preeclampsia *(prē-ē-klămp'-sē-ă).* A complication of pregnancy characterized by hypertension, edema, and/or proteinuria.

pregnancy *(prĕg'-nĕn-sē).* Condition in which there is a developing embryo in the body. In women, pregnancy from conception to delivery is about 266 days.

abdominal p. Pregnancy in which the fertilized ovum develops in the abdominal cavity.

ectopic p. Any pregnancy in which the fertilized ovum develops outside of the uterine cavity.

false p. Absence of menstruation and presence of other signs of pregnancy, without conception and development of an embryo.

ovarian p. Pregnancy in which the fertilized ovum develops inside an ovary.

phantom p. False pregnancy due to psychogenic factors.

tubal p. Pregnancy in which the fertilized ovum develops within a uterine tube.

prehensile *(prē-hĕn'-sĭl).* Adapted for seizing or grasping.

prehension *(prē-hĕn'-shŭn).* The act of seizing or grasping.

premalignant *(prē-mĕ-lĭg'-nĕnt).* Precancerous.

Premarin *(prĕm'ĕ-rĭn). See* Appendix, Common Prescription and OTC Drugs: By Trade Name.

premature *(prē-mă-choor').* Occurring before the proper or appropriate time.

premedication *(prē-mĕd-ĭ-kā'-shŭn).* Use of medication to induce unconsciousness prior to administration of inhalation anesthesia.

premenstrual *(prē-mĕn'-stoo-ĕl).* Just before the start of a menstrual period.

premenstrual syndrome (PMS). A condition that typically occurs several days before onset of menstruation and is characterized by various symptoms, which may include irritability, anxiety, mood changes, headache, breast tenderness, water retention, and emotional tension.

premie *(prē'-mē).* Slang for a premature or preterm infant.

premonition *(prĕ-mĕ-nĭ'-shŭn).* A feeling of an impending event.

Prempro *(prĕm'-prō). See* Appendix, Common Prescription and OTC Drugs: By Trade Name.

prenatal *(prē'-nā-tĕl).* Before birth.

prenatal care. Care of the woman during pregnancy, including periodic examinations to ensure the health of the mother and fetus; instructions in nutritional needs, care of newborns, and delivery; and suggestions on how to deal with the discomforts of pregnancy.

preoperative *(prē-ŏp'-ĕr-ă-tĭv).* Care that precedes surgery.

prepubertal *(prē-pū'-bĕr-ăl).* Before puberty, the time during which secondary sex characteristics develop.

prepuce *(prē'-pūs).* A covering fold of skin, as in prepuce of clitoris and prepuce of penis.

presbyopia *(prĕs-bē-ō'-pē-ă).* Impaired vision due to advancing age. There is a loss of elasticity of the crystalline lens, which causes the near point of clear vision to move farther from the eye.

prescription *(prē-skrĭp'-shŭn).* Written directions for the preparation and administration of a drug.

presenile *(prē-sē'-nīl).* Referring to a condition that resembles senility but which occurs in early to middle life.

presentation *(prē-sĕn-tā'-shŭn).* In obstetrics, the position of the fetus that can be touched by an examining finger through the cervix or, during labor, what is bounded by the girdle of resistance.

breech p. Presentation of the buttocks or feet of the fetus during labor.

cephalic p. Presentation of any part of the head during labor.

vertex p. Presentation of the vertex of the head during labor.

pressor *(prĕs'-ŏr).* Something that tends to increase blood pressure.

pressure *(prĕsh'-ĕr).* Force per unit area.

abdominal p. The pressure between the viscera within the abdominal cavity.

arterial p. Blood pressure.

atmospheric p. The pressure exerted by the atmosphere. At sea level, the pressure is one *atmosphere.* Pressure decreases as altitude increases.

blood p. The pressure of blood against blood vessel walls.

pulse p. The difference between the diastolic and systolic pressures.

venous p. The blood pressure in a vein.

presystolic *(prē-sĭs-tŏl'-ĭk).* Before the systole (contraction phase) of the heart.

pretibial *(prē-tĭb'-ē-ăl).* In front of the tibia.

Prevacid *(prĕv'-ă-sĭd).* *See* Appendix, Common Prescription and OTC Drugs: By Trade Name.

preventive medicine. A proactive approach to the practice of medicine; it is designed to avoid disease and discomfort.

priapism *(prī'-ă-pĭz-ĕm).* Persistent, abnormal erection of the penis, characterized by pain and tenderness.

prickly heat *(prĭk'-lē).* The blockage of sweat pores caused by bacterial damage of cells. Characterized by tiny blisters which may or may not be noticed. As the sweat reaches into the skin, the area becomes painful and itchy.

Prilosec *(prī-lō-sĕk). See* Appendix, Common Prescription and OTC Drugs: By Trade Name.

primary care *(prī'-măr-ē).* The goals of primary care are to provide patients with wide-ranging care, both preventive and curative, over time, and to coördinate all the patient's care.

primidone *(prĭm'-ĭ-dōn).* An antiseizure drug used to treat epilepsy.

primipara *(prī-mĭp'-ă-ră).* A woman who has produced one infant who was at least 500 grams or of 20 weeks' gestation, and was born alive or dead.

primordial *(prī-mŏr'-dē-ăl).* Existing first or in a primitive or early form.

Prinivil *(prĭ'-nĭ-vĭl). See* Appendix, Common Prescription and OTC Drugs: By Trade Name.

proband *(prō'-bănd).* The first individual in a family who presents with a physical or mental disorder and causes a study of his or her heredity to determine if other family members have the same condition or carry it.

probe *(prōb).* A device used to explore the direction and depth of a sinus or wound or, in dentistry, the surface features of teeth.

probenecid *(prō-bĕn'-ĕ-sĭd).* A preparation that is useful in treating gout. Trade names: Benemid and Sk-Probenecid.

probiotic *(prō-bī-ŏ'-tĭk).* A microbe that can counter the destruction of helpful intestinal bacteria by antibiotics and help prevent antibiotic-associated diarrhea. Commonly used probiotics include strains of *Lactobacillus,* including *L. acidophilus.*

procainamide *(prō-kān'-ĕ-mīd).* A drug used to treat cardiac arrhythmias.

procaine *(prō'-kān).* A drug used as a local anesthetic, a spinal anesthetic, and a peripheral nerve block. Trade name: Novocain.

procarbazine *(prō-kăhr'-bĕ-zēn).* A drug used primarily in combination with several others to treat advanced Hodgkin's disease, but also used to treat non-Hodgkin's lymphoma, primary brain tumors and multiple myeloma.

Procardia *(prō-kăr'-dē-ă).* Trade name for preparations of nifedipine, which are used to treat stable angina pectoris, coronary insufficiency, and hypertension.

procedure *(prō-sē'-jĕr).* A series of steps used to achieve a desired result.

process *(prŏs'-ĕs).* 1. A projection, as of bone. 2. A series of events, steps, or procedures that lead to achievement of a specific goal.

alveolar p. The portion of bone in the maxilla or the mandible that surrounds and supports the teeth.

procreation *(prō-krē-ā'-shŭn).* The

process of bringing a new individual into the world.

Procrit *(prō'-krĭt).* Trade name for a preparation of epoetin alfa, which is used to treat anemia of various causes.

proctalgia *(prŏk-tăl'-jē).* Neuralgia of the lower rectum.

proctectomy *(prŏk-tĕk'-tĕ-mē).* Surgical removal of the rectum.

proctitis *(prŏk-tī'-tĭs).* Inflammation of the rectum.

proctocolitis *(prŏk-tō-kō-lī'-tĭs).* Inflammation of the colon and rectum.

proctologist *(prŏk-tŏl'-ō-jĭst).* A physician who specializes in proctology.

proctology *(prŏk-tŏl'-ĕ-jē).* The branch of medicine that deals with disorders of the rectum and anus.

proctoplasty *(prŏk'-tō-plăs-tē).* Plastic surgery of the rectum.

proctoscope *(prŏk'-tō-skōp).* An instrument with a light source, used to inspect the rectum.

prodromal *(prō-drō'-mĕl).* Something that indicates the onset of a disease or morbid condition.

profundaplasty *(prō-fŭn'-dĕ-plăs-tē).* Reconstruction of a blocked deep femoral artery.

progenitor *(prō-jĕn'-ĭ-tŏr).* An ancestor or parent.

progeny *(prŏj'-ĕ-nē).* Offspring or descendents.

progeria *(prō-jĕr'-ē-ĕ).* A syndrome characterized by precocious senility and death from coronary artery disease usually before 10 years of age.

progesterone *(prō-jĕs'-tĕ-rōn).* The main hormone of the body, excreted by the corpus luteum, placenta, and the adrenal cortex. It prepares the uterus for development of a fertilized ovum and maintains an optimal environment for the developing fetus.

progestin *(prō-jĕs'-tĭn).* Any one of several steroid hormones that have the effect of progesterone.

prognosis *(prŏg-nō'-sĭs).* A prediction as to the probable outcome of an attack of disease and an estimate as to the possibility of recovery.

prognosticate *(prŏg-nŏs'-tĭ-kāt).* To predict the probable outcome of an attack of disease.

projection *(prō-jĕk'-shĕn).* 1. The act of throwing forward. 2. In psychiatry, the process whereby an individual displaces his or her own unconscious feelings onto someone else.

prolactin *(prō-lăk'-tĭn).* A hormone secreted by the pituitary gland and which stimulates and sustains lactation.

prolapse *(prō-lăps').* A falling down or sinking of a body part.

proliferation *(prō-lĭf-ĕ-rā'-shŭn).* Rapid multiplication or reproduction of similar forms, such as cells and cysts.

prolific *(prō-lĭf'-ĭk).* Productive.

proline *(prō'-lēn).* A nonessential amino acid that is a major component of collagen.

promazine *(prō'-mĕ-zēn).* Generic name of a drug used as an antipsychotic.

promethazine *(prō-mĕth'-ă-zēn).*

See Appendix, Common Prescription and OTC Drugs: By Generic Name.

pronate *(prō'-nāt).* To place in a prone position.

prone *(prōn).* Lying face downward.

propagation *(prŏp-ĕ-gā'-shŭn).* Reproduction.

Propecia *(prō-pē'-shă).* Trade name for a preparation of finasteride, used to stimulate hair growth.

prophylactic *(prō-fĕ-lăk'-tĭk).* Any substance or regimen that helps to prevent or ward off something, especially disease.

prophylaxis *(prō-fĕ-lăk'-sĭs).* Action aimed at preventing disease.

propolis *(prō'-pō-lĭs).* A sticky substance that bees collect from buds and use to fix holes in their hive. Medicinally it is used to treat mouth and throat inflammation, tonsillitis, cold, flu and cough. It is a rich source of most vitamins and twenty-one amino acids.

propoxyphene *(prō-pŏk'-sĭ-fēn). See* Appendix, Common Prescription and OTC Drugs: By Generic Name.

propranolol *(prō-prăn'-ĕ-lŏl).* Generic name of a preparation that decreases cardiac rate and output, lowers blood pressure, and helps prevent migraine.

proprietary *(prō-prī'-ĕ-tăr-ē).* Protected against free competition by trademark, patent, or other means.

proprietary medicine. Any drug, chemical, or similar product that is protected by patent, trademark, copyright, or similar means.

proprioceptive *(prō-prē-ō-sĕp'-tĭv).* Receiving stimuli within muscles, tendons, and other tissues of the body.

proptosis *(prŏp-tō'-sĭs). See* exophthalmos.

Proscar *(prŏs'-kăhr).* Trade name for a preparation of finasteride, used to treat benign prostatic hyperplasia and male pattern hair loss.

prospective study *(prŏ-spĕk'-tĭv).* A study in which the subjects are identified and then followed for a period of time.

prostaglandin *(prŏs-tĕ-glăn'-dĭn).* A large group of unsaturated fatty acids that have various biological effects in the body, including fluid balance, platelet aggregation, blood flow, and gastrointestinal function.

prostate *(prŏs'-tāt).* A gland in males that surrounds the bladder neck and urethra and contributes a secretion to the seminal fluid.

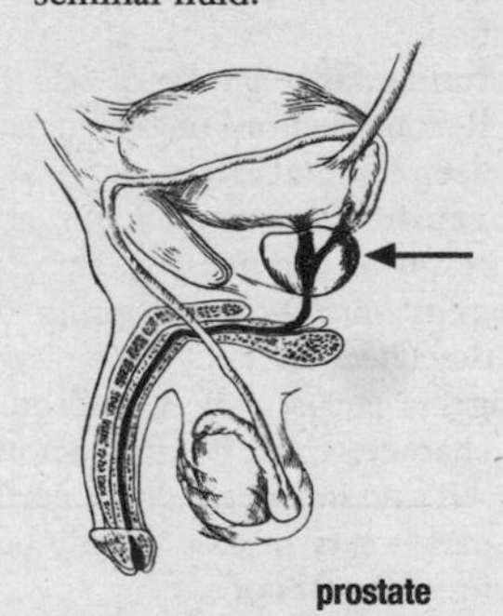

prostate

prostatectomy *(prŏs-tĕ-tĕk'-tĕ-mē).* Surgical removal of part or all of the prostate.
laser p. Removal of the prostate after it has been treated with a laser.
perineal p. Removal of the prostate through an incision in the perineum.
radical p. Removal of the prostate with its capsule and all the surrounding structures.
transvesical p. Removal of the prostate through an incision of the urinary bladder.

prostate specific antigen (PSA). A test used to help identify the possible presence of prostate cancer.

prostatic calcium *(prŏs-tăt'-ĭk).* A calcium stone in the prostate gland.

prostatic massage. Insertion of a finger into the rectum to massage the prostate and thus relieve congestion of the gland.

prostatism *(prŏs-tĕ-tĭz'-ĕm).* A benign enlargement of the prostate gland, frequently seen in older males.

prostatitis *(prŏs-tĕ-tī'-tĭs).* Inflammation of the prostate gland.

prosthesis *(prŏs-thē'-sĭs).* An artificial body part, such as an artificial leg or dentures.

prostration *(prŏs-strā'-shŭn).* Extreme exhaustion.

protean *(prō'-tē-ĕn).* Capable of assuming different shapes.

protease *(prō'-tēs).* An enzyme that can split a protein into peptides, from which the protein was originally made.

protease inhibitors. One of the anti-HIV drugs, designed to inhibit the activity of the enzyme protease and thus hinder virus reproduction.

protein *(prō'-tēn).* Any of a group of complex organic substances that contain carbon, hydrogen, oxygen, nitrogen, and usually sulfur. Protein is the main component of the protoplasm of all cells.

proteinosis *(prō-tēn-ō'-sĭs).* The accumulation of excess protein in the tissues.

proteinuria *(prō-tē-nū'-rē-ĕ).* The presence of an abnormally high level of protein in the blood.

Proteus *(prō'-tē-ŭs).* A genus of gram-negative bacteria of the family Enterobacteriaceae.

prothrombin *(prō-thrŏm'-bĭn).* A chemical in circulating blood that, through its interaction with thrombokinase, interacts with calcium salts to produce thrombin.

prothrombin time. A test used to determine clotting time, which is useful in evaluating the effect of anticoagulant drugs.

protocol *(prō'-tō-kŏl).* A detailed plan for an experiment, test, or procedure.

proton *(prō'-tŏn).* The positively charged particle of an atom, located in the nucleus.

Protonix *(prō'-tĕ-nĭks).* *See* Appendix, Common Prescription and OTC Drugs: By Trade Name.

protopathic sensibility *(prō-tō-păth'-ĭk).* The ability to feel or perceive basic stimulations such as pain, temperature, and some types of touch, which serves as a defensive agency against disease-causing changes in the tissues.

protoplasm *(prō'-tō-plăz-ĕm).* A thick, colloidal substance that makes up the physical foundation of all living activities.

prototype *(prō'-tō-tīp).* An original type or model from which subsequent models are developed.

Protozoa *(prō-tō-zō'-ă).* A subkingdom consisting of the simplest organisms of the animal kingdom. The one-celled organisms range in size from submicroscopic to macroscopic.

protrusion *(prō-troo'-zhĕn).* The state of being thrust forward or laterally.

protuberance *(prō-too'-bĕr-ĕns).* A projecting part.

proud flesh. Overgrowth of tissue in a wound that has not yet healed.

Proventil *(prō-vĕn'-tĭl). See* Appendix, Common Prescription and OTC Drugs: By Trade Name.

Provera *(prō-vĕr'-ă). See* Appendix, Common Prescription and OTC Drugs: By Trade Name.

proximal *(prŏk'-ĭ-mĕl).* Nearest; the opposite of *distal.*

Prozac *(prō'-zăk). See* Appendix, Common Prescription and OTC Drugs: By Trade Name.

prurigo *(proo-rī'-gō).* Any of various itchy skin eruptions characterized by dome-shaped lesions that have a small transient vesicle on top, followed by crusting. The cause is unknown.

pruritus *(proo-rī'-tĕs).* Itching. It can be the result of a drug reaction or be associated with cancer, parasites, aging, dry skin, contact skin reaction, liver or kidney disease, or unknown causes.

ani p. Intense chronic itching around the anus.

senilis p. Itching present in older adults, likely due to dry skin that occurs because of a decrease in sweat and sebum secretions, bathing too frequently, or both.

vulvae p. Intense itching of the external genitals of females.

PSA. Acronym for prostate specific antigen.

psammoma *(săm-ō'-mĕ).* Any tumor that is derived from fibrous tissue of the meninges or other structures associated with the brain. It is characterized by sandlike particles.

pseudoangina *(soo-dō-ĕn-jī'-nĕ).* A syndrome that occurs in nervous individuals, characterized by precordial pain, fatigue, and lassitude, but with no evidence of organic heart disease.

pseudocyst *(soo'-dō-sĭst).* An abnormal or dilated cavity that resembles a true cyst but is not lined with epithelium. Also called a *false cyst.*

pseudogout *(soo'-dō-gout).* Inflammation of the joints caused by deposits of calcium pyrophosphate crystals, resulting in arthritis, usually of the knees, hips, wrists, shoulders, and ankles. True gout is caused by a different type of crystal.

Pseudomonas *(soo-dō-mō'-nĕs).* A genus of gram-negative bacteria of the family Pseudomonadaceae, and consisting of several hundred species.

P. aeruginosa. The species that is the main cause of nosocomial infections, including *P. aeruginosa* pneumonia, urinary tract infections, wound infections, and abscesses.

P. putrefaciens. A species associated with spoilage in butter, meat, and marine foods.

P. vesicularis. A species associated with genitourinary tract infections.

pseudopregnancy *(soo-dō-prĕg'-nĕn-sē).* False pregnancy.

psittacosis *(sĭt-ĕ-kō'-sĭs).* A respiratory and systemic disease of wild and domestic birds, caused by infection with *Chlamydia psittaci,* and transmissible to humans, usually by inhaling dried bird excrement that contains the pathogen, handling an infected bird, or from the bite of an infected bird.

psoriasis *(sĕ-rī'-ĕ-sĭs).* A common, chronic skin disorder characterized by papules that coalesce to form plaques. If untreated, a silver-yellow scale develops. In about 5% of patients, arthritis develops as well.

psyche *(sī'-kē).* Everything that makes up the mind and its processes.

psychiatric *(sī-kē-ăt'-rĭk).* Referring to or within the realm of psychiatry.

psychiatrist *(sī-kī'-ĕ-trĭst).* A physician who specializes in psychiatry.

psychoanalysis *(sī-kō-ĕ-năl'-ĭ-sĭs).* Based on Freud's work, a theory of human mental phenomena and behaviors. It also refers to the therapeutic method based on his theory, which focuses on unconscious forces.

psychogenic *(sī-kō-jĕn'-ĭk).* Produced or caused by psychological factors.

psychologist *(sī-kŏl'-ō-jĭst).* An individual qualified in the specialty of psychology.

psychology *(sī-kŏl'-ō-jē).* The branch of science that deals with the mind and mental processes, especially in human and animal behavior.

abnormal p. The study of mental disorders and behavior disturbances.

clinical p. Use of psychologic knowledge and methods to treat individuals who have emotional, behavioral, mental, and developmental disorders.

developmental p. The study of behavioral changes that occur throughout the life span.

psychomotor *(sī-kō-mō'-tĕr).* Referring to or causing physical activity associated with one's thoughts, emotions, and desires.

psychopathic *(sī-kō-păth'ĭk).* Referring to psychopathy, especially to antisocial behavior.

psychopharmacology *(sī-kō-fărm-ă-kŏl'-ĕ-jē).* The study of the action and impact of drugs on psychological functioning and mental state.

psychophysiology *(sī-kō-fĭz-ē-ŏl'-ĕ-jē).* The study of the correlation of body and mind.

psychosexual *(sī-kō-sĕk'-shoo-ĕl).* Referring to the mental or emotional aspects of sex or sexuality.

psychosis *(sī-kō'-sĭs).* A mental disorder characterized by a loss of contact with reality as demonstrated by delusions, hallucinations, incoherent speech, or agitated behavior, usually occurring without the individual being aware of these behaviors.
alcoholic p. Psychosis associated with alcohol use and involving organic brain damage.
drug p. Any psychosis associated with drug use.
postpartum p. A psychotic episode that occurs during the postpartum period.
senile p. Paranoid or depressive delusions or hallucinations or other mental disorders caused mainly by deterioration of the brain in old age.

psychosomatic *(sī-kō-sō-măt'-ĭk).* Referring to the mind-body relationship; having physical symptoms that are emotional or mental in origin.

psychosurgery *(sī-kō-sĕr'-jĕr-ē).* Brain surgery performed for the treatment of psychiatric disorders.

psychotherapy *(sī-kō-thĕr'-ă-pē).* Treatment of mental conditions and behavioral disorders using verbal and nonverbal communication in an attempt to change maladaptive patterns of behavior and emotional disturbance and encourage personal growth.

psychotic *(sī-kŏ'-tĭk).* Referring to, characterized by, or caused by psychosis.

psychotropic drug *(sī-kō-trō'-fĭk).* Any drug that is capable of affecting the mind, emotions, and behavior. This category includes both legal drugs (e.g., lithium) and illicit drugs (e.g., cocaine).

psyllium *(sĭl'-ē-ŭm).* A plant of the genus *Plantago,* the seeds and husks of which are used as a mild bulk-forming laxative.

pterygium *(tĕr-ĭj'-ē-ŭm).* A winglike structure, especially referring to an abnormal fold of membrane in the eye that is attached to the cornea.

ptilosis *(tī-lō'-sĭs).* Loss of eyelashes.

ptomaine poisoning *(tō'-mān).* Poisoning with the nitrogenous organic substance that is formed by bacterial action on protein and amino acids.

ptosis *(tō'sĭs).* 1. Prolapse. 2. Drooping of the upper eyelid.

ptyalin *(tī'-ĕ-lĭn).* An enzyme in saliva that helps to break down starch into sugars.

ptyalism *(tī'ĕ-lĭz-ĕm).* Excessive salivation.

puberty *(pū'-bĕr-tē).* The period during which secondary sex characteristics begin to develop and males and females attain the capability of sexual reproduction.

precocious p. Onset of sexual maturation at an age earlier than normal: before age 8 in girls and age 9 in boys.

precocious p., central. Precocious puberty caused by premature hypothalamic-pituitary-gonadal maturation, characterized by increases in height and weight, development of the gonads, and short stature.

precocious p., neurogenic. Precocious puberty caused by a tumor, congenital defect, or trauma associated with the central nervous system.

pubescent *(pū-bĕs'-ĕnt).* Reaching sexual maturation.

pubic hair *(pū'-bĭk).* The hair that develops in the pubic region in both males and females.

pubis *(pū'-bĭs). See* **os.**

public health. An approach to medicine that involves the health of the entire community. A public health concern, for example, would be drinking water safety.

pudendal block *(pū'-dĕn-ăl).* A local anesthetic injection used to reduce the pain that occurs during the last part of labor and for an episiotomy.

pudendum *(pū-dĕn'-dŭm).* The external genitalia of humans, especially of females.

puerile *(pū'-ĕr-ĭl).* Referring to childhood or to children.

puerperal *(pū-ĕr'-pĕr-ĕl).* Referring to the puerperium.

puerperium *(pū-ĕr-pēr'-ē-ĕm).* The period from the end of the third stage of labor until involution of the uterus is complete, usually lasting 3 to 6 weeks.

pulmonary *(pool'-mō-năr-ē).* Referring to the lungs.

pulmonary edema. Abnormal accumulation of fluid in the lung tissues and air spaces characterized by intense shortness of breath and, in some forms, expectoration of frothy pink fluid.

pulmonary embolism. Closure of the pulmonary artery or one of its branches by an embolus.

pulmonary function test. A test used to measure the ability of the lungs to exchange air.

pulmonologist *(pool-mĕ-nŏl'-ĕ-jĭst).* An individual who specializes in the diagnosis and treatment of pulmonary conditions.

pulsatilla *(pŭl-sū-tē'-ă).* A homeopathic remedy prepared from the windflower plant (*Pulsatilla nigricans*). Symptoms that respond to pulsatilla include menstrual cramps, digestive problems, one-sided headache, chills, nosebleeds, sweating, and a bad taste in the mouth.

pulp *(pŭlp).* Any soft animal or vegetable tissue, such as dental pulp.

pulpectomy *(pĕl-pĕk'-tĕ-mē).* Root canal therapy in which dental pulp is completely removed from the pulp.

pulse *(pŭls).* The rhythmic expansion of an artery, which can be felt with the finger at various locations on the body, including the inside wrist.

abdominal p. The pulse over the abdominal aorta.
bigeminal p. A pulse in which two beats occur in rapid succession and are separated from the following pair by a longer interval.
bounding p. A strong pulse.
brachial p. The pulse that can be felt over the brachial artery at the inner aspect of the elbow.
carotid p. The pulse in the carotid artery.
femoral p. Pulse that can be felt over the femoral artery in the femoral triangle.
radial p. The pulse that is felt at the wrist.
thready p. Pulse that is very fine and scarcely perceptible.

pump *(pŭmp).* A device that draws or forces fluids or gases.
breast p. A manual or electric pump that abstracts milk from the breast.
insulin p. An externally worn pump that can provide insulin to subcutaneous tissues through a plastic tube.
stomach p. A pump that removes the contents from the stomach.

punch drunk. Boxer's dementia.

puncture *(pŭngk'-chĕr).* The act of penetrating or piercing with a pointed device.
lumbar p. A procedure that withdraws fluid from the subarachnoid space in the lumbar region of the back.
sternal p. Use of a needle to remove bone marrow from the sternum.
ventricular p. Puncture of a cerebral ventricle to withdraw fluid.

pupil *(pū'-pĭl).* The opening at the center of the iris of the eye, which is where light enters the eye.

pupillary *(pū'-pĭ-lăr-ē).* Referring to the pupil.

purgative *(pŭr'-gĕ-tĭv).* Causing emptying of the bowels; cathartic.

purpura *(pŭr'-pū-ră).* Any of various conditions characterized by bleeding into the skin, mucous membranes, internal organs, and other tissues.

purulent *(pūr'-ū-lĕnt).* Containing pus.

pus *(pŭs).* Liquid product of inflammation composed of albuminous substances, fluid, and white blood cells.

pustule *(pŭs'-tūl).* A collection of pus within or beneath the skin, often occurring in a hair follicle or sweat pore.

putrefaction *(pū-trĕ-făk'-shĕn).* Decomposition, especially of proteins, with enzymes, with the result being a foul-smelling substance.

Pycnogenol *(pĭk-năh'-gĕ-nŏl).* The trade name of a derivative of the French maritime pine tree that has been shown to be a potent antioxidant and a rich source of flavonoids. It is sometimes used to treat poor circulation, varicose veins, and fragile capillaries.

pyelitis *(pī-ĕ-lī'-tĭs).* Inflammation of the renal pelvis, often accompanied by pain and tenderness in the loins, an irritable bladder, bloody or purulent urine, and pain in the thigh when flexed.

pyelogram *(pī'-ĕ-lō-grăm).* X-ray of the kidney and ureter, especially one that shows the renal pelvis.

pyelonephritis *(pī-ĕ-lō-nĕ-frī'-tĭs).* Inflammation of the kidney and renal pelvis, due to a bacterial infection.

ascending p. Pyelonephritis caused by a urinary tract infection that spreads up the ureter into the kidney.

calculous p. Pyelonephritis caused by urinary calculi.

chronic p. Pyelonephritis caused by a recurring or progressive infection, and which may lead to chronic renal insufficiency.

pyelotomy *(pī-ĕ-lŏt'-ĕ-mē).* Incision of the renal pelvis.

pyemia *(pī-ē'-mē-ă).* A general septicemia (blood poisoning) characterized by fever, chills, sweating, jaundice, and abscesses that can appear on various parts of the body.

pyknic *(pĭk'-nĭk).* Having a short, thick, stocky body.

pylephlebitis *(pī-lē-flĕ-bī'-tĭs).* Inflammation of the portal vein, which usually results from intestinal disease.

pylorectomy *(pī-lō-rĕk'-tĕ-mē).* Excision of the pylorus.

pyloric stenosis *(pī-lŏr'-ĭk).* Narrowing of the pyloric opening.

pyloritis *(pī-lō-rī'-tĭs).* Inflammation of the pylorus.

pyloroplasty *(pī-lŏr'-ō-plăs-tē).* Surgical procedure to repair the pylorus, especially to increase the caliber of the pyloric opening.

pylorus *(pī-lŏr'-ŭs).* The lower orifice of the stomach opening into the duodenum.

pyoderma *(pī-ō-dĕr'-mă).* Any acute, purulent, inflammatory dermatitis caused by a bacterial infection.

pyometritis *(pī-ō-mē-trī'-tĭs).* Purulent inflammation of the uterus.

pyonephritis *(pī-ō-nĕf-rī'-tĭs).* Inflammation of the kidney, accompanied by pus.

pyosalpinx *(pī-ō-săl'-pĭnks).* Accumulation of pus in a uterine tube.

pyrexia *(pī-rĕk'-sē-ă).* Fever.

pyridoxine *(pĕr-ĭ-dŏk'-sēn).* One of the forms of vitamin B6 (the others are pyridoxal and pyridoxamine) and the one often associated with the supplement.

pyrogenic *(pī-rō-jĕn'-ĭk).* Causing fever.

pyuria *(pī-ū'-rē-ă).* The presence of pus in the urine.

Q

Q fever. An acute illness caused by *Coxiella burnetii,* which is contracted from animals. It is characterized by sudden onset of fever, headache, malaise, and pneumonia.

Qi *(kē).* According to traditional Chinese medicine, it is the life energy force that flows through the body and can be accessed at acupoints on the body. Also spelled *chi* and *ch'i.*

q.i.d. Latin: *quater in die,* meaning four times a day.

qigong *(chē-kŭng').* An ancient Chinese tradition that combines regulation of breath, posture, and the mind to promote well-being and health. It has proved useful in treating hypertension, asthma, arthritis, fatigue, insomnia, headache, and constipation.

QRS complex. A phase in heart action that is observable on an electrocardiogram.

quack *(kwăk).* Someone who fraudulently claims to have the ability to diagnose and provide effective treatment for health and medical conditions.

quadrant *(kwŏd'-rĕnt).* 1. One quarter of a circle. 2. Any one of four parts; e.g., the abdominal surface or field of vision.

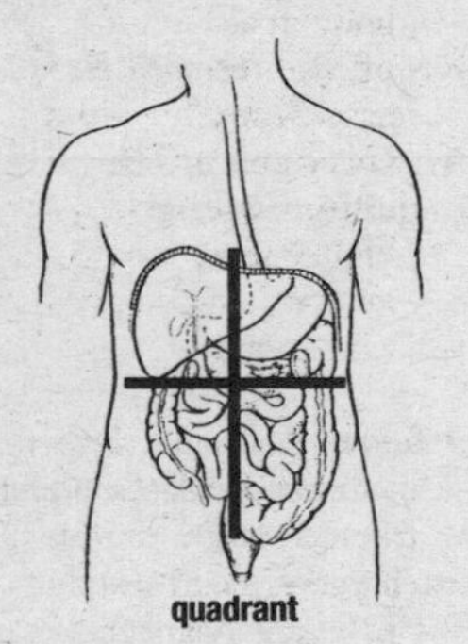

quadrant

quadriceps *(kwŏd'rĭ-sĕps).* 1. Four-headed. 2. Commonly used term to refer to the large muscles in the front of the thigh, which straighten the knee joint.

quadriplegia *(kwŏd-rĭ-plē'-jĕ).* Paralysis of all four limbs.

quadruplet *(kwŏd-roop'-lĕt).* One of four offspring produced in one gestation period.

qualitative *(kwăhl'-ĭ-tā-tĭv).* Referring to quality.

quantitative *(kwăhn'-tĭ-tā-tĭv).* Referring to or expressing a quantity.

quantum *(kwăhn'-tŭm). See* **theory, quantum.**

quarantine *(kwŏr'-ĕn-tēn).* Restriction of freedom of movement of seemingly healthy individuals who have been exposed to infectious disease. The quarantine period is typically the same as the maximum incubation period for the disease in question.

quickening *(kwĭk'-ĕn-ĭng).* The first observable movements of the fetus, which usually appear from the sixteenth to the twentieth week of pregnancy.

Quick's test *(kwĭks).* A test to help determine the adequacy of liver function.

quinapril *(kwĭn'ĕ-prĭl). See* Appendix, Common Prescription and OTC Drugs: By Generic Name.

quinine *(kwĭn'-īn).* A drug extracted from cinchona bark. It was once widely used to prevent and treat malaria, but is now used mainly to treat malaria that is resistant to other antimalarial medications.

quinsy *(kwĭn'-sē).* Abscess of the tonsil capsule caused by a bacterial infection.

quintuplet *(kwĭn-tŭp'-lĕt).* One of five offspring produced during a single gestation period.

quotidian *(kwō-tĭd'-ē-ĕn).* Occurring every day.

q.v. Latin for *quantum vis* (as much as you please); or *quod vide* (which see).

Q wave. On an electrocardiogram, a downward or negative wave that follows a P wave.

R

rabeprazole *(rĕ-bĕp'-rā-zōl). See* Appendix, Common Prescription and OTC Drugs: By Generic Name.

rabies *(rā'-bēz).* An acute infectious disease of the central nervous system that affects most mammals, including humans. Caused by a virus of the genus *Lyssavirus,* symptoms include pain, hyperexcitability, hallucinations, delirium, and bizarre behavior alternating with calmness, abnormal fear of fluids, convulsions, paralysis, and coma.

race *(rās).* A distinct ethnic group characterized by traits that are transmitted through the generations.

race-based drug. A drug designed to treat a specific racial group. The first such drug, BiDil, was reported in 2004 to reduce deaths from heart disease among African-Americans.

rad *(răd).* Acronym for radiation absorbed dose. A unit of measurement of the absorbed dose of ionizing radiation.

radiation *(rā-dē-ā'-shŭn).* The passage of energy through space. It may be transmitted by X-rays, ultraviolet rays, gamma rays, and other types of rays.

radiation burn. A burn caused by excessive exposure to X-rays, radium, the sun, or other sources.

radiation fibrosis. Scarring of the lungs caused by exposure to radiation.

radiation oncologist. An individual who specializes in the use of radiation as a treatment for cancer.

radiation sickness. Also known as acute radiation syndrome, it is caused by exposure to more than 50 rads of penetrating radiation to most or all of the body in a short time. Survivors of the Hiroshima and Nagasaki atomic bombs experienced this syndrome.

radiation therapy. The use of high-energy rays to stop the growth and reproduction of cancer cells.

radical hysterectomy. *See* **hysterectomy.**

radical mastectomy. *See* **mastectomy.**

radicular *(rĕ-dĭk'-ū-lĕr).* Referring to a radix or root.

radiculitis *(rĕ-dĭk-ū-lī'-tĭs).* Inflammation of the root of a spinal nerve.

radiculopathy *(rĕ-dĭk-ū-lŏp'-ĕ-thē).* Disease of the nerve roots.
cervical r. Radiculopathy of cervical nerve roots, often characterized by shoulder and neck pain.

radioactivity *(rā-dē-ō-ăk-tĭv'-ĭ-tē).* The ability of a substance to emit rays or particles from its nucleus.

radiograph *(rā'-dē-ō-grăf).* A film produced by radiography.

radiography *(rā-dē-ŏg'-ră-fē).* The production of film records (radiographs) of the body's internal structures by passing x-rays or gamma rays through the body to act on special film.

radioimmunoassay *(rā-dē-ō-ĭm-ū-nō-ăs'-ā).* A highly sensitive testing method to determine the concentration of substances, especially protein-bound hormones in blood plasma.

radioisotope *(rā-dē-ō-ī'-sō-tōp).* A radioactive form of an element.

radiologist *(rā-dē-ŏl'-ō-jĭst).* A physician who practices diagnosis and treatment by use of radiant energy.

radiology *(rā-dē-ŏl'-ō-jē).* The branch of medicine that involves the use of radioactive substances, including x-rays, radioactive isotopes, and ionizing radiation, to prevent, diagnose, and treat disease.

radiolucent *(rā-dē-ō-lū'-sĕnt).* Something that permits x-rays to pass through it.

radionecrosis *(rā-dē-ō-nĕ-krō'-sĭs).* Destruction of tissue due to exposure to radiation.

radiopaque *(rā-dē-ō-pāk').* Not penetrable by x-rays or other types of radiation.

radiosurgery *(rā-dē-ō-sĕr'-jĕr-ē).* Surgery in which tissue is destroyed using ionizing radiation rather than surgical incision.

radium *(rā'-dē-ŭm).* A radioactive element used to treat certain malignant diseases.

radius *(rā-dē-ŭs).* The outer and shorter bone of the forearm.

radon *(rā'-dŏn).* A heavy, colorless, radioactive element emitted by radium when it disintegrates. It is used to treat some types of cancer.

ragweed *(răg'-wēd).* Any of various plants of the genus *Ambrosia* whose pollen is a cause of hay fever.

rale *(rāl).* An intermittent sound that consists of a series of short nonmusical noises, heard primarily upon inhalation.

raloxifene *(răl-ŏk'-sĭ-fēn). See* Appendix, Common Prescription and OTC Drugs: By Generic Name.

ramipril *(răm'-ĭ-prĭl). See* Appendix, Common Prescription and OTC Drugs: By Generic Name.

ramus *(rā'-mŭs).* A branch or division of a forked structure.

random sampling *(răn'-dŏm).* A selection by chance.

range of motion. The range of flexion and extension through which a joint can be moved.

ranitidine *(rĕ-nĭ'-tĭ-dēn). See* Appendix, Common Prescription and OTC Drugs: By Generic Name.

ranula *(răn-ū'-lă).* A large cystic tumor located on the underside of the tongue, usually caused by obstruction of the ducts of the submaxillary or sublingual glands.

rape *(rāp).* Nonconsensual sexual penetration of an individual obtained by force or violence or when the victim is not capable of consent.

raphe *(rā'-fē).* The line of union of the halves of the symmetrical parts of an object or structure.

rapid eye movement sleep. Periods of sleep during which rapid eye movements occur. Typically people experience 3 to 5 periods of REM sleep per night, each period lasts from about 5 to 60 minutes or longer, and it is during REM sleep that dreams occur.

rarefaction of bone *(răr-ĕ-făk'-shĕn).* Loss of calcium in bone, which makes bone less dense and more brittle.

ratio *(rā'-shē-ō).* The relationship in number or degree between two things.

rationalization *(răsh-ŭn-ăl-ī-zā'-shŭn).* In psychology, a justification for an unreasonable or illogical idea or behavior in an attempt to make it appear reasonable.

Rauwolfia *(rou-wool'-fē-ĕ).* A large genus of tropical trees and shrubs, many species of which provide alkaloids that have medicinal value, including reserpine, which is used to treat high blood pressure.

ray *(rā).* A form of radiant energy that proceeds in a specific direction.

Raynaud's disease *(rā-nōz').* A peripheral vascular disorder that occurs most often in females between the ages of 18 and 30 and is characterized by abnormal constriction of the blood vessels in the hands and feet when they are exposed to cold or emotional stress.

RBC. Acronym for red blood cell.

RDI. Acronym for reference daily intake.

reaction *(rē-ăk'-shŭn).* A response; opposite action or counteraction.

hypersensitivity r. One in which the body responds with an exaggerated or inappropriate immune response to a substance that is foreign or that the body perceives to be foreign, resulting in tissue damage.

rebound r. A flare-up of symptoms when a drug is abruptly stopped.

rejection r. The body's destruction of a foreign body, including a structure or organ that has been transplanted from another person.

transfusion r. Symptoms that occur when the recipient of a blood transfusion receives an incorrectly matched transfusion or when the recipient has a hypersensitivity reaction to an element in the donor blood.

reagent *(rē-ā'-jĕnt).* A substance used to produce a chemical reaction for the purpose of detecting, producing, or measuring other substances.

rebound reaction. *See* **reaction.**

receptor *(rē-sĕp'-tĕr).* A molecular structure within a cell or on its surface, and which acts to receive an impulse.

recessive characteristics *(rē-sĕs'-ĭv).* Traits that are not apparent or active when inherited.

recidivation *(rē-sĭd-ĭ-vā'-shĕn).* Recurrence, relapse, or repetition of a condition or behavior pattern, especially a criminal act.

recombinant DNA. *See* **DNA.**

recrudescence *(rē-kroo-dĕs'-ĕns).* The recurrence of symptoms after a temporary abatement. Compared with a relapse, a recrudescence occurs after several days or weeks, while a relapse after several weeks or months.

rectocele *(rĕk'-tō-sēl).* Hernial protrusion of a portion of the rectum into the vagina.

rectosigmoidectomy *(rĕk-tō-sĭg-moi-dĕk'-tĕ-mē).* Surgical removal of the rectum and sigmoid colon.

rectovaginal *(rĕk-tō-văj'-ĭ-nĕl).* Referring to or communicating with the rectum and vagina.

rectum *(rĕk'-tŭm).* The lower portion of the large intestine, it measures about 5 inches in length.

recumbent position *(rē-kŭm'-bĕnt).* Lying down.

recuperation *(rē-koo-pĕr-ā'-shĕn).* Recovery of health.

recurrent disease *(rē-kŭr'-ĕnt).* A disease that returns after remission.

reduction *(rē-dŭk'-shĕn).* Restoration to a normal position, as in a fracture or hernia.

closed r. Reduction of a fracture or dislocation through manipulation rather than incision.

open r. Reduction of a fracture or dislocation through an incision into the site.

Reference Daily Intake (RDI). A set of dietary references based on the Recommended Dietary Allowances (RDAs) for essential vitamins and minerals and protein. "RDI" replaces "RDA."

referred pain. *See* **pain.**

reflex *(rē'-flĕks).* An involuntary response to a stimulus.

reflexology *(rē-flĕk-sŏl'-ĕ-jē).* A therapeutic technique based on the concept that there are sites on the hands or feet that correspond to the organs and organ systems of the body and that pressure applied to these sites can have a positive effect on the corresponding organ or system.

reflux *(rē'-flŭks).* A backward or return flow.

gastroesophageal r. Reflux of the stomach and duodenal contents into the esophagus. When chronic it is referred to as gastroesophageal reflux disease (GERD).

valvular r. The backflow of blood past a venous valve in the lower limbs, caused by venous insufficiency.

venous r. Any backflow of blood in the veins, usually occurring in the lower limbs, due to venous insufficiency.

refraction *(rē-frăk'-shĕn).* Deflection from a straight line, as when light rays pass through different densities.

refractory *(rē-frăk'-tō-rē).* Resistant to standard or ordinary treatment.

regeneration *(rē-jĕn-ĕr-ā'-shŭn).* The repair, restoration, or regrowth of a part, such as tissues.

regimen *(rĕj'-ĭ-mĕn).* A systematic plan of action, including regulation of diet, sleep, exercise, use of medication, and other activities to achieve a desired goal.

region *(rē'-jĕn).* A plane area with definite or less defined boundaries.

regression *(rē-grĕsh'-ĕn).* A return to a former state.

regurgitation *(rē-gŭr-jĭ-tā'-shĕn).* 1. To flow in the opposite direction from normal. 2. Vomiting.

rehabilitation *(rē-hă-bĭl-ĭ-tā'-shŭn).* A process of restoring levels of physical, psychosocial, vocational, and recreation activity to individuals after injury or illness.

rehydration *(rē-hī-drā'-shĕn).* To restore water or other fluids to the body or to a substance that has become dehydrated.

reiki *(rā'-kē).* An Eastern healing tradition that reportedly helps to rebalance the energy systems in the body that have become unbalanced. An unbalanced body often has physical symptoms, which can be relieved when a reiki practitioner channels energy through his/her hands to the recipient.

reishi *(rē'-shē).* A type of mushroom (*Ganoderma lucidum*) that is used by some health-care practitioners to treat arthritis pain. Some also believe it has anticancer properties.

relapse *(rē-lăps').* The return of a disease or other medical condition after it had apparently ceased.

REM. Acronym for rapid eye movement. *See* **sleep.**

remedial *(rĕ-mē'-dē-ĕl).* Acting as a remedy.

remedy *(rĕ'-mĕ-dē).* Anything that cures, helps alleviate, or prevents disease.

remineralization *(rē-mĭn-ĕr-ĕl-ĭ-zā'-shĕn).* To restore mineral elements, as of calcium salts to bones or teeth.

remission *(rē-mĭsh'-ĕn).* A reduction or elimination of symptoms of a disease. It also refers to the period of time during which reduction or elimination lasts.

renal *(rē'-nĕl).* Referring to the kidney.

renin *(rĕ'-nĭn).* A hormone released by the kidneys that has an effect on blood pressure.

renography *(rē-nŏg'-rĕ-fē).* X-ray of the kidney.

renovascular *(rē-nō-văs'-kū-lĕr).* Referring to or affecting the blood vessels of the kidney.

repaglinide *(rē-păg'-lĭ-nīd).* An oral hypoglycemic agent that stimulates the release of insulin from the pancreas. Often used along with metformin in the treatment of type II diabetes.

reperfusion *(rē-pĕr-fū'-shĕn).* Resumption of blood flow.

replantation *(rē-plăn-tā'-shĕn).* Replacement of an organ or other body part to the location from which it was removed or previously lost.

replication *(rĕp-lĭ-kā'-shĕn).* 1. Turning back of a part so as to form a duplication. 2. In genetics, the duplication of genetic material.

reposition *(rē-pō-zĭsh'-ŭn).* To restore an organ or tissue to its correct or original position.

repression *(rē-prĕsh'-ŭn).* In psychology, refusal to think about painful or disturbing ideas and suppressing them in the unconscious, where they continue to influence the thoughts and behavior of the individual.

reproduction *(rē-prō-dŭk'-shŭn).* Process by which organisms duplicate themselves.

resectable *(rē-sĕk'-tĕ-bĕl).* Capable of being resected.

resection *(rē-sĕk'-shĕn).* Removal of a part or all of an organ or other body part.

resectoscope *(rē-sĕk'-tō-skōp).* A device with an electrically activated wire loop, used for transurethral removal or biopsy of lesions of the prostate, bladder, or urethra.

reserpine *(rĕ-sĕr'-pēn).* An alkaloid isolated from the plant *Rauwolfia serpentine* and other *Rauwolfia* species and used to treat high blood pressure.

resolution *(rĕz-ō-loo'-shĕn).* 1. The cessation of a diseased or unhealthy state, as the disappearance or softening of inflammation. 2. The ability of the eye or a series of lenses to distinguish fine detail.

resonance *(rĕz'-ō-nĕns).* Quality of a sound produced by transmission of its vibrations to a cavity or hollow structure, such as the chest cavity.

resorption *(rē-sŏrp'-shŭn).* Act of removal by absorption, such as occurs when the body resorbs pus or exudates.

respiration *(rĕs-pĭr-ā'-shŭn).* The exchange of oxygen and carbon dioxide between the atmosphere and the cells of the body. Specifically, it involves taking in oxygen, use of the oxygen by the body's cells, and giving off carbon dioxide.

artificial r. Any man-made method of ventilation in which air is forced into and out of the lungs of an individual who has stopped breathing.

suppressed r. Respiration that occurs without any appreciable sound.

vicarious r. Increased action in one lung when that of the other lung is compromised.

respirator *(rĕs'-pĭr-ā-tŏr).* Ventilator.

respiratory acidosis. A condition in which the lungs are unable to expel a sufficient amount of carbon dioxide.

respiratory alkalosis. A condition in which there is an excessive amount of carbon dioxide exhaled by the lungs, as in hyperventilation.

respiratory arrest. Cessation of breathing.

respiratory distress syndrome. A syndrome of respiratory difficulty that affects newborns and is caused by a deficiency of a molecule called surfactant. It occurs in most newborns born before 37 weeks gestation.

respiratory failure. An inability of the lungs to transfer oxygen from inhaled air into the blood and transfer carbon dioxide from the blood into exhaled air.

respiratory rate. The number of breaths per minute. More precisely, the respiratory rate is determined by counting the number of times the chest rises or falls per minute.

respiratory system. The organs involved in respiration. In humans, these include the nose, pharynx, larynx, trachea, bronchi, and lungs.

respiratory therapy. Treatments and exercises designed to help individuals maintain and recover lung function, especially after surgery or for those who have cystic fibrosis.

restless legs syndrome. Condition characterized by an irresistible urge to move the legs and an internal itching sensation in the legs that occurs when sitting or lying down, especially just before sleep. Its cause is unknown, but may be due to inadequate circulation or a reaction to medication.

restraint *(rĭ-strānt').* Forcible confinement or control of a subject.

resuscitation *(rē-sŭs-ĭ-tā'-shĕn).* To restore life or consciousness to someone who is apparently dead.

retainer *(rē-tā'-nĕr).* In dentistry, a device for maintaining the teeth and jaws in position; also the part of a denture that joins the abutment tooth with a portion of a bridge, such as a crown.

retardation *(rē-tăhr-dā'-shĕn).* Delayed development.

mental r. A mental disorder characterized by significantly subaverage intellectual functioning. It is classified based on severity (see below).

mental r., mild. Condition in which the IQ is between 50–55 and 70, and the individual is capable of developing social and vocational skills sufficient for minimal self-support.

mental r., moderate. Condition in which the IQ is between 35–40 and 50–55 and the individual can benefit from vocational training and can perform personal care with moderate supervision.

mental r., profound. Condition in which the IQ is less than 20–25; the individual can achieve very limited self-care and requires constant supervision.

mental r., severe. Condition in which the IQ is between 20–25 and 35–40; the individual can be trained in basic hygiene skills and may learn to perform simple work under close supervision of an adult.

retching *(rĕch'-ĭng).* A strong involuntary effort to vomit.

rete *(rē'-tē).* A network, as a network of nerves or blood vessels.

retention *(rĭ-tĕn'-shŭn).* The act or process of holding or keeping something in place or keeping things within the body, such as urine or feces.

reticular *(rĭ-tĭk'-ū-lăr).* Resembling a network or mesh.

reticulocyte *(rĕ-tĭk'-ū-lō-sīt).* A red blood cell that contains a network of structures that represent an immature phase of development.

reticuloendothelial system *(rĕ-tĭk-ū-lō-ĕn-dō-thē'-lē-ăl).* A diverse system of cells throughout the body that have the power to ingest particulate matter (bacteria, colloidal particles). Some of those cells include macrophages, Kupffer cells of the liver, and reticular cells of lymphatic organs, among others.

reticuloendotheliosis *(rĕ-tĭk-ū-lō-ĕn-dō-thē-lī-ō'-sĭs).* Hyperplasia of the tissues that make up the reticuloendothelial system.

retina *(rĕt'-ĭ-nă).* The innermost of the three tunics of the eye. It is light sensitive, receives images formed by the lens, and is the structure upon which light rays come to a focus.

detached r. Separation of the inner layers of the retina from the pigment epithelium.

shot-silk r. A condition seen sometimes in young persons, in which the retina has an opalescent appearance.

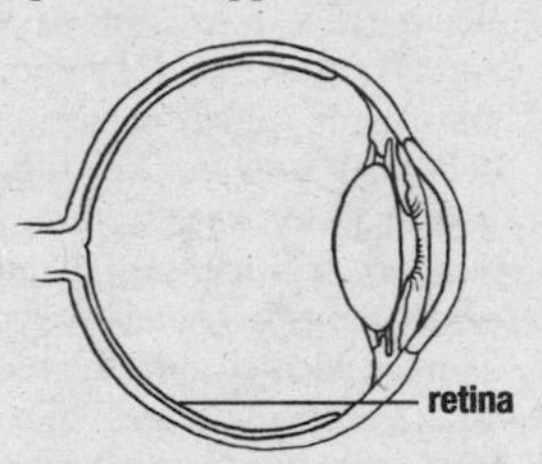

retinaculum *(rĕt-ĭ-năk'-ū-lĕm).* A structure that holds an organ or tissue in place.

retinal *(rĕt'-ĭ-nĕl).* Referring to the retina.

retinal detachment. *See* **retina.**

retinitis *(rĕt-ĭ-nī'-tĭs).* Inflammation of the retina.

retinoblastoma *(rĕt-ĭ-nō-blăs-tō'-mă).* A malignant congenital blastoma composed of tumor cells from retinoblasts, and seen in one or both eyes of children younger than five years of age.

retinopathy *(rĕt-ĭ-nŏp'-ă-thē).* Retinitis.

diabetic r. Retinopathy associated with diabetes.

hemorrhagic r. Retinopathy characterized by profuse bleeding in the retina.

renal r. Retinopathy associated with renal and hypertensive disorders.

retinoscope *(rĕt'-ĭ-nō-skōp).* An instrument for performing retinoscopy.

retinoscopy *(rĕt-ĭ-nŏs'-kĕ-pē).* A method for investigating, diagnosing, and evaluating refractive errors of the eye.

retraction *(rē-trăk'-shĕn).* The act of drawing back.

retractor *(rē-trăk'-tŏr).* An instrument that separates the edges of a wound and holds back underlying structures.

retrobulbar *(rĕt-rō-bŭl-bĕr).* Behind the eyeball.

retrocession *(rĕt-rō-sĕsh'-ŭn).* Backward displacement, especially of the uterus.

retroflexion *(rĕt-rō-flĕk'-shĕn).* The bending of an organ or part so that its top is turned posteriorly.

retrograde *(rĕt'-rō-grād).* Moving backward or against the usual flow.

retrogression *(rĕt-rō-grĕsh'-ĕn).* Return to a previous, less complex condition; deterioration.

retroplasia *(rĕt-rō-plā'-zhă).* Degeneration of a tissue or cell into a more simple or primitive type.

retrosternal *(rĕt-rō-stĕr'-nĕl).* Located or occurring behind the sternum.

retroversion *(rĕt-rō-vĕr'-shĕn).* Condition in which an entire organ or part tips in a backward direction.

Retrovir *(rĕt'-rŏ-vīr).* Trade name for preparations of zidovudine, used to treat HIV and AIDS.

retrovirus *(rĕt'-rō-vī-rĕs).* Any virus that belongs to the family Retroviridae.

retrusion *(rē-troo'-shĕn).* The state of being located posterior to the normal position.

reuptake *(rē'-ŭp-tāk).* The reabsorption of a secreted substance by the cell that originally produced and secreted it. The term is often associated with a type of medication that blocks the reuptake of serotonin; *see* **selective serotonin reuptake inhibitors.**

revascularization *(rē-văs-kū-lăr-ĭ-zā'-shĕn).* Restoration of the blood supply to a body part, or after a wound.

reversion *(rē-vĕr'-shĕn).* Returning to a prior condition.

revulsion *(rē-vŭl'-shĕn).* Drawing blood from one part of the body to another, as occurs in counterirritation.

Reye's syndrome *(rīz).* A rare, sometimes fatal disease of children, characterized by recurrent vomiting and changes in the liver and other viscera. Other symptoms may follow, including swelling of the brain, disturbances of consciousness, and seizures.

rhabdomyosarcoma *(răb-dō-mī-ō-săr-kō'-mă).* A very malignant neoplasm that originates in skeletal muscle.

rheography *(rē-ŏg'-rĕ-fē).* A recording of fluid flow.

rheumatic fever *(roo-măt'-ĭk).* An inflammatory, febrile disease that varies in severity, duration, and symptoms. It frequently results in heart or kidney disease.

rheumatism *(roo'-mă-tĭzm).* A general term for any of various disorders characterized by inflammation, degeneration, or metabolic derangement of connective tissue, especially the joints, muscles, bursae, tendons, and fibrous tissue, and accompanied by pain, stiffness, or limited motion.

rheumatoid arthritis *(roo'-mă-toid).* Form of arthritis in which there is inflammation of the joints, stiffness, swelling, deterioration of cartilage, and pain.

rheumatologist *(roo-mă-tŏl'-ō-jĭst).* A physician who specializes in the diagnosis and treatment of rheumatic disorders.

rhexis *(rēk'-sĭs).* Rupture of an organ, blood vessel, or tissue.

Rh factor. An antigen found in the red blood cells of most people: those with Rh factor are Rh positive; those without it are Rh negative. Blood transfusions must match donors for Rh status as well as for ABO blood group (*see* **ABO blood group**).

rhinitis *(rī-nī'-tĭs).* Inflammation of the mucous membrane of the nose.

allergic r. A general term that describes any allergic reaction of the nasal mucosa, which may occur seasonally or perennially.

atrophic r. A chronic form of rhinitis that is nonallergic and noninfectious, and characterized by wasting of the mucous membrane and dry areas with crusting.

perennial r. Allergic rhinitis that may occur intermittently or continuously around the year. It is caused by an allergen to which an individual is exposed to much or all of the time, such as dust, mites, certain foods, or pet dander.

rhinolaryngitis *(rī-nō-lăr-ĭn-jī'-tĭst).* Inflammation of the mucous membrane of the nose and larynx.

rhinopharyngitis *(rī-nō-făr-ĭn-jī'-tĭs).* *See* **nasopharyngitis.**

rhinophyma *(rī-nō-fī'-mĕ).* A bulbous enlarged red nose and puffy cheeks that may be accompanied by thick bumps on the lower half of the nose and nearby cheek areas. It occurs mainly in men as a complication of rosacea.

rhinoplasty *(rī-nō-plăs'-tē).* Plastic surgery of the nose, performed for cosmetic, reconstructive, or restorative reasons.

rhinorrhea *(rī-nō-rē'-ĕ).* Thin, watery discharge from the nose.

rhizotomy *(rī-zŏt'-ĕ-mē).* Interruption of a cranial or spinal nerve root.

RhoGAM *(rō'-găm).* Trade name for a preparation of Rh0 immunoglobulin, an injectable blood product used to protect an Rh-positive fetus from antibodies from its Rh-negative mother. This product is now given routinely to Rh-negative women after they give birth to Rh-positive infants to prevent the mother's immune system from reacting negatively to the Rh-positive blood of any subsequent fetus.

rhonchus *(rŏng'-kĕs).* A continuous, low-pitched, snorelike sound produced in the throat or bronchial tube, caused by partial obstructions such as secretions.

Rhus *(rŭs).* A genus of vines and shrubs of the family Anacardiaceae, many of which are poisonous (e.g., poison ivy, poison oak). Extracts of the leaves and twigs are used in the prevention and treatment of dermatitis associated with these plants.

rhythm *(rĭth'-ĕm).* A measured movement, or the recurrence of an activity or function at regular intervals.

ectopic r. A heart rhythm initiated outside the sinoatrial node.

gallop r. A disordered rhythm of the heart that resembles the canter of a horse, as it produces three or four heart sounds.

sinus r. Normal heart rhythm originating in the sinoatrial node.

rhythm method. A contraceptive approach in which sexual partners do not have sexual intercourse on the days of a woman's menstrual cycle when she could become pregnant, or use a barrier method (e.g., condom, diaphragm) on those days.

rhytidectomy *(rĭ-ĭ-dĕk'-tē-mē).* Plastic surgery to eliminate wrinkles from the face by removing loose or redundant tissue. The common term for this procedure is *face-lift.*

rib *(rĭb).* Any one of the arched bones, twelve on either side of the sternum, that form the major part of the thoracic skeleton.

riboflavin *(rī'-bō-flā-vĭn).* Vitamin B2, a water-soluble substance that is essential for humans, in that it is involved in metabolism, growth, and body size.

ribonucleic acid (RNA) *(rī-bō-noo-klē'-ĭk).* A nucleic acid that controls protein synthesis in all living cells and which plays a role in the transfer of genetic information.

ricin *(rī'-sĭn).* A potent protein toxin made from the byproduct of castor bean oil (*Ricinus communis*) processing. When inhaled it can produce severe respiratory symptoms and respiratory failure within 36–72 hours. When ingested, it causes severe gastrointestinal symptoms followed by death.

rickets *(rĭk'-ĕts).* A condition seen in children in which there is an inadequate deposition of lime salts in cartilage and bone, resulting in abnormally formed bones. Symptoms include bone pain, growth retardation, convulsions, osteomalacia, and fatigability.

rickettsial diseases *(rĭ-kĕt'-sē-ăl).* Conditions caused by an organism of the genus *Rickettsia.* Some of the most common diseases include Rocky Mountain spotted fever, endemic typhus, Q fever, and trench fever, among others.

rigidity *(rĭ-jĭd'-ĭ-tē).* Stiffness; an inability to bend or to be bent.

rigor mortis *(rĭ'-gŏr mŏr'-tĭs).* Stiffness that occurs in a dead body.

ringworm *(rĭng'-wĕrm).* Common name for tinea (in humans) or dermatophytosis (in other animals), so named because it causes ring-shaped lesions. *See,* for example, **tinea pedis.**

risk *(rĭsk).* The probability of suffering injury or other unfavorable consequences.

attributable r. The amount or proportion of incidence of disease or death in people who are exposed to a specific risk factor that can be attributed to exposure to that factor.

empiric r. The probability that a trait will occur or recur in a family, based on experience alone rather than on knowledge of the causative factor.

relative r. The ratio of the incidence rate among people with a given risk factor for a disease, death, or other outcome, to the incidence rate among those without it.

risk factors. Conditions (e.g., physical, environmental, psychological, genetic, chemical, among others) that increase the chances that a disease will develop.

Risperdal *(rĭs'-pĕr-dăl). See* Appendix, Common Prescription and OTC Drugs: By Trade Name.

risperidone *(rĭs-pĕr'-ĭ-dōn). See* Appendix, Common Prescription and OTC Drugs: By Generic Name.

Ritalin *(rĭt'-ĕ-lĭn).* Trade name for preparations of methylphenidate hydrochloride, used to treat attention deficit hyperactivity disorder (ADHD) and other behavioral disorders.

RN. Acronym for registered nurse.

RNA. Acronym for ribonucleic acid.

Rocky Mountain spotted fever. An infectious disease caused by the parasite *Rickettsia ricketsii,* which is transmitted by a wood tick. Symptoms include fever, headache, muscle ache, and a rash.

roentgenography *(rĕnt-gĕn-ŏg'-ră-fē).* The process of obtaining images using roentgen rays. The more common term is *radiography.*

roentgen rays *(rĕnt'-gĕn).* X-rays.

rofecoxib *(rō-fĕ-cŏk'-sĭb). See* Appendix, Common Prescription and OTC Drugs: By Generic Name.

Rogaine *(rō'-gān).* Trade name for preparations of minoxidil, used to treat high blood pressure and hair loss, in both males and females.

rolfing *(rŏlf'-ĭng).* A system of body education and soft tissue manipulation designed to bring the whole body into vertical alignment; used by athletes, dancers, and singers to improve flexibility and breathing. Also known as *structural integration.*

root *(root).* 1. The lowermost part of a plant or other structure. 2. The proximal end of a nerve. 3. The part of an organ that is implanted in tissues.

root canal. 1. The space in the root of the tooth that contains nerves. 2. Term used to describe the dental procedure in which the nerves are removed from the canal of a tooth.

Rorschach test *(rŏr'-shăhk).* A psychological test in which individuals are shown inkblots and asked to tell what they perceive in each one. The responses are used to determine what people read into the highly ambiguous shapes.

rosacea *(rō-zā'-shē-ĕ).* A chronic skin disease that usually involves the middle third of the face, characterized by persistent redness with acute episodes of edema, papules, and pustules.

rose hips. The fruit of the rose plant that is berrylike in structure. Nutritional supplements are made from rose hips and are high in vitamin C, as well as a good source of vitamins K and E.

rosiglitazone *(rō-sĭg-lĭt'-ĕ-zōn). See* Appendix, Common Prescription and OTC Drugs: By Generic Name.

rotator cuff *(rō'-tā-tŏr).* A group of four tendons that stabilize the shoulder joint. Each tendon is associated with a muscle that moves the shoulder in a specific direction.

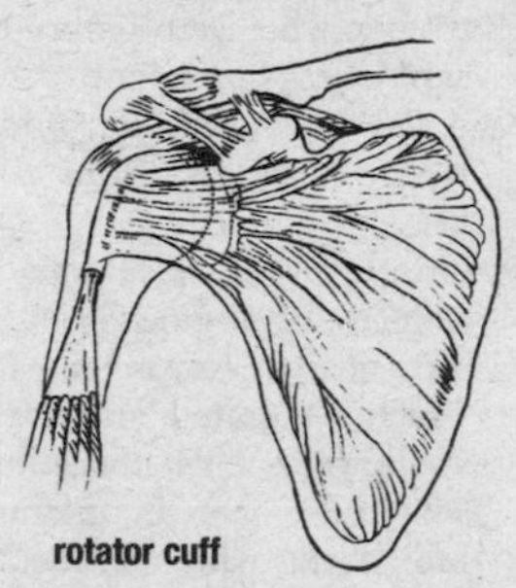

rotator cuff

rotavirus *(rō'-tă-vī-rŭs).* A group of viruses that are a major cause of acute infantile gastroenteritis and diarrhea in young children.

roughage *(rŭf'-ĕj).* Indigestible material such as fibers, cellulose, among others, in the diet.

rubefacient *(roo-bĕ-fā'-shĕnt).* Reddening of the skin, or an agent that reddens the skin.

rubella *(roo-bĕl'-ă).* An acute, usually benign, infectious disease caused by a togavirus. Characterized by a sore throat, slight cold, fever, enlargement of specific lymph nodes, and a pink rash that starts on the head and spreads throughout the body.

rubeola *(roo-bē'-ō-lĕ).* Measles.

rubor *(roo'-bŏr).* Redness, one of the primary signs of inflammation.

rudimentary *(roo-dĭ-mĕn'-tĕ-rē).* Imperfectly developed.

RU-486. Known as the French abortion pill or mifepristone, it is 95% effective at ending pregnancy. It can be used by women who are no more than seven weeks pregnant.

ruga *(roo'-gă).* A wrinkle, fold, or ridge, as of a mucous membrane.

rumination *(roo-mĭ-nā'-shĕn).* In humans, regurgitation of food after nearly every meal, part of which is vomited, and the rest swallowed. This condition is sometimes seen in infants or individuals who are mentally retarded.

rupture *(rŭp'-chĕr).* The forcible tearing or disruption of tissue.

rutin *(roo'-tĭn).* A bioflavonoid that is obtained from buckwheat and other plants.

S

Sabin vaccine *(sā'-bĭn).* A live, oral vaccine for the prevention of infantile paralysis.

sac *(săk).* A bag or pouch.

amniotic s. The sac formed by the amnion and commonly known as the *bag of waters.*

serous s. The sac composed of the pleura, pericardium, and peritoneum.

saccharin *(săk'-ĕ-rĭn).* A white crystalline compound that is several hundred times sweeter than sucrose and used as a sweetener in drug preparations.

sacculus *(săk'ū-lĕs).* A small sac or pouch.

sacralization *(să-krăl-ĭ-zā'-shŭn).* Fusion of the sacrum and the fifth lumbar vertebra.

sacrococcygeal *(să-krō-kŏk-sĭj'-ē-ăl).* Referring to the sacrum and coccyx.

sacroiliac *(să-krō-ĭl'-ē-ăk).* Referring to the sacrum and ilium or to the joint between them.

sacroiliitis *(să-krō-ĭl-ē-ī'-tĭs).* Inflammation in the sacroiliac joint.

sacrolumbar *(să-krō-lŭm'-bĕr).* Referring to the sacrum and the loin.

sacrospinal *(să-krō-spī'-nĕl).* Referring to the sacrum and the spine.

sacrum *(să'-krĕm).* The triangular bone located just below the lumbar vertebrae. It forms the base of the vertebral column.

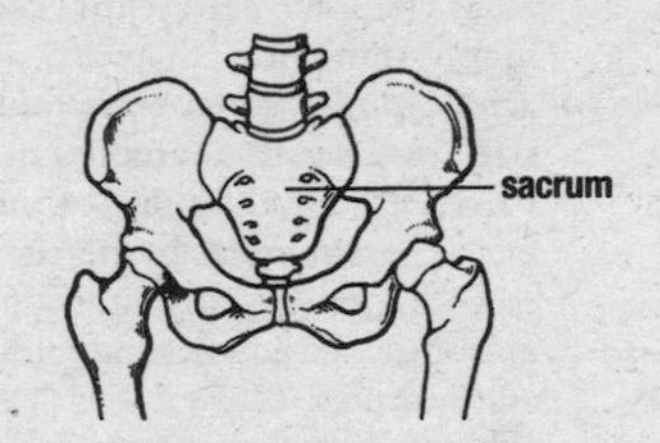

SAD. Acronym for seasonal affective disorder.

sadism *(sā'-dĭz-ĕm).* Deriving sexual pleasure from inflicting physical or psychological pain on another person.

safe sex. Protecting oneself from AIDS and other sexually transmitted diseases.

sagittal plane *(săg'-ĭ-tĕl).* A vertical plane through the longitudinal axis of the trunk dividing the body into two parts.

St. Anthony's fire. An infection caused by ergot, a consequence of the fungus *Clavideps purpurea* and characterized by an intensely painful burning sensation in the limbs and extremities.

St. John's wort. Any of several species of the genus *Hypericum. H. perforatum* specifically is used internally as a mild antidepressant and sedative and topically for skin inflammation, first-degree burns, and contusions.

St. Vitus' dance. Disease of the nervous system characterized by involuntary jerking motions.

salicylate drugs *(săl-ĭ-sĭl'-āt).* Any of a group of related compounds derived from salicylic acid, such as acetylsalicylic acid (aspirin), and used to treat pain, fever, and inflammation.

salicylic acid *(săl-ĭ-sĭl'-ĭk).* A substance obtained from the bark of the white willow and wintergreen leaves or prepared synthetically, to treat a variety of skin conditions such as acne, plantar warts, corns, and seborrheic dermatitis.

saline solution *(sā'-lēn).* A solution composed of sodium chloride and distilled water, often used as a fluid replacement to prevent or treat dehydration.

saliva *(sĕ-lī'-vă).* The clear, alkaline, slightly sticky secretion from the salivary glands and small mucous glands of the mouth. Saliva keeps the mouth moist, softens food, and contains a digestive enzyme that converts starch into maltose.

salivary glands *(săl'-ĭ-văr-ē).* Glands of the oral cavity that secrete saliva. There are three main glands—parotid, submandibular, and sublingual—and several minor glands.

salivation *(săl-ĭ-vā'-shŭn).* The secretion of saliva.

Salmeterol *(săl-mĕt'-ĕr-ōl). See* Appendix, Common Prescription and OTC Drugs: By Generic Name.

Salmonella *(săl-mō-nĕl'-ă).* A genus of bacteria that belongs to the family Enterobacteriaceae.

S. choleraesuis. Species that can cause septicemia.

S. enteritidis. Species that causes gastroenteritis and food poisoning.

S. typhimurium. Species isolated from individuals having acute gastroenteritis.

S. typhi. Species that causes typhoid fever.

salmonellosis *(săl-mō-nĕ-lō'-sĭs).* Infestation with *Salmonella* bacteria. Three forms of salmonellosis are typhoid fever, septicemia, and acute gastroenteritis.

salpingectomy *(săl-pĭn-jĕk'-tō-mē).* Surgical removal of the fallopian tube.

salpingitis *(săl-pĭn-jī'-tĭs).* Inflammation of the fallopian tube.

salpingocele *(săl-pĭng'-gō-sēl).* Hernia of a fallopian tube.

salpingocyesis *(săl-pĭng-ō-sī-ē'-sĭs).* Pregnancy in which the fetus begins to develop in a fallopian tube; also called a *tubal pregnancy.*

salpingography *(săl-pĭng-gŏg'-ră-fē).* X-ray image of the fallopian tubes into which a radiopaque dye is introduced.

salpingo-oophorectomy *(săl-pĭng-gō-ō-ŏf-ō-rĕk'-tō-mē).* Surgical removal of an ovary and fallopian tube.

salpingo-oophoritis *(săl-pĭng-ō-ō-ŏf-ō-rī'-tĭs).* Inflammation of a fallopian tube and ovary.

salpingoplasty *(săl-pĭng'-ō-plăs-tē).* Plastic surgery of the uterine tube.

salpinx *(săl'-pĭnks).* A tube.

salt *(săwlt).* Sodium chloride, or common table salt.

saltpeter *(săwlt-pē'-tĕr).* The common name for potassium nitrate.

sample *(săm'-pĕl).* A representative portion of a whole that demonstrates the characteristics of the whole.

sanguine *(săng'-gwĭn).* Having excessive blood.

sanitary *(săn'-ĭ-tăr-ē).* Referring to health or promoting health, especially pertaining to a clean environment.

sanity *(săn'-ĭ-tē).* Soundness of mind.

San Joaquin Valley fever *(săn wăh-kēn).* Primary coccidioidomycosis; so named because San Joaquin Valley is where the disease is especially prominent.

Sansert *(săn'-sĕrt).* Trade name for a preparation of methysergide maleate, used to treat migraine.

saphenous veins *(sĕ-fē'-nĕs).* Two major veins in the leg: the small and the great saphenous veins, the latter of which is the longest vein in the body.

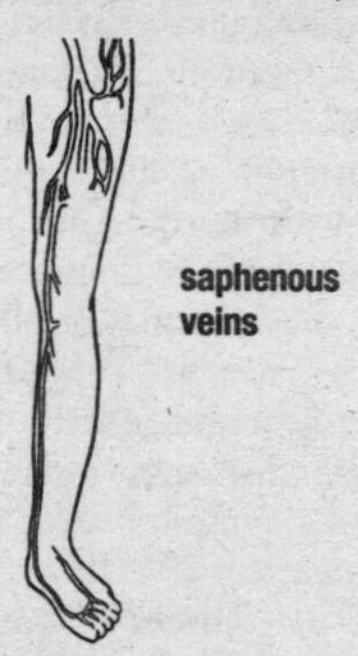

saponification *(sĕ-pŏn-ĭ-fĭ-kā'-shĕn).* Conversion into soap. Chemically, it is the splitting of fat by an alkali, resulting in glycerol and alkali salt of the fatty acid, the soap.

saprophyte *(săp'-rō-fīt).* An organism that lives on dead or decaying material.

sarcoidosis *(săhr-koi-dō'-sĭs).* A chronic, progressive, granulomatous reticulosis, characterized by hard tubercles that can appear in nearly any tissue or organ; hypercalcemia and hypergammaglobulinemia may also be present.

sarcolemma *(săhr-kō-lĕm'-ĕ).* The plasma membrane that covers every striated muscle fiber.

sarcoma *(săhr-kō-mĕ).* Any one of a group of tumors, usually malignant, that typically arise from connective tissue, but also from epithelial tissue.

embryonal s. Wilms' tumor.

Ewing's s. A highly malignant,

metastatic, primitive tumor of bone, usually occurring in long bones, ribs, and flat bones of children and adolescents.

giant cell s. A form of giant cell tumor of bone that develops malignant rather than transforming to malignancy.

sartorius muscle *(săr-tōr'-ē-ŭs).* A long muscle of the thigh that moves the thigh and helps flex the knee. It is the longest muscle in the body: it stretches from the calf to the pelvis.

satiety *(să-tī'-ĕ-tē).* Full gratification regarding appetite or thirst, with no desire to eat or drink.

saturated *(săt'-ū-rā-tĕd).* 1. In chemistry, having all the chemical bonds filled. 2. Holding all that can be absorbed, received, or combined.

saturated fat. A fat that is solid at room temperature. Most saturated fats come from animal products, such as butter, meat fat, and whole milk. These fats tend to raise cholesterol levels in the blood.

scabies *(skā'-bēs).* A contagious skin condition caused by the mite *Sarcoptes scabiei,* characterized by papules accompanied by intense itching frequently associated with eczema and bacterial infection.

scalp *(skălp).* The skin of the head, exclusive of the face and ears.

scalpel *(skăl'-pĕl).* A small surgical knife with a convex edge and thin blade.

scan *(skăn).* The image or data obtained from the examination of areas of the body by gathering information with a sensing device. For specific scans, *see* individual listings.

scaphoid *(skăf'-oyd).* Small bones in the wrist and ankle that are boat-shaped or hollowed.

scapula *(skăp'ū-lĕ).* The flat, triangular bone at the back of the shoulder; often referred to as the shoulder blade or "wingbone."

scar *(skăr).* A mark that remains after a wound or other injurious process heals.

scarification *(skăr-ĭ-fĭ-kā'-shĕn).* Making many superficial scratches or punctures on the skin, as for introducing the smallpox vaccine.

scarlet fever *(skăr'-lĕt).* An acute contagious disese characterized by sore throat, fever, scarlet rash, rapid pulse, and strawberry tongue.

scatology *(skă-tŏl'-ō-jē).* The scientific study of feces.

scatoma *(skă-tō'-mă). See* **stercoroma.**

Schick test *(shĭks).* A test for immunity against diphtheria.

Schilder's disease *(shĭl'-dĕrz).* A rare, fatal disease of the central nervous system characterized by adrenal atrophy and inflammation of the brain and the nerves leading from the brain.

Schimmelbusch's disease *(shĭm-ĕl-boosh'-ĕz).* Cystic disease of the breast.

schistocyte *(shĭs-tō-sīt).* A fragment of a red blood cell, often seen in the blood of hemolytic anemias.

schistosomiasis *(shĭs-tō-sō-mī'-ĕ-sĭs).* Infection with flukes of the genus *Schistosoma.* Symptoms depend on which species is causing the infection, but the areas most often affected are the intestinal tract, urinary tract, liver, and lungs.

schizoid *(skĭz'-oid).* 1. Condition characterized by extreme shyness, social withdrawal, introversion, and sensitivity. 2. A term loosely used to refer to traits associated with schizophrenia.

schizophrenia *(skĭz-ō-frĕ'-nē-ĕ).* A mental disorder or group of disorders comprising most major psychotic disorders. Characteristics include disturbances in thought, mood, sense of self, relationship to the world, and behavior.

ambulatory s. A mild form in which the patient can maintain him/herself in the community without hospitalization.

childhood s. Schizophrenia-like symptoms that first appear before puberty, characterized by autistic withdrawn behavior, failure to develop an identity separate from the mother's, and severe developmental immaturity.

paranoid s. A type of schizophrenia characterized by a preoccupation with one or more delusions or with frequent auditory hallucinations but without disorganized speech or disorganized or catatonic behavior.

Schwann's cell *(shwăhz).* A large nucleated cell whose membrane wraps around the axons of myelinated peripheral neurons and is the source of myelin.

schwannoma *(shwăh-nō'-mĕ).* A neoplasm that originates from Schwann cells of neurons.

sciatica *(sī-ăt'-ĭ-kĕ).* A syndrome characterized by pain that radiates from the back into the buttock and into the lower leg along the back or sides of the leg.

sciatic nerve. The largest nerve in the body, originating from the sacral plexus and going down the back of the thigh.

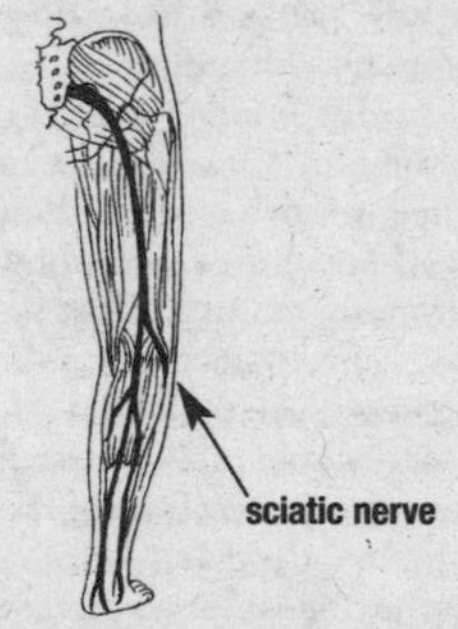

scintigraphy *(sĭn-tĭg'-rĕ-fē).* The production of two-dimensional images of the distribution of radioactivity in tissues after a radionuclide has been administered internally.

scintillation *(sĭn-tĭ-lā'-shŭn).* A visual sensation, as of seeing sparks.

scirrhoid *(skĭr'-oyd).* Referring to or similar to a hard carcinoma.

scirrhus *(skĭr'-ŭs).* A hard, cancerous tumor that arises from an overgrowth of fibrous tissue.

scission *(sĭzh'ŭn).* Dividing, splitting, or cutting.

sclera *(sklĕr'-ă).* A tough white fibrous tissue that covers the white of the eye.

scleral buckling *(sklĕr'-ĕl).* A surgical procedure in which tucks are placed in the sclera to shorten the eyeball.

sclerectoiridectomy *(sklĕ-rĕk-tō-ĭr-ĭ-dĕk'-tē-mē).* A surgical procedure in which a portion of the sclera and of the iris are removed to provide relief from glaucoma.

scleritis *(sklĕ-rī'-tĭs).* Inflammation of the sclera.

scleroderma *(sklĕr-ă-dĕr'-mă).* A chronic disease that causes hardening and thickening of the skin that can be localized (confined to the skin and subcutaneous tissue) or systemic (affecting the skin and internal organs).

sclerokeratitis *(sklĕ-rō-kĕr-ă-tī'-tĭs).* Inflammation of the sclera and of the cornea.

scleronychia *(sklĕ-rō-nĭk'-ē-ă).* Thickening and hardening of the nails.

sclerosing *(sklĕ-rōs'-ĭng).* Causing or undergoing sclerosis.

sclerosis *(sklĕ-rō'-sĭs).* Hardening of a part due to inflammation, increased development of connective tissue, or disease.

amyotrophic lateral s. A motor neuron disease characterized by progressive deterioration of neurons that give rise to motor cells of the brain and spinal cord. Also called *Lou Gehrig disease.*

multiple s. A chronic, slowly progressive disease of the central nervous system characterized by loss or destruction of the myelin covering of nerves and characterized by numerous symptoms, including unusual sensations, weakness, incoordination, visual disturbances, and tremor.

sclerotherapy *(sklĕr-ō-thĕr'-ĕ-pē).* Injection of a chemical irritant into a vein to produce inflammation and fibrosis, done to treat hemorrhoids, varicose veins, and esophageal varices.

sclerotic *(sklĕ-rŏ'-tĭk).* Hard or hardening.

sclerotomy *(sklĕ-rŏt'-ĕ-mē).* Surgical incision of the sclera.

scoliokyphosis *(skō-lē-ō-kī-fō'-sĭs).* Combined scoliosis and kyphosis curvature of the spine.

scoliosis *(skō-lē-ō'-sĭs).* Lateral curvature of the spine.

acquired s. Scoliosis that was not present at birth nor the result of a condition present at birth.

congenital s. Scoliosis present at birth or due to a condition present at birth.

scombroid poisoning *(skŏm'-broid).* Intoxication caused by eating raw or inadequately cooked fish belonging to the suborder *Scombroides*, which includes tuna, mackerel, albacore, and skipjack.

scopolamine *(skō-pŏl'-ă-mēn).* An anticholinergic alkaloid derived from plants in the nightshade family and used as a sedative.

scopophilia *(skō-pō-fĭl'-ē-ă).* Deriving sexual stimulation from looking at nudity and obscene pictures.

scopophobia *(skō-pō-fō'-bē-ă).* An abnormal fear of being seen.

scotoma *(skō-tō'-mă).* An area of lost or reduced vision within the visual field, surrounded by an area of normal or less suppressed vision.

screening *(skrēn'-ĭng).* Examination or testing of individuals to identify those who are well from those who have an ailment or undiagnosed disease, or who are at high risk.

scrofuloderma *(skrŏf-ū-lō-dĕr'-mĕ).* A tuberculous or nontuberculous bacterial infection that affects children and young adults in which the infection extends into the skin and is characterized by painless subcutaneous swellings that develop into abscesses, ulcers, and draining sinus tracts.

scrotum *(skrō'-tŭm).* The pouch that contains the testes and accessory organs.

scrub nurse *(skrŭb).* An operating room nurse who hands instruments to the surgeons and who has previously scrubbed his or her hands and has donned sterile surgical garb before entering the operating room.

scurvy *(skŭr'-vē).* A condition caused by a deficiency of vitamin C (ascorbic acid) and characterized by weakness, anemia, spongy gums, hard muscles in the lower limbs, and a tendency to experience mucocutaneous bleeding.

scybalum *(sĭb'-ĕ-lĕm).* A dry, hard mass of fecal matter in the intestinal tract.

seasickness *(sē'-sĭk-nĕs).* A disorder caused by the motion of a boat or by riding in a car, bus, train, or elevator, or can refer to a similar condition experienced by air travelers. Also known as *motion sickness.*

seasonal affective disorder *(sē'-sŏn-ĕl).* A type of depression that tends to occur as the amount of daylight decreases in fall and winter. Symptoms include excessive eating, excessive sleeping, and weight gain during fall and winter months and complete remission during spring and summer.

sebaceous *(sĕ-bā'-shĕs).* Referring to sebum.

seborrhea *(sĕ-bŏr-ē'-ă).* A chronic inflammatory disease of the skin characterized by excessive secretion of sebum and the accumulation of scales and frequently crusted, itchy patches. It most often affects the scalp, face, ears, eyebrows, eyelids, and genitalia.

seborrheic dermatitis. *See* **dermatitis.**

sebum *(sē'-bŭm).* The thick, semifluid secretion of the sebaceous glands.

Seconal *(sĕk'-ō-năhl).* Trade name for preparations of secobarbital sodium, a short-acting sedative often used presurgically along with anesthesia.

secondhand smoke *(sĕk'-ŏnd-*

hănd). Environmental tobacco smoke that is inhaled passively or involuntarily by people who are not smoking. Such smoke is produced from the burning of cigarettes, cigars, and pipes, and from exhaled smoke, and was classified by the US government as a known human carcinogen in 2000.

secretagogue *(sē-krēt'-ĕ-gŏg).* A substance that stimulates secretion.

secretary's knee *(sĕk'-rĕ-tăr-ēz).* The most common cause of chronic knee pain, also known as patellofemoral syndrome. The syndrome is characterized by vague discomfort of the inner knee, which is aggravated by activity or prolonged sitting.

secrete *(sē-krēt').* To separate from the blood, a living organism, or a gland.

secretin *(sē-krē'-tĭn).* A hormone secreted by the mucosa of the duodenum and upper jejunum; it stimulates the release of pancreatic juice by the pancreas and bile by the liver.

secretion *(sē-krē'-shĕn).* The process whereby cells of glands produce certain substances from the blood. Also refers to the substance that is produced by glands.

section *(sĕk'-shĕn).* 1. An act of cutting. 2. A segment of an organ.

secundines *(sē-kŭn'-dīnz).* Afterbirth.

sedative *(sĕd'-ă-tĭv).* A substance that exerts a soothing or calming effect.

sediment *(sĕd'-ĭ-mĕnt).* The substance that settles at the bottom of a liquid.

segment *(sĕg'-mĕnt).* A portion of a larger structure or body.

segmentation *(sĕg-mĕn-tā'-shĕn).* Division into parts that are similar.

seizure *(sē'-zhŭr).* 1. A sudden recurrence or attack of a disease. 2. A single episode of epilepsy.
absence s. A seizure involving a sudden momentary break in consciousness of activity or thought, usually accompanied by clonic movements.
atonic s. An absence seizure characterized by a sudden loss of muscle tone.
complex partial s. A type of partial seizure associated with a temporal lobe disease and characterized by impaired consciousness and automatisms, followed by amnesia for them.
generalized tonic-clonic s. Seizure of grand mal epilepsy, consisting of a loss of consciousness and generalized tonic convulsions followed by clonic convulsions.
myoclonic s. Seizure characterized by a brief episode of myoclonus with immediate recovery and usually without loss of consciousness.
tonic s. Seizure characterized by tonic (persistent, involuntary contractions) but not clonic convulsions.

selective estrogen-receptor modulator (SERM) *(sē-lĕk'-tĭv).* A man-made estrogen that possesses many but not all of the actions of estrogen. Generally, a SERM helps prevent bone loss and reduces serum cholesterol levels, but does not stimulate the endometrial lining of the uterus.

selegiline hydrochloride *(sĕ-lĕj'-ĕ-lēn).* An inhibitor of monoamine oxidase type B used along with levodopa and carbidopa to treat Parkinson's disease.

self-hypnosis *(sĕlf-hīp-nō'-sĭs).* The act or process of hypnotizing oneself.

self-limited *(sĕlf-lĭm'-ĭ-tĕd).* Limited by its own nature and not by outside influences. Often used to refer to a disease that runs a definite limited course.

semen *(sē'-mĕn).* The thick, whitish secretion of the reproductive organs of males, consisting of spermatozoa in plasma, secretions from the prostate, seminal vesicles, and other glands, epithelial cells, and other substances.

semicircular canals *(sĕm-ē-sŭr'-kū-lăr).* Passages that form part of the inner ear.

semicomatose *(sĕm-ē-kō'-mĕ-tōs).* A stupor from which an individual can be aroused.

semilunar *(sĕm-ē-loo'-nĕr).* Resembling a half-moon.

seminal *(sĕm'-ĭ-năl).* Referring to semen or seed.

seminal vesicles. The two saclike structures in the male that lie behind the bladder near the prostate and are connected to the ductus deferens on each side. They secrete a thick fluid that is part of the semen.

semination *(sĕm-ĭ-nā'-shŭn).* The introduction of semen into the female genital tract, either through sexual intercourse or artificially.

seminoma *(sĕm-ĭ-nō'-mă).* A tumor of the testis.

seminuria *(sĕ-mĭn-ū'-rē-ă).* The presence of semen in the urine.

semirecumbent position *(sĕm-ē-rē-kŭm'-bĕnt).* Reclining but not completely recumbent.

semitendinosus muscle *(sĕm-ē-tĕn-dĭn-ō'-sĭs).* A spindle-shaped muscle of the posterior and inner part of the thigh.

senescence *(sē-nĕs'-ĕns).* The process of growing old.

senile *(sē'-nīl).* Referring to or characteristic of old age.

senile psychosis. *See* **psychosis.**

senility *(sĕ-nĭl'-ĭ-tē).* The physical and mental deterioration associated with old age.

senna *(sĕn'-ă).* Any plant of the genus *Cassia*; the dried leaves of *C. acutifolia* or *C. augustifolia* are used as a cathartic.

senopia *(sĕ-nō'-pē-ĕ).* A decrease in presbyopia in the elderly.

sensation *(sĕn-sā'-shŭn).* An impression or awareness of conditions within or outside the body resulting from stimulation of sensory receptors.

sense *(sĕns).* To perceive through a sense organ.

sensibility *(sĕn-sĭ-bĭl'-ĭ-tē).* The capacity to receive and respond to stimulation.

sensitivity *(sĕn-sĭ-tĭv'-ĭ-tē).* A term used when assessing the value of a diagnostic test, clinical observation, or procedure. It is determined by the number of true positive results divided by the total number of people with the disease.

sensorium *(sĕn-sŏr'-ē-ŭm).* All the parts of the brain as a whole that receive, process, and interpret sensory stimuli.

sensory deprivation *(sĕn'-sō-rē).* Forced elimination or absence of usual or accustomed sensory stimulation. Prolonged sensory deprivation can result in severe mental changes, including auditory and visual hallucinations, depression, and anxiety.

sensory integration. The skill and performance needed to develop and coordinate sensory input, motor output, and sensory feedback, which includes, in part, sensory awareness, balance, motor coordination, body integration, and other components.

sentient *(sĕn'-shē-ĕnt).* Capable of feeling or perceiving sensation.

sepsis *(sĕp'-sĭs).* A disease state that is usually characterized by fever and which results from the presence of microorganisms or their toxic products in the blood.

septal *(sĕp'-tăl).* Concerning a septum.

septectomy *(sĕp-tĕk'-tō-mē).* Excision of a septum, especially the nasal septum.

septic *(sĕp'-tĭk).* Produced or caused by decomposition by microorganisms.

septicemia *(sĕp-tĭ-sē'-mē-ă).* Presence of disease-causing bacteria in the blood. Symptoms and signs usually include chills, fever, abscesses, and pustules.

septoplasty *(sĕp'-tō-plăs-tē).* Surgical reconstruction of the nasal septum.

septotomy *(sĕp-tŏt'-ĕ-mē).* Incision of the nasal septum.

septum *(sĕp'-tŭm).* Generally, a dividing wall or partition. Commonly used to refer to the nasal septum.

sequelae *(sĕ-kwĕl'-ĕ).* Any lesion or condition that follows or is a result of a disease.

sequestration *(sē-kwĕs-trā'-shĕn).* 1. The formation of a sequestrum. 2. Isolation of a patient for quarantine or treatment.

sequestrum *(sē-kwĕs'-trŭm).* Any tissue that becomes sequestered. 2. A piece of dead bone that separates from sound bone during necrosis.

Serevent *(sĕr'-ĕ-vĕnt).* Trade name for a preparation of salmeterol xinafoate.

serioscopy *(sē-rē-ŏs'-kĕ-pē).* X-ray visualization of the body in a series of parallel planes using multiple exposures.

seroconversion *(sĕr-ō-kŏn-vĕr'-shŭn).* The development of detectable antibodies in the blood that target an infectious agent.

serofibrinous exudates *(sĕr-ō-fī'-brĭn-ĕs).* Composed of both serum and fibrin.

seronegative *(sĕr-ō-nĕg'-ă-tĭv).* Showing negative results on serological examination; showing a lack of antibody.

seropositive *(sĕr-ō-pŏs'-ĭ-tĭv).* Showing positive results on serological examinations; showing a high level of antibody.

seroma *(sēr-ō'-mă).* A tumorlike accumulation of serum in the tissues.

serosa *(sēr-ō'-să).* The membrane that lines the exterior walls of various body cavities.

serosanguineous *(sĕr-ō-săng-gwĭn'-ē-ĕs).* Referring to or containing both blood and serum.

serositis *(sĕr-ō-sī'-tĭs).* Inflammation of a serous membrane.

serotonin *(sĕr-ō-tō'-nĭn).* A chemical, 5-hydroxytryptamine (5-HT), present in many body tissues, including the intestinal mucosa, pineal body, and the central nervous system. It plays many roles, including neurotransmitter, stimulator of smooth muscle, and precursor of melatonin.

serous *(sēr'-ĕs).* Referring to or resembling serum.

serpiginous *(sĕr-pĭj'-ĭ-nĕs).* Having a very uneven or wavy margin.

serration *(sĕ-rā'-shĕn).* A formation or structure that has teeth like those of a saw.

sertaline *(sĕr'-tră-lēn).* *See* Appendix, Common Prescription and OTC Drugs: By Generic Name.

serum *(sēr'-ŭm).* The clear portion of any body fluid, such as blood or lymph.

serum sickness. A hypersensitivity reaction that may occur several days or weeks after receiving antisera or following certain drug treatments. Symptoms include fever, painful joints, swollen lymph nodes, and rash.

sesamoid *(sĕs'-ă-moyd).* Referring to a small nodular bone embedded in a joint capsule or tendon.

sessile *(sĕs'-ĭl).* Attached by a base, not by a stalk.

sex *(sĕks).* The characteristics that differentiate males and females among most animals and plants, based on the type of gametes produced by the individuals.

sex chromosomes. Chromosomes associated with determination of sex.

sex-linked disorder. A medical condition that is controlled by genes on the sex-linked chromosomes.

sexology *(sĕks-ŏl'-ō-jē).* The scientific study of sexuality.

sexual dysfunction. Inadequate enjoyment or a failure to enjoy sexual activity, which may be caused by physical and/or psychological factors.

sexually transmitted disease (STD). A disease acquired as the result of sexual intercourse or oral sexual activity with an infected individual.

shaken baby syndrome *(shā'-kĕn).* Injuries caused by violently shaking an infant, with those injuries especially affecting the head. This syndrome is the most common cause of infant death from head injuries.

shaking palsy. A disease that affects the basal ganglion in the brain and is characterized by progressive rigid tremors, unusual gait, muscular contractions, and weakness.

shank *(shăngk)*. The tibia or the region of the leg that extends from the knee to the ankle. Also known as the *shin.*

sheath *(shēth)*. Something that covers or encloses an organ or body part.

shiatsu *(shē-ăt'-sū)*. A form of energy therapy based on traditional Chinese medicine. Practitioners use their fingers, elbows, knees, and feet to apply pressure to specific points and gently stretch the limbs to release any blocked energy.

Shigella *(shĭ-gĕl'-lă)*. A genus of gram-negative bacteria that belongs to the family *Enterobacteriaceae* and contains species that cause digestive disorders.

S. boydii. A species that causes acute diarrhea in humans.

S. dysenteriae. A species that causes severe dysentery.

S. flexneri. A species that causes severe dysentery.

S. sonnei. A species that causes a mild but common form of bacillary dysentery.

shigellosis *(shĭ-gĕl-ō'-sĭs)*. Any condition produced by an infection with organisms of the genus *Shigella.*

shin *(shĭn)*. The front portion of the tibia.

shingles *(shĭng'-gĕlz)*. Appearance of inflammatory, herpetic vesicles on the torso along a peripheral nerve and occasionally elsewhere on the body. Also known as *herpes zoster.*

shin splints. Pain in the front region of the tibia, usually following vigorous exercise and caused by ischemia of the muscles and tiny tears in the tissues.

shock *(shŏk)*. A syndrome in which the peripheral blood flow is insufficient to return adequate blood to the heart and to other organs to ensure their normal function.

anaphylactic s. Shock due to an allergic reaction.

endotoxin s. Septic shock caused by the release of endotoxins by gram-negative bacteria.

insulin s. A hypoglycemic reaction to an overdose of insulin, strenuous exercise, or lack of food in an insulin-dependent diabetic. Initial symptoms include tremor, dizziness, cool moist skin, hunger, and rapid heartbeat.

septic s. Shock associated with an overwhelming infection, usually caused by gram-negative bacteria. Initial symptoms include chills, fever, increased cardiac output, warm flushed skin, and mild hypotension.

short-term memory. *See* **memory.**

shoulder *(shōl'-dĕr)*. The junction of the upper limb and the trunk.

shunt *(shŭnt)*. 1. To divert or bypass something. 2. A passageway between two natural channels, especially between blood vessels.

sialadenitis *(sī-ĕl-ăd-ĕ-nī'-tĭs).* Inflammation of a salivary gland.

sialadenoma *(sī-ĕl-ăd-ĕ-nō'-mĕ).* A benign tumor of the salivary glands.

sialagogue *(sī-ăl'-ĕ-gŏg).* A substance that promotes the flow of saliva.

sialoadenectomy *(sī-ĕ-lō-ăd-ĕ-nŏt'-ĕ-mē).* Incision and drainage of a salivary gland.

sialography *(sī-ĕ-lŏg'-rĕ-fē).* X-ray that demonstrates the salivary ducts with the help of injected opaque substances.

sialolithotomy *(sī-ĕ-lō-lĭ-thŏt'-ĕ-mē).* Incision of a salivary gland or duct to remove a calculus.

sialosemelology *(sī-ĕ-lō-sĕ-mī-ŏl'-ĕ-jē).* Analysis of the saliva to determine the physiologic status of the patient, especially concerning metabolism.

sibling *(sĭb'-lĭng).* Any of two or more offspring of the same parents.

sick building syndrome *(sĭk).* A condition caused by exposure to noxious substances that have a negative impact on individuals who work in a building that houses many people who work in close proximity to each other. Symptoms of the syndrome can include anxiety, hyperventilation, cramps, muscle twitches, and severe breathlessness.

sickle cell anemia *(sĭk'-ĕl).* *See* **anemia.**

sickle cell crisis. An episode characterized by fever and severe pain in the joints and the abdomen and occurring when the sickled cells hinder the transport of oxygen and obstruct blood flow in the capillaries.

sick sinus syndrome. Failure of the sinus node to regulate the heartbeat properly, resulting in either a fast or slow heartbeat.

side effect *(sīdĕ-fĕct).* The effect or action of a drug other than that desired. Usually side effects are negative, such as nausea, insomnia, headache, and vomiting.

siderinuria *(sĭd-ĕr-ĭ-nū'-rē-ĕ).* Excretion of iron in the urine.

sideropenia *(sĭd-ĕr-ō-pē'-nē-ĕ).* Iron deficiency.

siderosis *(sĭd-ĕr-ō'-sĭs).* A benign type of pneumoconiosis caused by the inhalation of iron particles.

SIDS. Acronym for sudden infant death syndrome.

sigmoid *(sĭg'-moid).* The lower colon; short for "sigmoid colon."

sigmoidectomy *(sĭg-moi-dĕk'-tĕ-mē).* Removal of the sigmoid colon.

sigmoiditis *(sĭg-moid-ī'-tĭs).* Inflammation of the sigmoid colon.

sigmoidoscope *(sĭg-moi'-dō-skōp).* A flexible or rigid endoscope with a light for examining the sigmoid colon.

sigmoidostomy *(sĭg-moi-dŏs'-tĕ-mē).* Surgical formation of an artificial opening from the body surface into the sigmoid colon.

sign *(sīn).* Any objective evidence of a disease. Such evidence is observable as opposed to a subjective sensation (*see also* **symptom).**

sildenafil *(sĭl-dĕn'-ĕ-fĭl). See* Appendix, Common Prescription and OTC Drugs: By Generic Name.

silent ischemia. *See* ischemia.

silicosis *(sĭl-ĭ-kō'-sĭs).* Pneumoconiosis caused by inhalation of the dust of sand, stone, or flint that contains silica, resulting in the formation of nodules in the lungs.

silver nitrate *(sĭl'vĕr).* A powerful germicide used as an antiseptic and astringent for infections of the skin and mucous membranes.

simethicone *(sī-mĕth'-ĭ-kōn).* A medication, this is composed of dimethicones and silicon dioxide and is used as an antiflatulent.

simulation *(sĭm-ū-lā'-shĕn).* The act of counterfeiting a disease.

simvastatin *(sĭm'-vĕ-stăt-ĭn). See* Appendix, Common Prescription and OTC Drugs: By Generic Name.

sinew *(sĭn'-ū).* The tendon of a muscle.

Singulair *(sĭng'-ū-lār). See* Appendix, Common Prescription and OTC Drugs: By Trade Name.

sinistrality *(sĭn-ĭs-trăl'-ĭ-tē).* Development of the right side of the brain to dominate over the left side, as seen in left-handed individuals.

sinoatrial *(sī-nō-ā'-trē-ĕl).* Referring to the sinus venosus and the atrium of the heart.

sinoatrial node. One of the main elements in the cardiac conduction system, which controls heart rate. The SA node is the heart's natural pacemaker.

sinuous *(sĭn'-ū-ĕs).* Winding or bending in and out.

sinus *(sī'-nŭs).* A cavity, space, or channel.

ethmoid s. Air cavities in the ethmoid bone, above and on the sides of the nose.

frontal s. A cavity in the frontal bone on each side of the midline above the nasal bridge.

maxillary s. A cavity in the maxillary bone, located on each side of the nose.

paranasal s. The air cavities in the cranial bones, including the ethmoidal, frontal, maxillary, and sphenoidal sinuses.

sphenoid s. Air sinuses that occupy the sphenoid bone and connect with the nasal cavity.

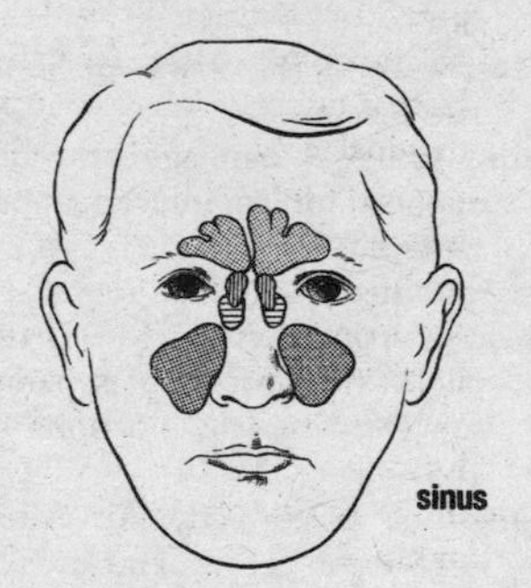

sinus

sinusitis *(sī-nūs-ī'-tĭs).* Inflammation of a sinus, especially a paranasal sinus.

sinusotomy *(sī-nŭs-ŏt'-ō-mē).* Surgically incising a sinus.

sinus rhythm. The normal heart rhythm that begins at the sinoatrial node.

siphonage *(sī'-fŭn-ĭj).* Use of a siphon to drain a body cavity.

sitz bath *(sĭtz).* Bath in which one sits with water (usually warm) above and covering the hips, usually with the legs out of the water.

skeleton *(skĕl'-ĕ-tĕn).* The bony framework of the body.

Skene's glands *(skēnz).* The small glands located on either side of the urethra in females. In acute gonorrhea, these glands are usually infected.

skin *(skĭn).* The outer covering of the body, consisting of the dermis and the epidermis. The skin is the largest organ of the body.

skin tabs. Projections of skin near the anus, often seen after hemorrhoids have been removed.

skull *(skŭl).* The skeleton of the head.

sleepwalking *(slēp'wawk-ĭng).* Somnambulism.

sling *(slĭng).* A bandage or other support for an injured upper extremity.

slipped disk. *See* **disk.**

slough *(slŭf).* Necrotic (dead) tissue in the process of separating from healthy portions of the body.

smallpox *(smăwl'-pŏks).* An acute, highly contagious, and usually fatal infectious disease caused by an orthopoxvirus. Symptoms include fever and distinctive skin eruptions.

smear *(smēr).* A specimen for study under a microscope, prepared by spreading the material to be studied across a glass slide.

smegma *(smĕg'-mă).* The secretion of sebaceous glands, especially the cheeselike secretion found mainly beneath the prepuce.

smooth muscle (smooth). The type of muscle that contracts and relaxes involuntarily, such as the type in blood vessels, the intestines, and the bladder.

Snellen test *(snĕl'-ĕn).* A test to determine visual acuity.

social medicine. *See* **medicine.**

socket *(sŏk'ĕt).* A hollow or depression into which a corresponding part fits.

dry s. A painful condition that frequently occurs after a tooth has been extracted, in which the bone is exposed in the socket due to disintegration or loss of the blood clot.

eye s. The orbit.

sodium *(sō'-dē-ŭm).* A soft, white, alkaline metallic element, the salts of which are commonly used in medicine. Sodium salts are found in body fluids and are needed to preserve a balance between calcium and potassium to maintain normal heart action.

sodomy *(sŏd'-ĕ-mē).* Anal intercourse.

solar plexus *(sō'-lĕr).* A nerve center in the upper abdomen that contains nerves that supply the stomach, liver, gallbladder, pancreas, and other structures.

soleus muscle *(sōl'-ē-ŭs).* One of the two large muscles that make up the calf of the leg.

solipsism *(sŏl'-ĭp-sĭz-ĕm).* The belief that the world exists only in a person's mind or that it consists solely of the person and his/her own experiences.

soluble *(sŏl'-ū-bĕl).* Susceptible of being dissolved.

solution *(sĕ-loo'-shĕn).* A mixture that contains one or more dissolved substances (solutes) in a dissolving medium (solvent). The solute may be a gas, liquid, or solid, and the solvent is usually liquid, but may be solid.

isotonic s. A solution that has the same strength as the dissolved substance encountered in body tissues.

saturated s. A solution in which the solvent has taken up all the solutes that it can hold.

solvent *(sŏl'-vĕnt).* A substance that dissolves or is capable of dissolving.

Soma *(sō'-mă).* *See* Appendix, Common Prescription and OTC Drugs: By Trade Name.

somatic *(sō-măt'-ĭk).* Referring to or characteristic of the body, or soma.

somatization *(sō-mĕ-tĭ-zā'-shĕn).* In psychiatry, the conversion of mental experiences into bodily symptoms.

somatoform *(sō-măt'-ō-fŏrm).* Referring to physical symptoms that cannot be attributed to an organic disease and appear to be psychological in origin.

somatotype *(sō-măt'-ō-tīp).* A certain body type or build based on physical characteristics. *See* **ectomorph, endomorph,** and **mesomorph.**

somnambulism *(sŏm-năm'-bū-lĭz-ĕm).* Sleepwalking; rising from bed and walking about or engaging in behaviors during an apparent state of sleep.

somnifacient *(sŏm-nĭ-fā'-shĕnt).* Hypnotic.

somnolence *(sŏm'-nĕ-lĕns).* Excess drowsiness or sleepiness.

sonography *(sĕ-nŏg'-rĕ-fē).* Ultrasonography.

soporific *(sŏp-ō-rĭf'-ĭk).* Inducing or causing profound sleep.

sorbitol *(sŏr'-bĭ-tŏl).* 1. A type of sugar alcohol and a precursor of the fructose in seminal plasma. 2. A sweetening agent in foods as well as an official sweetening agent in pharmaceutical preparations.

sordes *(sŏr'-dēz).* The accumulation of food, bacteria, and other debris that collects on the teeth and lips during a prolonged fever.

sotalol *(sō'-tĕ-lŏl).* A beta-adrenergic blocking agent used to treat life-threatening cardiac arrhythmias.

sound *(sound).* Auditory sensations produced by vibrations and measured in decibels.

auscultatory s's. Sounds heard on auscultation, including breath sounds, heart sounds, and Korotkoff sounds.

breath s's. Auscultatory sounds

heard associated with an individual's breathing.

heart s's. Sounds heard over the heart region in auscultation and produced by the functioning of the heart.

Korotkoff s's. Sounds heard during auscultatory determination of blood pressure.

spasm *(spăz'-ĕm).* A sudden, violent, involuntary muscle contraction accompanied by pain and transient dysfunction of the affected area or part.

clonic s. A spasm characterized by alternating contraction and relaxation of the involved muscle or muscles.

tonoclonic s. A convulsive twitching of the muscles.

tonic s. Spasm due to a poison.

spastic *(spăs'-tĭk).* Characterized by or referring to spasms.

spasticity *(spăs-tĭs'-ĭ-tē).* The state of being spastic.

species *(spē'-shēz).* A taxonomic category below genus and higher than subspecies, and made up of individuals who possess common characteristics that distinguish them from other categories of individuals of the same taxonomic level.

specific *(spĕ-sĭf'-ĭk).* Limited in effect, application, and so on to a certain function, structure, and so on.

specific gravity. The weight of a volume of a substance compared with the weight of an equal volume of another substance used as a standard. Water is the standard used for solids and liquid.

speculum *(spĕk'-ū-lĕm).* A device that exposes the interior of a cavity or passage by enlarging the opening.

speech center *(spēch).* The region of the brain that controls speech.

speech pathologist. An individual who has been trained to improve the communication skills of children and adults who have language and speech difficulties due to physiological disturbances, defective articulation, or dialect.

speed *(spēd).* The street name for the illicit drug methamphetamine.

sperm *(spĕrm).* The fluid, containing spermatozoa, that is ejaculated from the penis of the male. Also called *spermatozoon.*

spermatic cord *(spĕr-mă'-tĭk).* The tubular structure that carries semen and leads from the testicle to the seminal vesicle.

spermatocele *(spĕr'-mĕ-tō-sēl).* A cystic tumor containing spermatozoa that develops on the epididymis.

spermatogenesis *(spĕr-măt-ō-jĕn'-ĕ-sĭs).* The process of forming spermatozoa.

spermaturia *(spĕr-mă-tū'-rē-ă).* Semen discharged into the urine.

spermatozoon *(spĕr-mĕ-tō-zō'-ŏn).* A mature male germ cell and the element of semen that fertilizes the ovum. Also called *sperm.*

sperm count. An estimate of the number of sperm in an ejaculation.

sperm donor. A male who contributes his sperm to impregnate a female who does not have a sexual partner or whose husband/partner is not able to contribute viable sperm.

spermicide *(spĕr'-mī-sīd).* A substance that kills spermatozoa.

sphenoiditis *(sfē-noy-dī'-tĭs).* Inflammation of the sphenoidal sinus.

sphenoid sinus *(sfē'-noyd). See* **sinus.**

spherocytes *(sfē'-rō-sīt).* A red blood cell that takes on a spheroid shape.

spherocytosis *(sfē-rō-sī-tō'-sĭs).* Condition in which red blood cells assume a spheroid shape, which occurs in some hemolytic anemias.

sphincter *(sfĭngk'-tĕr).* A ringlike band of muscle that closes or constricts an opening.

anal s. Sphincter that closes the anus.

cardiac s. Plain muscle near the esophagus at cardiac opening into the stomach.

pyloric s. A thickening of the muscular wall around the pyloric opening.

sphincterismus *(sfĭngk-tĕr-ĭz'-mŭs).* A spasm of the anal sphincter.

sphincterotomy *(sfĭngk-tĕr-ĕk'-tĕ-mē).* Excision of a sphincter.

sphygmomanometer *(sfĭg-mō-mĕ-nŏm'-ĕ-tĕr).* An instrument used to measure blood pressure in the arteries.

spica *(spī'-kă).* A figure-eight bandage that crosses at the shoulder or hip.

spicule *(spĭk'-ūl).* A needlelike body.

spider veins *(spī'-dĕr).* A group of widened veins that are visible through the skin's surface. Also known as *spider telangiectasia.*

spina bifida *(spī'-nă bĭ'-fĭ-dă).* A congenital defect in the walls of the spinal canal caused by the failure of the laminae of the vertebrae to join.

spinal canal *(spī'-năl).* The canal of the vertebral column that contains the spinal cord.

spinal column *(spī'-năl).* The vertebral column that encloses the spinal cord and 33 vertebrae.

spinal cord. The column of nervous tissue that extends from the medulla to the second lumbar vertebra in the spinal canal.

spinal fusion. Surgery performed to immobilize adjacent vertebra.

spinalis muscle *(spī-nā'-lĭs).* A muscle that is attached to the spinal portion of a vertebra.

spinal tap. *See* **lumbar puncture.**

spine *(spīn).* The spinal column, consisting of 33 vertebrae.

cervical s. The first seven vertebrae, which are in the neck.

dorsal s. The portion of the spine below the cervical spine, of twelve vertebrae. Also known as the thoracic spine.

lumbar s. The portion of the

spine in the lower back, consisting of five vertebrae.
sacrococcygeal s. The portion of the spine consisting of five fused sacral vertebrae and four fused coccygeal vertebrae.

spirometer *(spī-rŏm'-ē-tĕr).* An instrument used to measure the air inhaled into and exhaled out of the lungs.

splanchnic *(splănk'-nĭk).* Referring to the viscera.

splanchnicectomy *(splănk-nĭ-sĕk'-tĕ-mē).* Removal of one or more splanchnic nerves to treat hypertension or intractable pain.

spleen *(splēn).* A large glandlike, ductless organ located in the upper part of the abdominal cavity. It disintegrates red blood cells and frees hemoglobin, which the liver converts into bilirubin; it also produces lymphocytes and plasma cells.

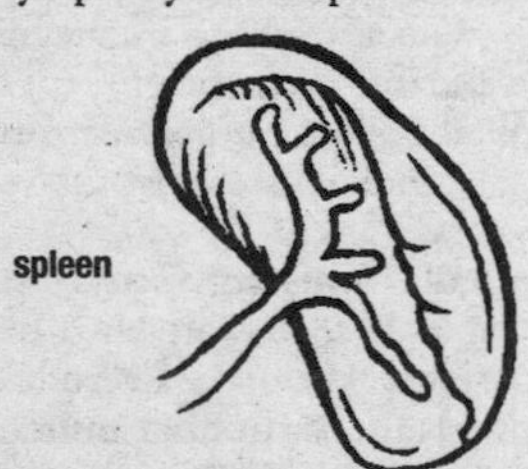

splenectomy *(splē-nĕk'-tō-mē).* Surgical removal of the spleen.

splenic infarct *(splē'-nĭk).* A clot of a vessel of the spleen.

splenitis *(splē-nī'-tĭs).* Inflammation of the spleen.

splenohepatomegaly *(splē-nō-hĕp-ă-tō-mĕg'-ă-lē).* Enlargement of the spleen and liver.

splenomegaly *(splē-nō-mĕg'-ă-lē).* Enlargement of the spleen.

splenoportography *(splē-nō-pŏr-tŏg'-ră-fē).* X-ray study of the spleen and portal vein after injection of a radiopaque dye into the spleen.

splenorenal shunt *(splē-nō-rē'-năl).* The joining together of the splenic vein to the renal vein, which allows blood from the portal system to enter general venous circulation.

splenorrhaphy *(splē-nŏr'-ă-fē).* Suture of a wound of the spleen.

splint *(splĭnt).* An appliance made of wood, metal, bone, or plaster of paris, used to fix, join, or protect an injured body part.

spondylarthritis *(spŏn-dĕl-ăhr-thrī'-tĭs).* Arthritis of the spine.

spondylitis *(spŏn-dĕ-lī'-tĭs).* Inflammation of the vertebrae.
ankylosing s. A form of degenerative joint disease that affects the spine; it mainly afflicts young people.
traumatic s. Spondylitis that occurs as a result of injury to the vertebrae.

spondylolisthesis *(spŏn-dĕ-lō-lĭs-thē'-sĭs).* Forward displacement of one vertebra over another, usually involving the fifth lumbar over the sacrum or the fourth lumbar over the fifth.

spondylosis *(spŏn-dĭ-lō'-sĭs).* Immobility and fusion of a vertebral joint.

sponge *(spŭnj).* The fibrous skeleton of certain marine animals, used primarily as an absorbent.

spongioblast *(spŭn-jē-ō-blăst).* A cell that develops from embryonic neural tube and later is transformed into other types of cells.

spongioblastoma *(spŭn-jē-ō-blăs-tō'-mă).* A glioma of the brain that arises from spongioblasts.

spongiocyte *(spŭn'-jē-ō-sīt).* A neuroglial cell.

spontaneous abortion *(spŏn-tā'-nē-ŭs).* See **abortion.**

sporadic *(spĕ-răd'-ĭk).* Occurring in a random manner.

spore *(spŏr).* A reproductive cell produced by plants and some protozoons.

sporotrichosis *(spŏr-ō-trī-kō'-sĭs).* A chronic fungal infection usually involving the skin and superficial lymph nodes.

spotting *(spŏt-ĭng).* Appearance of blood-tinged discharge from the vagina, usually occurring between menstrual periods or at the beginning of labor.

sprain *(sprān).* Injury or trauma to a joint that causes pain and disability.

sprue *(sproo).* A chronic form of malabsorption syndrome, characterized by weakness, weight loss, and various digestive disorders, especially impaired absorption of glucose, vitamins, and fats.

spur *(spŭr).* A pointed projection.

sputum *(spū'-tŭm).* Substance expelled by coughing or clearing the throat.

squamous *(skwă'-mŭs).* Scalelike.

squint *(skwĭnt).* Visual abnormality in which the right and left visual axes do not focus simultaneously toward a given point.

stab *(stăb).* To pierce with a knife.

stabilization procedure *(stā-bĭl-ĭ-zā'-shĕn).* A process used to create a stable state.

stable *(stā'-bĕl).* Steady, firm.

stage *(stāj).* A period or distinct phase in the course of a disease, the course of any biological process, or the life cycle of an organism.

anal s. In psychoanalytic theory, the second stage of psychosexual development, occurring between the ages of 1 and 3 years. During this stage, the anal zone is the center of the infant's activities, concerns, and interests.

oral s. In psychoanalytic theory the first stage of psychosexual development, lasting from birth to about 18 months. During this stage, the oral region is the center of the infant's needs and experiences.

phallic s. In psychoanalytic theory, the third stage in psychosexual development, determined to be from ages 2 or 3 years to ages 5 to 6 years. During this stage, sexual curiosity and experiences are centered on the penis in boys and the clitoris in girls.

stagnation *(stăg-nā'-shĕn).* The cessation of motion of any fluid in the body.

stain *(stān).* 1. Any discoloration. 2. A dye or pigment used to color samples used for microscopic examination.

stalk *(stăwk).* An elongated structure that usually supports or serves to attach a structure or organ.

stamina *(stăm'-ĭ-nă).* Strength; endurance.

standard deviation (SD) *(stăn'-dĕrd).* In statistics, it is the measure of dispersion or variability in a distribution.

stannous fluoride *(stăn'-ŭs).* A fluoride compound used in toothpaste to help prevent cavities.

stapedectomy *(stā-pē-dĕk'-tō-mē).* Removal of the stapes to improve hearing, especially in cases of otosclerosis.

stapes *(stā'-pēz).* A small bone in the middle ear that articulates with the incus. Also known as a *stirrup.*

Staphcillin *(stăf-sĭl'-ĭn).* Trade name for methicillin sodium, used to treat a variety of bacterial infections.

staphylectomy *(stăf-ĭ-lĕk'-tō-mē).* Amputation of the uvula.

staphylococcemia *(stăf-ĭl-ō-kŏk-sē'-mē-ă).* The presence of staphylococci in the blood.

Staphylococcus *(stăf-ĭ-kō-kŏk'-ŭs).* A genus of gram-positive micrococci belonging to the family Micrococcaceae.

S. aureus. A species commonly found on the skin and mucous membranes and a cause of boils, carbuncles, and internal abscesses.

S. epidermidis. A species commonly associated with abscesses, infected wounds, peritonitis, and subacute bacterial endocarditis.

S. haemolyticus. A variety that may be associated with infections of the urinary tract, wounds, and the conjunctiva.

stasis *(stā'-sĭs).* A cessation or decrease in the flow of blood or other body fluids.

statistical significance *(stă-tĭs'-tĭ-kăl).* In statistics, a conclusion that the event investigated has a certain probability of being due to chance, but the probability is so insignificant that it is presumed that the event did not occur due to chance alone.

status *(stă'-tŭs).* A state or condition. It is especially used to refer to a morbid condition.

STD. Acronym for sexually transmitted disease.

steady state *(stĕ'-dē).* The condition of the metabolic needs of a body system, such as the muscles receiving nutrients at the same rate as the energy being put out.

steatitis *(stē-ă-tī'-tĭs).* Inflammation of fat tissue.

steatolysis *(stē-ă-tŏl'-ĭ-sĭs).* The process by which fats are emulsified and then hydrolyzed to fatty acids and glycerine as the preparatory steps to absorption.

steatoma *(stē-ă-tō'-mă).* 1. Sebaceous cyst. 2. A benign tumor composed of fat cells.

steatorrhea *(stē-ă-tō-rē'-ă).* Excessive amounts of fats present in the feces, as occurs in malabsorption syndromes.

stem cell *(stĕm).* A master cell that has the ability to develop into any one of the body's more than 200 cell types. All stem cells are unspecialized and retain the ability to divide throughout life and to result in cells that can become highly specialized and replace cells that die or are lost.

stenosed *(stĕ-nōzd').* Narrow or constricted.

stenosis *(stĕ-nō'-sĭs).* An abnormal narrowing of a canal or duct.

aortic s. Narrowing of the opening of the aortic value or of the supravalvular or subvalvular regions.

mitral s. Narrowing of the left atrioventricular opening.

pyloric s. Obstruction of the pyloric opening of the stomach

tricuspid s. Narrowing of the tricuspid opening of the heart.

stent *(stĕnt).* 1. A device used to keep a skin graft in place. 2. A device used to provide support for tubular structures that are being joined.

stercoroma *(stĕr-kĕ-rō'-mă).* A large accumulation of fecal matter that forms a tumorlike mass in the rectum.

stereoanesthesia *(stĕr-ē-ō-ăn-ĕs-thē'-zhĕ).* An inability to identify by touch the size, weight, texture, and form of objects.

stereotactic *(stĕr-ē-ō-tăk'-tĭk).* Refers to the precise positioning in three-dimensional space. For example, in a stereotactic needle biopsy, the site to be biopsied is identified three-dimensionally, the data is entered into a computer, and the computer calculates the information and positions the needle.

sterile *(stĕr'-ĭl).* Unable to produce offspring.

sterility *(stĕ-rĭl'-ĭ-tē).* An inability to either conceive (females) or to induce conception (males).

sterilization *(stĕr-ĭl-ĭ-zā'-shĕn).* 1. Any procedure, such as castration, vasectomy, or salpingectomy, that renders an individual incapable of reproduction. 2. The total destruction or elimination of all living microorganisms using physical, mechanical, radiation, or chemical methods.

sternalgia *(stĕr-năl'-jĕ).* Pain in the sternum.

sternoclavicular *(stĕr-nō-klĕ-vĭk'-ū-lĕr).* Referring to the sternum and clavicle.

sternotomy *(stĕr-nŏt'-ĕ-mē).* A surgical procedure in which the sternum is cut.

sternum *(stĕr'-nŭm).* The narrow, flat bone in the middle of the thorax in front. It consists of three parts: the manubrium, the gladiolus, and the xiphoid process.

steroid *(stĕr'-oyd).* An organic compound that contains a specific ring system in its nucleus: perhydrocyclopentanophenanthrene. Some examples of steroids include progesterone, sex hormones, cholesterol, and bile acids, among others.

stethoscope *(stĕth'-ō-skōp).* An instrument that allows one to listen to internal body sounds, such as those emitted by the heart, lungs, digestive tract, and other systems and organs.

stevia *(stē'-vē-ă).* A South American shrub whose leaves have been used for centuries by natives as a sweetener. Stevioside, the main ingredient in stevia, is nearly calorie-free and hundreds of times sweeter than table sugar.

stigma *(stĭg'-mă).* A physical or mental mark or peculiarity that can assist in the identification or diagnosis of a disease.

stilbestrol *(stĭl-bĕs'-trŏl).* Diethylstilbestrol.

stillbirth *(stĭl'bĭrth).* Delivery of a dead child.

stillborn *(stĭl'-bŏrn).* Born dead.

Still's disease *(stĭlz).* A variety of chronic polyarthritis that affects children and characterized by enlarged lymph nodes, rash, and irregular fever. Also called *juvenile rheumatoid arthritis.*

stimulant *(stĭm-ū-lĕnt).* Any substance that produces stimulation, especially by causing tension on muscle fibers through the nervous system. Stimulants are usually classified according to the organ upon which they act.

central s. Stimulant of the central nervous system.

cerebral s. Stimulant that increases the activities of the brain.

general s. Stimulant that acts upon the entire body.

spinal s. Stimulant that acts upon and through the spinal cord.

topical s. A local stimulant.

stimulation *(stĭm-ū-lā'-shĕn).* The process of being stimulating or the condition of being stimulated.

stimulus *(stĭm'-ū-lŭs).* Any agent or factor capable of influencing living protoplasm directly or of initiating an impulse in a nerve.

stippling *(stĭp'-lĭng).* The appearance of fine dark or light spots. This can be seen, for example, in the retina in certain diseases.

stitch *(stĭch).* 1. A suture; a single loop of suture material that is passed through the flesh by a needle. 2. A popular term for a sudden, sharp pain, usually in the side.

stoma *(stō'-mă).* A small opening, mouth, or pore. It often refers to such an opening created in the abdominal wall by colostomy, ileostomy, or a similar operation.

stomach *(stŭm'-ĕk).* An expandable, saclike portion of the alimentary canal below the esophagus that serves as an organ of digestion.

stomatitis *(stō-mă-tī'-tĭs).* Inflammation of the mouth.

allergic s. Stomatitis caused by allergens or as part of an allergic condition.

denture s. Condition seen in some individuals with new dentures or with ill-fitting ones.

herpetic s. Herpes simplex that involves the oral mucosa and lips and characterized by vesicles that rupture and result in painful ulcers.

stomatology *(stō-mă-tŏl'-ĕ-jē).* Branch of medicine that concerns the mouth and its structure, diseases, and functions.

stone *(stōn).* 1. Calculus. 2. In Great Britain, the equivalent of 14 pounds or about 6.4 kilograms.

stool *(stool).* Feces.

fatty s. Feces that contain fat, seen in malabsorption conditions and diseases of the pancreas.

mucous s. Feces that contain a large amount of mucus, seen in mucous colitis or intestinal inflammation.

tarry s. Black feces that contain blood; seen when there has been bleeding in the stomach or duodenum.

storm *(stŏrm).* A sudden outburst of symptoms of a disease.

thyroid s. A complication of thyrotoxicosis that if not treated is usually fatal. Symptoms include sudden onset of fever, sweating, rapid heartbeat, pulmonary edema, tremors, and restlessness.

strabismus *(stră-bĭz'-mŭs).* An eye disorder in which the optic axes are unable to be directed to the same object.

strain *(strān).* 1. Excessive use of a body part to the point that it is injured. 2. A group of organisms within a species or variety that is characterized by a certain quality. Often used to refer to a *strain of bacteria.*

high-jumper's s. Strain of the muscles of the thigh seen in high jumpers.

resistant s. A strain of organisms that is resistant to the effects of substances used to control them, such as antibiotics or insecticides.

straitjacket *(strāt'-jăk-ĕt).* A cloth jacket with extra-long arms that allow the patient's arms to be wrapped around him/her and thus restrain movement.

strangulation *(străng-gū-lā-shĕn).* Compression or constriction of a part of the body so that it causes a cessation of breathing (constriction of the throat) or the passage of materials (e.g., constriction of the bowel).

strangury *(străng-gū-rē).* Slow, painful urination caused by spasm of the urethra and bladder.

stratification *(străt-ĭ-fĭ-kā'-shĕn).* The process of arranging something in layers.

stratum *(stră'tŭm).* A layer.

strep throat *(strĕp thrōt).* An infection of the pharynx and tonsils caused by streptococcus. Characteristics can include high fever, malaise, swollen lymph glands, and chills.

streptococcemia *(strĕp-tō-kŏk-sē'-mē-ă).* The presence of streptococci in the blood.

Streptococcus *(strĕp-tō-kŏk'-ŭs).* A genus of gram-positive bacteria in the family *Streptococcaceae.*

S. bovis. A species found in the bovine alimentary tract and occasionally in human feces, and thus can be associated with human endocarditis.

S. mitis. A species found in the normal human upper respiratory tract and can be associated with subacute bacterial endocarditis.

S. pneumoniae. A species that is the most common cause of lobar pneumonia; also causes meningitis, septicemia, peritonitis, and empyema.

S. pyogenes. A species that causes septic sore throat, scarlet fever, rheumatic fever, and other conditions.

streptokinase *(strĕp-tō-kī'-nās)*. A protein produced by beta-hemolytic streptococci and used as a thrombolytic agent to treat acute coronary arterial thrombosis, acute pulmonary embolism, deep venous thrombosis, and acute arterial thromboembolism or thrombosis.

streptomycin *(strĕp-tō-mī'-sĭn)*. An aminoglycoside antibiotic effective against a wide variety of aerobic gram-negative bacilli and some gram-positive bacteria. Its use is now limited because many resistant strains have developed.

streptothricosis *(strĕp-tō-thrī-kō'-sĭs). See* **actinomycosis.**

stress *(strĕs)*. 1. Forcibly exerted pressure or influence. 2. A condition of physical or emotional strain caused by adverse stimuli.

stress incontinence. *See* **incontinence.**

stretch marks *(strĕch)*. Lines on the abdominal skin that frequently appear in women during the later months of pregnancy.

stria *(strī-ă)*. *(pl.* striae). A band, streak, line, or stripe of differing color or texture than the surrounding tissue.

striated muscle *(strī'-āt-ĕd)*. Voluntary muscle whose fibers are marked by transverse bands into striations.

stricture *(strĭk'-chŭr)*. An abnormal narrowing; stenosis.

stridor *(strī'-dŏr)*. A harsh, high-pitched breath sound, such as may be heard when the larynx fails to open sufficiently during breathing.

stroke *(strōk)*. A sudden, severe attack in which there is a loss of consciousness followed by paralysis caused by bleeding in the brain, development of an embolus or thrombus that blocks an artery, or the rupture of an extracerebral artery.

stroma *(strō'-mă)*. The tissue that supports an organ, as compared with tissue that is involved in its function.

strontium-90 *(strŏn'-shē-ĕm)*. A radioactive isotope of strontium that is used in the treatment of various benign ophthalmologic conditions.

structure *(strŭk'-chĕr)*. The arrangement of the components that make up an organ or body part.

struma *(stroo'-mă)*. Goiter.

stunt *(stŭnt)*. To retard or hinder the growth of something.

stupefactive *(stoo-pĕ-făk'-tĭv).* Producing stupor.

stupor *(stoo'-pĕr).* A reduced level of consciousness or awareness in which an individual responds only after vigorous stimulation.

stuttering *(stŭt'-ĕr-ĭng).* A speech disorder characterized by spasmodic repetition of the same syllable or words, prolongation of sounds, and long pauses.

stye *(stī).* A purulent, inflammatory staphylococcal infection that affects one or more sebaceous glands of the eyelids. Also called a *hordeolum.*

stylet *(stī'-lĕt).* A wire that is passed through a catheter or cannula to make it rigid or to remove material from inside the lumen.

styptic *(stĭp'-tĭk).* Astringent.

subacute *(sŭb-ĕ-kūt').* Used to describe the course of a disease, being between acute and chronic, with some features of acute disease.

subarachnoid *(sŭb-ĕ-răk'-noid).* Between the arachnoid membrane and the pia mater in the brain.

subarachnoid hemorrhage. *See* **hemorrhage.**

subarachnoid space. The space beneath the arachnoid membrane.

subclinical disease *(sŭb-klĭn'-ĭ-kĕl).* A disease whose signs and symptoms are so slight or mild that they are not apparent or detectable by clinical examination or tests.

subcostal *(sŭb-kŏs'-tĕl).* Beneath a rib or all of the ribs.

subcutaneous *(sŭb-kū-tā'-nē-ŭs).* Beneath the skin.

subdeltoid *(sŭb-dĕl'-toid).* Beneath the deltoid muscle.

subdural *(sŭb-doo'-rĕl).* Between the dura mater and the arachnoid membrane in the brain.

subdural hematoma. *See* **hematoma.**

subjective *(sŭb-jĕk'-tĭv).* Refers to symptoms an individual claims to be experiencing but which cannot be observed or ascertained by others, including the examining physician.

sublethal *(sŭb-lē'-thĕl).* Insufficient to cause death.

subliminal *(sŭb-lĭm'-ĭ-nĕl).* Below the threshold, or limen, of sensation; or below normal consciousness.

sublingual *(sŭb-lĭng'-gwĕl).* Beneath the tongue.

subluxation *(sŭb-lĕk-sā'-shĕn).* 1. An incomplete or partial dislocation, as of a bone. 2. In chiropractic, any mechanical impediment to nerve function.

submandibular *(sŭb-măn-dĭb'-ū-lĕr).* Beneath the mandible.

submaxillary *(sŭb-măk'-sĭ-lăr-ē).* Beneath the maxilla.

submental *(sŭb-mĕn'-tĕl).* Under the chin.

submucosa *(sŭb-mū-kō'-să).* The layer of connective tissue below the mucous membrane.

subperiosteal *(sŭb-pĕr-ĭ-tō-nē'-ăl).* Beneath the peritoneum.

subscapular *(sŭb-skăp'-ū-lăr).* Beneath the shoulder blade.

substance P. A peptide composed of amino acids found in nerve cells throughout the body and in some endocrine cells in the gut.

substernal *(sŭb-stĕr'-năl).* Beneath the sternum.

substrate *(sŭb'-strāt).* A substance upon which an enzyme acts.

subungual *(sŭb-ŭng'-gwĕl).* Under the fingernail or toenail.

suckle *(sŭk'-ĕl).* To breast-feed an infant.

sucrose *(soo'-krōs).* A derivative of sugar cane, sugar beets, and sorghum, composed of glucose and fructose and used as a sweetener.

Sudafed *(soo'-dĕ-fĕd).* Trade name for preparations that contain pseudoephedrine hydrochloride and used to treat nasal congestion.

sudamen *(soo-dā'-mĕn).* A white vesicle that develops due to retention of sweat in the pores of the skin.

Sudeck's atrophy *(soo'-dĕks).* Degeneration of bone following a severe injury.

sudoresis *(sū-dō-rē'-sĭs).* Profuse sweating.

suffocate *(sŭf'-ō-kāt).* To block or stop respiration; smother.

suffusion *(sĕ-fū'-zhĕn).* Spreading a bodily fluid into surrounding tissues.

suicide *(soo'-ĭ-sīd).* The taking of one's own life.

sulcus *(sŭl'-kŭs).* A slight depression, groove, or fissure, especially in the brain.

sulfa drugs *(sŭl'-fĕ).* Drugs that contain sulfanilamide or other sulfonamides. They are used to inhibit the growth of certain infectious bacteria.

sulfide *(sŭl'-fīd).* Any compound of sulfur with an element or base.

sulfmethemoglobin *(sŭlf-mĕt-hē'-mō-glō-bĭn).* Hemoglobin that has a sulfur atom, which makes it ineffective in transporting oxygen.

sulfonamides *(sŭl-fŏn'-ĕ-mīdz).* The sulfa drugs, which are used to treat various bacterial infections. Because many strains of bacteria are resistant to sulfonamides, when they are used they are usually combined with another sulfonamide or other antimicrobial agents.

sulfur *(sŭl'-fŭr).* An element necessary for human health, as it is required for the synthesis of proteins.

sumatriptan *(sū-mă-trĭp'-tĕn).* *See* Appendix, Common Prescription and OTC Drugs: By Generic Name.

sunburn *(sŭn'-bĕrn).* Injury to the skin due to excessive exposure to sunlight. It is characterized by redness, tenderness, and, depending on the severity of the burn, blistering.

sundowning *(sŭn'-doun-ĭng).* A condition sometimes seen in older individuals who have dementia or other mental disorders, it is characterized by confusion, agitation, other disruptive behaviors, and an inability to remain asleep, all seen only or significantly worsening at night.

sunstroke *(sŭn'-strōk)*. A condition produced by excessive exposure to the sun and characterized by high body temperature (greater than 105°), convulsions, headache, cessation of sweating, confusion, fast pulse, and coma.

superego *(soo-pĕr-ē'-gō)*. In psychoanalytic theory, the part of the personality that monitors ego functioning and is associated with self-criticism, ethics, and moral standards.

superficial *(soo-pĕr-fĭsh'-ĕl)*. Limited or referring to the surface.

superinfection *(soo-pĕr-ĭn-fĕk'-shŭn)*. A new infection that occurs in an individual who already has a preexisting infection. Such conditions can be difficult to treat when the new infection is resistant to the drugs used to treat the initial infection.

superior *(soo-pē'-rē-ŏr)*. Higher than or situated above; better than.

superiority complex. An exaggerated conviction that one is superior to others.

supernumerary *(soo-pĕr-noo'-mĕr-ăr-ē)*. Referring to anything that is in excess of the regular or normal number.

supersaturate *(sū-pĕr-săch'-ĕr-āt)*. To add more of a substance to a solution than the latter can hold permanently.

supinate *(soo'-pĭ-nāt)*. To assume or place in a supine position.

supine *(soo'-pīn)*. Lying on one's back with the face upward.

suppository *(sŭ-pŏz'-ĭ-tō-rē)*. A semisolid substance, usually containing medication, that is introduced into the rectum, vagina, or urethra, where it dissolves.

suppuration *(sūp-ū-rā'-shŭn)*. The process of pus formation.

suppurative *(sūp'-ū-ră-tĭv)*. Producing or involved with the development of pus.

surfactant *(sŭr-făk'-tănt)*. A substance that reduces surface tension. Examples include oils and detergents.

surgeon *(sŭr'-jŭn)*. A medical practitioner who specializes in surgery.

surgery *(sŭr'-jĕr-ē)*. The branch of medicine that treats injuries, physical abnormalities, and diseases using manual or operative approaches.

susceptibility *(sŭ-sĕp-tĭ-bĭl'-ĭ-tē)*. The state of being readily acted upon or affected; for example, a tendency to develop a disease if exposed to it.

susceptible *(sŭ-sĕp'-tĭ-bl)*. Capable of being readily acted upon; not having immunity.

suspension *(sŭs-pĕn'-shŭn)*. A state of temporary cessation of an activity or process.

suspensory *(sŭs-pĕn'-sō-rē)*. Holding up a part, such as a muscle or bone.

suture *(sū'chūr)*. A loop of thread or similar material used to secure the edges of a surgical or accidental wound. Also called a *stitch*.

Swan-Ganz catheter. *See* **catheter.**

swayback *(swā'-băk)*. Abnormal concavity in the curvature of the lumbar and cervical spine; also called *lordosis*.

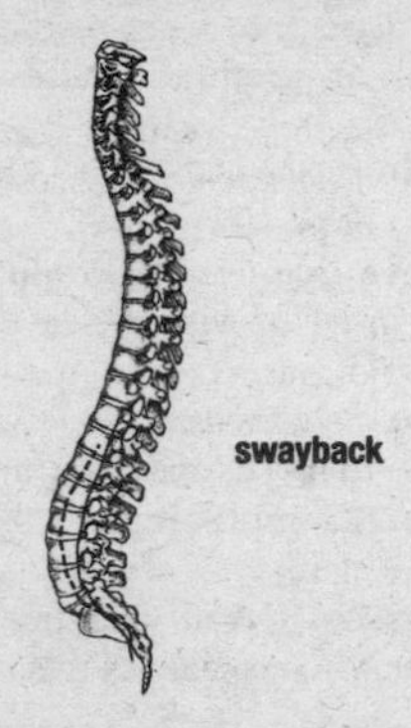
swayback

sweat glands *(swĕt)*. Simple, tubular glands located nearly everywhere on the body's surface, especially the palms of the hands and soles of the feet.

symbiosis *(sĭm-bē-ō'-sĭs)*. When two organisms of different species live together in close association. If neither organism is harmed, it is called *commensalism;* if both are benefited, it is called *mutualism;* if one is harmed and one benefits, it is called *parasitism*.

symmetry *(sim'-ĕt-rē)*. Corresponding in shape, size, and position of parts on opposite sides of the body.

sympathectomy *(sĭm-pă-thĕk'-tō-mē)*. Surgical removal of part (e.g., a nerve, plexus, ganglion, or several ganglia) of the sympathetic part of the autonomic nervous system.

sympathetic nervous system *(sĭm-pă-thĕt'-ĭk)*. A large part of the autonomic nervous system, it is made up of the nerves, ganglia, and plexuses that supply the involuntary muscles.

sympathetic ophthalmia. Inflammation of the uveal tract in one eye because of a similar inflammatory process in the other eye.

symphysis *(sĭm'-fĭ-sĭs)*. A line of fusion between two bones that are separate during early development.

symptom *(sĭm'-tŭm)*. Any observable change in the body or how it functions that indicates the presence of disease or other medical abnormality.

symptomatology *(sĭmp-tō-mă-tŏl'-ō-jē)*. The science of symptoms and the systematic discussion of symptoms.

symptom complex. All the symptoms of a disease; a syndrome.

synapse *(sĭn'-ăps)*. The point of junction between neurons where an impulse is transmitted from one neuron to another.

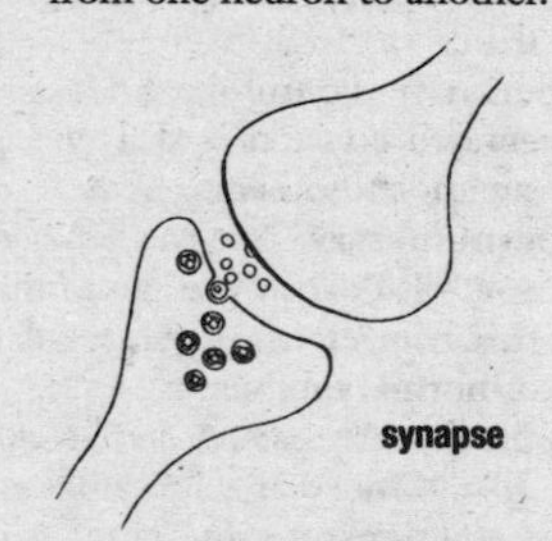
synapse

synchondrosis *(sĭn-kŏn-drō'-sĭs)*. A union between two bones formed by cartilage. It is usually temporary, as the cartilage is usually converted into bone before adulthood.

synchronism *(sĭn'-krō-nĭzm).* Occurrence at the same time.
syncope *(sĭn'-kū-pē).* Fainting.
syndrome *(sĭn'-drŭm).* A set of symptoms that occur at the same time.
synechia *(sĭn'-ĕk-ē-ă).* Adhesion of one body part to another, especially adhesion of the iris to the cornea.
synergetic *(sĭn-ĕr-jĕt'-ĭk).* Acting in harmony.
synergism *(sĭn'-ĕr-jĭzm).* Cooperation between two or more structures (e.g., organs) or drugs.
synergy *(sĭn'-ĕr-jē).* The action of two or more organs, substances, or agents working together.
synesthesia *(sĭn-ĕs-thē'-zē-ă).* A feeling or sensation in one part of the body from a stimulus applied to another part.
synovectomy *(sĭn-ō-vĕk'-tō-mē).* Surgical removal of a synovial membrane.
synovial fluid *(sĭn-ō'-vē-ăl).* Clear fluid secreted by the synovial membrane and lubricates the joint.
synovial membrane. Membrane that lines the capsule of a joint.
synovioma *(sĭn-ō-vē-ō'-mă).* A tumor that develops from a synovial membrane.
synovitis *(sĭn-nō-vī'-tĭs).* Inflammation of a synovial membrane.
synthesis *(sĭn'-thĕ-sĭs).* The union of elements to produce compounds.
Synthroid *(sĭn'-throyd). See* Appendix, Common Prescription and OTC Drugs: By Trade Name.
syntropy *(sĭn'-trō-pē).* Turning or pointing in the same direction.
syphilis *(sĭf'-ĭ-lĭs).* An infectious disease caused by *Treponema pallidum*, which is usually transmitted by sexual contact or acquired in utero.
syringe *(sĭr-ĭnj').* An instrument used to inject or withdraw liquids from a cavity or tissues.
syringotomy *(sĭr-ĭn-gŏt'-tō-mē).* Surgery for the incision of a fistula.
system *(sĭs'-tĕm).* An organized grouping of related parts or structures.
systemic *(sĭs-tĕm'-ĭk).* Affecting or referring to several parts of the body rather than just one of the parts.
systole *(sĭs'-tō-lē).* The period during the heart cycle in which the heart contracts, and is then followed by relaxation (diastole).
systolic blood pressure *(sĭs-tŏl'-ĭk).* The highest blood pressure reading produced when the heart contracts. It is the first number in a blood pressure measurement.

T

tabes *(tā'-bēz).* Wasting away of a body part.
tablet *(tăb'-lĕt).* A solid dosage form of a medication, which may be compressed or molded.
tachography *(tĕ-kŏg'-rĕ-fē).* A recording of the speed of blood flow.

tachycardia *(tăk-ĭ-kăhr'-dē-ē).* An excessively rapid heart rate, usually defined as greater than 100 beats per minute in an adult.

tachypnea *(tăk-ĭp-nē'-ĕ).* Excessive rapid breathing.

tactile *(tăk'-tĭl).* Referring to touch.

Taenia *(tē'-nē-ă).* A genus of tapeworms of the family Taeniidae.

taeniafuge *(tē'-nē-ĕ-fūj).* A substance that expels tapeworms.

tag *(tăg).* A small flap, appendage, or polyp.

Tagamet *(tăg'-ĕ-mĕt).* Trade name for preparations of cimetidine, used to treat peptic ulcer, symptoms of hyperactivity, and gastroesophageal reflux.

tai chi *(tī chē).* An ancient Chinese system of postures and slow, graceful movements that helps create inner and outer balance and harmony, improve cardiovascular and respiratory functioning, and relieves stress.

taint *(tānt).* To spoil or cause putrefaction.

talcosis *(tăl-kō'-sĭs).* Respiratory disease caused by inhalation of talc in the body.

talipes *(tăl'-ĭ-pēz).* Any of various deformities of the foot.

talus *(tā'-lĕs).* Also called the *ankle,* it is the bone that articulates with the tibia and fibula to form the ankle joint.

tamoxifen *(tĕ-mŏk'-sĭ-fĕn).* A nonsteroidal, antiestrogen that is used in the prevention and treatment of breast cancer.

tampon *(tăm'-pŏn).* A plug or pack made of cotton or other material and used in surgery or medical procedures to control bleeding or to absorb secretions.

tamponade *(tăm-pŏn-ād').* Surgical use of a tampon.

cardiac t. Acute compression of the heart caused by an increase in pressure caused by accumulation of blood or fluid in the pericardium, trauma, or progressive effusion.

tamsulosin *(tăm-soo'-lō-sĭn). See* Appendix, Common Prescription and OTC Drugs: By Generic Name.

tapeworm *(tāp'-wĕrm).* Any flatworm of the class *Cestoidea,* many of which are intestinal parasites. The eggs usually enter the body via raw or poorly cooked beef and can cause symptoms including abdominal pain, fatigue, weight loss, and diarrhea.

tapping *(tăp'-ĭng).* Insertion of a needle into the body to remove fluids, such as blood, pus, or serum.

tardive dyskinesia *(tăr'-dĭv dĭs-kĭ-nēsh'-ă).* A neurological syndrome characterized by repetitive, involuntary movements such as grimacing, lip smacking, rapid movements of the arms, legs, and trunk, and tongue protrusion. Caused by long-term use of neuroleptics (drugs used to treat psychiatric, neurological, and gastrointestinal disorders).

target cell *(tăr'-gĕt).* A cell with a specific receptor for an antibody, antigen, or hormone.

tarsalgia *(tăhr-săl'-jē).* Pain in the foot or ankle.

tarsus *(tähr'-sŭs).* The area of the articulation between the foot and the leg.

tartar *(tăhr'-tĕr).* Deposits that gather around the base of teeth; also known as *plaque* or *dental calculus.*

taste *(tāst).* A sense that depends on sense organs called taste buds, on the surface of the tongue, which, when stimulated, produces one or more of the four basic taste sensations: sweet, bitter, sour, and salty.

taurine *(tăw'-rēn).* A derivative of the amino acid cysteine, it is present in bile and is also thought to be a central nervous system neurotransmitter.

Taxol *(tăk'-sŏl).* Trade name for a preparation of paclitaxel, used to treat advanced ovarian or breast cancer, non-small cell lung cancer, and AIDS-related Kaposi's sarcoma.

Tay-Sach's disease *(tā săks).* An inherited, fatal disease that occurs almost exclusively among Ashkenazi Jews; characterized by onset between 3 to 6 months of age, blindness, brain deterioration, convulsions, red spots on the macula, and death by age 5 years.

TB. Abbreviation for tuberculosis.

T-cells. A type of white blood cell that plays a major role in the immune system. Some have the ability to kill tumor cells. Also known as *T lymphocytes.*

tear duct *(tēr).* The passage that carries the secretions of the lacrimal (tear) glands.

teat *(tēt).* The nipple of the mammary gland.

technetium-99 *(tĕk-nē'-shē-ŭm).* An isotope of the metallic element technetium and the most commonly used radionuclide in nuclear medicine.

technique *(tĕk-nēk').* The systematic procedure or process by which a task is completed.

teeth *(tēth). See* **tooth.**

Tegretol *(tĕg'-rĕ-tŏl).* Trade name for preparations of carbamazepine, used to treat the pain associated with trigeminal neuralgia and certain seizures associated with epilepsy.

telalgia *(tĕl-ăl'-jĕ).* Referred pain.

telangiectasia, telangiectasis *(tĕl-ăn-jē-ĕk-tā'-zhĕ) (tĕl-ăn-jē-ĕk'-tĕ-sĭs).* Permanent dilation of small blood vessels, resulting in red lesions in the skin or mucous membranes.

telangitis *(tĕl-ăn-jē-ī'-tĭs).* Inflammation of the capillaries; also called *capillaritis.*

telecardiography *(tĕl-ĕ-kăhr-dē-ŏg'-rĕ-fē).* Recording an electrocardiogram by transmitting the impulses to a location at a distance from the patient.

telediagnosis *(tĕl-ĕ-dī-ĕg-nō'-sĭs).* Determining a diagnosis at a location remote from the patient based on the electronic transmission of health information.

telemedicine *(tĕl-ĕ-mĕd'-ĭ-sĭn).* A branch of telehealth in which the diagnosis and treatment of medical conditions are done at a great distance through methods such as videoconferencing or rapid transmission of digital files, all facilitated by healthcare professionals at each location.

telesurgery *(tĕl-ĕ-sŭr'-jĕr-ē).* Surgical procedures performed at a distance using advanced robotic and computer technology.

temperature *(tĕm'-pĕr-ĕ-chĕr).* The degree of heat or cold.

basal body t. Temperature of the body when it is at absolute rest.

core t. The temperature of structures deep within the body.

normal t. The temperature of a healthy human body, usually about 98.6° F measured orally.

room t. The ordinary temperature of a room, 65°–80° F.

temple *(tĕm'-pĕl).* The region of the head in front of and slightly above the ear.

temporal *(tĕm'-pĕ-rĕl).* 1. Referring to the lateral area of the head. 2. Referring to time; temporary.

temporomandibular joint *(tĕm-pŏ-rŏ-măn-dĭb'-ū-lăr).* The junction of the lower jawbone (the mandible) and the temporal bone, which is immediately in front of the ear.

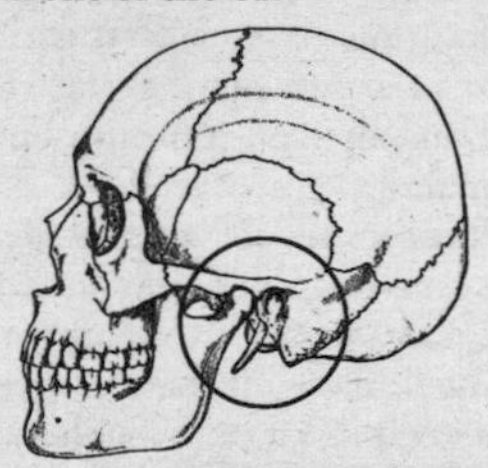

temporomandibular joint

temporomandibular joint syndrome. A disorder of the temporomandibular joint, characterized by pain, usually in front of the ear(s), headache, and sometimes neck pain and spasm. Grinding and clenching the teeth due to stress are common causes; trauma, arthritis, and poor dental work are others.

tenaculum *(tĕ-năk'-ū-lĕm).* A hook-like instrument used to seize and hold tissues.

tenderness *(tĕn'-dĕr-nĕs).* Abnormal sensitiveness to pressure or touch.

tendinitis *(tĕn-dĭ-nī'-tĭs).* Inflammation of tendons and where tendons and muscles attach.

tendinoplasty *(tĕn-dĭ-nō-plăs'-tē).* *See* **tenoplasty.**

tendon *(tĕn'-dŏn).* A type of connective tissue by which a muscle is attached.

tendonitis *(tĕn-dŏn-ī'-tĭs).* Inflamed tendon, usually due to excessive overuse. *See also* **tendinitis.**

tenesmus *(tĕ-nĕz'mĕs).* Straining, especially when referring to painful straining in urination or to pass stools.

tennis elbow *(tĕn'-ĭs).* An injury to the tendon that is attached to the outer part of the elbow caused by repetitive twisting of the forearm or wrist. Characterized by inflammation of the extensor tendon.

tenodesis *(tĕ-nŏd'-ĕ-sĭs).* Suturing of the end of a tendon to a bone.

tenoplasty *(tĕn'-ō-plăs-tē).* Surgical repair of a defect in a tendon.

Tenormin *(tĕn'-ŏr-mĭn). See* Appendix, Common Prescription and OTC Drugs: By Trade Name.

tenorrhaphy *(tĕ-nŏr'-ĕ-fē).* Suturing of a divided tendon.

tenosynovitis *(tĕn-ō-sĭn-ō-vī'-tĭs).* Inflammation of a tendon sheath.

tenotomy *(tĕ-nŏt'-ĕ-mē).* Surgical cutting of any tendon.

tenovaginitis *(tĕn-ō-văj-ĭ-nī'-tĭs).* Inflammation of a tendon sheath; also called *tenosynovitis.*

tension *(tĕn'-shŭn).* The state of being stretched or strained.

tensile strength *(tĕn'-sĭl).* The maximum amount of stress a substance can maintain without rupturing.

tentorium *(tĕn-tō'-rē-ŭm).* A tent-like structure or part.

tepid *(tĕ'-pĭd).* Lukewarm.

teratocarcinoma *(tĕr-ŏ-tō-kăhr-sĭ-nō'-mă).* A malignant neoplasm that occurs most often in the testis and which contains elements of teratoma along with those of embryonal carcinoma or choriocarcinoma.

teratogen *(tĕr'-ĕ-tō-jĕn).* Any substance that can hinder the development of an embryo or fetus, cause a birth defect, or stop a pregnancy.

teratoma *(tĕr-ĕ-tō'-mă).* A type of germ (reproductive) cell tumor composed of different types of tissue from one or more of the three germ cell layers.

term *(tĕrm).* 1. A limit or boundary. 2. A specified period of time.

terminal *(tĕr'-mĭ-nĕl).* Referring to or forming an end.

tertiary medical care *(tĕr'-shē-ăr-ē).* Health care provided by specialists who have had patients referred to them by other physicians.

test *(test).* A procedure done to aid in making a diagnosis. *See* individual names of tests.

testis *(tĕs'-tĭs)* (*pl.* testes). Either of the oval-shaped glands, located in the scrotum, where spermatozoa are produced and from which the hormone testosterone is secreted.

abdominal t. An undescended testis located in the abdominal cavity.

ascending t. A testis that is located in the scrotum in an infant male but later ascends.

undescended t. A testis that fails to descend into the scrotum.

test meal. Consumption of a meal, which is then removed from the stomach via a tube after a specified number of minutes or hours, for the purpose of analyzing how the stomach is functioning.

testosterone *(tĕs-tŏs'-tĕ-rōn).* A hormone produced in the testes that is responsible for male characteristics; it also stimulates skeletal muscle and influences metabolism. This hormone is also produced by the adrenal gland in both males and females.

test tube baby. An infant born as the result of fertilization that has occurred outside the female's body. Eggs that have been removed from a female are combined with sperm and cultured. If fertilization occurs, the egg (or eggs) is implanted into the woman's uterus.

tetanus *(tĕt'-ĕ-nĕs).* An acute, often fatal infectious disease caused by the bacillus *Clostridium tetani,* which usually invades the body through a puncture wound. Characteristics include muscular contractions, lockjaw, respiratory spasm, seizures, and paralysis.

tetany *(tĕt'-ĕ-nē).* A nervous condition characterized by intermittent tonic (persistent, involuntary) spasms that involve the extremities.

tetracycline *(tĕt-rĕ-sī'-klēn).* Any of a group of related broad-spectrum antibiotics effective against aerobic and anaerobic gram-positive and gram-negative bacteria, as well as some protozoa.

tetralogy of Fallot *(tĕ-trăl'-ĕ-jē).* A combination of four heart defects that are present at birth and make up about 10 percent of all congenital heart disease. The four defects include a hole between the left and right ventricles, pulmonary stenosis, enlargement and thickening of the right ventricle, and an aorta that straddles the septum.

thalamus *(thăl'-ĕ-mĕs).* A region in the posterior part of the brain consisting of three parts, each of which contains cells that serve as relay centers for sensory signals.

thalassemia *(thăl-mĕ-sē'-mē-ĕ).* A classification of hereditary anemias seen in people who live along the Mediterranean and in Southeast Asia. Severity varies; along with anemia some characteristics may include an enlarged heart and/or spleen, jaundice, fatigue, and leg ulcers.

thalidomide *(thĕ-lĭd'-ō-mīd).* A sedative and hypnotic commonly used in Europe in the 1950s and 1960s until it was discovered to cause severe congenital defects in the fetus. It is now used to treat leprosy.

thallium-201 *(thăl'-ē-ĕm).* A radioactive isotope of thallium that is used in the form of thallous chloride to aid in diagnostic imaging.

thanatoid *(thăn'-ĕ-toid).* Resembling death.

thanatology *(thăn-ĕ-tŏl'-ĕ-jē).* The medicolegal study of death and conditions that affect dead bodies.

thanatophobia *(thăn-ĕ-tō-fō'-bē-ĕ).* An irrational fear of death.

theca *(thē'-kĕ).* A sheath or enclosing case.

thecitis *(thē-sī'-tĭs).* Inflammation of the sheath of a tendon; also called *tenosynovitis.*

thecoma *(thē-kō'-mĕ).* A theca cell tumor.

thelalgia *(thē-lăl'-jĕ).* Pain in the nipple.

thelerethism *(thĕ-lĕr'-ĕ-thĭz-ĕm).*

Protrusion or erection of the nipple.

thelitis *(thē-lī'-tĭs).* Inflammation of the nipple.

thenar *(thē'-nĕr).* The pad of flesh on the palm at the base of the thumb.

theobromine *(thē-ō-brō'-mīn).* A white crystalline alkaloid found in cocoa or made synthetically, it is used as a diuretic, a smooth muscle relaxant, a vasodilator, and a myocardial stimulant.

theophylline *(thē-ŏf'-ĕ-līn).* A compound found in tea leaves and also prepared synthetically; its salts and derivatives act as smooth muscle relaxants, central nervous system and heart muscle stimulants, and bronchodilators.

theory *(thē'-rē).* An assumption that is based on specific evidence or observations but lacking definitive scientific proof.

therapeutics *(thĕr-ĕ-pū'-tīks).* The branch of medical science that is involved with the treatment of disease.

therapeutic touch. An alternative medicine practice based on the premise that the therapist possesses a healing force that can be passed along to a patient when the therapist passes his/her hands over the patient and identifies energy imbalances in the patient.

therapist *(thĕr'-ĕ-pĭst).* An individual who is skilled in the treatment of disease or a particular type of treatment (e.g., physical therapy, speech therapy).

therapy *(thĕr'-ĕ-pē).* The treatment of disease.

endocrine t. Treatment of disease using hormones.

group t. A form of psychotherapy in which a group of people are led by a group leader, who is usually a therapist.

maintenance t. Therapy for chronically ill individuals that is designed to keep the medical condition at its present level and prevent any worsening.

occupational t. The therapeutic use of work, play, and self-care activities designed to improve a patient's function, enhance development, and prevent disability.

physical t. Treatment using physical means, such as exercise and specially designed movements, rather than surgical or medical approaches to help patients disabled by disease, injury, or pain.

replacement t. Treatment designed to replace deficiencies in the body by administering appropriate substances.

speech t. The use of special techniques to help correct speech and language disorders.

thermal *(thĕr'-mĕl).* Referring to or characterized by heat.

thermesthesia *(thĕrm-ĕs-thē'-zhē).* A sense of temperature.

thermoanesthesia *(thĕr-mō-ăn-ĕs-thē'-zhĕ).* An inability to recognize sensations of heat and cold.

thermocautery *(thĕr-mō-kăw'-tĕr-ē).* Cauterization using a hot wire or point; *see also* **electrocautery.**

thermodilution *(thĕr'-mō-dī-loo'-shĕn).* A method of measuring blood flow by injecting a cool or cold substance, such as a saline solution, into the cardiovascular system and measuring the temperature over time.

thermography *(thĕr-mŏg'-rĕ-fē).* Use of an infrared camera to photograph the surface temperatures of the body, sometimes used to diagnose underlying disease processes, such as breast tumors.

thermoplegia *(thĕr-mō-plē'-jĕ).* Heat stroke.

thermostabile *(thĕr-mō-stā'-bĭl).* The quality of withstanding the effects of heat without undergoing changes.

thermotaxis *(thĕr-mō-tăk'-sĭs).* The normal adjustment of body temperature.

thermotherapy *(thĕr-mō-thĕr'-ĕ-pē).* The application of heat to treat disease.

thiamine *(thī'-ĕ-mĭn).* Vitamin B1, a water-soluble compound, is essential for the normal metabolism of carbohydrates and fats. It is found especially in wheat germ, pork, legumes, nuts, whole grains, and organ meats.

thiazides *(thī'-ĕ-zīds).* A group of compounds that act as diuretics and are used for the treatment of edema caused by congestive heart failure or chronic hepatic or renal disease.

thigh *(thī).* The portion of the lower extremity extending from the hip to the top of the knee.

thimerosal *(thĭ-mĕr'-ō-sĕl).* A mercury-containing preservative used in some vaccines and other products since the 1930s. In 1999, use of thimerosal was reduced or eliminated in vaccines as a precautionary measure against the possibility that it causes autism or other behavioral disorders.

thoracentesis *(thŏr-ă-sĕn-tē'-sĭs).* Surgical puncture of the chest wall, usually with a large-bore needle, to remove fluids.

thoracic *(thĕ-răs'-ĭk).* Referring to or concerning the thorax (chest).

thoracic duct. The primary lymph duct that receives lymph from all of the body except the right side of the head, neck, and thorax and right upper extremity.

thoracoscope *(thĕr-ăk'-ō-skōp).* An endoscope used to examine the pleural cavity. The scope is inserted into the cavity through an incision in an intercostal space.

thoracoscopy *(thŏr-ĕ-kŏs'-kĕ-pē).* An examination of the pleural cavity through an endoscope to aid in diagnosis.

thoracostomy *(thŏr-ĕ-kŏs'-tĕ-mē).* Surgical creation of an opening in the chest wall in order to drain fluids.

thorax *(thŏr'-ăks).* The chest.

Thorazine *(thŏr'-ĕ-zēn).* Trade

name for preparations of chlorpromazine, used to treat severe behavioral disorders in children.

thready pulse *(thrĕd'-ē).* *See* **pulse.**

threonine *(thrē'-ō-nīn).* An essential amino acid required for optimal growth in infants and for nitrogen balance in adults.

threshold *(thrĕsh'-ōld).* The minimum amount of input needed to produce a result or cause an event to occur.

auditory t. The slightest perceptible sound.

pain t. *See* **pain threshold.**

throat *(thrōt).* The pharynx.

thrombectomy *(thrŏm-bĕk'-tĕ-mē).* Surgical removal of a thrombus from a blood vessel.

thrombin *(thrŏm'-bĭn).* An enzyme formed from prothrombin, it converts fibrinogen to fibrin, which is the basis of a blood clot.

thromboangiitis obliterans *(thrŏm-bō-ăn-jē-ī'-tĭs).* A chronic, recurring, inflammatory disease that primarily affects the peripheral arteries and veins in the extremities. Symptoms include severe pain in a leg or foot that is worse at night; blue, cold extremity; thrombosis; and reduced sense of heat and cold.

thromboarteritis *(thrŏm-bō-ăr-tĕ-rī'-tĭs).* Inflammation of an artery in connection with thrombosis.

thrombocytopenia *(thrŏm-bō-sī-tō-pē'-nē-ă).* Abnormal decrease in the number of blood platelets.

thrombocytosis *(thrŏm-bō-sī-tō'-sĭs).* An increase in the number of circulating platelets.

thromboembolism *(thrŏm-bō-ĕm'-bō-lĭz-ĕm).* Blockage of a blood vessel with thrombotic material that has been transported to the site from another location from which it broke away.

thrombolysis *(thrŏm-bŏl'-ĭ-sĭs).* The dissolution or breaking up of a thrombus.

thrombopenia *(thrŏm-bō-pē'-nē-ă).* *See* **thromobocytopenia.**

thrombophilia *(thrŏm-bō-fĭl'-ē-ă).* The tendency to form thromboses (blood clots). This tendency can be inherited or acquired.

thrombophlebitis *(thrŏm-bō-flĕ-bī'-tĭs).* Inflammation of a vein (phlebitis) associated with the formation of a thrombus.

thromboplastin *(thrŏm-bō-plăs'-tĭn).* A blood coagulation factor (factor III) present in both blood and tissues.

thrombosis *(thrŏm-bō'-sĭs).* The formation, development, or presence of a thrombus.

cerebral t. Thrombosis of a cerebral vessel, which may lead to a cerebral infarction.

coronary t. Development of a thrombus in a coronary artery, usually seen in atherosclerosis and associated with myocardial infarction or sudden death.

thromboxane *(thrŏm-bŏk'-sān).* A substance manufactured by platelets that causes blood clotting and constriction of blood vessels. There are two types of thromboxanes, and both are derived from arachidonic acid.

thrombus *(thrŏm'-bŭs).* A blood clot in a blood vessel or in the heart.

thrush *(thrŭsh).* Candidiasis that affects the oral mucosa. It is characterized by white plaques composed of a soft curdlike substance and usually affects sick infants, individuals in poor health, and immunocompromised patients.

thymectomy *(thī-mĕk'-tĕ-mē).* Surgical removal of the thymus.

thymic *(thī'-mĭk).* Referring to the thymus.

thymol *(thī'-mŏl).* A phenol obtained from thyme oil or other volatile oils and used as an antiseptic, antibacterial, and antifungal.

thymoma *(thī-mō'-mĕ).* A tumor derived from the thymus.

thymus *(thī'-mŭs).* A lymphoid organ located in the center of the chest immediately behind the sternum. The thymus is where lymphocytes mature, reproduce, and become T-cells.

thyroid *(thī'-roid).* A gland located in the neck, in front of and partially surrounding the upper rings of the trachea. It secretes two hormones, triiodothyronine and tetraiodothyronine.

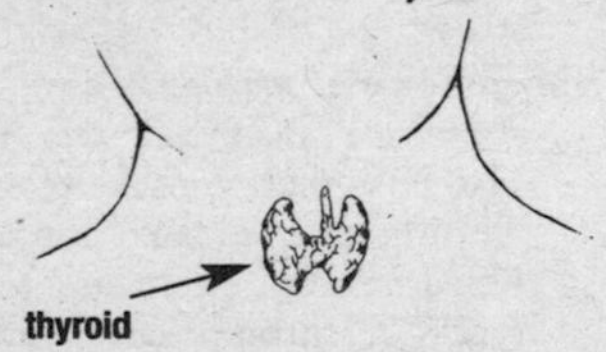

thyroid crisis. *See* **thyroid storm.**

thyroidectomy *(thī-roid-ĕk'-tĕ-mē).* Surgical removal of the thyroid gland.

thyroiditis *(thī-roid-ī'-tĭs).* Inflammation of the thyroid gland.

thyroid-stimulating hormone. A hormone secreted by the pituitary gland that stimulates the thyroid gland.

thyroid storm. A complication of thyrotoxicosis that if not treated can be fatal. Characteristics include sudden onset of fever, rapid heartbeat, sweating, tremors, restlessness, pulmonary edema or congestive heart failure.

thyrotoxicosis *(thī-rō-tŏks-ĭ-kō'-sĭs).* A toxic condition caused by hyperactivity of the thyroid gland.

thyrotropin *(thī-rŏt'-rĕ-pĭn).* A hormone that promotes the growth of, maintains, and stimulates the hormonal secretions of the thyroid gland.

thyroxine *(thī-rŏk'-sĭn).* The main hormone secreted by the thyroid gland. Its primary function is to increase the rate of cell metabolism; it also plays a role in the central nervous system and regulates other functions.

TIA. Acronym for transient ischemic attack.

tibia *(tĭb'-ē-ĕ).* The shin bone, or the inner and larger bone of the leg below the knee.

tic *(tĭk).* An involuntary, repetitive, compulsive movement or vocalization that may be psychological or neurological in origin. Occurrence is exacerbated by stress and reduced during sleep or engrossing activities.

convulsive t. A facial spasm.

diaphragmatic t. Spasmodic twitching of the diaphragm.

habit t. Any tic that is psychological in origin.

tic douloureux. Pressure on or degeneration of the trigeminal nerve, resulting in neuralgia and severe, stabbing pain that radiates from the angle of the jaw along one or more of the three branches associated with the nerve.

tick *(tĭk).* A blood-sucking parasite that can transmit serious diseases to humans.

tick fever. Any one of several diseases caused by a tick bite.

t.i.d. Latin for "ter in die" (three times a day); typically used to indicate how often a medication or therapy should be administered.

tidal drainage *(tī'-dăl).* A device that allows the automatic drainage of urine from a paralyzed bladder.

tilt-table test. A test designed to detect postural hypotension, a condition that results when changing from a horizontal position to a more vertical one. The test involves placing a patient on a table, which is tilted upward while blood pressure and pulse are measured and symptoms recorded.

timolol *(tī'-mō-lōl).* A beta-adrenergic blocking agent used to reduce intraocular pressure in glaucoma and ocular hypertension.

Tinactin *(tĭn-ăk'-tĭn).* Trade name for preparation of timolol maleate, used to reduce intraocular pressure in glaucoma and ocular hypertension.

tincture *(tĭnk'-chĕr).* An alcohol- or alcohol-and-water-based solution prepared from vegetable materials or chemical substances.

tinea *(tĭn'-ē-ă).* Any fungal skin disease that can occur on various parts of the body. The name indicates the part affected.

t. barbae. Tinea of the bearded area of the face and neck.

t. capitis. Tinea that affects the scalp and may also affect the eyebrows and eyelashes.

t. corporis. Tinea of the body, characterized by red, slightly elevated scaly patches that are itchy.

t. cruris. Tinea in the groin or perineal area that may spread to nearby areas.

t. pedis. Tinea that involves the feet. Also called *athlete's foot.*

t. unguium. Tinea that involves the nails, usually caused by both a fungus and bacteria.

tinnitus *(tĭn-ī'-tĕs).* A sound in the ears, such as ringing, buzzing, roaring, or clicking.

tissue *(tĭsh'-oo).* An aggregation of similar cells, such as brain cells, muscle cells.

connective t. Tissue that binds together and supports various structures of the body.

erectile t. Spongy tissue that contains spaces that fill with blood, causing it to expand and harden. Such tissue is found in the penis, clitoris, and the nipples.

fibrous t. Connective tissue that consists primarily of fibers.

granulation t. The new tissue that forms as wounds heal.

tissue plasminogen activator (tPA). An enzyme that helps dissolve blood clots. It is used in the treatment of heart attack and stroke.

tissue typing. Techniques used to determine the compatibility of tissues to be used in transplants and grafts with the recipient's tissues and cells.

titration *(tī-trā'-shŭn).* An estimation of the concentration of a chemical solution by adding known amounts of a liquid reagent until a given change in color is reached.

tobramycin *(tō-brĕ-mī'-sĭn).* An aminoglycoside antibiotic that is effective against many aerobic gram-negative and some gram-positive bacteria. It is used externally to treat eye infections and internally to treat *Pseudomonas aeruginosa* infections in patients with cystic fibrosis.

tocography *(tō-kŏg'-rĕ-fē).* A graphic recording of uterine contractions.

tocopherol *(tō-kŏf'-ĕr-ŏl).* A generic term for vitamin E and several chemically related compounds, most of which have the biological activity of vitamin E.

Tofranil *(tō-frā'-nīl).* Trade name for a preparation of imipraine pamoate, used to treat depression.

tolbutamide *(tŏl-būt'-ă-mīd).* An oral hypoglycemic drug used to treat type II diabetes. Trade name: Orinase.

tolerance *(tŏl'-ĕr-ăns).* A capacity for enduring a great amount of a substance without experiencing an adverse reaction and demonstrating a decreased sensitivity to subsequent doses of the same substance.

tomography *(tō'-mō-grăf).* An instrument that moves an x-ray source in one direction as the film is moved in the opposite direction, resulting in a focused image of tissue on one plane while blurring or eliminating details in other planes; *see also* **computed tomography; positron emission tomography.**

tomography *(tō-mŏg'-rĕ-fē).* Recording of internal body structures using a tomography.

tongue *(tŭng).* The muscular organ that arises from the floor of the mouth and back of the throat. It is the organ of speech and taste; it also aids in chewing.

bifid t. A birth defect in which the front of the tongue is split into two parts.

coated t. A tongue covered with a white or yellow layer of bacteria, debris, and other material that can be easily removed by scraping.

hairy t. A benign condition characterized by a furlike appearance that is the result of enlargement of the filiform papillae (minute projections) on the tongue.

strawberry t. Condition in which the tongue is dark red and the surface resembles that of a strawberry. Seen in conditions such as *Staphylococcus aureus* infection and streptococcal pharyngitis.

tonic *(tŏn'-ĭk).* 1. Producing and restoring normal tone. 2. When referring to seizures, the type that is characterized by continuous tension.

tonicity *(tō-nĭs'-ĭ-tē).* The state of tissue tension or tone.

tonometer *(tōn-ŏm'-ĕ-tĕr).* An instrument that measures pressure or tension, especially one that measures intraocular pressure.

tonsil *(tŏn'-sĭl).* A small, rounded mass of lymphoid tissue located in the mouth near the back of the tongue.

tonsillectomy *(tŏn-sĭl-ĕk'-tō-mē).* Surgical removal of the tonsils.

tonsillitis *(tŏn-sĭl-ī'-tĭs).* Inflammation of the tonsils.

tonus *(tō'-nŭs).* A slight, continuous contraction of a muscle, which in skeletal muscles serves to maintain posture and help return blood to the heart.

tooth (tooth) Any of the hard calcified structures set in the upper and lower jaws and used for chewing.

bicuspid t. *See* **premolar tooth.**

canine t. The tooth immediately lateral to the second incisor. It has the longest and most powerful root of all the teeth.

deciduous t. The 20 teeth that appear first and then are shed and replaced by the permanent teeth.

impacted t. A tooth that is prevented from emerging by a physical barrier.

incisor t. One of the two most frontal teeth in each jaw.

molar t. The most posterior teeth on either side in each jaw. There are a total of eight molars: two on each side, both upper and lower jaw.

premolar t. The permanent teeth between the canines and the molars. There are two on each side of each jaw.

wisdom t. The third molar and the last of the permanent teeth to emerge, usually between the ages of 17 and 21.

tophus *(tō'-fŭs).* A chalky substance that often forms around the joints in bone, bursae, cartilage, subcutaneous tissue, and in the external ear, resulting in chronic inflammation.

topography *(tō-pŏl'-ră-fē).* The description of an anatomical area or special part of the body.

toponeurosis *(tō-pō-nū-rō'-sĭs).* Neurosis that affects a limited area.

Toprol *(tŏp'-rŏl). See* Appendix, Common Prescription and OTC Drugs: By Trade Name.

torpid *(tŏr'-pĭd).* Sluggish; not acting with normal vitality.

torpor *(tŏr'-pŏr).* Lack of response to ordinary or normal stimuli.

torsion *(tŏr'-shŭn).* The process or act of twisting, turning, or rotating about an axis.

torso *(tŏr'-sō).* The trunk of the body.

torticollis *(tŏr-tĭ-kŏl'-ĭs).* An acquired or congenital condition in which the neck muscles undergo spasmodic contractions, causing the head and neck to be placed in an unnatural position.

torus *(tōr'ŭs).* A rounded swelling or elevation.

total parenteral nutrition (TPN). The provision of a patient's total caloric needs through intravenous feeding when a patient is unable to consume food by mouth.

tourniquet *(toor'-nĭ-kĕt).* An instrument that compresses a blood vessel when it is placed around an extremity to control blood circulation and to prevent blood flow to and from the distal area.

toxemia *(tŏks-ē'-mē-ă).* Any condition that is caused by the spread of toxins, including endotoxins and exotoxins, in the bloodstream.

toxic *(tŏks'-ĭk).* Poisonous.

toxicity *(tŏks-ĭs'-ĭ-tē).* The quality of being poisonous.

toxicology *(tŏks-ĭ-kŏl'-ō-jē).* The study and science of poisons, their actions, and the treatment of the conditions that they cause.

toxicosis *(tŏks-ĭ-kō'-sĭs).* Any disease condition that is caused by poisoning.

toxic shock syndrome. A sometimes fatal disease in which toxins are produced by certain strains of the bacterium *Staphylococcus aureus* or by group A *Streptococcus.* Characteristics include high fever of sudden onset, vomiting, diarrhea, muscle pain, hypotension, and a sunburnlike rash that peels.

toxin *(tŏks'-ĭn).* A poison, often used to refer to a protein produced by certain plants, animals, and bacteria that is highly poisonous to other living organisms.

toxoid *(tŏks'-oyd).* A toxin that has been modified to destroy its toxicity but it is still able to induce formation of antibodies.

toxoplasmosis *(tŏks-ō-plăs-mō'-sĭs).* A disease caused by infection with the protozoa *Toxoplasma gondii.* Symptoms may be very mild or may be more severe and include malaise, muscle pain, pneumonitis, hepatitis, and encephalitis.

trabecula *(tra-bĕk'-ū-lă).* A supportive or anchoring strand of connective tissue; for example, one that extends from a capsule into the interior of an organ.

trabeculectomy *(trĕ-bĕk-ū-lĕk'-tē-mē).* Surgical procedure in which a fistula is created between the anterior chamber of the eye and the subconjunctival space by removing part of the trabecular meshwork. Performed in patients who have glaucoma.

tracer *(trās'-ĕr).* A radioactive particle or particles that is introduced into the body and detected as it travels. Tracers are used in absorption and excretion studies and to study the distribution of substances in the body.

trachea *(trā'-kē-ĕ).* A cylindrical tube about 4.5 inches long that is lined with mucous membrane and extends from the larynx to the bronchial tubes. Also known as the *windpipe.*

tracheitis *(trā-kē-ī'-tĭs).* Inflammation of the trachea. It may be associated with bronchitis and laryngitis.

tracheobronchitis *(trā-kē-ō-brŏng-kī'-tĭs).* Referring to the trachea and the bronchial tubes.

tracheoesophageal fistula *(trā-kē-ō-ē-sŏf'-ă-jē-ĕl).* A birth defect in which there is an abnormal passage between the trachea and the esophagus.

tracheotomy *(trā-kē-ŏt'-ō-mē).* Surgical creation of an opening into the trachea through the neck. Also called a *tracheostomy.*

trachoma *(trā-kō'-mă).* A chronic, infectious disease of the cornea and conjunctiva, characterized by photophobia, pain, and excessive tearing.

tracing *(trās'-ĭng).* A recording of movements or actions; for example, a tracing of heart action seen on an electrocardiogram.

tract *(trăkt).* 1. A bundle of nerve fibers that have the same origin, function, and end point. 2. A number of organs that serve a common function.

alimentary t. The part of the digestive system made up of the esophagus, stomach, and small and large intestines.

biliary t. The organs, ducts, and other structures that are involved in the secretion, storage, and delivery of bile into the duodenum.

gastrointestinal t. The stomach and intestines.

motor t. Any bundle of nerve fibers that carry motor signals from the central nervous system to a muscle.

respiratory t. The organs and structures by which ventilation and gas exchange between air and the blood is accomplished. The main structures include the nose, larynx, trachea, bronchi, bronchioles, and lungs.

traction *(trăk'-shŭn).* The process of drawing or exerting a pulling force.

Bryant's t. Overhead vertical traction to treat a fracture of the femoral shaft.

cervical t. Traction applied to the neck, usually involving a sling placed under the chin and behind the head. Used to treat osteoarthritis, rheumatoid arthritis, and other conditions that affect the cervical spine.

tragus *(trā'-gŭs).* The small projection of cartilage that points backward over the opening of the external ear.

trance *(trăns).* A state of altered consciousness in which an individual experiences enhanced focal awareness and reduced peripheral awareness.

tranquilizer *(trăn-kwĭ-līz'-ĕr).* A drug that produces a calming, soothing effect. It usually refers to minor tranquilizers, or antianxiety agents, rather than major tranquilizers, which are now referred to as antipsychotic agents.

transaminases *(trăns-ăm'-ĭ-nāzēs).* A general term for a group of enzymes that help transform one amino acid into another amino acid.

transducer *(trăns-dū'-sĕr).* A receptor or a device that translates one form of energy into another; for example, the translation of pressure or temperature into an electrical signal.

transduction *(trăns-dŭk'-shŭn).* The transfer of genetic material from one cell to another, which results in a change in the genetic constitution of the second organism.

trans fat. A substance created through a chemical process called hydrogenation, in which liquid oils are solidified to be used in foods to increase shelf life and flavor stability. Ingestion of trans fats is associated with heart disease, obesity, and cancer.

transferase *(trăns'-fĕr-ās).* A class of enzymes that transfer a chemical group from one compound to another compound.

transference *(trăns-fĕr'-ĕns).* In psychotherapy, the unconscious tendency to assign to others in one's current life emotions and attitudes associated with important people in one's early life, especially parents.

transformation *(trăns-fŏr-mā'-shĕn).* Change of structure or form.

transfusion *(trăns-fū'-zhĕn).* The introduction of whole blood or blood components directly into the bloodstream.

exchange t. Repetitive withdrawal of small amounts of blood and replacement with donor blood until much of the blood has been exchanged.

immediate t. Transfer of blood from one person to another without using an intermediate container or anticoagulant. Also called *direct t.*

indirect t. Transfer of blood from a donor to a container, and then to the recipient.

placental t. Returning to a newborn some of the blood contained in the fetal placenta.

transitional cell carcinoma *(trăn-zĭ'-shĕn-ăl).* A type of cancer that develops in the lining of the ureter, bladder, or renal pelvis.

transmigration *(trăsn-mī-grā'-shĕn).* Wandering away from, across, or through, especially a change of place from one side of the body to the other.

transmissible *(trăns-mĭs'-ă-bĕl).* Capable of being carried from one person to another, such as an infection.

transmutation *(trăns-mū-tā'-shŭn).* 1. Evolutionary change of one species into another. 2. The change of a chemical element into another.

transparent *(trăns-păr'-ĕnt).* Allowing the passage of light rays, so that objects can be seen through the substance.

transpiration *(trăn-spĭ-rā'-shĕn).* The release of air, sweat, or vapor through the skin.

transplant *(trăns-plănt).* An organ or tissue taken from the body to be grafted or implanted into another area of the same body or into another individual.

transposition *(trăns-pō-zĭ'-shĕn).* A transfer of position from one site to another.

transsexualism *(trăn-sĕks'-ū-ă-lĭzm).* A persistent intense desire to change one's physical gender. People who practice transsexualism may dress and act as individuals of the opposite sex and take hormones or undergo surgery to develop the desired secondary sex characteristics.

transsexual surgery. Surgical procedures that alter the anatomical sex of an individual whose psychological gender is not in agreement with his/her anatomical sexual characteristics.

transudate *(trăns'-ū-dāt).* The fluid that passes through a membrane, especially one that passes through capillary walls.

transvaginal *(trăns-văj'-ĭ-nĕl).* Performed through the vagina.

transverse *(trăns-vĕrs').* Placed crosswise.

transverse colon. *See* **colon.**

trapezius *(tră-pē'-zē-ŭs).* A flat, triangular muscle that covers the back of the neck and shoulder.

trauma *(trăw'mă).* Physical or psychological injury or damage.

traumatology *(trăw-mĕ-tŏl'-ĕ-jē).* The branch of surgery that deals with wounds and any disabilities that arise from injuries.

traveler's diarrhea. *See* **diarrhea.**

trazodone *(trā'-zō-dōn). See* Appendix, Common Prescription and OTC Drugs: By Generic Name.

tremor *(trĕm'-ĕr).* An involuntary shaking or quivering.

action t. An involuntary, rhythmic, oscillatory motion of a body part during voluntary movement.

essential t. A hereditary tremor that begins as a fine rapid tremor of the hands, followed by tremor of the head, tongue, limbs, and trunk, and is exacerbated by emotional factors.

parkinsonian t. The resting tremor commonly seen with parkinsonism, characterized by slow, regular movements of the hands and frequently the face, jaw, neck, and lower limbs. It can be exacerbated by fatigue, cold, and emotional stress.

pill-rolling t. A parkinsonian

tremor of the hand characterized by flexion and extension of the fingers and thumb.

resting t. A tremor that occurs when a limb or other body part is at rest.

tremulous *(trĕm'-ū-lĕs)*. Characterized or referring to tremors.

trench mouth *(trĕnch)*. A painful condition in which there is ulceration of the mucous membranes of the mouth and pharynx.

trephination *(trĕf-ĭ-nā'-shĕn)*. The process of cutting out a piece of bone with a surgical instrument called a trephine. The procedure usually involves cutting into the skull.

tretinoin *(trĕt'-ĭ-noin)*. All-trans-retinoic acid, a substance applied topically to treat acne vulgaris. A trade name is Retin-A.

triad *(trī'-ăd)*. 1. A group of three objects or entities. 2. A trivalent element.

Charcot's t. 1. A symptom complex consisting of nystagmus (eye twitching), tremor, and staccato speech, once thought to be a sign of multiple sclerosis. 2. A symptom complex of biliary colic, jaundice, and fever and chills, associated with intermittent cholangitis.

Virchow's t. Three features that predispose one to vascular thrombosis: changes in the vascular wall, changes in blood-flow pattern, and changes in the blood constituents.

Whipple's t. Three features of insulin-producing tumors: Spontaneous hypoglycemia with blood glucose levels less than 50 mg/100 mL; vasomotor or central nervous system symptoms; and symptom relief with administration of intravenous glucose.

triage *(trē'ăhzh)*. The sorting and prioritizing of patients for treatment in either an emergency or nonemergency situation.

triamcinolone *(trī-ăm-sĭn'-ĕ-lōn)*. *See* Appendix, Common Prescription and OTC Drugs: By Generic Name.

triceps *(trī'-sĕps)*. Typically refers to the muscles in the back of the upper arms that help extend the elbow.

trichiasis *(trĭ-kī'-ĕ-sĭs)*. A condition of ingrowing hairs around an opening or of ingrowing eyelashes.

trichinosis *(trĭk-ĭ-nō'-sĭs)*. A disease caused by infection with *Trichinella spiralis*, which occurs after eating undercooked contaminated meat. Symptoms include diarrhea, nausea, colic, and fever, followed by stiffness, swollen muscles, sweating, and insomnia.

trichobezoar *(trĭk-ō-bē'-zŏr)*. A hard mass composed of hairs that forms within the stomach or intestines.

trichoesthesia *(trĭk-ō-ĕs-thē'-zhĕ)*. The perception that a hair on the skin has been touched, which is caused by stimulation of a hair follicle receptor.

trichology *(trī-kŏl'-ĕ-jē).* The study of hair and its care, treatment, and diseases that affect it.

Trichomoniasis vaginalis *(trĭk-ō-mō-nī'-ĕ-sĭs vă-jĭ-năl'-ĭs).* An infection with *Trichomonas vaginalis,* seen in both males and females and usually transmitted by sexual intercourse. Symptoms in females can include severe vaginitis; in males, urethritis, an enlarged prostate, and epididymitis.

trichomycosis *(trĭk-ō-mī-kō'-sĭs).* Any disease of the hair that is caused by a fungus.

trichophagia *(trĭk-ō-fā'-jĕ).* The habit of eating hair.

trichosis *(trī-kō'-sĭs).* Any disease or abnormal growth of hair.

tricuspid *(trī-kŭs'-pĭd).* 1. Having three cusps or points. 2. Referring to the tricuspid valve; *see* **valve.**

trifocal lenses *(trī'-fō-kăl).* A lens consisting of three parts: one each for near, intermediate, and distant vision.

trigeminal nerve *(trī-jĕm'-ĭ-nĕl).* The fifth cranial nerve, which supplies the face.

trigeminal neuralgia. *See* **neuralgia.**

trigger finger *(trĭg'-ĕr).* Condition in which flexion or extension of a digit is temporarily interrupted, then completed with a jerk.

trigger point. A site of muscle tension caused by pain from the spine and connected tissue that when pressed feels like a hard nodule.

triglyceride *(trī-glĭs'-ĕr-īd).* A combination of glycerol with three of five different fatty acids. Elevated levels contribute to arteriosclerosis.

triiodothyronine *(trī-ī-ō-dō-thī'-rō-nēn).* An iodine-containing hormone secreted by the thyroid in amounts smaller than those of thyroxine. Most triiodothyronine is produced by the deiodination of thyroxine in the liver and other tissues; *see also* **thyroxine.**

trimester *(trī-mĕs'-tĕr).* A period of three months; usually refers to pregnancy.

Trimox *(trī'-mŏks).* *See* Appendix, Common Prescription and OTC Drugs: By Trade Name.

triplet *(trĭp'-lĕt).* One of three infants produced in one gestation and one birth.

trisomy *(trī'-sō-mē).* In genetics, having three homologous chromosomes per cell instead of two. Depending on which chromosome is affected, children are born with severe deformities and/or mental retardation. Down's syndrome is a trisomy condition.

trocar *(trō'-kăhr).* A sharp-pointed surgical instrument with a cannula, used to puncture the wall of a body cavity and withdraw fluid.

trochanter *(trō-kăn'-tĕr).* A bony prominence or projection below the neck of the femur (thighbone).

troche *(trō'-kē).* A small medicated tablet, or lozenge, usually used to treat a sore or inflamed throat.

trochlear nerve *(trōk'-lē-ĕr).* The fourth cranial nerve; it supplies the muscle that goes to the upper, outer portion of the eyeball.

trophic *(trŏf'ĭk).* Referring to nutrition or nourishment.

trophology *(trō-fŏl'-ĕ-jē).* The science of nutrition.

tropism *(trō'-pĭz-ĕm).* An involuntary movement, turning, or growth of an organism or part of an organism in response to an external stimulus.

trunk *(trŭngk).* The main part of the body to which the limbs and head are attached. Also called a *torso.*

truss *(trŭs).* A cloth or metallic device used to retain a hernia reduced within the abdominal cavity.

trypanosomiasis *(trī-păn-ō-sō-mī'-ĕ-sĭs).* Infection with protozoa of the genus *Trypanosoma,* which causes sleeping sickness.

trypsin *(trĭp'-sĭn).* A substance, secreted by the pancreas, that plays an important role in protein digestion.

tryptophan *(trĭp'-tō-făn).* An essential amino acid that is necessary for optimal growth in infants and for nitrogen balance in adults.

tubal ligation *(tŭ'-bĕl lī-gā'-shĕn).* A surgical procedure in which the uterine tubes are severed, crushed, or constricted, rendering the female sterile.

tubal pregnancy. *See* **pregnancy.**

tubercle *(too'-bĕr-kĕl).* 1. A small, round, gray lesion characteristic of tuberculosis. 2. A nodule, such as a rounded prominence on a bone.

tuberculin test *(too-bĕr'ū-lĭn).* A skin test for tuberculosis that uses various types of tuberculin and application methods.

tuberculosis *(too-bĕr-kū-lō'-sĭs).* Any of various infectious diseases caused by species of *Mycobacterium* and characterized by the formation of tubercles and tissue death.

basel t. Tuberculosis that is located in the lower part of the affected lung.

disseminated t. The spread of tubercle bacilli from the primary site to other parts of the body.

primary t. Pulmonary tuberculosis when an individual is first infected. This form is often asymptomatic.

pulmonary t. Infection of the lungs by *Mycobacterium tuberculosis.* It may spread to other organs via the blood or lymph. Symptoms may include weight loss, fatigue, night sweats, chest pain, and wasting.

tuberosity *(too-bĕ-rŏs'-ĭ-tē).* An elevation or protuberance.

tuboplasty *(too'-bō-plăs-tē).* Surgical plastic repair of a tube, such as the uterine tube (*see* **salpingoplasty**).

tubule *(too'-būl).* A small tube.

tularemia *(too-lĕ-rē'-mē-ĕ).* An infectious, plaguelike disease caused by infection with the bacillus *Francisella tularensis.* It is transmitted by the bites of deer flies, fleas, and ticks, ingestion of contaminated food or water, or contact with contaminated animals or their products.

tumefaction *(too-mĕ-făk'-shĕn).* Swelling.

tumor *(too'mĕr).* 1. Swelling. 2. A new growth of tissue in which the reproduction of cells is uncontrolled and progressive. Also called a *neoplasm.*

benign t. A tumor that lacks the ability to spread beyond its fibrous capsule; not malignant.

malignant t. A tumor that has the capability to invade and spread.

mixed t. A tumor that contains more than one type of neoplastic tissue, especially a benign or malignant mixed tumor of the salivary glands.

Schwann's t. *See* **schwannoma.**

theca cell t. A fibroidlike tumor of the ovary that may be the result of excessive production of estrogen.

Wilm's t. A rapidly developing malignant mixed tumor of the kidneys. It typically affects children before age 5 years, but it may be seen in the fetus and among the elderly.

tumorigenesis *(too-mĕr-ĭ-jĕn'-ĕ-sĭs).* The production of tumors.

tumorous *(too'-mĕr-ĕs).* Neoplastic; tumorlike.

tunic *(too'-nĭk).* A sheath or membrane that covers an organ or structure.

tunica vaginalis *(too'-nĭ-kă).* Serous membrane that surrounds the front and sides of the testicle.

tunnel vision. *See* **vision.**

turgid *(tŭr'-jĭd).* Swollen.

turmeric *(tĕr'-mĕr-ĭk).* The rhizome of *Curcuma longa,* a member of the ginger family. Used medicinally for the treatment of indigestion; reportedly has anticancer and anti-inflammatory properties.

Turner's syndrome *(tĕr'-nĕrz).* A chromosome disorder in females characterized by the absence of part or all of the second sex chromosome in some or all cells. It occurs in 1 out of every 2,500 to 3,000 female births. Characteristics include webbing of the neck, delayed growth, cardiovascular abnormalities, broad chest, among others.

tussis *(tŭs'-ĭs).* A cough.

twilight sleep *(twī'-līt slēp).* A state of partial anesthesia in which the sense of pain has been significantly reduced by an injection of morphine and scopolamine.

Tylenol *(tī'-lĕ-nŏl).* *See* Appendix, Common Prescription and OTC Drugs: By Trade Name.

tympanic membrane *(tĭm-păn'-ĭk).* Referring to the tympanum.

tympanic nerve. The nerve that supplies the middle ear, the Eustachian tube, and the mastoid bone.

tympanites *(tĭm-păn-ī'-tēz).* Distension of the stomach caused by excessive gas in the intestinal tract, as is seen in peritonitis.

tympanoplasty *(tĭm-păn-ō-plăs'-tē).* Any of several surgical procedures designed to cure a chronic inflammatory process in the middle ear or to restore hearing, including restoration of the drum membrane.

tympanum *(tĭm'-păn-ŭm).* The eardrum.

typhilitis *(tĭf'-lī-tĭs).* Inflammation of the cecum.

typhoid fever *(tī'-foid).* An acute febrile disease caused by *Salmonella typhi* and usually transmitted by ingesting contaminated food and water. Symptoms and signs include fever, malaise, rash, abdominal pain, enlarged spleen, delirium, and leucopenia.

typhus fever *(tī'-fŭs).* Any of a group of acute infectious diseases caused by *Rickettsia* and characterized by severe headache, chills, back and limb pain, high fever, stupor, and bluish spots.

typing *(tīp-ĭng).* Determination of the category to which an object, individual, or other entity belongs; for example, blood type or tissue type.

tyrosine *(tī'-rō-sēn).* An amino acid present in many proteins, especially casein.

Tzanck test *(tsănk).* Analysis of the fluid from a blister (bulla) to identify Tzanck cells, which are characteristic of varicella (chicken pox), herpes simplex, herpes zoster, and pemphigus vulgaris.

U

ubiquinone *(ū-bĭ-kwĭ-nōn').* A fat-soluble substance present in virtually all cells, where it performs several functions. Also known as *coenzyme Q.*

ulcer *(ŭl'sĕr).* A lesion or open sore on the skin or mucous membrane, caused by superficial loss of tissue, usually accompanied by inflammation, and which may become infected.

decubitus u. Ulcer that develops in pressure areas of skin, especially over a bony site such as the hip bone or heels. Usually seen in bedridden individuals or people otherwise immobilized. Also known as a *bedsore.*

duodenal u. Ulcer that develops on the mucosa of the duodenum, caused by irritation from gastric juices.

gastric u. Ulcer that develops in the stomach; also known as a *peptic ulcer.*

indolent u. Nearly painless ulcer that usually forms on the leg.

penetrating u. An ulcer that extends deep into the tissues of an organ.

peptic u. Term for an ulcer that develops on the mucosa of the stomach or duodenum.

perforated u. Ulcer that extends through the entire thickness of a body part, such as the intestine or foot.

varicose u. Ulcer, especially of the lower extremities, associated with varicose veins.

ulcerative colitis *(ŭl'-sĕ-rā-tĭv). See* **colitis.**

ulitis *(ū-lī'-tĭs).* Gingivitis.

ulna *(ŭl'-nĕ).* The inner and larger bone of the forearm, it is located on the side opposite the thumb.

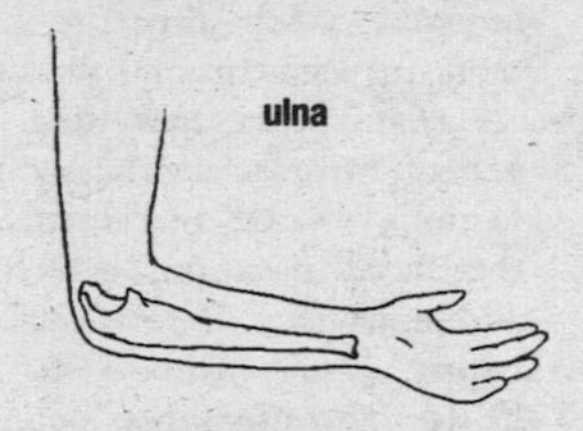

Ultram *(ŭl'-trăm).* Trade name for a preparation of tramadol hydrocholoride, an opioid analgesic, used to treat moderate to severe pain following surgical procedures.

ultrasonic *(ŭl-tră-sŏn'-ĭk).* Referring to sound waves that are beyond the upper limit the human ear can perceive (greater than 20,000 hertz).

ultrasonography *(ŭl-trĕ-sĕ-nŏg'-rĕ-fē).* A method used to visualize the deep structure of the body by recording the echoes of pulses of ultrasonic waves that are focused into the tissues.
continuous wave Doppler u. Method that uses two transducers and is capable of recording signals with very high velocities, such as seen in severely stenotic valves.
Doppler u. Method used to record the velocities of moving objects, such as cardiovascular blood flow.
duplex u. Method that combines real-time with Doppler ultrasonography.
pulsed wave Doppler u. Method that uses one transducer and is capable of determining where signals originate but cannot record high velocity signals.
real-time u. A series of ultrasound images produced in rapid succession so that the monitor shows the motion of an organ or body part.

ultraviolet radiation *(ŭl-trĕ-vī'-ĕ-lĕt).* Rays beyond the violet end of the spectrum. These rays can induce sunburn and tanning of the skin plus produce vitamin D2 (ergocalciferol) by interacting with ergosterol in the skin.

umbilical cord *(ŭm-bĭl'-ĭ-kăl).* The attachment that connects the fetus with the placenta and through which flows blood from the mother. The cord is surgically severed after delivery of the child.

umbilical hernia. A hernia that develops in the region of the umbilicus.

umbilicus *(ŭm-bĭl-ĭ'-kŭs).* A depressed point in the middle of the abdomen or the scar that marks the former site where the umbilical cord was attached to the body.

unconditioned reflex *(ŭn-kŏn-dĭsh'-ŭnd).* An inborn or natural reflex.

unconscious *(ŭn-kŏn'-shŭs).* Lacking awareness of the environment; incapable of responding to sensory stimuli and having subjective experiences.

unction *(ŭngk'-shĕn).* Ointment.

underbite *(ŭn'-dĕr-bīt).* A popular name for retrognathism.

undescended testicle. *See* **testis.**

undifferentiated *(ŭn-dĭf-ĕr-ĕn'-shē-ā-tĕd)*. A loss of differentiation of cells toward a more embryonic type or toward a malignant state.

undulant *(ŭn'-jĕ-lĕnt)*. Characterized by wavelike fluctuations.

unguentum *(ĕng-gwĕn'-tĕm)*. Ointment.

unicellular *(ū-nĭ-sĕl'-ū-lĕr)*. Composed of a single cell, as the bacteria.

unilateral *(ū-nĭ-lăt'-ĕr-ăl)*. Affecting only one side.

unilocular *(ū-nĭ-lŏk'-ū-lĕr)*. Having only one cavity or compartment.

union *(ūn'-yĕn)*. The process of joining two or more things into one part, as a broken bone.

United States Pharmocopeia (USP). A compendium of standards for drugs, published by the United States Pharmacopeial Convention, Inc., and revised periodically.

universal recipient *(ū-nĭ-vĕr'-săl)*. An individual who has blood type AB, Rh positive, and whose serum is compatible with the cells of the other ABO blood.

unsaturated *(ŭn-săt'-ū-rāt-ĕd)*. Capable of dissolving or absorbing to a greater degree.

upper airway obstruction. Condition of the respiratory system in which normal function is hindered by an obstruction in the upper portion of the airway, such as the mouth, larynx, or nose.

upper GI (gastrointestinal) series. A series of X-rays of the esophagus, stomach, and small intestine taken after the patient drinks a barium solution.

upper respiratory infection. A general term for nearly any infectious disease that involves the nasal passages, bronchi, and pharynx.

uranium *(ū-rā'-nē-ŭm)*. A hard, heavy radioactive metallic element.

uranoplasty *(ū-rē-nō-plăs'-tē)*. *See* **palatoplasty.**

urea *(ū-rē'-ă)*. A compound formed in the liver and excreted by the kidney. It is the main end product of protein catabolism and makes up about half of all urinary solids. Elevated levels suggest kidney problems or urinary tract obstruction.

uremia *(ū-rē'-mē-ă)*. Condition characterized by an excess of urea, creatinine, and other nitrogenous products in the blood.

ureter *(ū-rē'-tĕr)*. A fibromuscular tube that carries urine from the kidney to the bladder.

ureterectomy *(ū-rē-tĕr-ĕk'-tĕ-mē)*. Surgical removal of all or part of a ureter.

ureteritis *(ū-rē-tĕr-ī'-tĭs)*. Inflammation of a ureter.

ureterocele *(ū-rē'-tĕr-ō-sēl)*. Cystlike dilatation of the ureter near where it opens into the bladder.

ureterolithiasis *(ū-rē-tĕr-ō-lĭ-thī'-ĕ-sĭs)*. The formation or presence of calculi in the ureter.

ureterolithotomy *(ū-rē-tĕr-ō-lĭ-*

thŏt'-ĕ-mē). Surgical removal of a calculus from the ureter.

ureterostomy *(ū-rē-tĕr-ŏs'-tĕ-mē).* Any of various channels artificially created to divert urine, leaving all or part of the ureter intact.

ureterotomy *(ū-rĕ-tĕr-ŏt'-ĕ-mē).* Surgical incision of a ureter.

uretero-ureterostomy *(ū-rē-tĕr-ō-ū-rē-tĕr-ŏs'-tĕ-mē).* Reestablishment of a passage between the ends of a divided ureter.

ureterovesicostomy *(ū-rē-tĕr-ō-vĕs-ĭ-kŏs'-tĕ-mē).* Reimplantation of a ureter into the bladder.

urethra *(ū-rē'-thră).* The canal that carries urine from the bladder to the exterior of the body.

urethritis *(ŭ-rē-thrī'-tĭs).* Inflammation of the urethra.

urethrocele *(ū-rē'-thrō-sēl).* Pouchlike protrusion of the urethral wall in a female.

urethroplasty *(ū-rē'-thrō-plăs-tē).* Reparative surgery of the urethra.

urethroscope *(ū-rē'-thrō-skōp).* Device used to examine the interior of the urethra.

uric acid *(ū'-rĭk).* An acid that occurs as an end product of purine metabolism. It is a common constituent of urinary and renal calculi and of gouty concretions.

uricemia *(ū-rĭ-sē'-mē-ă).* Excess uric acid in the blood.

urinalysis *(ū-rĭ-năl'-ĭ-sĭs).* Chemical, physical, or microscopic analysis of the urine.

urinary *(ū'-rĭ-năr-ē).* Referring to, secreting, or containing urine.

urinary bladder. Vessel for urine excreted by the kidneys.

urinary calculi. Concretions that form in the urinary passages.

urinary system. Consisting of the kidneys, ureters, bladder, and urethra.

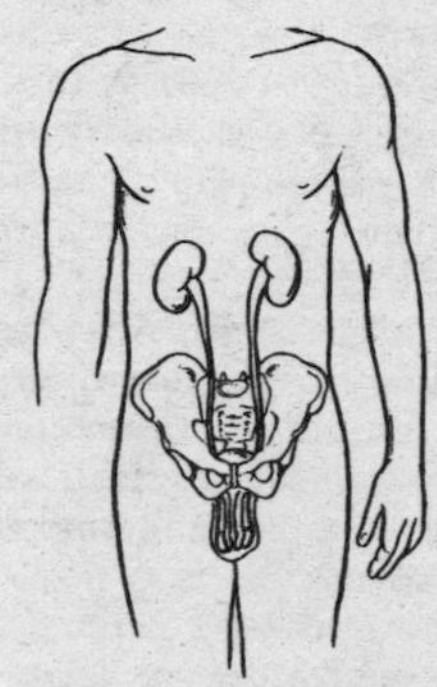

urinary system (male)

urination *(ū-ĭ-nā'-shŭn).* The act of voiding urine.

urine *(ū'-ĭn).* The fluid secreted from the blood by the kidneys, stored in the bladder, and then eliminated through the urethra.

urinemia *(ū-rĕ-nē'-mē-ă).* Accumulation in the blood of substances such as urea, which are normally excreted in the urine.

urobilinogen *(ū-rō-bĭ-lĭn'-ō-jĕn).* A colorless substance formed in the intestines when bilirubin is reduced.

urobilinogenuria *(ū-rō-bĭ-lĭn-ō-jĕ-nū'-rē-ă).* Presence of excessive urobilinogen in the urine, which can occur in liver dysfunction or the jaundice associated with hemolytic anemia.

urodynamic *(ū-rō-dī-năm'-ĭk).* Referring to the flow and motion of liquids in the urinary tract.

urogastrone *(ū-rō-găs'-trōn)*. Human epidermal growth factor.

urogenital *(ū-rō-jĕn'-ĭ-tĕl)*. Referring to the urinary and reproductive organs. *See* **genitourinary.**

urography *(ū-rŏg'-ră-fē)*. X-rays of any part of the urinary tract after a radiopaque contrast substance has been introduced.

urokinase *(ū-rō-kī'-nās)*. An enzyme obtained from human urine and used in the treatment of acute coronary arterial thrombosis and acute pulmonary embolism.

urologist *(ū-rŏl'-ĕ-jĭst)*. A physician who specializes in urology.

urology *(ū-rŏl'-ĕ-jē)*. The medical specialty that deals with the urinary tract in both males and females, and with the genital organs in males.

urosepsis *(ū-rō-sĕp'-sĭs)*. A condition characterized by chills, fever, hypotension, and occasionally altered mental state, caused by an infection that has moved from the urinary tract to the bloodstream.

urticaria *(ŭr-tĭ-kă'-rē-ĕ)*. Hives. Different types are named for the factor that causes them.

aquagenic u. Urticaria caused by exposure of the skin to ordinary water.

cold u. Urticaria caused by exposure to cold, and which may progress to angioedema.

heat u. Form of urticaria caused by application of heat to the skin or by exposure to high temperatures. Symptoms may include cramps, weakness, flushing, and salivation.

solar u. Urticaria that develops rapidly after brief exposure to sunlight.

usher syndrome *(ŭsh'-ĕr)*. A genetic disorder characterized by hearing impairment and retinitis pigmentosa, an eye disorder in which vision worsens over time. More than 50 percent of all deaf-blind people have usher syndrome.

uterine *(ū'-tĕr-ĭn)*. Referring to the uterus.

uterography *(ū-tĕr-ŏg'-ră-fē)*. X-ray of the uterus.

uterosacral *(ū-tĕr-ō-să'-krĕl)*. Referring to the uterus and sacrum.

uterus *(ū'-tĕr-ŭs)*. An organ of the female reproductive system that contains and nourishes the embryo and fetus from the time the fertilized egg implants in the uterus to the time of birth.

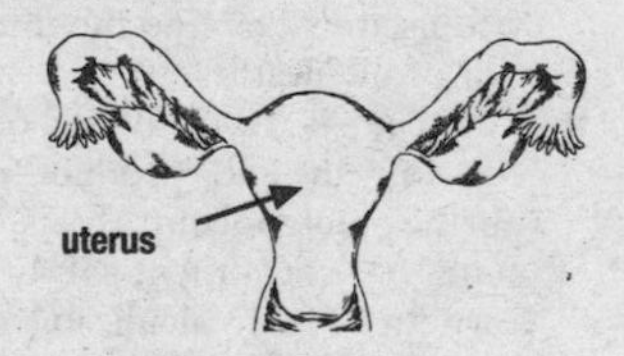

utricle *(ū-trĭk'-ĕl)*. 1. Any tiny sac. 2. One of two sacs in the bony vestibule of the inner ear that communicates with other parts of the ear and helps with changes in position.

uvea *(ū'-vē-ă)*. The layer of the eye located immediately beneath the sclera. It consists of the iris, ciliary body, and choroids, and forms the pigmented layer.

uveitis *(ū-vē-ī'-tĭs).* Inflammation of the iris, ciliary body, and choroids, or the entire uvea.

uveoparotitis *(ū-vē-ō-păr-ĕ-tī'-tĭs).* Inflammation of the parotid gland and uveitis.

uvula *(ū'-vū-lĕ).* The soft structure that hangs from the edge of the soft palate above the root of the tongue.

uvulitis *(ū-vū-lī'-tĭs).* Inflammation of the uvula.

uvulotomy *(ū-vū-lŏt'-ĕ-mē).* Surgical removal of the uvula.

V

vaccination *(văk-sĭ-nā'-shŭn).* The act of giving a vaccine to establish resistance to a specific infectious disease.

vaccine *(văk-sēn').* Any preparation given for the purpose of establishing resistance to an infectious disease. Most are administered via injection, but oral or nasal spray are also utilized.

attenuated v. Live pathogens that have lost their ability to cause disease but which are still capable of inducing an immune response against virulent forms of the pathogen.

autogenous v. One that is prepared from a culture of the patient's own bacteria.

killed v. Vaccine prepared from dead microorganisms, used for strains that are highly virulent.

live v. One prepared from living, weakened (attenuated) organisms.

polyvalent v. One that has been prepared from two or more strains of the same species or microorganism.

vacuole *(văk'-ū-ōl).* A minute space in any tissue or a clear space within a cell.

vagal *(vā'-gĕl).* Referring to the vagus nerve.

vagina *(vĕ-jī'-nă).* The musculomembranous tube that forms the passageway between the uterus and the vulva.

vaginectomy *(văj-ĭn-ĕk'-tĕ-mē).* Surgical removal of part or all of the vagina.

vaginismus *(văj-ĭn-ĭz'-mŭs).* Painful spasm of the vagina that can interfere with or prevent intercourse.

vaginitis *(văj-ĭn-ī'-tĭs).* Inflammation of the vagina.

atrophic v. Vaginitis that follows menopause. Also called *senile v.*

candidal v. Vaginitis caused by infection with *candida albicans.*

***Trichomonas vaginalis* v.** Vaginitis associated with or caused by *T. vaginalis.* It may be spread by sexual intercourse and thus is classified as a sexually transmitted disease.

vaginoplasty *(vă-jī-nō-plăs'-tē).* Plastic surgery of the vagina.

vaginosis *(văj-ĭ-nō'-sĭs).* A disease of the vagina.

bacterial v. Inflammation of the vagina caused by infection, frequently with *Gardnerella vaginalis.*

vagitis *(vă-jī'-tĭs).* Inflammation of the vagus nerve.

vagolysis *(vā-gŏl'-ĭ-sĭs).* Surgical destruction of the vagus nerve.

vagus nerve *(vā'-gŭs).* The tenth cranial nerve, it has both motor and sensory functions and supplies the pharynx, larynx, trachea, lungs, heart, and gastrointestinal tract.

valerian *(vă-lĕr'-ē-ăn).* An herb, *Valeriana officinalis,* used as a sedative and for menopausal symptoms.

valgus *(văl'-gĕs).* Referring to a body part that is twisted or turned away from the midline of the body.

valine *(văl'-ēn).* A nutritionally essential amino acid.

Valium *(văl'-ē-ŭm). See* Appendix, Common Prescription and OTC Drugs: By Trade Name.

valley fever *(văl'-ē).* Coccidioidomycosis, a condition caused by the fungus *Coccidioides immitis,* and seen primarily in the southwestern United States, northwestern Mexico, and parts of Central and South America. Characterized by flu-like symptoms.

valproate sodium *(văl-prō'-āt).* An anticonvulsant used to treat seizure disorders. Similar to its parent compound, valproic acid.

valproic acid *(văl-prō'-ĭk).* An anticonvulsant used to treat seizure disorders.

Valsalva's maneuver *(văhl-săhl'-vĕz).* An attempt to forcibly exhale with the mouth, nose, and glottis closed. This maneuver can slow the pulse, increase intrathoracic pressure, and decrease return of blood to the heart.

valsartan *(văl-săhr'-tăn). See* Appendix, Common Prescription and OTC Drugs: By Generic Name.

valve *(vălv).* A fold of the membrane that lines a canal or another hollow organ.

aortic v. Valve between the left ventricle and the ascending aorta of the heart.

mitral v. The valve that closes the left atrium and left ventricle of the heart.

pulmonary v. The valve that separates the pulmonary artery and right ventricle.

tricuspid v. The valve that closes the opening between the right atrium and right ventricle of the heart.

valvotomy *(văl-vŏt'-ō-mē).* A procedure that involves cutting through a stenosed (narrowed) cardiac valve to relieve obstruction and allow blood flow.

valvulitis *(văl-vū-lī'-tĭs).* Inflammation of a valve, especially a heart valve.

vanadium *(vă-nā'-dē-ŭm).* A metallic element and one that is important in human health, as a deficiency can result in abnormal bone growth and elevate cholesterol and triglyceride levels.

vancomycin *(văn-kō-mī'-sĭn).* An antibiotic effective against gram-negative organisms. Trade name: Vancocin.

vanillin *(vă-nĭl'-ĭn).* Substance that is derived from vanilla and used as a flavoring agent and in pharmaceutical preparations.

vapor *(vā'-pŏr).* 1. The gaseous state of a substance. 2. A medicinal substance that is inhaled.

variant *(văr'-ē-ănt).* Having the tendency to change, show diversity, differ from others, or not conform.

varicella *(văr-ĭ-sĕl'-ă).* An acute contagious disease caused by the varicella-zoster virus and usually occurring in children. Also known as *chicken pox.* It is characterized by an eruption of papules, which become vesicles and then pustules.

varicocele *(văr'-ĭ-kō-sēl).* A condition characterized by abnormal enlargement of the veins of the spermatic cord. Symptoms include a full feeling in the scrotum and dull ache along the cord.

varicocelectomy *(văr-ĭ-kō-sē-lĕk'-tĕ-mē).* Surgical removal of part of the scrotal sac to relieve varicocele.

varicose veins *(văr'-ĭ-kōs).* Enlarged, twisted, superficial veins that most commonly occur in the lower extremities and the esophagus, but may appear anywhere. Caused by faulty valves in the veins, which hinder blood flow.

varicotomy *(văr-ĭ-kŏt'-ō-mē).* Surgical removal of a varicose vein.

variola *(vĕ-rī'-ō-lă).* An acute, contagious viral disease characterized by skin eruptions that pass through phases of macules, papules, vesicles, pustules, and crusts. Also known as *smallpox.*

varix *(vă'-rĭks)* (*pl.* varices). 1. A dilated vein. 2. An enlarged and twisted vein, lymphatic vessel, or artery.

varus *(văr'-ŭs).* Referring to something that is bent or twisted inward toward the midline of the limb or body. Incorrectly used interchangeably with *valgus.*

Vascor *(văs'-kŏr).* Trade name for a preparation of bepridil hydrochloride, used to treat chronic angina pectoris.

vascularization *(văs-kū-lăr-ĭ-zā'-shŭn).* The formation of new blood vessels in a body part.

vascular surgery. Surgery performed on any part of the vascular system.

vascular system *(văs'-kū-lăr).* The body system that consists of the blood vessels, heart, and lymphatics, plus the pulmonary and portal systems considered collectively.

vasculitis *(văs-kū-lī'-tĭs).* Inflammation of a blood or lymph vessel; *see also* **angiitis.**

vas deferens *(văs dĕf'-ĕr-ĕnz).* The secretory duct of the testicle, it transports sperm from each testis to the prostatic urethra. Also known as *ductus deferens.*

vasectomy *(văs-ĕk'-tĕ-mē).* Surgical removal of all or part of the vas deferens, usually performed to make a male sterile.

Vaseline *(văs-ō-lēn').* Trade name

for petroleum jelly; used as a topical ointment to treat minor skin irritation and as a barrier against moisture.

vasoconstriction *(văs-ō-kŏn-strĭk'-shŭn).* Reduction in the caliber of blood vessels.

vasoconstrictor *(văs-ō-kŏn-strĭk'-tŏr).* 1. A substance that causes blood vessels to narrow (constrict). 2. A nerve that, when stimulated, causes blood vessels to constrict.

vasodepression *(văs-ō-dē-prĕs'-shŭn).* Reduction in blood vessel tone with vasodilation, resulting in a decline in blood pressure.

vasodilatation *(văs-ō-dī-lă-tā'-shŭn).* Widening of the caliber of blood vessels.

vasomotor *(văs-ō-mō'-tĕr).* Referring to the nerves that have motor control over blood vessel walls, causing them to constrict or dilate.

vasopressin *(văs-ō-prĕs'-ĭn).* A hormone, produced by the pituitary gland of healthy domestic animals or made synthetically, that has antidiuretic effects and elevates blood pressure.

vasopressor *(văs-ō-prĕs'-ŏr).* A substance that causes vasoconstriction and raises blood pressure.

vasospasm *(văs'-ō-spăzm).* Spasm of a blood vessel.

Vasotec *(văs'-ō-tĕk). See* Appendix, Common Prescription and OTC Drugs: By Trade Name.

vasovagal *(văs-ō-văg'-ăl).* Referring to the action of the vagus nerve on blood vessels.

vasovasostomy *(văs-ō-vă-sŏs'-tō-mē).* Surgical rejoining of a previously severed ductus deferens, done to reverse a vasectomy or correct an obstruction.

VDRL. An acronym for Venereal Disease Research Laboratory.

vector *(vĕk'-tŏr).* A carrier, especially an animal, that transfers an infectious agent from one host to another.

Veetids *(vē'-tĭds). See* Appendix, Common Prescription and OTC Drugs: By Trade Name.

vegan *(vĕj'-ăn, vē'-găn).* A vegetarian who omits all animal foods from his or her diet.

vegetarian *(vĕj-ĕ-tăr'-ē-ăn).* Individual whose diet does not include animal flesh, such as beef, pork, mutton/lamb, veal, fish, and fowl, and which may or may not include other animal products, such as eggs and dairy.

lacto v. Individual who consumes dairy products but not eggs or animal flesh.

lacto-ovo v. Individual who consumes dairy products and eggs but not animal flesh.

vegetative *(vĕj'-ĕ-tā-tĭv).* Functioning unconsciously; a state of greatly impaired consciousness.

vehicle *(vē'-ĭ-kl).* 1. A substance that serves as a medium to carry active ingredients of a medication. 2. A substance (e.g., food, dust, clothing) by or upon which an infectious agent is passed from the infected host to a susceptible recipient.

vein *(vān)*. A blood vessel that carries blood toward the heart. All veins except the pulmonary vein carry unoxygenated blood to the heart.

vellicate *(věl'-ĭ-kāt)*. To contract or twitch spasmodically; especially refers to this action by muscles.

Velpeau's bandage *(věl-pōz')*. A bandage designed to immobilize the shoulder, forearm, and arm.

vena *(vē'-nă)*. Vein.

vena cava inferior. The main vein that drains the lower portion of the body and terminates in the right atrium of the heart.

vena cava superior. The main vein that drains the upper portion of the body and ultimately empties into the right atrium of the heart.

venereal disease *(vĕ-nē'-rē-ăl)*. Disease acquired usually as a result of sexual intercourse with an individual who is afflicted. Some of the diseases include gonorrhea, syphilis, chlamydiosis, AIDS, genital herpes, and genital warts.

venipuncture *(věn'-ē-pŭnk-chŭr)*. Puncture of a vein, usually to obtain a blood sample or to inject a substance.

venlafaxine *(věn-lě-făk'-shěn)*. *See* Appendix, Common Prescription and OTC Drugs: By Generic Name.

venogram *(vē'-nō-grăm)*. A radiograph of the veins.

venom *(věn'-ŏm)*. A poisonous fluid secreted by snakes, scorpions, and some insects and transmitted by bites or stings.

venospasm *(vē'-nō-spăzm)*. Contraction of a vein, which may occur after an irritating substance is injected into a vein.

venostasis *(vē-nō-stā'-sĭs)*. *See* **phlebostasis.**

venous *(vē'-nŭs)*. Referring to one or more veins.

venous return. The blood that returns to the heart via the coronary sinus and great veins.

venous thrombosis. *See* **thrombosis.**

ventilation *(věn-tĭ-lā'-shŭn)*. 1. Circulation of fresh air in a given space, usually a room, and removal of foul air. 2. The movement of gases into and out of lungs. 3. The oxygenation of blood.

continuous positive-pressure v. Method of ventilation in which a device administers air or oxygen to the lungs under continuous pressure.

manual v. Use of a gas-filled bag to force gases into a patient's lungs.

ventilator *(věn'-tĭ-lā-tŏr)*. A device, manually or mechanically driven, that provides artificial circulation of air/oxygen for the lungs.

ventral *(věn'-trăl)*. Referring to the belly.

ventricle *(věn'-trĭk-ĕl)*. 1. Either of two lower chambers of the heart. 2. One of the cavities of the brain.

left v. The lower chamber on the left side of the heart that receives blood from the left atrium and sends it into the aorta.

right v. The lower chamber on the right side of the heart that receives blood from the right atrium and sends it into the pulmonary artery.

ventricular fibrillation *(vĕn-trĭk'-ū-lăr). See* **fibrillation.**

ventriculography *(vĕn-trĭk-ū-lŏg'-rĕ-fē).* An x-ray method that allows visualization of the size and shape of the ventricles of the brain by injecting air to show the cerebrospinal fluid that normally fills these cavities.

ventriculo-peritoneal shunt. *See* **shunt.**

venule *(vĕn'-ūl).* A tiny vein continuous with a capillary.

verapamil *See* Appendix, Common Prescription and OTC Drugs: By Generic Name.

vermicide *(vĕr'-mĭ-sīd).* A substance that destroys worms.

vermifuge *(vĕr'-mĭ-fūj).* A substance that expels intestinal worms.

vermin *(vĕr'-mĭn).* Insects or small animals, such as lice or rats, that cause disease or that are destructive or annoying.

vernix caseosa *(vĕr'-nĭks kā-sē-ō'-să).* An oily substance that covers the fetus and protects it in utero.

verruca *(vĕr-rū'-kă).* A flesh-colored tumor of the epidermis of the skin, caused by a papillomavirus. Also referred to as a wart.

verruca plana. A flat or slightly raised wart, seen most commonly on the face of the young.

verruca vulgaris. The common wart, most often seen on the backs of hands and fingers of young people.

version *(vĕr'-shŭn).* Changing the position of a fetus in the uterus, either naturally or by a doctor to facilitate delivery.

vertebra *(vĕr'-tĕ-bră).* Any one of the thirty-three bony segments that make up the spinal column.

basilar v. The lowest of the lumbar vertebrae.

cervical v. The seven vertebrae of the neck.

false v. The fusion of the sacral and coccygeal vertebrae.

lumbar v. The five vertebrae between the thoracic vertebrae and the sacrum.

sacral v. The five fused vertebrae that form the sacrum.

thoracic v. The twelve vertebrae that connect the ribs and form part of the back wall of the thorax.

vertebrate *(vĕr'-tĕ-brāt).* Having a vertebral column.

vertex *(vĕr'-tĕks).* The top of the head.

vertex presentation. The most common position of an unborn child during delivery, with the back of the head appearing at the vaginal opening.

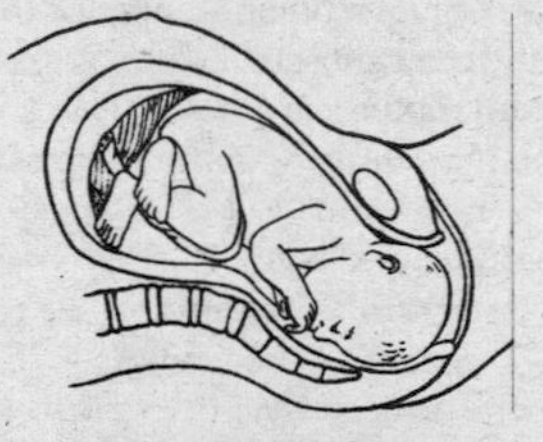

vertex presentation

vertigo *(vĕr'-tĭ-gō)*. The sensation of moving around in space or that objects are moving around you. It can be caused by various situations, such as middle ear disease, infectious disease, use of alcohol or certain drugs, or postural hypotension.

benign paroxysmal positional v. A brief, recurring disorder of the labyrinth of the inner ear characterized by sudden attacks of vertigo and nystagmus when the head is in a certain position or moving in a certain direction.

central v. Vertigo caused by a central nervous system disease.

cerebral v. Vertigo caused by a brain disease.

essential v. Vertigo that has an unknown cause.

height v. Vertigo experienced when someone looks down from a great height or looks up at, say, a high building.

horizontal v. Vertigo that occurs when an individual is lying down.

ocular v. Vertigo caused by an eye condition.

vestibular v. Vertigo caused by a disease or dysfunction of the vestibular mechanism.

vesical *(vĕs'-ĭ-kăl)*. Referring to or shaped like a bladder.

vesication *(vĕs-ĭ-kā'-shŭn)*. The process of blistering.

vesicle *(vĕs'-ĭ-kl)*. A small (less than 0.5 cm) blisterlike elevation of the skin that contains fluid.

vesiculation *(vĕ-sĭk-ū-lā'-shŭn)*. The formation of vesicles.

vesiculitis *(vĕ-sĭk-ū-lī'-tĭs)*. Inflammation of any vesicle, especially of a seminal vesicle.

vessel *(vĕs'-ĕl)*. A structure that conveys or contains fluid.

vestibular *(vĕs-tĭb'-ū-lăr)*. Referring to a vestibule, especially in the ear.

vestibule *(vĕs'-tĭ-būl)*. A small space at the entrance to a canal.

vestige *(vĕs'-tĭj)*. A primitive or undeveloped structure that had a function in a previous stage of the species or individual development. The appendix is an example of a vestige.

viable *(vī'-ă-bl)*. Capable of living. When referring to a fetus, it usually means one that has reached twenty gestational weeks and at least 500 grams in weight.

Viagra *(vī-ăg'-ră)*. *See* Appendix, Common Prescription and OTC Drugs: By Trade Name.

vibramycin *(vī-bră-mī'-sĭn)*. Trade name for doxycycline, a broad-spectrum antibacterial drug used to treat a variety of infections.

Vibrio *(vĭb'-rē-ō)*. A genus of gram-negative bacteria, several of which cause diseases in humans.

V. alginolyticus. Species associated with ear and wound infections and bacteremia.

V. cholerae. Species that cause cholera.

V. fluvialis. Species associated with diarrheal disease.

V. furnissili. Species associated with diarrheal disease and gastroenteritis.

vibrissae *(vī-brĭs'-ē)*. The hairs that grow in the vestibular area of the nasal cavity.

Vicodin *(vī'-kō-dĭn)*. *See* Appendix, Common Prescription and OTC Drugs: By Trade Name.

vikriti *(vĭk'-rĭ-tē)*. In ayurveda, a disordered physical constitution, the result of an imbalance of the doshas.

villus *(vĭl'-ŭs)* (*pl.* villi). A short, hairlike projection from the surface of a cell or mucous membrane.

vinblastine *(vĭn-blăs'-tĭn)*. A substance derived from the herb vinca rosea and used to treat Hodgkin's disease, leukemia, and other neoplastic diseases. Trade name: Velban.

vinca rosea *(vĭn'-kă rō'-să)*. A species of myrtle that contains two substances, vinblastine and vincristine, which are useful against cancer.

Vincent's angina *(vĭn'-sĕnts ăn-jī'-nă)*. *See* **trench mouth.**

vincristine *(vĭn-krĭs'-tĭn)*. A substance derived from the herb vinca rosea. It is similar to vinblastine and also used to treat various neoplastic diseases, except it is more useful than vinblastine when treating some conditions, such as acute leukemia.

viral hepatitis. *See* hepatitis.

viremia *(vī-rē'-mē-ă)*. The presence of viruses in the bloodstream.

virgin *(vĕr'-jĭn)*. An individual who has never had sexual intercourse.

virile *(vĭr'-ĭl)*. Referring to the male sex; have male characteristics.

virilism *(vĭr'-ĭl-ĭzm)*. The presence or development of male secondary characteristics in a woman or in a prepubertal male.

virology *(vī-rŏl'-ō-jē)*. The study of viruses and viral diseases.

virulent *(vĭr'-ū-lĕnt)*. Extremely poisonous.

virus *(vī'-rŭs)*. A term for a group of infectious agents which, in most cases, are smaller than bacteria, rely on the nutrients inside of cells for reproductive and metabolic needs, and usually are not visible through a light microscope, but can be seen through an electron microscope. *See also* **names of individual viruses.**

attenuated v. A variant strain of a disease-causing virus, modified so that it produces protective antibodies without producing the disease.

helper v. A virus that allows a defective virus to be present in the same cell to reproduce.

viscera *(vĭs'-ĕr-ă)*. Internal organs included within a cavity; especially refers to the abdominal organs.

visceral *(vĭs'-ĕr-ăl)*. Referring to the viscera.

visceroptosis *(vĭs-ĕr-ŏp-tō'-sĭs)*. Descent of the viscera from their normal positions in the body.

viscid *(vĭs'-ĭd)*. Sticky, glutinous.

viscosity *(vĭs-kŏs'-ĭ-tē)*. 1. A state of being gummy or sticky. 2. Resistance of a fluid to change its shape or form.

viscus *(vĭs'-kŭs)*. Any internal organ that is located within a cavity, such as the thoracic or abdominal cavity.

vision *(vĭzh'-ŭn).* The act of seeing external objects.

binocular v. Seeing a single image by both eyes simultaneously.

double v. Seeing of one object as two; *see* **diplopia.**

field of v. The space within which an object can be seen while the eye remains fixed on one point. Also referred to as *visual field.*

halo v. Condition in which colored or lighted rings are seen around lights.

low v. Condition in which there is significant visual impairment but also a significant amount of remaining vision.

night v. The ability to see when there is low or limited light.

peripheral v. Sight that is the result of light falling on the retina outside of the macular field; what can be seen out of the sides of the eye.

tunnel v. Constriction of the visual field, as though an individual were looking through a tube.

visualization *(vĭzh-ū-ăl-ĭ-zā'-shŭn).* The act of seeing or sensing an image of an object.

vital *(vī'-tĕl).* Referring to life.

vital capacity. The volume of air that can be exhaled following a full inhalation.

vital signs. The basic signs of life: blood pressure, body temperature, heartbeat, and respiration.

vitamins *(vī'-tă-mĭns).* Organic substances, other than proteins, carbohydrates, fats, minerals, or organic salts, that are necessary for healthy metabolism, growth, and development.

vitamin A. A fat-soluble vitamin that is essential for growth and development, healthy epithelial tissues, and growth and development of teeth and bones.

vitamin B complex. A vitamin supplement that typically contains thiamin, riboflavin, niacin, pyridoxine, biotin, folic acid, and cyanocobalamin. *See* the individual vitamins.

vitamin C. Ascorbic acid; a substance necessary for tissue integrity and important in maintaining a healthy immune system.

vitamin D. One of several fat-soluble vitamins that helps prevent rickets; also essential in calcium and phosphorus metabolism and normal development of bones and teeth.

vitamin E. An essential, fat-soluble vitamin found in various oils, whole grains, and animal tissue.

vitamin K. A fat-soluble substance that aids in blood coagulation and the production of prothrombin. It is found in fats, oats, wheat, and rye.

vitiation *(vĭzh'-ē-ā-shŭn).* A change that reduces efficiency or that causes contamination or injury, or impaired use.

vitiligo *(vĭt-ĭ-lī'-gō).* A condition characterized by nonpigmented white patches of various sizes and usually bordered by pigmented areas. The cause is unknown.

vitreous *(vĭt'-rē-ĕs).* 1. Resembling glass. 2. Referring to the vitreous body of the eye.

vitreous body. A transparent, jellylike substance that fills the cavity of the eyeball behind the lens.

vitreous humor. The clear, jellylike substance that fills the interstices of the stroma of the vitreous body.

viviparous *(vī-vĭp'-ĕr-ĕs).* Giving birth to live young.

vivisection *(vĭv-ĭ-sĕk'-shŭn).* Any cutting procedure performed upon a living animal, often under anesthesia, for purposes of experimentation.

VLDL. Acronym for very-low-density lipoprotein; *see also* **lipoprotein.**

vocal cords *(vō'-kăl).* Two thin folds of tissue within the larynx that vibrate when air travels between them, resulting in sounds that form the basis of speech.

voice *(vŏys).* The sound produced by the vibration of the vocal cords in humans.

void *(voyd).* To empty or evacuate the bladder or bowels.

volume *(vŏl'-ūm).* The space occupied by a substance.

expiratory reserve v. The maximal amount of air that can be forced from the lungs after normal exhalation.

inspiratory reserve v. The maximal amount of air that can be inhaled after the end of normal inhalation.

packed cell v. The volume of packed red blood cells in a sample of blood.

stroke v. The amount of blood expelled by a ventricle in one contraction.

tidal v. The volume of air inhaled and exhaled in one normal respiratory cycle.

volvulus *(vŏl'-vū-lŭs).* A twisting of the bowel upon itself, resulting in obstruction.

vomit *(vŏm'-ĭt).* Material ejected from the stomach through the mouth.

bilious v. Vomit that contains large amounts of bile, indicating bowel obstruction.

coffee-ground v. Vomit that looks like coffee grounds because blood is mixed in with the stomach contents. It is associated with bleeding of the stomach.

vomiting. Ejection through the mouth of the stomach contents and, in some cases, intestinal contents. It may result from intake of toxins, brain tumor, diseases of the stomach, motion sickness, migraine, pregnancy, intestinal obstruction, and other causes.

cyclic v. Periodic and recurring bouts of vomiting often associated with individuals who have a nervous temperament.

dry v. Nausea without vomiting.

induced v. Vomiting brought

on by the administration of certain substances, such as syrup of ipecac, or by physically stimulating the pharynx.
projectile v. Vomiting with great force.

von Willebrand's disease *(fŏn vĭl'-ĕ-brăhnt).* A congenital bleeding disease that manifests at an early age and is characterized by nosebleeds, easy bruising, and excess blood loss during menstruation.

voracious *(vŏr-ā'-shŭs).* Having an unsatiable appetite.

vortex *(vŏr'-tĕks).* A structure that has a spiral appearance or pattern.

voyeurism *(voi'-yĕr-ĭz-ĕm).* A paraphilia in which a person experiences sexual gratification by observing unsuspecting nude individuals or those engaged in sexual activities.

vulva *(vŭl'-vă).* The area of female external genitalia consisting of the labia majora, labia minora, clitoris, vestibule of the vagina, vaginal opening, and bulb of the vestibule, mons pubis, and the greater and lesser vestibular glands.

vulvectomy *(vŭl-vĕk'-tĕ-mē).* Surgical removal of the vulva.

vulvitis *(vŭl-vī'-tĭs).* Inflammation of the vulva.

vulvovaginitis *(vŭl-vō-văj-ĭ-nī'-tĭs).* Inflammation of the vulva and vagina, or of the vulvovaginal glands. It can have various causes, such as infections by bacteria, yeasts, viruses, or parasites, or by tight-fitting underclothes or chemical irritation.

W

WAIS. Acronym for Wechsler Adult Intelligence Scale.

waist *(wāst).* The portion of the trunk between the ribs and the pelvis.

wall. The substance that surrounds, encloses, or limits a cavity, cell, blood vessel, or other anatomical structure.
cell w. The outer membrane of some animal and plant cells.
chest w. The bony and muscular structures that make up the outer framework of the thorax and move during breathing.

walleye *(wăwl'-ī).* An eye that has a light-colored or white iris.

ward. A large room in a hospital in which there are beds for several or many patients, or such a room set aside for patients who have the same condition.

warfarin *(wŏr'-fĕr-ĭn).* *See* Appendix, Common Prescription and OTC Drugs: By Generic Name.

wart *(wŏrt).* *See* **verruca.**

Wassermann test *(văhs'-ĕr-măhn).* A test used in the diagnosis of syphilis.

wasting *(wāst'-ĭng).* A gradual loss of strength or size; emaciation.

water balance. The mechanism that maintains a balance between the amount of fluid taken into the body and the amount that is excreted.

water blister. A blister with clear watery contents that is not purulent (does not contain pus) and is not sanguineous (does not contain blood). One that is more than 5 mm in diameter with thin walls and is full of watery fluid; is called a bulla or a bleb.

wave *(wāv).* 1. A vibrating or undulating motion. 2. An oscillation seen on the recording of an electrocardiogram or other graphic records of physiological activity.

alpha w's. Brain waves seen on an electroencephalogram, with a frequency of 8 to 13 per second. They are typical of a normal individual in a quiet, waking state.

beta w's. Brain waves seen on an electroencephalogram, with a frequency of 18 to 30 per second. They are typical during periods of high activity of the nervous system.

brain w's. Electrical impulses produced by the brain.

delta w's. Brain waves seen on an electroencephalogram, with a frequency of less than 3.5 per second. They occur during deep sleep, in infants, and in people who have serious brain disorders.

theta w's. Brain waves seen on an electroencephalogram, with a frequency of 4 to 7 per second. They occur primarily in children but are seen in adults who are under extreme stress.

WBC. Acronym for white blood cell.

weaning *(wēn'-ĭng).* 1. To gradually accustom an infant to cessation of breast-feeding and to accept other forms of nourishment. 2. The gradual discontinuation of dependency on a form of therapy.

weight *(wāt).* The gravitational force exerted upon an object, usually by the earth. The weight of an object can vary depending on the force of gravity.

atomic w. Weight of an atom of an element compared with that of oxygen.

birth w. In humans, the weight of an infant taken within less than the first 60 completed minutes after birth.

equivalent w. Weight of a chemical that is equal to and will replace a hydrogen atom in a chemical reaction.

molecular w. The sum of all the atomic weights of all the elements in a molecule.

Wellbutrin *(wĕl'-bū-trĭn).* *See* Appendix, Common Prescription and OTC Drugs: By Trade Name.

wens *(wĕnz).* A cyst that develops when fatty secretions from a sebaceous gland are retained.

Wernicke's syndrome *(vĕr'-nĭ-kĕz).* Condition seen in old age in which individuals experience loss of memory and disorientation with confabulation.

Wertheim operation *(vĕrt'-hīmz).* Surgical removal of the entire uterus, fallopian tubes, ovaries, and the surrounding tissue, sometimes done to treat cancer of the uterus or cervix.

western blot technique. A method for analyzing and identifying protein antigens.

West Nile encephalitis. A mild, febrile disease caused by the West Nile virus and transmitted by mosquitoes. Symptoms may include drowsiness, severe frontal headache, rash, abdominal pain, loss of appetite, nausea, and lymph node disease.

wet brain. Condition characterized by an increased amount of cerebrospinal fluid and swelling of the meninges of the brain, caused by alcoholism.

Wharton's duct *(hwăr'-tŏnz).* Duct of the submandibular salivary gland that secretes saliva into the mouth at the side of the frenulum linguae, which is under the tongue.

wheal *(hwēl).* An itchy, round, evanescent eruption of the skin that is white in the center with a pale red periphery.

wheat germ oil. An oil obtained from the germ of wheat seed, *Triticum aestivum,* one of the richest sources of natural vitamin E.

wheeze *(hwēz).* To breathe with difficulty and with a whistling, sighing, squeaking, or puffing sound due to narrowing of a respiratory passageway.

whey *(hwāy).* The watery material that remains after separation of the casein in milk.

whiplash injury *(hwĭp'-lăsh).* A general term used to describe injury to the cervical vertebrae and nearby tissues due to a sudden jerking or rapid backward-and-forward movement of the head in relationship to the vertebrae.

Whipple's disease *(hwĭp'-ĕlz).* Intestinal disease characterized by weight loss, abnormal skin pigmentation, fatty stools, loss of strength, chronic arthritis, and other signs of malabsorption.

whitlow *(hwĭt'-lō).* Inflammation at the end of a toe or finger accompanied by pus.

herpetic w. Painful herpes simplex virus infection of a finger, often accompanied by swollen lymph glands.

WHO. Acronym for World Health Organization.

whooping cough *(hūp'-ĭng).* An acute infectious disease caused by *Bordetella pertussis* characterized by recurrent coughing spasms that end in a whooping sound.

Wilm's tumor *(vĭlmz).* A rapidly developing malignant tumor of the kidney, usually seen in children.

Wilson's disease *(wĭl'-sŏnz).* A hereditary condition characterized by an accumulation of copper in various organs and increased absorption of copper by the intestines. A green ring in the outer margin of the cornea is a sign of the disease.

windburn *(wĭnd'-bĕrn).* Redness of the face due to exposure to wind.

windpipe *(wĭnd'-pīp). See* **trachea.**

wisdom teeth. *See* **tooth.**

witch hazel *(wich' ha'zel).* Extract prepared from the bark of the shrub *Hamamelis virginiana,* used externally to treat contusions, headache, and noninflammatory hemorrhoids.

witch's milk. Milk secreted by a newborn's breast, stimulated by the lactating hormone from its mother.

withdrawal *(wĭth-drăw'-ăl).* Generally refers to cessation of administration or use of a drug, especially a narcotic, alcohol, or nicotine, to which an individual has become either physically or psychologically addicted.

womb *(woom).* Female organ that contains, protects, and nourishes a fetus; uterus.

workup *(wĕrk'-ŭp).* The process of obtaining all the necessary information to diagnose and treat a patient. Elements include family and personal medical history, social and job history, physical examination, and lab, x-ray, and any medical procedure results.

wound *(woond).* Trauma to any tissue in the body, especially any caused by physical means and which breaks the integrity of the tissue.

aseptic w. One that does not become infected.

incised w. One made with a sharp instrument, resulting in a clean cut.

nonpenetrating w. A wound, especially within the abdomen or thorax, that does not disrupt the surface of the body.

penetrating w. One that breaks the integrity of the skin and enters underlying tissue or a body cavity.

puncture w. Wound made by a sharp-pointed instrument, such as a needle or ice pick.

septic w. One that becomes infected.

subcutaneous w. A wound that extends below the skin into subcutaneous tissue but not into bone or organs.

wrist *(rĭst).* The region that lies between the hand and the forearm, consisting of the carpal bones and associated soft tissue.

wristdrop *(rĭst'-drŏp).* Paralysis of the extensor muscles in the wrist and fingers, leaving the hand flexed at the wrist and incapable of being extended.

writer's cramp *(rīt'-ĕrz krămp).* A cramp that affects the muscles of the thumb and two adjacent fingers, due to excessive writing.

wryneck *(rī'-nĕk).* Condition in which one or more neck muscles is contracted, producing an abnormal position for the head. *See* **torticollis.**

X

Xalatan *(xăl'-ă-tăn). See* Appendix, Common Prescription and OTC Drugs: By Trade Name.

Xanax *(xă'-năks).* *See* Appendix, Common Prescription and OTC Drugs: By Trade Name.

xanthelasma *(zăn-thĕl-ăz'-mă).* Flat or slightly raised yellowish tumor that usually appears on the upper and lower eyelids, and is seen most often in the elderly.

xanthochromia *(zăn-thō-kōr'-mē-ă).* Patchy yellow discoloration of the skin resembling jaundice.

xanthoderma *(zăn-thō-dĕr'-mă).* Any yellowish discoloration of the skin.

xanthogranuloma *(zăn-thō-grăn-ū-lō'-mă).* A tumor with characteristics of an infectious granuloma and a xanthoma, occurring most commonly in women.

xanthoma *(zăn-thō'-mă)* (*pl.* xanthomata). A yellow flat or slightly elevated nodule or plaque usually seen on the eyelids.

diabetic x. An eruption of xanthoma associated with severe diabetes.

disseminatum x. Characterized by an eruption of xanthomata throughout the body. Also called *Hand-Schuller-Christian disease.*

xanthomatosis *(zăn-thō-mă-tō'-sĭs).* Widespread eruption of xanthomata, especially on the knees and elbows, and often affecting the mucous membranes. Also called *xanthoma multiplex.*

xanthopsia *(zăn-thŏp'-sē-ă).* A condition in which objects appear yellow; it may occur in cases of jaundice or digitalis poisoning.

xanthosis *(zăn-thō'-sĭs).* A condition in which the skin is yellowish due to ingestion of excessive amounts of carrots, yellow squash, egg yolks, or other foods that contain carotenoids.

X chromosome. *See* **chromosome.**

xenogenic *(zē-nō-jĕn'-ĭk).* Originating outside of the organism or obtained from a foreign substance that has been introduced into the organism.

xenograft *(zēn'-ō-grăft).* Surgical tissue graft obtained from one species to be given to a different species.

xenology *(zē-nŏl'-ō-jē).* The study of parasites and their relationships to one another and to their hosts.

xenophobia *(zē-nō-fō'-bē-ă).* An abnormal fear of strangers.

xeroderma *(zĕr-ō-dĕr'-mă).* A condition characterized by excessively dry skin due to a slight increase of the horny layer and reduced moisture content of the skin.

xeromammography *(zĕr-ō-mă-mŏg'-ră-fē).* Examination of the breast using xeroradiography.

xeromenia *(zĕr-ō-mē'-nē-ă).* A condition in which a woman expresses the symptoms of menstruation without experiencing menstrual flow.

xerophthalmia *(zĕr-ŏf-thăl'-mē-ă).* Excessive dryness of the conjunctiva and cornea, followed by development of a horny layer (keratinization), all due to chronic conjunctivitis or a vitamin A deficiency.

xeroradiography *(zĕr-ō-rā-dē-ŏg'-ră-fē).* A radiographic method in which, instead of film, plates covered with an electronically powdered substances are used to produce images.

xerosis *(zĕr-ō'-sĭs).* Abnormal dryness of the skin, mucous membranes, or conjunctiva.

xerostomia *(zĕr-ō-stō'-mē-ă).* Dry mouth, due to reduced amount of secretion from the salivary glands. May be caused by diabetes, hysteria, acute infections, certain drugs, and other causes.

xiphoid process *(zĭf'-oyd).* The lowest portion of the sternum. There are some abdominal muscles attached to it but not ribs.

X-linked. Referring to characteristics that are transmitted by genes on the X chromosome.

X-rays. High-energy electromagnetic waves with the ability to penetrate most solid matter to some degree and to act on photographic film, making them useful in diagnosis and treatment.

XX chromosome. *See* **chromosome.**

XXY chromosome. *See* **chromosome.**

XY chromosome. *See* **chromosome.**

xylitol *(zī'-lĭ-tŏl).* A type of sugar similar to sucrose that may be used in place of sucrose as a sweetener, especially in diabetic diets.

Y

YAG laser. *See* **laser.**

Yang. *See* **yin-yang.**

yarrow *(yăr'-ō).* A preparation of the plant *Achillea millefolium* used to treat loss of appetite, indigestion, and conditions of the liver and gallbladder.

yaws *(yăws).* An infectious tropical disease caused by *Treponema pertenue* and characterized by crusting ulcers on the extremities of rheumatism.

Y chromosome. *See* **chromosome.**

years of life lost. Number of years a person is estimated to have remained alive if he or she had not contracted a specific disease.

yeast *(yēst).* A general term for a true fungus of the Saccharomycetaceae family. It is a one-celled organism that is capable of fermenting carbohydrates.

Brewer's y. A byproduct of beer brewing.

cultivated y. A type of yeast grown in culture and used in making breads, beer, and so on.

yellow enzymes *(yĕl'-ō).* A group of respiratory enzymes that catalyze reactions in the body, permitting cells to breathe. These reactions are called oxidation-reduction reactions.

yellow fever. An acute infectious disease characterized by jaundice, stomach tenderness, fever, hemorrhage, and vomiting.

yerba santa *(yĕr'-bă săhn'-tă).* An herbal tea made from the dried leaves of mata, a South American evergreen tree.

Yersinia *(yĕr-sĭn'-ē-ă).* A genus of gram-negative bacteria in the Enterobacteriacea family.

Y. enterocolitica. A species that causes yersiniosis in humans.

Y. pseudotuberculosis. A species that causes pseudotuberculosis in birds and rodents, and occasionally in humans.

yersiniosis *(yĕr-sĭn-ē-ō'-sĭs).* An infection with *Yersinia enterocolitica,* characterized by gastroenteritis and swollen lymph nodes in children, and by arthritis, septicemia, and erythema nodosum in adults.

yin-yang *(yĭn'-yăng').* In ancient Chinese philosophy, the concept of two opposing and complementary influences that form the basis of and control all nature. The goal of Chinese medicine is to achieve a balance between them.

yin-yang

yoga *(yō'-gă).* A system of beliefs and practices whose goal is to achieve union between self and an ultimate or universal self or consciousness.

hatha y. A yoga practice best known in the West, it consists of more than one thousand asanas (postures) designed to promote physical and mental well-being to achieve self-transformation.

Iyengar y. A type of yoga that emphasizes correct body alignment in the asanas.

yohimbine *(yō-hĭm'-bēn).* The active ingredient of yohimbe, the bark of *Corynanthe yohimbi.* It causes a rise in blood pressure, irritability, tremor, nausea, and vomiting, but is also used as a sexual stimulant.

yolk *(yōk).* A substance stored in the egg (ovum) that provides nutrition to a developing embryo.

Z

zafirlukast *(zĕ-fĭr'-loo-kăst).* A drug used as an antiasthmatic.

Zantac *(zăn'-tăk).* *See* Appendix, Common Prescription and OTC Drugs: By Trade Name.

Zeis' glands *(tsīs).* Oil-secreting glands of the eyelid. Each gland is associated with an eyelash.

zero balancing. A type of bodywork that combines Western and Eastern concepts and uses gentle manipulation at specific areas of the body to align the body's energy with its structure for the purpose of reducing stress and to promote well-being.

Zestril *(zĕs'-trĭl). See* Appendix, Common Prescription and OTC Drugs: By Trade Name.

zinc *(zĭngk).* A metallic element and an essential bioelement in the body. It is also a cofactor in many proteins.

zinc oxide. A substance used as a skin protective in ointments and in dusting powders.

zinc peroxide. A yellow-white powder used in anti-infective preparations.

zinc sulfate. A substance used as an astringent in the treatment of gonorrhea, conjunctivitis, and various skin diseases.

Zithromax *(zĭth'-rō-măx). See* Appendix, Common Prescription and OTC Drugs: By Trade Name.

Zocor *(zō'-kŏr). See* Appendix, Common Prescription and OTC Drugs: By Trade Name.

Zollinger-Ellison syndrome *(zōl'-ĭn-jĕr ĕl'-ĭ-sŏn).* A condition caused by noninsulin secreting tumors of the pancreas, which release excessive amounts of gastrin, which stimulate the stomach to secrete large amounts of hydrochloric acid and pepsin, resulting in the development of peptic ulcers.

Zoloft *(zō'-lŏft). See* Appendix, Common Prescription and OTC Drugs: By Trade Name.

zolpidem *(zōl-pĭ'-dĕm). See* Appendix, Common Prescription and OTC Drugs: By Generic Name.

zone *(zōn).* An area or belt.

comfort z. The range of temperature, humidity, and, if applicable, sun and wind exposure in which an individual working at a specific rate and in specific clothing is comfortable.

erogenous z. An area of the body that may produce sexual arousal when stimulated.

zoology *(zō-ŏl'-ĕ-jē).* The scientific study of animal life.

zoonoses *(zō-ō-nō'-sēs).* Diseases that are transmitted from animals to humans under natural conditions.

zooplasty *(zō'-ō-plăst-ē).* Transplantation of animal tissue to humans.

zoster *(zŏs'-tĕr).* Acute inflammatory disease in which vesicles gather along different regions of the skin.

herpes z. Infection caused by a herpesvirus and characterized by eruption of groups of vesicles on one side of the body along the course of a nerve caused by inflammation of the ganglia and dorsal nerve roots.

Zyban *(zī'-băn).* Trade name for a preparation of bupropion hydrochloride, used to treat depression and for smoke cessation.

zygote *(zī'-gŏt).* The cell that is produced when two gametes unite; a fertilized egg.

Zyprexa *(zī-prĕk'-să).* Trade name for a preparation of olanzapine, used to treat schizophrenia and bipolar disorder.

Zyrtec *(zīr'-tĕk).* Trade name for preparations of certirizine hydrochloride, used to treat allergies.

Part 2

Appendix—Practical Health Information Every Consumer Should Know

1

SIGNS AND SYMPTOMS: WHAT DO THEY MEAN?

A well-informed health-care consumer has the best chance of getting the quality health care he or she needs and deserves. Part of the information you need includes an understanding of the signs and symptoms associated with various medical conditions you or your loved ones may experience. The following list brings together brief, descriptive explanations of dozens of signs and symptoms. These explanations are not meant to provide you with a diagnosis: that is the task of your physician. However, they can help you identify the disorders that may be associated with certain signs and symptoms. (**NOTE:** A **sign** is objective evidence of disease—one that can be detected by someone other than the person who is affected. Examples include a rash, external bleeding, hair loss, and swollen ankles. In contrast, a **symptom** is subjective: it is experienced by the individual and consists of sensations only the patient can perceive. Fatigue, neck pain, and anxiety are examples of symptoms.)

Alopecia

- Hair loss, or alopecia, may occur at the top and front of the head, which is characteristic of male pattern baldness; in patches, which is called alopecia areata; or involve the entire head, which is known as alopecia capitis totalis.
- Alopecia may be caused by and/or be a warning sign of systemic lupus erythematosus, chemotherapy, medication use, or radiation therapy.

Bad Breath

- A foul odor to the breath. Also referred to as halitosis.
- Bad breath can be caused by and/or be a warning sign of an abscessed tooth, bronchiectasis, esophageal cancer, gingivitis, impacted teeth, intestinal obstruction, lung abscess, pharyngitis, renal failure, sinusitis, Sjogren's syndrome.

Bloating

- Bloating is a subjective feeling that the abdomen is larger than normal. This is in contrast to *distention*, in which the abdomen is actually larger than normal.
- Bloating can be caused by and/or be a warning symptom of celiac disease, dyspepsia, irritable bowel syndrome, dietary fat, or lactose intolerance.

Blood in Sputum

- Bloody mucus or spitting up blood (also referred to as hemoptysis) can occur in the presence of common lung and airway infection, such as acute bronchitis or pneumonia.
- Blood in sputum can be caused by and/or be a warning sign of amyloidosis, bronchiectasis, bronchitis (acute), deep vein thrombosis, fungal infection of the lung, lung abscess, lung cancer, lung trauma, pneumonia, pulmonary embolism, tuberculosis.

Blood in Stool

- Blood in the stool may be visible (bright red or maroon in color) or occult (invisible to the naked eye). Blood that remains in the intestine for some time can turn stools black.
- Blood in stool can be caused by and/or be a warning sign of anal fissure, colon cancer, colon polyps, Crohn's disease, diverticulosis, hemorrhoids, peptic ulcer disease, and stomach cancer.

Blood in Urine

- Blood in urine (also referred to as hematuria) is always abnormal, may or may not be accompanied by pain, and should be evaluated by a health-care practitioner.
- Blood in urine can be caused by and/or be a warning sign of appendicitis (acute), bladder cancer, cystitis, endometriosis, high blood pressure, kidney cancer, kidney stone, penile cancer, polycystic kidney disease, prostate cancer, systemic lupus erythematosus, urinary tract infection, excessive exercise/trauma, vasculitis.

Blood in Vomit

- Bloody vomit (also referred to as hematemesis) should be evaluated by a health-care practitioner.
- Blood in vomit can be caused by and/or be a warning sign of esophageal cancer, nonsteroidal anti-inflammatory drug (NSAID) use, peptic ulcer disease, stomach cancer.

Cold Feet/Hands

- Cold sensations to the hands and feet that can range from mild to painful.
- Cold feet and/or hands can be caused by and/or be a warning sign of exposure to cold, frostbite, hypothyroidism, poor circulation, peripheral neuropathy, peripheral vascular disease, Raynaud's phenomenon.

Constipation

- Medically speaking, constipation is defined as fewer than three bowel movements per week, and severe constipation is considered to be less than one bowel movement per week. Most people are irregular, however, and what is typical for one person may be atypical for another.
- Constipation can be caused by and/or be a warning symptom of adrenal insufficiency, anal fissure, colorectal cancer, depression, diverticulosis, endometriosis, fibroids (uterine), hemorrhoids, hypothyroidism, intestinal obstruction,

irritable bowel syndrome, multiple sclerosis, Parkinson's disease, proctitis, scleroderma.

Convulsion

- A manifestation of uncontrolled electrical activity of the brain. The type of convulsion (also known as a seizure) and symptoms depends on where the abnormal electrical activity occurs in the brain, the patient's age, and overall health.
- Convulsions can be caused by and/or be a warning sign of alcohol abuse, Alzheimer's disease, brain tumor, fever, head injury, infection, kidney failure, lead poisoning, stroke, systemic lupus erythematosus, toxoplasmosis.

Cough

- Cough is a rapid expulsion of air from the lungs in an effort to clear the airways of mucus, fluid, or other materials. It may be acute (lasting less than three weeks) or chronic (greater than three weeks).
- Cough can be caused by and/or be a warning sign of asthma, chemical irritation, chronic obstructive pulmonary disease, chronic rhinitis, common cold, congestive heart failure, croup, emphysema, esophageal cancer, gastroesophageal reflux disease, laryngitis, lung cancer, measles, pleurisy, pneumonia, post-nasal drip, pulmonary edema, sarcoidosis, sinusitis, systemic lupus erythematosus, tuberculosis, whooping cough.

Cramps (abdominal)

- Abdominal pain generally refers to pain that originates from organs within the abdominal cavity (stomach, small intestine, colon, liver, gallbladder, and pancreas), although pain may be felt in the abdomen from other organs, such as the kidneys and uterus.
- Abdominal cramps can be caused by and/or be a warning symptom of abdominal aortic aneurysm, appendicitis, colic (infants), colon cancer, constipation, diverticulosis, ectopic

pregnancy, endometriosis, gallstones, gastroenteritis, hernia, inflammatory bowel disease, irritable bowel syndrome, kidney stones, lactose intolerance, liver cancer, ovarian cancer, ovarian cysts, pancreatitis, peptic ulcer disease, stomach cancer, ulcerative colitis.

Diarrhea

- Diarrhea is abnormally frequent or abnormally soft bowel movements. Acute diarrhea (lasting 3 to 5 days) is usually caused by a viral infection; prolonged diarrhea (lasting more than 4 to 6 weeks) is usually due to a gastrointestinal disease.
- Diarrhea can be caused by and/or be a warning sign of Addison's disease, adrenal insufficiency, appendicitis (acute), celiac disease, colon cancer, Crohn's disease, dehydration, diabetes mellitus, endometriosis, food poisoning, gastroenteritis, HIV, hyperthyroidism, inflammatory bowel syndrome, intestinal obstruction, irritable bowel disease, lactose intolerance, rotavirus, pancreatic cancer, ulcerative colitis.

Dizziness

- Dizziness is an indistinct term used to describe various conditions such as light-headedness, unsteadiness, and vertigo.
- Dizziness can be caused by and/or be a symptom of anemia, benign paroxysmal positional vertigo, brain tumor, cardiac arrhythmias, cardiomyopathy, deafness, disturbances of the vestibular (balance) system of the inner ear, earache, hypertension, medications (some), Meniere's disease, migraine, motion sickness, otosclerosis, Paget's disease, stroke, viral infection, vision disturbances.

Dry Eye

- Dryness of the eyes can occur as a normal response to weather (especially dry, windy, warm weather) or use of certain medications.

- Dry eye can be caused by and/or be a sign of conjunctivitis, corneal ulcer or infection, Sjogren's syndrome.

Earache

- Ear pain can occur within the ear or in the ear canal, or affect the outer ear. Acute middle ear infection (inflammation of the middle ear, or otitis media) is common among children. Earache also occurs when the ear canal becomes infected (otitis externa) or when the outer ear (pinna) becomes inflamed.
- In addition to otitis media and otitis externa, earache can be caused by and/or be a symptom of barotitis, injury/trauma, laryngitis, pharyngitis, Ramsay Hunt syndrome, relapsing polychondritis, sinusitis, temporomandibular joint syndrome, tonsillitis, tooth abscess, tooth decay.

Edema

- Leg swelling generally occurs when there is an abnormal accumulation of fluid in the lower extremities.
- Edema can be caused by and/or be a sign of Baker's cyst, cellulitis, cirrhosis, congestive heart failure, eosinophilic fasciitis, kidney failure, leg vein obstruction, medication use, phlebitis, salt retention, scleroderma, trauma.

Erectile Dysfunction

- The consistent inability to sustain an erection sufficient for sexual intercourse or an inability to achieve ejaculation, or both.
- Erectile dysfunction can be caused by and/or be a sign of adrenal insufficiency, alcohol abuse, atherosclerosis, coronary artery disease, Cushing's syndrome, depression, diabetes, high blood pressure, hypothyroidism, kidney failure, medications, multiple sclerosis, peripheral vascular disease, prostate cancer, prostatitis, renal failure, smoking, stress.

Fainting

- A temporary loss of consciousness (either partial or complete) with an interruption of awareness is known as fainting, or syncope, and is caused by a temporary reduction in blood flow and supply of oxygen to the brain.
- Fainting can be caused by and/or be a sign of anaphylaxis, anemia, aortic stenosis, cardiac arrhythmias, cardiomyopathy, congestive heart failure, dehydration, diabetes, heatstroke, high altitudes, hypoglycemia, hypothermia, low blood pressure, meningitis, migraine, Parkinson's disease, postural hypotension, pulmonary hypertension, stroke.

Fatigue

- Fatigue is a reduced ability to work or to accomplish tasks following physical or mental activity.
- Fatigue can be caused by and/or be a symptom of Addison's disease, adrenal insufficiency, anemia, cancer, congestive heart failure, chronic fatigue syndrome, chronic obstructive pulmonary disease, dehydration, depression, diabetes mellitus, fibromyalgia, hepatitis (acute viral and chronic), hormone imbalance, hyperthyroidism, infection, insomnia, jet lag, kidney failure, leukemia, liver disease, Lyme disease, multiple sclerosis, myasthenia gravis, myocarditis, pericarditis, pneumonia, renal failure, sarcoidosis, stress, surgery, systemic lupus erythematosus, trauma, vasculitis.

Fever

- A body temperature greater than 100.4° F and part of the body's defense mechanism against disease-producing organisms. Fever accompanied by troubling symptoms should be evaluated by a health-care practitioner.
- Fever can be caused by and/or be a sign of dozens of conditions, some of which include allergy, appendicitis (acute), cancer, chicken pox, chronic fatigue syndrome, common cold, Crohn's disease, croup, endocarditis, flu, gastroenteritis, German measles, HIV, infection (bacterial, parasitic,

viral), infectious mononucleosis, inflammatory bowel disease, leukemias, Lyme disease, malaria, meningitis, pelvic inflammatory disease, pericarditis, peritonitis, pleurisy, pneumonia, sarcoidosis, sinusitis, sore throat, staph infection, strep throat, toxoplasmosis, typhoid fever, urinary tract infection, and viral hepatitis.

Gastroesophageal Reflux (heartburn)

- The regurgitation of the contents of the stomach and first portion of the small intestine (duodenum) into the esophagus, which produces an uncomfortable burning, warm feeling behind the breastbone that may rise as high as the neck. It typically occurs after eating, when lying down, or while sleeping.
- Gastroesophageal reflux can be caused by and/or be a warning sign for eating fatty, spicy, or acidic foods; excessive alcohol consumption, excessive caffeine consumption; hiatal hernia, scleroderma, smoking.

Headache

- Pain in the head that can arise from a variety of causes. Headache can be primary (not associated with a disease) or secondary (associated with disease).
- Headache can be caused by and/or be a symptom of anemia, bacterial infection, brain aneurysm, brain tumor, caffeine, chronic fatigue syndrome, cough, fibromyalgia, heatstroke, giant cell arteritis, glaucoma, hypoglycemia, Lyme disease, medication, meningitis, meningococcemia, multiple myeloma, Paget's disease, renal failure, sarcoidosis, sinusitis, spondylosis (cervical), stroke, systemic lupus erythematosus, trauma to the head or neck, temporomandibular joint disorder, toothache, vasculitis, and viral infection.

Incontinence (urine)

- An involuntary loss of urine that results from a loss of voluntary control over the muscles around the ureter opening.

- Urinary incontinence can be caused by and/or be a sign of Alzheimer's disease, bladder infection, irritable bladder, kidney infection, medications (some), multiple sclerosis, pregnancy, prostate cancer, spinal tumor, stroke.

Itching

- An uncomfortable skin sensation that can occur locally or generally on the body and that makes an individual want to scratch.
- Itching can be caused by and/or be a sign of anemia, chicken pox, dry skin, eczema, food allergy, hay fever, head lice, hemorrhoids, hives, jaundice, kidney failure, leukemia, psoriasis, rash, ringworm, renal failure, scabies, STDs, shingles, tinea cruris (jock itch), vaginitis, varicose veins.

Jaundice

- An abnormally high blood level of bilirubin, a bile pigment, which causes the skin and whites of the eyes to yellow. In some cases urine also turns dark.
- Jaundice can be caused by and/or be a sign of alcohol abuse, cirrhosis, gallstones, hepatitis B or C, HIV, liver cancer, medications (some), pancreatic cancer, pancreatitis, pregnancy, primary biliary cirrhosis, sickle cell disease, viral hepatitis.

Loss of Appetite

- Loss of appetite, or anorexia, can be a temporary or persistent condition, the latter of which should be evaluated by a health-care practitioner.
- Loss of appetite can be caused by and/or be a symptom of Addison's disease, anemia, bipolar disorder, chronic inflammatory diseases, colorectal cancer, Crohn's disease, dementia, depression, gastroesophageal reflux disease, glomerulonephritis (acute, chronic), hepatitis, liver tumors, loss of sense of taste, lymphoma, medications (some), pelvic inflammatory disease, peptic ulcer

disease, stomach cancer, stroke, systemic lupus erythematosus.

Nausea/Vomiting

- Nausea is the sensation that there is a need to vomit. Both nausea and vomiting can have physical or psychological sources.
- Nausea and vomiting can be caused by and/or be signs of Addison's disease, adrenal insufficiency, alcohol abuse, anemia, appendicitis, brain tumor, bulimia, chemotherapy, concussion, congestive heart failure, depression, diabetes, esophageal cancer, food poisoning, gallbladder disorders, gastroenteritis, gastroesophageal reflux disease, glaucoma, headache/migraine, heart attack, hyperparathyroidism, influenza, irritable bowel syndrome, kidney failure, leukemias, medications (some), Meniere's disease, ovarian cancer, pancreatic cancer, peptic ulcer disease, peritonitis, pregnancy, renal calculi, renal failure, vestibular balance disorders, viral hepatitis.

Nosebleed

- Many nosebleeds occur spontaneously, triggered by environmental factors or a medical condition. In most cases they are not serious and can be stopped with proper first aid.
- Nosebleed can be caused by and/or be a sign of alcohol abuse, anemia, dry nasal membranes (common in dry climates or during winter months when the air is warm and dry from heating systems), hemophilia, hypertension, infection, leukemias, medications (those that prevent normal blood clotting, such as warfarin, aspirin), platelet function disorders, rhinitis, sinusitis, trauma.

Numbness (paresthesia)

- Numbness of the fingers and toes usually results from conditions that affect the nerves and/or blood flow that supply these body parts. Numbness is frequently accompanied by tingling.

- Numbness can be caused by and/or be a symptom of anemia, brain tumor, carpal tunnel syndrome, degenerative disc disease, diabetes mellitus, hyperventilation, multiple sclerosis, peripheral neuropathy, peripheral vascular disease, Raynaud's phenomenon, renal failure, sciatica, shingles, spinal cord trauma, tarsal tunnel syndrome, transient ischemic attack, vasculitis.

Pain (abdominal)

- Generally refers to pain that originates from organs within the abdominal cavity, including the stomach, colon, small intestine, liver, gallbladder, and pancreas. Occasionally abdominal pain may be "referred pain," meaning it arises from organs that are close to but not within the abdominal cavity, such as the kidneys, uterus, or lower lungs.
- Abdominal pain can be caused by and/or be a symptom of abdominal aortic aneurysm, appendicitis, colic (in infants), colon cancer, constipation, diverticulosis, endometriosis, gallstones, gastroenteritis, hernia, inflammatory bowel disease, irritable bowel syndrome, kidney stone, lactose intolerance, liver cancer, ovarian cancer, ovarian cysts, pancreatic cancer, pancreatitis, peptic ulcer disease, peritonitis, stomach cancer, ulcerative colitis.

Pain (back)

- Pain in the lower back can be associated with the lumbar spine, spinal cord and nerves, disks between the vertebrae, ligaments around the spine and disks, muscles of the lower back, and internal organs in the abdominal and pelvic areas, while pain in the upper back can be the result of disorders of the aorta, chest tumors, and inflammation of the spine.
- Back pain can be caused by and/or be a symptom of abdominal aortic aneurysm, ankylosing spondylitis, bladder stones, bone cancer, cervical cancer, degenerative disk disease, endometriosis, fibromyalgia, fibroids, fractured bone, gallbladder disorders, infection of the spine or pelvis, kid-

ney stones, muscle spasm, obesity, osteoarthritis, Paget's disease, pancreatic cancer, pregnancy, sciatica, scoliosis, shingles, spinal stenosis, sprain, strain, uterine prolapse, viral infections.

Pain (breast)

- Many women experience breast pain or discomfort associated with their menses, although there are also other less common causes of breast pain. Breast pain is also referred to as mastodynia, mastalgia, or mammalgia.
- Breast pain can be caused by and/or be a symptom of benign breast tumors, blocked milk duct, breast cancer, breast infection (mastitis, abscess), certain medications, pregnancy, premenstrual hormonal fluctuations.

Pain (chest)

- Chest pain can have many causes, most of them serious, and thus should not be ignored. Seek immediate medical attention.
- Chest pain can be caused by and/or be a symptom of angina, aortic stenosis, emotional stress, gallstones, gastroesophageal reflux disease, heart attack, peptic ulcer disease, pericarditis, pleurisy, pneumonia, pneumothorax, pulmonary embolism, pulmonary hypertension, shingles.

Pain (head; *see* headache)

Pain (joint)

- Joint pain can be caused by an injury to any of the structures that surround a joint or be associated with inflammation or infection. *Arthralgia* is another name for joint pain.
- Joint pain can be caused by and/or be a symptom of aseptic necrosis, bone tumor, bursitis, gout, hepatitis (acute viral), infectious arthritis, Lyme disease, osteoarthritis, osteochondritis dissecans, Paget's disease, rheumatoid arthri-

tis, Sjogren's syndrome, sickle cell anemia, sprain, steroid withdrawal, systemic lupus erythematosus, tendonitis, torn cartilage.

Pain (knee)

- The knee is one of the most commonly injured joints, making knee pain a common occurrence as well. Knee pain can also be the result of disease.
- Knee pain can be caused by and/or be a symptom of ankylosing spondylitis, aseptic necrosis, bone cancer, bursitis, Crohn's disease, juvenile arthritis, Kawasaki disease, Lyme disease, obesity, osteoarthritis, osteochondritis dissecans, reactive arthritis, rheumatoid arthritis, sickle cell anemia, sprain, Still's disease, systemic lupus erythematosus, tendonitis, torn cartilage, ulcerative colitis.

Pain (neck)

- Neck pain can arise from conditions that affect any structure in the neck, including the seven vertebrae that surround the spinal cord, neck muscles, blood vessels, lymph glands, thyroid gland, esophagus, larynx, and trachea.
- Neck pain can be caused by and/or be a symptom of ankylosing spondylitis, carotidynia, degenerative disk disease, fracture, meningitis, osteoarthritis, osteomyelitis, polymyalgia rheumatica, rheumatoid arthritis, sciatica, shingles, sprain injury (e.g., whiplash), strep throat, thyroid inflammation, viral infection.

Palpitations

- Palpitations are uncomfortable sensations arising from an irregular or forceful heartbeat. They may or may not be associated with heart disease.
- Palpitations can be caused by and/or be a sign of alcohol abuse, anemia, aortic stenosis, arrhythmia, atrial fibrillation, bradycardia, depression, fever, hypertension, hyperthyroidism, hypoglycemia, Lyme disease, medications (some),

mitral valve prolapse, paroxysmal atrial tachycardia, smoking, stress.

Rash

- Rash is an inflammation of the skin and can also include eruptions and red, scaly, itchy patches.
- Rash can be caused by and/or be a sign of anaphylaxis, eczema, Lyme disease, measles, meningitis, psoriasis, rheumatic fever, rheumatoid arthritis, ringworm, rosacea, scabies, shingles, systemic lupus erythematosus, typhoid fever.

Ringing in the Ears

- The perception of buzzing, roaring, clicking, or other abnormal ear noises when no sound is present. Also known as tinnitus. Hearing loss often accompanies tinnitus.
- Ringing in the ears can be caused by and/or be a symptom of aging, anemia, aneurysm, concussion, ear infection, Eustachian tube problems, fluid in the ears, hypertension, labyrinthitis, loud noises, medications (some), Meniere's disease, otosclerosis, Paget's disease, perforated eardrum.

Shortness of Breath

- Shortness of breath, also referred to as dyspnea, can occur when you are at rest or engaged in physical activity, and it may occur gradually or suddenly. These characteristics are important to tell your health-care practitioner so he/she can better detect the cause of dyspnea.
- Shortness of breath can be caused by and/or be a sign of asthma, anxiety, bronchitis, congestive heart failure, croup, emphysema, epiglottitis, goiter, heart attack, leukemia, lung cancer, myocardial infarction, muscular dystrophy, multiple sclerosis, obesity, pericarditis, pleurisy, pneumonia, pulmonary edema, pulmonary embolism, thyroid cancer, thyroid nodules.

Sore Throat

- Throat soreness may be linked with a variety of inflammatory conditions and is usually associated with painful swallowing.
- Sore throat can be caused by and/or be a symptom of bacterial infection, chemical irritation, chemotherapy, dry air, chronic fatigue syndrome, common cold, epiglottitis, gingivitis, inflamed pharynx, inflamed tonsils, inflamed larynx, infectious laryngitis, mononucleosis, pharyngitis, roseola, sinusitis, Sjogren's syndrome, strep throat, tonsillitis, viral infection.

Swollen Glands (lymphadenopathy)

- Enlargement of the lymph glands is common and can occur in various parts of the body, including the chest, neck, armpits, and groin.
- Swollen glands can be caused by and/or be a sign of bacterial, viral, fungal, or parasitic infections; alcohol abuse; bulimia; cancer; chronic fatigue syndrome; connective tissue disease; Hodgkin's disease; HIV; hyperthyroidism; mononucleosis (infectious); mumps; non-Hodgkin's lymphoma; pharyngitis; salivary gland disorders; sarcoidosis; Sjogren's syndrome; strep throat; thyroid cancer; thyroiditis; tonsillitis; tooth abscess.

Urination: Burning/Painful

- A feeling of discomfort, pain, and/or burning when urinating.
- Painful urination can be caused by and/or be a symptom of bacterial infection of the urinary tract affecting the bladder (cystitis) or kidney (pyelonephritis); bladder cancer; chlamydia; epididymitis; gonorrhea; irritation from chemicals in bubble baths, soaps, douches, and spermicides; kidney stone; prostatitis; sexually transmitted diseases; vaginitis; vulvitis.

Watery Eye

- Although tearing of the eye is normal and desirable to maintain a healthy eye, excessive tearing may indicate an underlying health condition.
- Watery or tearing eyes can be caused by and/or be a sign of allergy, Bell's palsy, common cold, corneal ulcers and infections, headache (cluster), medications (some), tear duct blockage, trauma/injury, uveitis.

Weight Gain (unintentional)

- Weight gain can result from an increase in fat, muscle mass, or body fluid.
- Unintentional weight can be caused by and/or be a sign of acromegaly, bipolar disorder, congestive heart failure, Cushing's syndrome, depression, hypothyroidism, kidney failure, lack of exercise, medications (some), pregnancy, renal failure (acute).

Weight Loss (unintentional)

- Weight loss can be intentional (dieting, exercise) or unintentional (e.g., associated with illness), the latter of which should be checked by a health-care practitioner.
- Unintentional weight loss can be caused by and/or be a sign of Addison's disease, adrenal insufficiency, AIDS/HIV, alcohol abuse, anemia, anorexia nervosa, bulimia, cancer, Crohn's disease, dehydration, dementia, depression, diabetes, hyperthyroidism, leukemias, lymphoma (Hodgkin's), malabsorption, medications (some), ovarian cancer, parasites, pancreatic cancer, pancreatitis, peptic ulcer, rheumatoid arthritis (juvenile), scleroderma, stomach cancer, systemic lupus erythematosus, tapeworm infection, thyroiditis, tuberculosis, ulcerative colitis.

2

WHAT YOU NEED TO KNOW ABOUT COMMON MEDICAL PROCEDURES

Unlike a medical test, which is a method or examination used to identify the nature of a symptom or medical condition, a medical procedure is a specific way to accomplish a desired result through use of an approach or tool. The following are brief explanations of medical procedures that are among some of the most common ones you or a loved one may encounter. Advance knowledge of a procedure can help alleviate any concerns or nervousness you may have about it and stimulate questions you may want to ask your health-care practitioner.

Amniocentesis

- **What it is:** A procedure in which a long, thin needle is used to remove fluid from the amniotic sac during pregnancy for detection of possible abnormalities of a fetus. Evaluation of the fluid can identify lung maturity, presence of infection, and evidence of spina bifida, as well as provide genetic information and chromosome analysis.
- **How to prepare:** No special preparation is required.
- **What to expect:** The physician will sterilize the insertion site and insert a long, thin-bore needle into the abdomen under ultrasound guidance. Local anesthesia may or may not be used. Approximately 30 cc (2 tablespoons) of amniotic fluid will be removed for analysis. Information obtained from evaluation of amniotic fluid can help women make informed decisions regarding their pregnancy and infant. You may experience some faintness or abdominal cramping.

Angiography

- **What it is:** A procedure in which a catheter is used to inject dye into a blood vessel, which allows the arteries to be viewed on x-ray. Angiography can be used to identify blockages and other problems that hinder blood flow in the legs, abdomen, adrenal glands, brain, or heart. Another name for this procedure is arteriography.
- **How to prepare:** If a sedative will be given before the procedure, no food or liquid (except small amounts of water needed to take medication) should be consumed 4 to 8 hours before the angiography. You should remove all jewelry and empty your bladder before the procedure. Tell your health-care practitioner if you are allergic to shellfish or iodine or if you've ever had an adverse reaction to contrast media.
- **What to expect:** The physician will sterilize the site where the catheter (a very small, thin tube) will be inserted into a blood vessel in the groin or arm. An incision will be made, the catheter inserted and advanced to the desired site, and a dye will be injected through the catheter. The procedure is painless, but you may feel some warmth or burning when the dye is injected. If you will be going home the same day, you will need to arrange transportation. Possible side effects of the procedure include an allergic reaction to the dye, which could cause a drop in blood pressure, breathing difficulties, or loss of consciousness.

Arthrocentesis

- **What it is:** A procedure whereby a sterile needle and syringe are used to remove fluid from a joint for examination in a lab.
- **How to prepare:** No special preparation is required unless glucose levels in the synovial fluid will be measured. In such cases, you should avoid food and fluids for 6 to 12 hours before the procedure (except for small amounts needed to take medication).
- **What to expect:** The physician will sterilize the skin over the joint either by injection, a topical freezing liquid, or

both, and a needle with syringe will be inserted into the joint. In some cases, medication is injected into the joint after the fluid has been removed. A bandage will be applied over the injection point. Results of analysis of the fluid can help uncover the causes of joint swelling or arthritis, including infection, gout, and rheumatoid disease. Complications of arthrocentesis are uncommon but may include bruising, minor bleeding into the joint, or loss of pigment in the skin that was injected. Rarely, joint infection occurs. The affected joint may be sore or swollen for several days.

Arthroscopy

- **What it is:** A surgical procedure in which a small tube is inserted through tiny incisions made into an affected joint as a way to diagnose or treat joint problems.
- **How to prepare:** Preoperative evaluation usually includes a physical examination, blood tests, and a urinalysis. Individuals who have a history of heart or lung problems or who are older than 50 usually are asked to have a chest x-ray and electrocardiogram. Do not consume food or fluids (except small amounts of water if needed to take medication) 12 hours before the procedure.
- **What to expect:** A local or general anesthetic will be given, depending on the site of the arthroscopy. An arthroscope (a small tube containing optical lenses and fibers) will be inserted through tiny incisions (approximately ¼ inch long) into the affected joints. The arthroscope is connected to a video camera, which allows the physician to see the interior of the joint on a monitor. After the procedure, the treated area may feel stiff or sore for a few days.

Balloon Angioplasty

- **What it is:** A nonsurgical procedure that relieves narrowing and obstruction of the coronary arteries. Also known as percutaneous transluminal coronary angioplasty (PTCA).
- **How to prepare:** Do not consume food or liquids (except

small amounts of water if needed to take medication) after midnight prior to the day of the procedure.

- **What to expect:** You will be mildly sedated, and the catheter insertion site will be sterilized. A small balloon catheter will be inserted into an artery in the arm or groin and advanced to the blocked area in the coronary artery. The balloon will then be inflated to enlarge the narrowed section. You may experience mild discomfort at the incision site and minor chest pain when the balloon is inflated. Bed rest is recommended for 6 to 12 hours post-procedure. When successful, this procedure can relieve angina pain and reduce or stop a heart attack without the patient resorting to open heart bypass surgery.

Blood Transfusion

- **What it is:** The transfer of blood or blood products from one individual into another individual's bloodstream.
- **How to prepare:** No special preparation is required of the donor or the donoree. However, if you are donating blood for transfusion to a specific individual, it should be drawn at least two working days prior to its transfusion to allow time for the blood to be tested.
- **What to expect:** If you are receiving a blood transfusion, the blood is delivered through an intravenous (IV) line directly into the veins. You will be monitored for any adverse reactions. Allergic reactions include swelling, rash, itchiness, fever, headache, and dizziness. Once the transfusion is complete, a compress bandage will be placed over the IV site.

Breast Biopsy, Fine-needle Aspiration

- **What it is:** Insertion of a thin needle on a syringe to withdraw fluid and/or tissue from a breast abnormality. It is usually used to drain fluid from benign, fluid-filled cysts.
- **How to prepare:** No specific preparation is required.
- **What to expect:** A long, thin needle will be inserted through the breast into the abnormality. Because the nee-

dles used for this procedure are very thin, use of a local anesthetic is usually not required, but may be offered. There is generally little or no discomfort associated with this procedure. If the sample looks suspicious or the abnormality turns out to be solid, your physician will recommend another biopsy be performed.

Bronchoscopy

- **What it is:** Use of a bronchoscope (a flexible tube with a camera and tiny light) to examine a patient's lungs and airways, including the trachea and larynx. Lung tissue samples can be biopsied using this approach. Bronchoscopy is helpful in the diagnosis of conditions such as lung cancer, pneumonia, and tuberculosis.
- **How to prepare:** Do not consume any food or liquid after midnight prior to the procedure, except required medications taken with small amounts of water. Medications that increase the risk of bleeding (e.g., aspirin, warfarin) and nonsteroidal anti-inflammatory products should be discontinued at some point before the procedure, according to physician's orders. Dentures, contact lenses, and glasses must be removed before the procedure.
- **What to expect:** An intravenous catheter is inserted for delivery of medication and fluids, and a local anesthetic is administered to the nose and back of the throat to ease passage of the flexible bronchoscope. Once the scope is in place, additional anesthetic is sprayed to help ensure comfort. If a rigid bronchoscope is used, general anesthesia is needed. A flexible bronchoscope rarely causes any discomfort or pain, but you may feel an urge to cough.

Cervical Biopsy

- **What it is:** A procedure in which tissue samples are taken from the cervix and evaluated for the presence of disease or other abnormalities.
- **How to prepare:** No special preparation is required. It is suggested, however, that you empty your bladder and

bowels before the procedure and abstain from sexual intercourse or douching for 24 hours before the biopsy.

- **What to expect:** The procedure is painless, although you may feel a pinch when tissue samples are taken. Some women experience mild cramping during the biopsy as well as after, and in both instances the pain may be relieved or eliminated by taking slow, deep breaths.

Colonoscopy

- **What it is:** Examination of the inside of the colon using a thin, lighted instrument called a colonoscope, which is inserted into the rectum. It can be used to diagnose colon cancer, inflammatory bowel disease, and other conditions.
- **How to prepare:** Your physician will advise you on how to thoroughly clean your bowel before the procedure, which will include the use of enemas, avoidance of solid foods for 2 to 3 days before the procedure, and use of laxatives. Use of any iron-containing preparations should be stopped several weeks before the procedure. If you have valvular heart disease, you may be given antibiotics before and after the procedure to prevent infection.
- **What to expect:** A rectal exam is usually done before the procedure, to make sure there are no blockages. You will receive a sedative and pain medication to help you relax. You may feel pressure as the colonoscope moves inside your bowel, and mild cramping and gas pains are common, both during the procedure and after. You can reduce discomfort by taking slow, deep breaths.

Colposcopy

- **What it is:** Use of a thin tube equipped with a magnifying lens and a light to examine the vagina and cervix for abnormalities.
- **How to prepare:** Avoid use of vaginal creams or medications for 24 hours before the procedure, and empty your bladder just before it begins.

- **What to expect:** You will lie on your back with your feet resting in stirrups. The physician will insert a speculum to open up the vagina, the cervix will be cleansed with a solution, and a colposcope will be passed through the speculum. Abnormal areas will be noted and a biopsy may be performed. If a biopsy is done, you should avoid strenuous activity for up to 24 hours after the procedure. You may also experience some cramping and/or slight bleeding post-procedure if a biopsy was performed.

Cystoscopy

- **What it is:** Examination of the bladder and urethra using a thin, lighted device (cystoscope) that is inserted into the urethra. The cystoscope removes tissue samples, which can be examined for diagnosis of bladder cancer, cystitis, enlarged prostate, incontinence, and other conditions.
- **How to prepare:** No special preparation is required. However, you should arrange for transportation home after the procedure.
- **What to expect:** A local anesthetic is applied, and the cystoscope is advanced through the urethra into the bladder. You will feel a strong urge to urinate when the fluid passes into the bladder. You may also feel a pinch if the physician takes a tissue sample. Once the cystoscope is removed, your urethra may be sore and you may experience burning during urination for a few days.

Endometrial Biopsy

- **What it is:** A procedure in which a tissue sample is taken from the endometrium (inside lining of the uterus) and examined for abnormalities.
- **How to prepare:** No special preparation is required.
- **What to expect:** This procedure can be done with or without anesthesia. After the physician performs a pelvic examination, the cervix will be cleaned with an antiseptic. A small, hollow tube will be passed into the uterus and gentle suction will remove a small tissue sample. Some

women experience cramping when the sample is being taken and for a short time after the procedure.

Hemodialysis

- **What it is:** A method of removing toxic substances from the blood when the kidneys can no longer do so. Hemodialysis involves circulating the blood through various filters and typically is done three times per week.
- **How to prepare:** You should follow the dietary and medication usage guidelines set by your physician.
- **What to expect:** Since each dialysis session lasts several hours and you are required to remain quiet, you should be prepared to occupy your time in a comfortable manner.

Laparoscopy

- **What it is:** A procedure in which a thin, lighted tube (laparoscope) is passed through the abdominal wall to examine the inside of the abdomen and the pelvic region. Tissue samples or organs may be removed or other procedures may be performed during laparoscopy.
- **How to prepare:** Do not consume any food or liquid (except for small amounts of water needed to take medication) for 8 hours before the procedure.
- **What to expect:** The procedure is done under general anesthesia. A small incision will be made near the navel and a needle will be inserted to inject carbon dioxide gas, which enlarges the surgical area. The physician will also insert a laparoscope to examine the internal organs. Additional incisions may be made if organs or tissue samples must be removed, or if other procedures must be performed. In some cases, a drain may be left in one or more incisions to allow removal of accumulated fluid. You may experience throbbing and pain at the incision sites after the procedure. Some people have shoulder pain or an urgent need to urinate post-procedure.

Lithotripsy

- **What it is:** A procedure used to break up stones that can form in the gallbladder, ureter, kidney, or bladder. In most cases, high-energy shock waves are used to break up stones; laser is an alternative when shock waves are not feasible.
- **How to prepare:** Do not consume food or liquids at least 6 hours before the procedure (except for small amounts of water needed to take medication).
- **What to expect:** You will receive some form of painkiller and sedative before the procedure begins. No incisions are required if shock wave lithotripsy is done. However, if laser lithotripsy is performed, your physician will need to make an incision and insert an endoscope through the urinary tract. If you have kidney or urinary stones treated, you will need to drink lots of water for several weeks after the procedure to help pass the stone fragments. Some people experience pain and nausea during this time. A small amount of blood in the urine is common for a few days to weeks post-procedure. Your physician may prescribe an antibiotic or anti-inflammatory after your procedure.

Lumbar Puncture

- **What it is:** A procedure in which a needle is inserted into the lower spinal column to collect cerebrospinal fluid or to administer anticancer drugs. This procedure is also known as a spinal tap.
- **How to prepare:** Be prepared to remain in the hospital for at least 6 to 8 hours after the procedure, during which time you must remain lying flat.
- **What to expect:** You will lie on your side with your knees tucked up against your abdomen, or you will be asked to sit while bending forward. A local anesthetic will be injected into the lower spine, and the spinal needle will be inserted between the lumbar vertebrae. Once the cerebrospinal fluid sample is collected, the collection site will be cleaned and bandaged.

Paracentesis

- **What it is:** A procedure in which a needle is inserted through the abdominal wall to obtain a fluid sample for analysis. Paracentesis is also known as an abdominal tap.
- **How to prepare:** No special preparation is required. However, tell your health-care practitioner if you are allergic to any medications or anesthetics, if you have any bleeding problems, or if you are pregnant.
- **What to expect:** Paracentesis is an outpatient procedure. The puncture site will be cleaned, and you will be given a local anesthetic. The needle will be inserted and fluid drawn into a syringe. In some cases a tiny incision is needed to insert the needle. Once the needle has been withdrawn, a dressing will be applied and several stitches made if an incision was required.

Pericardiocentesis

- **What it is:** Insertion of a needle into the pericardial sac (the membrane that surrounds the heart) to obtain a fluid sample for analysis or to remove fluid that is compressing the heart.
- **How to prepare:** You should refrain from food and fluids at least 6 hours before the procedure (except for small amounts of water needed to take medication).
- **What to expect:** The insertion site will be cleaned and a local anesthetic given. Insertion of the needle will be just below the sternum (breastbone) and may be guided by electrocardiographic leads attached to the needle or by an echocardiogram. When the needle reaches the pericardial sac, a wire will be inserted and the needle removed and replaced with a catheter. Fluid will be withdrawn through the catheter. You may experience a feeling of pressure or chest pain during the procedure.

Sigmoidoscopy

- **What it is:** Examination of the lower colon using a thin, lighted instrument called a sigmoidoscope. Tissue samples obtained during the examination are used to identify the cause of diarrhea and/or to diagnose inflammatory bowel disease, cancer, and other conditions.

- **How to prepare:** Consume only clear liquids for 12 to 48 hours before the procedure, according to your health-care practitioner's instructions. You may be asked to self-administer an enema before the sigmoidoscopy.
- **What to expect:** You will lie on your left side with your knees drawn up toward your chest. The physician will do a digital rectal examination to check for any blockages and to dilate the anus. This will be followed by insertion of the sigmoidoscope into the rectum and the passage of air into the scope. You will feel pressure and a feeling of the need to defecate during the procedure, as well as some cramping and bloating. Once the procedure is done, you will expel some gas.

Thoracentesis

- **What it is:** Use of a needle to withdraw fluid from the space between the lining of the outside of the lungs (pleura) and the chest wall. A thoracentesis is done to remove accumulated fluid and/or to identify the reason for it.
- **How to prepare:** No special preparation is required. Your physician may order a chest x-ray before and after the procedure.
- **What to expect:** The injection site on your chest or back will be sterilized with a solution and a local anesthetic injected into the area. Fluid will then be withdrawn through a needle that will be placed through the skin of the chest wall. You will experience a stinging sensation when the anesthetic is injected, and you may feel some pressure when the needle is inserted into the chest wall.

3

WHAT YOU NEED TO KNOW ABOUT COMMON MEDICAL TESTS

Physicians have hundreds of tests at their disposal to help them make more accurate diagnoses, verify their suspicions, and assist them in choosing the most appropriate treatment course. The following is a brief explanation of some of the more common tests you or your loved ones may encounter. Advance knowledge of a test can help alleviate any concerns or nervousness you may have about it and stimulate you to ask questions of your health-care practitioner.

Ambulatory Electrocardiography Monitoring

- A noninvasive, painless test that detects abnormal heart rhythms.
- Used to evaluate chest pain, the efficacy of antiarrhythmic drug therapy, a patient's cardiac status after a heart attack, the status of a patient's pacemaker, and central nervous system and breathing symptoms.

Amsler Grid

- A checkerboard-pattern grid individuals can use at home to help identify possible problems with the macular area of the retina.
- Used to detect blind or partially blind spots in the macula.

Antinuclear Antibody Test

- A sensitive screening test that analyzes a blood sample for the presence of antinuclear antibodies, which are found in patients whose immune system may be predisposed to cause inflammation against their own tissues.
- Used to detect autoimmune diseases, such as rheumatoid arthritis, lupus, and multiple sclerosis.

Apgar Test

- A quick observational evaluation process performed immediately after birth to determine a newborn's physical condition and how well he or she tolerated the birthing process.
- Factors considered include respiration, skin color, muscle tone, reflexes, and heart rate. A score of 8 to 10 is normal; lower scores indicate the infant needs help adjusting to his/her new environment.

Audiology Test

- Detects how well an individual can hear. The normal range of hearing is 0 to 85 decibels (dB). Exposure to levels greater than 85 dB can cause hearing loss.
- Used to detect severity of hearing loss.

Barium Enema

- Involves introduction of a thick, chalky liquid (barium sulfate) into the rectum, used as an enema. Once the liquid is in place, x-rays are taken.
- Used to help confirm a diagnosis of inflammatory disease of the lower bowel, colon cancer, polyps, and physical changes in the large intestine.

Barium Swallow

- Involves swallowing a thick, chalky liquid (barium sulfate) and using various monitoring devices to follow the progress of the barium.

- Facilitates the diagnosis of polyps, tumors, ulcers, strictures, and other problems that can affect the upper gastrointestinal tract.

Blood Culture

- A blood sample is taken and placed in a container that encourages organisms to grow.
- Used to help detect suspected disease-causing organisms in the bloodstream.

Blood Pressure Test

- A measurement of the force of the blood against the artery walls as it flows through the body. The upper or first number (systolic) is the pressure when the heart is contracted; the lower or second number (diastolic) is the pressure when the heart is at rest.
- Used as an indicator of risk for heart attack, stroke, kidney failure, and other medical conditions.

Bone Density Testing

- A noninvasive test that can be done using several different devices: (1) dual-energy x-ray absorptiometry (DEXA), which checks density of the wrist, vertebrae, and hip; (2) dual-photon absorptiometry, or DPA, checks the spine and hip; and (3) single-photon absorptiometry, or SPA, checks the wrist or heel.
- Used to monitor loss of bone density, such as that associated with osteoporosis; to detect bone tumors; and to identify infection.

Cerebrospinal Fluid Analysis

- A sample of cerebrospinal fluid is obtained during a lumbar puncture procedure (*see* page 429) through a syringe.
- Used to help identify meningitis, tumors, bleeding around the spinal cord or brain, or any obstructions in fluid circulation.

CHEM-20

- A blood sample is taken and analyzed for 20 different factors, including levels of specific proteins, enzymes, minerals, sugars, and metabolites.
- Used to help physicians identify various problems with specific body functions.

Chest Radiography

- An x-ray of the chest, used for diagnostic purposes and also as part of a preadmission test before surgery.
- Used to detect problems with the lungs, such as breathing difficulties, tumors, pneumonia, and collapsed lung.

Cholesterol Testing

- A blood sample is taken and analyzed for the levels of total cholesterol; levels of high-density and low-density lipoprotein cholesterol also may be done.
- Used to help evaluate an individual's risk for coronary artery disease, atherosclerosis, and sudden death. This test is usually a part of a Chem-20.

Complete Blood Count (CBC)

- A routine screening test whose results may indicate a need for additional tests. Factors tested for include number of red blood cells, number of white blood cells, total hemoglobin level, percentage of blood composed of cells (hematocrit), number of platelets, mean corpuscular hemoglobin, mean corpuscular hemoglobin concentration, and mean corpuscular volume.
- Used to help diagnose and monitor various diseases; for example, a low hematocrit indicates anemia and a high white blood cell count indicates leukemia or infection.

Computerized Axial Tomography (CAT)

- A noninvasive diagnostic test that produces three-dimensional images of various body parts.
- Used to help evaluate vascular structures, tumors, the brain, chest, and abdomen; and to guide physicians during various procedures such as biopsies and placement of tubes.

C-Reactive Protein Test

- A blood sample is taken to identify the amount of C-reactive protein in the bloodstream.
- Used to identify inflammatory diseases, such as rheumatoid arthritis and rheumatic fever; also helps physicians monitor how patients are responding to therapy for these and other inflammatory conditions.

Echocardiography

- A test that uses sound waves to create a moving picture of the heart. It is also referred to as a Doppler ultrasound of the heart.
- Used to evaluate blood circulation, valve disorders, heart function, and abnormal heart sounds; to detect excess fluid around the heart; and to measure the size of the heart's chambers. It is also used to monitor individuals who have undergone bypass surgery or arterial reconstruction.

Echoencephalography

- A noninvasive diagnostic test in which ultrasound beams are sent through tissue and the reflected sound waves are transformed into electrical signals that are viewed on a screen.
- Used to identify the size and position of brain structures and to evaluate suspected disease of the brain and spinal cord.

Electrocardiography (ECG)

- A noninvasive diagnostic test in which electrical signals from the heart are obtained through electrodes attached to the chest and printed on graph paper.
- Used to help identify heart damage and electrolyte abnormalities that have an impact on the heart muscle; also is useful in determining how well a patient's pacemaker is functioning.

Electroencephalography (EEG)

- A noninvasive test that produces a graphic recording of the electrical currents that are produced by brain activity. The recording is obtained as a result of the electrodes that are placed on the scalp and whose lead wires are attached to an electrical device.
- Used to help diagnose epilepsy, brain injuries, and tumors.

Electromyography (EMG)

- A noninvasive diagnostic test in which electrical signals produced by muscle are measured.
- Used to help diagnose neuromuscular diseases, such as amyotrophic lateral sclerosis (ALS), muscular dystrophy, and nerve disorders.

ELISA (Enzyme-Linked Immunoabsorbant Assay)

- A blood test that can be used to help detect the presence of disease-causing organisms or antibodies.
- Used to help diagnose HIV, Lyme disease, chlamydia, mumps, rubella, blood clots, allergies, immune system diseases, and other conditions.

Erythrocyte Sedimentation Rate

- A common blood test that measures the rate at which red blood cells (erythrocytes) settle in a tube over a given period of time. The normal rate for males is 0–15 millimeters per hour, 0–20 for females, and may be slightly greater in the elderly.

- Used to detect and monitor inflammation and conditions associated with inflammation. The greater the inflammation, the greater the sedimentation rate.

Exercise Electrocardiography

- Also known as a stress test, it is the most commonly used cardiac stress test. It involves exercising on a treadmill according to a specific protocol while the patient's electrocardiogram, heart rhythm, blood pressure, and heart rate are monitored continuously.
- Used to help detect any obstruction to blood flow or abnormal rhythms, which are characteristics of coronary artery disease.

Fasting Plasma Glucose

- A blood test in which a sample is taken after an individual has fasted for 12 hours. Also known as fasting blood sugar.
- Used to detect abnormalities in sugar (glucose) metabolism and to screen for diabetes mellitus.

Fecal Occult Blood Test

- Special chemical tests performed on stool samples to detect occult (not visible) blood.
- Used to identify colon polyps or cancers that cause chronic blood loss from the colon. Individuals who have a positive result on this test are then referred for a colonoscopy (*see* page 426).

Glomerular Filtration Rate (GFR)

- A urine sample and a blood sample are taken to determine the amount of fluid that the kidneys filter per minute. Also known as a creatinine clearance test.
- Used to evaluate kidney function.

Homocysteine Test

- A blood sample is obtained and used to measure levels of this protein in the blood. Elevated levels are believed to cause narrowing and hardening of the arteries.
- Used by some physicians to screen for elevated homocysteine levels in patients with early onset of heart attack, stroke, or other symptoms related to atherosclerosis, or individuals at risk for these conditions.

Intravenous Pyelogram

- X-rays are taken of the urinary tract both before and after injection of a special dye.
- Used to evaluate the structure and function of the urinary tract and to detect any abnormalities, such as tumors, kidney stones, blockages, and other problems.

Lactose Intolerance

- A test in which individuals drink a liquid that contains lactose (a sugar) and then have several blood samples drawn over a two-hour period to measure blood sugar levels.
- Used to determine how well the body digests lactose. If lactose is not completely metabolized, blood glucose levels fail to rise, which indicates lactose intolerance.

Liver Scan

- An imaging test in which a radioactive substance is injected into the veins, after which a special x-ray camera is used to take images of the liver.
- Used to identify tumors and other abnormalities of the liver and spleen and to monitor the function of a diseased liver.

Magnetic Resonance Imaging (MRI)

- The use of magnetic fields and radio waves to evaluate various parts of the body. In some cases, a dye is injected into a vein to allow physicians a better view of the internal structures.

- Used to determine the location and size of tumors and to confirm diagnoses.

Mammography

- A mammogram is an x-ray that can visualize normal and abnormal structures within the breasts.
- Used to identify cysts, calcifications, tumors, and other abnormalities.

Obstetric Screening Profile

- A battery of tests done on blood samples obtained from pregnant women. The tests include complete blood count, differential white cell count, hepatitis B detection, blood typing, antibody screen, test for syphilis, and test for immunity to rubella to determine baseline values.
- Used to determine a woman's baseline health status.

Ophthalmoscopy

- A painless test in which an ophthalmoscope is used to examine the back of the eye.
- Used to evaluate the retina, blood vessels, and nerves in the eye to help detect eye disorders.

Pap (Papanicolaou) Smear

- A procedure in which a swab is used to obtain a cell sample from a woman's cervix, and the sample is then smeared on a microscope slide for examination.
- Used to look for premalignant or malignant changes.

Positron Emission Tomography (PET)

- A highly specialized imaging technique that uses short-lived radioactive substances to produce three-dimensional images showing how those substances function within the body.

- Used to study metabolic activity or body function, which CT and MRI cannot do. It can detect active tumor tissue, identify the causes of seizures and dementia, and assess the results of coronary artery bypass surgery.

Pregnancy Test (Urine and Blood)

- Blood and urine samples are taken to detect the amount of human chorionic gonadotropin (HCG). Such pregnancy testing kits are available over-the-counter.
- Used to determine if a woman is pregnant. Levels of HCG are detectable within 10 days after fertilization.

Prostate-Specific Antigen (PSA) Test

- A blood sample is taken and subjected to a very sensitive detection method known as monoclonal antibody technique to measure levels of prostate-specific antigen (PSA).
- Used to screen for prostate cancer or to monitor PSA levels in men who have the disease. Abnormal results (greater than 4 mg/mL but less than 10 mg/mL) on screening indicates further testing is needed. As a monitoring test, an abnormal result indicates a recurrence of prostate cancer.

Radio-Allergosorbent Test (RAST)

- An allergy test in which a sample of blood is taken and mixed with substances known to trigger allergies.
- Used to measure the level of allergy antibodies in the blood. A negative result on the test indicates that you probably don't have a true allergy, although there is a small chance that an allergy exists. Positive or elevated results typically indicate an allergy, even though you may not have a physical response when exposed to the allergen.

Rh Typing

- A blood sample is taken to type an individual's blood, especially new mothers'.
- Used to determine whether new mothers require an Rh-immunoglobulin injection, or whether a donor's blood is compatible before a transfusion is performed.

Rheumatoid Factor (RF)

- A blood sample is obtained to look for the presence of rheumatoid factor, which is found in about 80 percent of adults who have rheumatoid arthritis.
- Used to help in the diagnosis of rheumatoid arthritis. Rheumatoid factor is also present in individuals who have other conditions, such as systemic lupus erythematosus, infectious hepatitis, syphilis, parasites, tuberculosis, liver disease, sarcoidosis.

Skin Biopsy

- A test that involves removal of a small skin tissue sample, under local anesthesia, for examination under a microscope. The sample may be obtained using a sharp scalpel to remove the outer layer of a lesion; a punch, which removes a circular core of tissue from the center of a lesion; or excision, which removes the entire lesion.
- Used to identify benign and malignant growths, or to diagnose bacterial and fungal infections or inflammatory and autoimmune skin conditions.

Spirometry

- This test involves breathing into a plastic tube that is part of a machine called a spirometer, which measures the volume of air inhaled and exhaled.
- Used to determine whether the airways are narrowed and helps monitor the effectiveness of therapies for lung conditions.

Sputum Culture

- A sputum sample is obtained through deep coughing or, if necessary, through the use of suction or bronchoscopy.
- Used to help verify a diagnosis of respiratory disorders such as pneumonia, bronchitis, and tuberculosis.

Throat Culture

- A sterile swab is used to collect a sample from the back of the throat.
- Used to isolate and identify bacteria that can infect the tonsils, throat, and pharynx. It may also be used as a screening tool to identify asymptomatic individuals who may have *Neisseria meningitis*.

Thyroid Function Test

- A blood sample is taken and used to measure the levels of various hormones to determine the status of the thyroid.
- Used to show levels of total thyroxine (T4), total T3 resin uptake, and thyroid-stimulating hormone, and thus helps determine whether an individual has hypothyroidism, hyperthyroidism, or other thyroid function abnormalities.

Ultrasound

- This imaging approach involves sending high-frequency sound waves into the body, which send back various signals based on the density of the tissue being examined. The images can be seen on a computer screen.
- Used to examine and evaluate many different body parts and functions: for example, detection of blood clots, abnormalities of the heart valves, kidney stones, gallstones, cysts, tumors, or abscesses; evaluation of the size and structure of various organs; and as a guide when performing fluid withdrawal from the chest, lungs, or around the heart.

Urinalysis

- A urine sample is tested using a microscope or by dipping reagent strips into the sample and noting any chemical changes to the strips, which indicate various conditions.
- Used to identify urinary tract infections, kidney infections, and diabetes.

Visual Acuity

- An eye test performed as part of a routine eye examination.
- Used to evaluate distant and near vision and to identify how nearsighted or farsighted an individual may be.

Visual Field Test

- An eye test performed to measure an individual's complete scope of vision for both eyes.
- Used to detect any signs of glaucoma damage or diseases of the optic nerve or retina or patterns of vision loss.

4

26 MEDICAL CONDITIONS: WHAT TO ASK YOUR DOCTOR

The following offers health-care consumers a concise explanation of some of the more common medical conditions that affect individuals. It is in the form of Q&A, with questions you should ask your health-care provider and brief answers to each query. We hope this Q&A will prompt you to make further inquiries of your doctor should you or a loved one have any of these medical conditions.

Alzheimer's Disease

- **What is it?** A progressive disease of the brain characterized by impaired memory and dysfunction of at least one other thinking function (e.g., language, perception of reality). Many experts believe this disease is a result of an accumulation of beta-amyloid protein in the brain, which causes nerve cell death.
- **Who gets it?** Ten percent of individuals older than age 65 and 50 percent of those older than 85 have Alzheimer's disease. In a small percentage of cases, in which a genetic basis has been identified, the disease develops before age 50.
- **What are the symptoms/warning signs?** The Alzheimer's Association lists the following: (1) memory loss that affects job skills; (2) difficulty performing familiar tasks; (3) language difficulties; (4) disorientation to place and time; (5) poor judgment; (6) misplacing objects; (7) difficulty with abstract thinking; (8) behavior or mood changes; (9) personality changes; (10) loss of initiative.

- **What causes it?** Except in rare cases in which a mutated gene has been identified as the cause of the disease, the vast majority of cases are caused by multiple factors, including but not limited to abnormally low levels of some neurotransmitters such as acetylcholine and serotonin, environmental toxins, history of stroke or heart disease, high-fat diet, lack of exercise, and gender (women are at higher risk than men).
- **How is it diagnosed?** A combination of history, physical examination, laboratory tests, psychometric tests, and other studies.
- **What is the prognosis?** The rate and severity of decline can vary greatly among individuals. Some people have the disease for up to five years, while others have it for up to 20 years. Infection is the most common cause of death.
- **How is it treated?** There is no cure for Alzheimer's disease. Medications are available to help manage behavioral and psychiatric manifestations and to help slow the progression of mental decline.

Asthma

- **What is it?** A chronic disease characterized by inflammation of the bronchial airways, which causes swelling and narrowing of the airways and results in breathing difficulties.
- **Who gets it?** Asthma affects people of all ages, but it is more common among children. In fact, it is the most common chronic condition affecting children: 1 in every 15 children has asthma. Five percent of adults in North America have asthma.
- **What are the symptoms/warning signs?** Shortness of breath, especially at night or with exertion; wheezing when breathing out; cough that may be chronic, that is usually worse at night and in early morning, and that often occurs after exercise or when exposed to dry, cold air; and tightness in the chest.
- **What causes it?** Generally caused by allergy to one or more substances or irritants. Hereditary predisposition may be a factor.

- **How is it diagnosed?** A physical examination, allergy skin tests, chest x-rays, and pulmonary function tests are typically conducted. Blood tests also may be done.
- **What is the prognosis?** Fifty percent of children with asthma outgrow it as they get older.
- **How is it treated?** Treatments include use of bronchodilator inhalers, oral bronchodilator and/or antileukotriene medications, inhaled corticosteroidal drugs, and use of a peak flow meter if indicated by your physician.

Breast Cancer

- **What is it?** The growth of malignant (cancerous) cells in the breast.
- **Who gets it?** Primarily women, with a progressively increased risk among women older than 50; it is rare among men (less than 1% of all cases).
- **What are the symptoms/warning signs?** A lump or swelling, usually painless, in the breast or in the underarm region; a change in the appearance of the breast, such as dippling, creasing, scaliness, or indentations; changes in the nipple, including abnormal discharge; pain or discomfort in the breast.
- **What causes it?** The exact cause is unknown, but risk factors include age greater than 50, family history of breast cancer (especially among immediate or near-immediate family members), having had no children or children late in life, onset of menstruation before age 11, menopause after one's early 50s, presence of nonmalignant tumors and cysts in the breast, a diet high in animal fat, use of oral contraceptives and/or hormone replacement therapy.
- **How is it diagnosed?** Mammography, breast examination, biopsy of a lump; if needed, ultrasound, thermography, or computer tomography.
- **What is the prognosis?** Depending on the stage at which the cancer is treated, five-year survival can range from 95% at stage 0 to less than 10% at stage 4.
- **How is it treated?** Surgery to remove the lump and/or

surrounding tissue (if indicated), accompanied by (if needed) radiation, chemotherapy, and/or hormone therapy.

Cataracts

- **What is it?** A loss of the transparency in the lens of the eye.
- **Who gets it?** About 75% of people older than 60 exhibit some signs of cataracts.
- **What are the symptoms/warning signs?** Fuzzy, foggy, or filmy vision; difficulty seeing at night, especially when driving; difficulty with glare from the sun or lamps; seeing halos around lights; double vision in one eye; decreased sensitivity to contrast; frequent changes in eyeglass or contact lens prescriptions.
- **What causes it?** Aging is the single greatest risk factor, especially the cumulative exposure to ultraviolet rays from the sun; also exposure to radiation, eye inflammation, physical injury to the eye, low serum calcium levels, prolonged use of corticosteroids.
- **How is it diagnosed?** Eye examination by an ophthalmologist, including a slit lamp examination.
- **What is the prognosis?** Surgery is successful in 95% of cases.
- **How is it treated?** The only treatment is removal of the cataract. Most people are then fitted with an artificial lens (intraocular lens); contact lenses or cataract glasses are options when lens replacement is not possible.

Chronic Fatigue Syndrome

- **What is it?** A term used to describe a group of symptoms that includes, primarily, extreme and chronic fatigue and weakness, among other symptoms.
- **Who gets it?** Most often affects women ages 30 to 50.
- **What are the symptoms/warning signs?** In addition to the primary symptom of extreme, debilitating fatigue, other symptoms include chronic sore throat, aching and painful muscles and joints, low-grade fever, depression,

painful lymph nodes, irritability, memory loss, sleep difficulties, headache, and confusion.

- **What causes it?** The cause is unknown. Some possibilities include inflammation of pathways of the nervous system, which indicates an autoimmune response; a virus; abnormally low levels of certain hormones, such as cortisol and corticotropin-releasing hormone; stress, genetics, and prior illnesses may also play a role.
- **How is it diagnosed?** Patients must exhibit the primary symptom of extreme, debilitating fatigue for at least 6 months, plus tests should be done to rule out other conditions that mimic chronic fatigue syndrome, such as anemia, infections, heart disease, thyroid disease, and others.
- **What is the prognosis?** Prognosis is variable and difficult to determine. Some people recover completely within 6 to 12 months; others take longer, while some never fully regain their former level of health.
- **How is it treated?** Over-the-counter pain relievers, regular moderate exercise, eating a healthy diet, getting lots of rest, low-dose antidepressants, and psychological counseling are recommended.

Colorectal Cancer

- **What is it?** The growth of malignant (cancerous) cells in the colon or rectum.
- **Who gets it?** Occurs most often in men and women older than 50.
- **What are the symptoms/warning signs?** Often there are no warning signs in the early stages of the disease. Indications include changes in bowel habits that last longer than 10 days, bloody or black stools, loss of appetite, unexplained weight loss, fatigue, intestinal obstruction.
- **What causes it?** Risk factors include a family history of colorectal, breast, or endometrial cancer; personal or family history of colon polyps or ulcerative colitis; a diet high in animal fat.
- **How is it diagnosed?** Digital rectal exam; sigmoidoscopy

or colonoscopy; blood and stool samples; in some cases a barium enema may be ordered.

- **What is the prognosis?** Five-year survival depends on the stage of the disease. Stage I has a 90% five-year survival rate; stage II, 75–85%; stage III, 40–60%; stage IV, median survival is 1 to 2 years.
- **How is it treated?** Surgery is necessary to remove the tumor. This is followed by chemotherapy. Radiation therapy is rarely used.

Coronary Artery Disease

- **What is it?** Narrowing of the coronary arteries, which are the blood vessels that supply blood to the heart.
- **Who gets it?** The lifetime risk of developing coronary artery disease after age 40 is 49% for men and 32% for women.
- **What are the symptoms/warning signs?** There are no symptoms in the early stages of the disease. Eventual symptoms/signs include chest pain or squeezing, burning, or heaviness in the chest that lasts 30 seconds to 5 minutes, accompanied by pain or discomfort that radiates down the arm, into the neck, or along the jaw. Also shortness of breath, dizziness, choking sensation.
- **What causes it?** Accumulation of plaque in the arterial walls (atherosclerosis). Risk factors include smoking, high blood cholesterol, high blood pressure, diabetes mellitus, obesity, sedentary lifestyle, family history of premature heart attack.
- **How is it diagnosed?** Diagnosis may include patient history and physical examination, an electrocardiogram, chest x-rays, stress testing, blood tests, echocardiogram, coronary angiography.
- **What is the prognosis?** Prognosis depends on several factors. Individuals who take their medications, eat a healthy diet, and exercise regularly as instructed by their healthcare practitioners generally do well.
- **How is it treated?** Nitrates can be prescribed to relieve or prevent chest pain (angina); other medications that may be

prescribed include beta-blockers, ACE inhibitors, calcium channel blockers, anticoagulants, vasodilators. Preventive measures should be followed, including low-fat diet, regular physical exercise, stress management, an avoidance of excessive alcohol intake and substances that raise blood pressure, such as tobacco, caffeine, diet pills, and nasal decongestants.

Crohn's Disease

- **What is it?** Chronic inflammation of the intestinal tract, most often occurring in the end portion of the small intestine or the colon.
- **Who gets it?** Can affect any age, but most often occurs in adolescents and early adulthood.
- **What are the symptoms/warning signs?** Abdominal spasms or cramps, persistent watery diarrhea, anal fissures, nausea, fever, fatigue, loss of appetite and weight, blood in the stool, systemic conditions such as joint pain, inflamed eyes, and skin lesions.
- **What causes it?** The only known risk factors are a family history of the disease, Jewish ancestry, and smoking. Food allergies, viral or bacterial infections, autoimmune disorders, and lymphatic obstruction may play a role.
- **How is it diagnosed?** Patient history, physical examination, blood tests, upper gastrointestinal series of x-rays, barium enema, sigmoidoscopy or colonoscopy. A biopsy of the colon may be taken during the sigmoidoscopy or colonoscopy.
- **What is the prognosis?** Crohn's disease is chronic, characterized by periods of improvement followed by deterioration. There is an increased risk of small bowel or colorectal cancer associated with this disease.
- **How is it treated?** Mild episodes can be treated with over-the-counter antidiarrheal medications and pain relievers; anti-inflammatory drugs, antibiotics and/or immunosuppressive drugs may be used. Nutritional supplementation is recommended.

Diabetes

- **What is it?** A metabolic disorder characterized by abnormally high levels of glucose (sugar) in the bloodstream and little or no production of insulin by the pancreas.
- **Who gets it?** There are two main types of diabetes: type I, which typically develops in individuals younger than 30; and type II, which tends to affect people older than 40.
- **What are the symptoms/warning signs?** Excessive and frequent urination, increased appetite, increased thirst, unintentional weight loss, blurry vision, weakness, fatigue, numbness and tingling in the hands and feet, chronic or recurring bladder, skin, and/or gum infections.
- **What causes it?** Type I diabetes is an autoimmune disorder in which the immune system attacks the pancreas and causes cessation of insulin production. In type II diabetes, risk factors include obesity and family history of the disease.
- **How is it diagnosed?** Patient history, physical examination, fasting blood tests, glucose tolerance testing, urinalysis.
- **What is the prognosis?** The risk of long-term complications can be reduced in patients who maintain strict control of their blood glucose levels and blood pressure, as well as adhere to dietary and exercise guidelines set by their physician.
- **How is it treated?** Type I diabetes is treated with daily injections of insulin; type II can be treated with a combination of diet, exercise, and weight loss, with oral medications used as needed.

Endometriosis

- **What is it?** A benign condition in which endometrial tissue attaches itself to reproductive or abdominal organs, which swell during menstruation and may also develop into abnormal tissue, both of which can cause significant pain.
- **Who gets it?** It is most common among women ages 25 to 40.

- **What are the symptoms/warning signs?** Pain in the vagina, lower abdomen, and/or lower back that begins just prior to monthly menstruation, continues through the period, and worsens just after the flow ceases; vaginal pain during sexual intercourse, diarrhea, constipation, painful bowel movements, bleeding from the rectum, blood in the urine during menstruation, infertility.
- **What causes it?** The cause is unknown, although hereditary and hormonal changes appear to play a part.
- **How is it diagnosed?** Pelvic examination, although a definitive diagnosis requires that a tissue sample be taken and/or direct visualization using laparoscopy be performed.
- **What is the prognosis?** Among women who have undergone conservative surgery for endometriosis because they were believed to be infertile, infertility rates are very low among women who have mild endometriosis, about 40 to 50 percent among those with moderate disease, and 60 to 70 percent among those with a severe condition.
- **How is it treated?** Depending on the severity, options include over-the-counter pain medications, prescription drugs designed to stop menstruation in an attempt to shrink abnormal endometrial tissue, surgical removal of the tissue, hysterectomy.

Genital Herpes

- **What is it?** A highly contagious viral infection characterized by episodes of painful sores that develop on the genitals.
- **Who gets it?** Statistics vary, but research shows that 25% of people 25 to 45 in the United States have been exposed to infection with herpes simplex virus type 2 (HSV-2).
- **What are the symptoms/warning signs?** Most people with genital herpes have no symptoms. When they do occur, they include watery blisters in the genital area that break and form painful ulcers; painful urination in women; fever, loss of appetite, enlarged lymph nodes in the groin, fatigue, headache.

- **What causes it?** Herpes simplex virus type 2 is the most common cause of genital herpes. Recurrence of episodes may be triggered by sexual intercourse, menstrual periods, stress, fatigue, exposure to ultraviolet light, and injury.
- **How is it diagnosed?** Patient history, physical examination, culture of fluid from blisters.
- **What is the prognosis?** Genital herpes can recur at any time: as seldom as once a year or so often it seems to be continuous. There is no cure.
- **How is it treated?** The goal of treatment is relieve symptoms and includes use of antiviral drugs and/or painkillers for pain. Warm baths may relieve discomfort during an outbreak.

Glaucoma

- **What is it?** A group of diseases characterized by increased pressure within the eye that can damage the optic nerve and result in partial or total vision loss. There are two forms: open-angle, which accounts for about 90% of cases and is a slow, progressive condition; and closed-angle, characterized by rapid deterioration of vision and eye pain.
- **Who gets it?** It is most commonly seen in people older than 40 years and is most prevalent among African-Americans or in individuals with a family history of the disease.
- **What are the symptoms/warning signs?** For open-angle glaucoma, gradual loss of peripheral vision, characterized by blind spots. For closed-angle glaucoma, severe eye pain, nausea and vomiting, blurry vision, and the appearance of halos around lights.
- **What causes it?** It is the result of an abnormal accumulation of the fluid (aqueous humor) inside the eyeball.
- **How is it diagnosed?** Tests include visual acuity, dilated eye examination, tonometry, and ultrasonic waves to measure cornea thickness.
- **What is the prognosis?** Untreated closed-angle glaucoma results in permanent vision loss once symptoms appear, but vision can be preserved with prompt treatment.

Untreated open-angle glaucoma can progress to blindness within 20 to 25 years. However, early intervention can prevent further loss.

- **How is it treated?** Treatments include medications, laser trabeculoplasty, conventional surgery, or a combination of these approaches.

Hepatitis C

- **What is it?** Inflammation of the liver associated with infection by hepatitis C virus.
- **Who gets it?** Those at greatest risk include individuals who have sex with multiple partners and/or with persons who have hepatitis C, health-care workers, individuals who receive blood or tissue products from a donor who has hepatitis C, street drug users who share needles with someone who has hepatitis C, individuals who have been on long-term kidney dialysis.
- **What are the symptoms/warning signs?** Jaundice, abdominal pain, fatigue, loss of appetite, nausea, vomiting, low-grade fever, dark urine, itching, bleeding from the esophagus, pale or clay-colored stools.
- **What causes it?** Infection with hepatitis C virus, which is usually spread through contact with blood products—use of IV drugs, sharing needles, getting a blood transfusion before 1992, accidentally being stuck with an infected needle.
- **How is it diagnosed?** Physical examination, patient history, blood tests, liver biopsy.
- **What is the prognosis?** At least 80% of people with acute hepatitis C later develop chronic liver infection, 20 to 30% develop cirrhosis, and up to 5% develop liver cancer.
- **How is it treated?** Several variations of prescription medications called interferon are available. Although they aren't a cure, they do relieve symptoms in about 25 percent of people who take it, and they may help prevent future liver problems.

Herpes Zoster (Shingles)

- **What is it?** A painful, often lingering disorder caused by varicella zoster, the same virus that causes chicken pox.
- **Who gets it?** Shingles usually strikes people older than 50 and only those who have had chicken pox in the past.
- **What are the symptoms/warning signs?** The initial indications include a sensation of tingling followed by pain and often accompanied by fever and headache. These are followed by severe pain and itching along with appearance of a rash consisting of fluid-filled blisters on one side of the torso, arms, legs, or face. This lasts from 1 to 4 weeks, although the pain can persist for months or years after the skin has cleared.
- **What causes it?** Exactly why the varicella zoster virus becomes reactivated and causes shingles is unknown, although it is generally believed that it occurs when the immune system is weakened by factors such as age, illness, stress, or use of immunosuppressant drugs.
- **How is it diagnosed?** The appearance of the rash, a prior history of chicken pox, culture of the skin lesions.
- **What is the prognosis?** Shingles usually clears in several weeks and rarely recurs. If motor nerves were involved, temporary or permanent nerve palsy may occur. Long-term (years) nerve pain occurs in 50% of people older than 60 who have had shingles. An outbreak that has caused eye lesions may lead to permanent blindness.
- **How is it treated?** Over-the-counter pain relievers, acyclovir and similar medications, corticosteroids, cold compresses.

High Blood Pressure

- **What is it?** A condition characterized by a persistent increase in the force that the blood exerts upon arterial walls. Systolic blood pressure (the "top" number of a blood pressure measurement) represents the pressure exerted when the heart beats. Diastolic blood pressure (the "bottom" number) represents the pressure when the heart is at rest. Blood pressure consistently measuring greater than 140 systolic and/or 90 diastolic is high blood pressure.

- **Who gets it?** It can affect people of any age; however, more than 50 percent of men and women age 55 and older have hypertension. According to the American Heart Association (2004), 50 million Americans have high blood pressure.
- **What are the symptoms/warning signs?** Symptoms are rare with many cases of high blood pressure, but if it is dangerously high, symptoms may include headache, dizziness, ringing in the ears, palpitations, nosebleeds, numbness or tingling in the feet or hands, drowsiness, confusion.
- **What causes it?** Risk factors include family history, race (incidence is twice as great among blacks as among whites), emotional stress, alcohol use, cigarette smoking, high-sodium diet.
- **How is it diagnosed?** Accurate measurement of abnormally high blood pressure on at least three different occasions over a period of a week or longer.
- **What is the prognosis?** It is manageable with lifelong treatment.
- **How is it treated?** Diet, exercise, stress management, use of medications such as angiotensin-converting enzyme (ACE) inhibitors, angiotensin receptor blockers (ARBs), alpha-blockers, beta-blockers, calcium channel blockers, diuretics.

Hyperthyroidism

- **What is it?** A metabolic imbalance caused by overproduction of thyroid hormone.
- **Who gets it?** It most often affects individuals ages 30 to 40 and is five times more prevalent among women than men.
- **What are the symptoms/warning signs?** Weight loss despite an increase in appetite and food consumption, anxiety, insomnia, tremors of the fingers or tongue, increased sweating and intolerance to heat, watery bulging eyes, muscle weakness, swelling in the neck, palpitations.
- **What causes it?** It can be caused by overproduction of

thyroid hormone by an enlarged gland, a condition known as Grave's disease. Less often, it is caused by a single nodule that forms on the gland and produces excessive amounts of thyroid hormone.

- **How is it diagnosed?** Patient history, physical examination, blood tests that show high levels of thyroid hormones and low levels of thyroid-stimulating hormone, thyroid scan.
- **What is the prognosis?** With frequent medical supervision, prognosis can be good. Hyperthyroidism increases the risk of osteoporosis and of thyroid "storm," an acute increase in the severity of symptoms brought on by stress or infection.
- **How is it treated?** Oral radioactive iodine, which gradually destroys parts of the gland so it will produce less hormone, is the preferred treatment approach. Other drugs can be given to inhibit production, while surgery is reserved for patients with extreme thyroid enlargement.

Hypothyroidism

- **What is it?** Condition in which the thyroid gland produces insufficient amounts of thyroid hormone.
- **Who gets it?** It can affect individuals of any age, but it is most prevalent among women older than 50.
- **What are the symptoms/warning signs?** Unusual weight gain, fatigue, intolerance to cold, slowed movement, constipation, dry flaky skin, muscle cramps, weakness, deepened voice, lack of interest in sex, puffiness around the eyes, dry hair or hair loss.
- **What causes it?** In some cases it is an autoimmune response whereby the immune system attacks the thyroid (Hashimoto's disease); it may also be caused by underproduction of thyroid-stimulating hormone, use of certain medications (e.g., lithium), surgical removal of the thyroid, or a birth defect.
- **How is it diagnosed?** Physical examination, patient history, blood tests to determine levels of thyroid hormone and thyroid-stimulating hormone.

- **What is the prognosis?** With regular treatment, a normal lifestyle can be maintained.
- **How is it treated?** Lifelong hormone replacement therapy with thyroxine is usually needed.

Irritable Bowel Syndrome

- **What is it?** A complex gastrointestinal disorder of the lower intestinal tract characterized by intermittent bouts of constipation or diarrhea, along with pain and other symptoms. It is the most common gastrointestinal disorder in the United States.
- **Who gets it?** Although it affects both sexes, it is more common among women than men and usually begins in adolescence or early adulthood.
- **What are the symptoms/warning signs?** Diarrhea, constipation, or alternating episodes of both, abdominal pain, bloating, cramps, excess gas, nausea, a feeling that the bowels will not empty completely.
- **What causes it?** The cause is unknown, but emotional stress and consumption of certain foods (high-fat foods, dairy products, gas-producing foods) appear to be contributory factors.
- **How is it diagnosed?** Patient history, physical examination; barium enema and/or sigmoidoscopy or colonoscopy may be used.
- **What is the prognosis?** This is usually a chronic, lifetime condition, but symptoms can be alleviated with judicious treatment.
- **How is it treated?** Low-fat diet, increased intake of high-fiber foods (does not work for everyone), stress management techniques (e.g., meditation, biofeedback, counseling), regular exercise; antidiarrheal medications, low-dose antidepressants, bulk-forming agents, and/or antispasmodics can prove helpful.

Lung Cancer

- **What is it?** The growth of malignant cells in the lungs.
- **Who gets it?** It is the number one cause of death from cancer in the United States for both women and men. More than 90% of cases are caused by smoking.
- **What are the symptoms/warning signs?** Cough that may be accompanied by bloody sputum, fatigue, chest pain, weight loss, clubbed fingers, hoarseness, swollen neck and face, swallowing difficulty, wheezing and shortness of breath.
- **What causes it?** Smoking is responsible for 90% of lung cancer cases. Other causes include air pollutants, exposure to secondhand smoke and other toxic substances, chronic bronchitis.
- **How is it diagnosed?** Patient history, physical examination, sputum sample, chest x-rays, magnetic resonance imaging, computed tomography, biopsy of cancerous tissue.
- **What is the prognosis?** Prognosis depends on the type and stages of the disease. For non-small-cell carcinoma, stages I and II can usually be cured in about 50% of cases; stage III cure rates are much lower, and stage IV is rarely curable. For small-cell carcinoma, cure rates are generally 25% or lower.
- **How is it treated?** Surgical removal of cancerous tissue, radiation therapy and/or chemotherapy.

Migraine

- **What is it?** A throbbing, severely painful headache that is usually confined to one side of the head and which is often debilitating.
- **Who gets it?** It can affect people of any age, but it usually first appears during childhood or adolescence and is more common among women than men.
- **What are the symptoms/warning signs?** Severe, throbbing pain, usually beginning on one side of the head and lasting up to two days; nausea and vomiting; sensitivity to light; early warning symptoms (an aura) in some people include visual disturbances, dizziness, numbness on one side of the face, weakness of an arm or leg.

- **What causes it?** The exact cause is unknown, but a popular theory is that abnormal constriction of arteries that supply the brain and scalp, followed by dilation, is a major factor. Other causes include heredity, stress, menstrual periods, use of oral contraceptives, consumption of certain foods and/or additives (e.g, nitrites, tyramine).
- **How is it diagnosed?** Patient history, physical examination, brain scans (computed tomography, magnetic resonance imaging).
- **What is the prognosis?** Every person responds differently to treatment. Often it requires individuals try different medications along with other techniques to manage the pain and the frequency of attacks.
- **How is it treated?** Medications such as over-the-counter painkillers, beta-blockers, calcium channel blockers, antidepressants, and anticonvulsants; stress management techniques (e.g., biofeedback, meditation, self-hypnosis).

Osteoarthritis

- **What is it?** Also known as degenerative joint disease, it is a chronic condition characterized by the gradual deterioration of cartilage within the joints.
- **Who gets it?** Nearly all men and women have some degree of osteoarthritis in one or more joints by the age of 60.
- **What are the symptoms/warning signs?** Joint pain that grows worse with movement and is relieved by rest, joint stiffness, especially in the morning, overgrowth on the joints on the fingers, changes in gait.
- **What causes it?** The result of wear and tear of the joints associated with aging. Obesity increases the risk of developing osteoarthritis.
- **How is it diagnosed?** Patient history, physical examination, x-rays.
- **What is the prognosis?** Treatment can greatly improve symptoms and improve quality of life. There is no cure.
- **How is it treated?** Pain medications, such as acetaminophen and nonsteroidal anti-inflammatory drugs; regular

exercise under the guidance of a health-care professional; hydrotherapy; glucosamine and chondroitin supplements; injections of corticosteroids.

Osteoporosis

- **What is it?** Condition characterized by loss of bone mass that leaves bones susceptible to fracture.
- **Who gets it?** It is most common among people older than 70, and it affects women four times more often than men.
- **What are the symptoms/warning signs?** Lower back pain, gradual loss of height, stooping posture, fractures of the wrist, hip, or vertebrae.
- **What causes it?** Although some bone loss occurs naturally with aging, factors that augment that loss include reduction in estrogen levels, sedentary lifestyle, smoking, excessive alcohol use, being underweight, dietary calcium deficiency, and heredity.
- **How is it diagnosed?** Patient history, physical examination, bone density scan, urine or blood tests, bone biopsy, x-rays or other imaging scans.
- **What is the prognosis?** Progression of the disease can sometimes be slowed or stopped with treatment.
- **How is it treated?** Bisphosphonate therapy helps build bone and slow bone resorption, calcitonin can slow bone loss, weight-bearing and/or resistance exercises to help preserve function, adequate intake of calcium and vitamin D for bone health, over-the-counter painkillers for pain.

Parkinson's Disease

- **What is it?** A degenerative disease of the nervous system in which there is progressive death of nerve cells in the gray matter of the brain.
- **Who gets it?** It typically appears between ages 55 and 70 and is slightly more prevalent among men.
- **What are the symptoms/warning signs?** Slowness of movement, rhythmic tremor of the hands that lessens or stops with movement and sleep, muscle stiffness, hesitant

gait with shuffling steps, stooped posture, loss of balance, difficulty swallowing, drooling, expressionless voice, bowel or bladder dysfunction.

- **What causes it?** The cause is unknown, but both genetic and environmental factors, along with aging, are believed to play major roles.
- **How is it diagnosed?** Patient history, physical examination by a neurologist familiar with Parkinson's disease.
- **What is the prognosis?** Most people respond to medications to some degree; however, the extent of relief varies greatly and side effects can be severe.
- **How is it treated?** Various medications can be used to help increase the levels of dopamine in the brain. Levodopa is the primary therapy given; others include selegiline, anticholinergic drugs, amantadine, dopamine agonists, and catechol-O-methyltransferase inhibitors.

Prostate Cancer

- **What is it?** The growth of malignant cells in the prostate gland.
- **Who gets it?** It affects between 8 and 10 percent of men. African-Americans have a 1.5 times greater incidence of prostate cancer than white Americans.
- **What are the symptoms/warning signs?** Frequent or urgent need to urinate, dribbling, painful urination, blood in the urine, painful or bloody ejaculation, erectile dysfunction, lower back or pelvic pain.
- **What causes it?** The cause is unknown, but age (incidence increases faster with age than any other form of cancer), family history, and race (African-Americans are at greater risk) are strong risk factors. High dietary fat and elevated testosterone levels may also play a role.
- **How is it diagnosed?** Patient history, physical examination including digital rectal exam, blood tests [prostate specific antigen (PSA)].
- **What is the prognosis?** Prognosis depends on the severity of the cancer, age and overall health of the patient, and whether the cancer has just appeared or is recurring.

- **How is it treated?** Watchful waiting (careful monitoring of the cancer to identify if and when intervention may be necessary), surgery to remove cancer tissue, radiation therapy, hormone therapy, chemotherapy.

Rheumatoid Arthritis

- **What is it?** A chronic, systemic inflammatory disorder that can affect joints throughout the body.
- **Who gets it?** It typically develops between the ages of 20 and 50 and is three times more prevalent among women than men.
- **What are the symptoms/warning signs?** Early symptoms that precede obvious joint problems include fatigue, weakness, low-grade fever, loss of appetite, weight loss. These are followed by red, swollen, painful joints that may be warm; red, painless skin lumps.
- **What causes it?** The cause is unknown. However, it is believed to be an autoimmune disorder (in which the body attacks its own cells and tissues) with genetic, hormonal, and/or infectious factors playing a role.
- **How is it diagnosed?** Patient history, physical examination, blood tests for rheumatoid factors, x-rays of affected joints.
- **What is the prognosis?** Often the disease can be controlled with a combination of treatments, although having the disease shortens expected lifespan by about 3 to 7 years. After 10 to 15 years, about 10% of affected individuals are severely disabled.
- **How is it treated?** A combination of therapies, including disease-modifying, antirheumatic drugs (DMARDs), nonsteroidal anti-inflammatory drugs (NSAIDs), antimalarial drugs, penicillamine, sulfasalazine, physical therapy, oral corticosteroids, tumor necrosis factor inhibitors, topical medications containing capsaicin, camphor, or menthol.

Systemic Lupus Erythematosus

- **What is it?** A chronic inflammatory condition that can affect many organ systems throughout the body. It is often referred to simply as lupus.
- **Who gets it?** Women are 8 to 10 times more likely to develop lupus than are men. It can develop at any age, but onset is usually between the ages of 15 and 35.
- **What are the symptoms/warning signs?** Fever, fatigue, loss of appetite, weight loss, abdominal pain, nausea, vomiting, headache, joint pain, rash over both cheeks and bridge of the nose, blurry vision, unusual bruising or bleeding, jaundice, dark urine, enlarged glands, swelling of the abdomen and ankles, tingling or pain in the muscles, stiffness, spasms, seizures, emotional depression. Severe symptoms develop in some and include kidney disease, myocarditis, hemolytic anemia, seizures.
- **What causes it?** Lupus is an autoimmune disorder. The exact causes are not known, but heredity, sex hormones, exposure to ultraviolet radiation, infections, and/or stress appear to play a role.
- **How is it diagnosed?** Patient history, physical examination, blood tests for autoimmune antibodies, urine tests for excessive protein and red blood cell levels, lumbar puncture, computed tomography or magnetic resonance imaging.
- **What is the prognosis?** The 10-year survival rate for lupus is greater than 85%. The worst prognosis is for individuals who have severe involvement of the brain, heart, kidneys, and lungs.
- **How is it treated?** Mild disease can be treated with nonsteroidal anti-inflammatory drugs (NSAIDs), antimalarial drugs and low-dose corticosteroids; severe conditions require more aggressive therapies.

5

COMMON PRESCRIPTION AND OTC DRUGS: BY TRADE NAME

The following is a list of the top 100+ most often prescribed prescription drugs in the United States, along with their generic names, type of drug, and the uses for which they are most often ordered. Your health-care practitioner may prescribe any of these drugs for indications other than those given here. Whenever you get a prescription from your health-care practitioner, always be sure to tell him or her about any other prescription, over-the-counter, or recreational drugs you may be taking, as well as herbal and nutritional supplements, as they may cause undesirable reactions.

TRADE NAME	GENERIC NAME	TYPE OF DRUG	COMMON USES
Accupril	Quinapril	Antihypertensive	Hypertension, congestive heart failure
Aciphex	Rabeprazole	Proton pump inhibitor	Gastroesophageal reflux
Actos	Pioglitazone	Antidiabetic	Noninsulin-dependent diabetes
Adderall	Amphetamine	CNS stimulant	Hyperactivity in children and adults
Advair	Salmetrerol/ fluticasone	Bronchodilator	Bronchial asthma, COPD
Allegra	Fenofenadine	Antihistamine	Allergies, asthma

TRADE NAME	GENERIC NAME	TYPE OF DRUG	COMMON USES
Altace	Ramipril	Antihypertensive	Hypertension, congestive heart failure
Ambien	Zolpidem	Sedative	Insomnia
Amoxil	Amoxicillin	Antibiotic	Various infections
Ativan	Lorazepam	Antianxiety	Anxiety, insomnia
Augmentin	Amoxicillin/ clavulanate	Antibiotic	Various infections, especially upper respiratory
Avandia	Rosiglitazone	Antidiabetic	Noninsulin-dependent diabetes
Azmacort	Triamcinolone	Corticosteroid (inhaled)	Asthma, COPD
Benadryl	Diphenhydramine	Antihistamine	Allergy, nausea/ vomiting, common cold
Calan	Verapamil	Calcium channel blocker	Angina, hypertension, cardiac dysrhythmia
Cardizem	Diltiazem	Calcium channel blocker	Angina, hypertension, cardiac dysrhythmia
Cardura	Doxazosin	Antihypertensive	Benign prostatic hyperplasia, hypertension
Catapres	Clonidine	Antihypertensive	Hypertension, migraine prevention
Celebrex	Celecoxib	NSAID/ COX-2 inhibitor	Arthritis

TRADE NAME	GENERIC NAME	TYPE OF DRUG	COMMON USES
Celexa	Citalopram	Antidepressant (SSRI)	Depression, anxiety
Cipro	Ciprofloxacin	Antibacterial	Various bacterial infections (e.g., urinary tract infections, conjunctivitis)
Clarinex	Desloratadine	Antihistamine	Allergies, asthma symptoms
Claritin	Loratadine	Antihistamine	Allergies, asthma symptoms
Coumadin	Warfarin	Anticoagulant	Prevention of blood clots
Cozaar	Losartan	Antihypertensive	Hypertension
Darvocet	Propoxyphene w/ acetaminophen	Analgesic (narcotic)	Mild to moderate pain
Deltasone	Prednisone	Corticosteroid	Many: adrenal insufficiency, inflammation, dermatitis, asthma, hemolytic anemia, others
Depakote	Divalproex	Anticonvulsant	Epilepsy
Desyrel	Trazodone	Antidepressant	Depression
Diflucan	Fluconazole	Antifungal	Candidiasis, cryptococcal meningitis
Digitek	Digoxin	Antiarrhythmic	Atrial fibrillation, congestive heart failure
Diovan	Valsartan	Antihypertensive	Hypertension, edema

TRADE NAME	GENERIC NAME	TYPE OF DRUG	COMMON USES
Dyazide	Hydrocholorothiazide	Antihypertensive	Hypertension, edema
Effexor	Venlafaxine	Antidepressant	Depression
Elavil	Amitriptyline	Antidepressant	Depression, migraine prevention
Evista	Raloxifene	Antiosteoporosis agent	Osteoporosis
Flexeril	Cyclobenzaprine	Antispasmodic	Muscle spasms
Flomax	Tamsulosin	Alpha-adrenergic blocker	Urinary obstruction
Flonase	Fluticasone	Corticosteroid (inhaled)	Allergic rhinitis
Flovent	Fluticasone propionate	Corticosteroid (inhaled)	Asthma, COPD
Fosamax	Alendronate	Antiosteoporosis agent	Osteoporosis, Paget's disease
Glucophage	Metformin	Antidiabetic	Noninsulin-dependent diabetes
Glucotrol XL	Glipizide	Antidiabetic	Noninsulin-dependent diabetes
Humulin N	Human insulin NPH	Blood glucose regulator	Diabetes mellitus
Imitrex	Sumatriptan	Antimigraine	Migraine
Indur	Isosorbide mononitrate	Antianginal	Angina
K-Dur	Potassium chloride	Mineral supplement	Hypokalemia, muscle cramps

TRADE NAME	GENERIC NAME	TYPE OF DRUG	COMMON USES
Lanoxin	Digoxin	Antiarrhythmic	Atrial fibrillation, congestive heart failure
Lasix	Furosemide	Antihypertensive, diuretic	Edema, hypertension, congestive heart failure
Levaquin	Levofloxacin	Antibiotic	Infections (e.g., skin, urinary tract, respiratory)
Levoxyl	Levothyroxine	Thyroid medication	Hypothyroidism, goiter
Librium	Chlordiazepoxide	Antianxiety	Anxiety
Lipitor	Atorvastatin	Antihyperlipidemic	Hypercholesterolemia, hyperlipidemia
Lopid	Gemfibrozil	Antihyperlipidemic	Hypercholesterolemia, hyperlipidemia
Lopressor	Metoprolol	Beta-blocker	Angina, hypertension
Lotensin	Benazepril	Antihypertensive	Congestive heart failure, hypertension
Lotrel	Amlodipine/ benazepril	Antihypertensive, ACE inhibitor	Hypertension
Micronase	Glyburide	Antidiabetic	Noninsulin-dependent diabetes
Motrin	Ibuprofen	NSAID	Arthritis, mild to moderate pain, inflammation

TRADE NAME	GENERIC NAME	TYPE OF DRUG	COMMON USES
Naproxyn	Naproxen	NSAID	Arthritis, mild to moderate pain, inflammation
Nasonex	Mometasone	Adrenocorticoid	Nasal allergy, nasal polyps
Neurontin	Gabapentin	Anticonvulsant	Epilepsy, pain
Nexium	Esomeprazole	Proton pump inhibitor	Gastroesophageal reflux
Norvasc	Amlodipine	Calcium channel blocker	Angina, hypertension
Ortho-Tri-Cyclen	Norgestimate/ Ethinyl estradil	Female sex hormone	Prevents pregnancy
Oxycontin	Oxycodone	Analgesic (narcotic)	Moderate to severe pain
Paxil	Paroxetine	Antidepressant (SSRI)	Depression, anxiety, PTSD
Percocet	Oxycodone w/ acetaminophen	Analgesic (narcotic)	Moderate to severe pain
Plavix	Clopidogrel	Antiplatelet	Prevent blood clots
Pravachol	Pravastatin	Antihyperlipidemic	Hypercholesterolemia, hyperlipidemia
Premarin	Conjugated estrogens	Female sex hormones	Relieve menopausal symptoms
Prempro	Conjugated estrogens	Female sex hormones	Relieve menopausal symptoms
Prevacid	Lansoprazole	Antigastric agent	Acid/peptic disorders, duodenal ulcer, erosive esophagitis

TRADE NAME	GENERIC NAME	TYPE OF DRUG	COMMON USES
Prilosec	Omeprazole	Antigastric agent	Acid/peptic disorders, duodenal ulcer, erosive esophagitis
Prinivil	Lisinopril	Antihypertensive	Hypertension
Phenergan	Promethazine	Antiemetic, antihistamine	Allergies, motion sickness, nausea/vomiting
Protonix	Pantoprazole	Proton pump inhibitor	Gastroesophageal reflux
Proventil	Albuterol	Antiasthmatic	Acute asthma symptoms, COPD
Provera	Medroxyprogesterone	Progestin	Amenorrhea, endometrial cancer, contraception
Prozac	Fluoxetine	Antidepressant (SSRI)	Depression, obsessive/compulsive disorder
Risperdal	Risperidone	Antipsychotic	Psychosis, schizophrenia
Singulair	Montelukast	Antiasthmatic	Mild to moderate asthma
Soma	Carisoprodol	Antispasmodic	Muscle spasms, especially back pain
Synthroid	Levothyroxine	Thyroid medication	Hypothyroidism, goiter
Tenormin	Atenolol	Antianginal, beta-blocker	Angina, hypertension, myocardial infarction
Toprol-XL	Metoprolol	Beta-blocker	Angina, hypertension

TRADE NAME	GENERIC NAME	TYPE OF DRUG	COMMON USES
Trimox	Amoxicillin	Antibiotic	Infections, especially upper respiratory, urinary tract
Tylenol	Acetaminophen	Analgesic	Mild to moderate pain, fever
Tylenol #3	Acetaminophen w/codeine	Analgesic (narcotic)	Moderate to moderately severe pain
Valium	Diazepam	Antianxiety	Anxiety, insomnia
Vasotec	Enalapril	Antihyper-tensive	Hypertension, con-gestive heart failure
Veetids	Pencillin VK	Antibiotic	Infections, especially upper respiratory
Viagra	Sildenafil citrate	Anti-impotence agent	Erectile dysfunction
Vicodin	Hydrocodone w/ acetaminophen	Analgesic (narcotic)	Moderate to severe pain
Wellbutrin	Bupropion	Antidepressant	Depression, smoking cessation
Xalatan	Latanoprost	Antiglaucoma agent	Glaucoma
Xanax	Alprazolam	Antianxiety	Anxiety, panic disorders
Zantac	Ranitidine	Histamine H2 receptor	Duodenal and gastric ulcers, heartburn
Zestril	Lisinopril	Antihyper-tensive	Congestive heart failure, hypertension

TRADE NAME	GENERIC NAME	TYPE OF DRUG	COMMON USES
Zithromax	Azithromycin	Antibiotic	Infections, especially upper respiratory
Zocor	Simvastatin	Antihyper-lipidemic	Hypercholes-terolemia, hyperlipidemia
Zoloft	Sertraline	Antidepressant (SSRI)	Depression, anxiety, panic disorder
Zyprexa	Olanzepine	Antipsychotic	Psychosis, schizo-phrenia
Zyrtec	Cetirizine	Antihistamine	Allergic rhinitis, some asthma symptoms

ACE = angiotension-converting enzyme
CNS = central nervous system
COPD = chronic obstructive pulmonary disease
NPH = neutral protamine Hagedorn
NSAID = nonsteroidal anti-inflammatory drug
PTSD = post-traumatic stress disorder
SSRI = selective serotonin reuptake inhibitor

6

COMMON PRESCRIPTION AND OTC DRUGS: BY GENERIC NAME

The following is a list of the top 100+ most often purchased over-the-counter drugs in the United States, along with their trade name(s), type of drug, and the uses for which they are most often taken. Before taking any over-the-counter drug, it is recommended that you consult your health-care practitioner about the safety of using the drug if you are currently also using any prescription or recreational drugs, as well as herbal and nutritional supplements, as they may cause undesirable reactions.

GENERIC NAME	TRADE NAME	TYPE OF DRUG	COMMON USES
Acetaminophen	Tylenol, Datril	Analgesic	Mild to moderate pain, fever
Acetaminophen w/ codeine	Tylenol #3	Analgesic (narcotic)	Moderate to moderately severe pain
Albuterol	Proventil	Antiasthmatic	Acute asthma symptoms, COPD
Alendronate	Fosamax	Anti-osteoporosis agent	Osteoporosis, Paget's disease
Allopurinol	Aloprim, Lopurin	Antigout	Gout
Alprazolam	Xanax	Antianxiety	Anxiety, panic disorders

GENERIC NAME	TRADE NAME	TYPE OF DRUG	COMMON USES
Amitriptyline	Elavil	Antidepressant	Depression, migraine prevention
Amlodipine /benazepril	Lotrel	Antihypertensive, ACE inhibitor	Hypertension
Amlodipine	Norvasc	Calcium channel blocker	Angina, hypertension
Amoxicillin	Amoxil, Trimox	Antibiotic	Infections, especially upper respiratory, urinary tract
Amoxicillin /clavulanate	Augmentin	Antibiotic	Various infections, especially upper respiratory
Amphetamine	Adderall	CNS stimulant	Hyperactivity in children and adults
Aspirin	Various	Analgesic, anti-inflammatory	Pain, inflammation, fever; prevent blood clots
Atenolol	Tenormin	Antianginal, beta blocker	Angina, hypertension, myocardial infarction
Atorvastatin	Lipitor	Antihyperlipidemic	Hypercholesterolemia, hyperlipidemia
Azithromycin	Zithromax	Antibiotic	Infections, especially upper respiratory
Benazepril	Lotensin	Antihypertensive	Congestive heart failure, hypertension
Bupropion	Wellbutrin	Antidepressant	Depression, smoking cessation

GENERIC NAME	TRADE NAME	TYPE OF DRUG	COMMON USES
Carisoprodol	Soma	Antispasmodic	Muscle spasms, especially back pain
Celecoxib	Celebrex	NSAID/ COX-2 inhibitor	Arthritis
Cephalexin	Keflex	Antibiotic	Various infections
Cetirizine	Zyrtec	Antihistamine	Allergic rhinitis, some asthma symptoms
Chlordiaze-poxide	Librium	Antianxiety	Anxiety
Ciprofloxacin	Cipro	Antibacterial	Various bacterial infections (e.g., urinary tract infections, conjunctivitis)
Citalopram	Celexa	Antidepressant (SSRI)	Depression, anxiety
Clonazepam	Klonopin	Anticonvulsant	Seizure disorders, anxiety
Clonidine	Catapres	Antihyper-tensive	Hypertension, migraine prevention
Clopidogrel	Plavix	Antiplatelet	Prevent blood clots
Conjugated estrogens	Premarin, Prempro	Female sex hormones	Relieve menopausal symptoms
Cyclobenza-prine	Flexeril	Antispasmodic	Muscle spasms
Deslorata-dine	Clarinex	Antihistamine	Allergies, asthma symptoms
Diazepam	Valium	Antianxiety	Anxiety, insomnia

GENERIC NAME	TRADE NAME	TYPE OF DRUG	COMMON USES
Digoxin	Digitek, Lanoxin	Antiarrhythmic	Atrial fibrillation, congestive heart failure
Diltiazem	Cardizem	Calcium channel blocker	Angina, hypertension, cardiac dysrhythmia
Diphenhydramine	Benadryl	Antihistamine	Allergy, nausea/vomiting, common cold
Divalproex	Depakote	Anticonvulsant	Epilepsy
Doxazosin	Cardura	Antihypertensive	Benign prostatic hyperplasia, hypertension
Doxycycline	Vibramycin	Antibiotic	Infections, especially upper respiratory
Enalapril	Vasotec	Antihypertensive	Hypertension, congestive heart failure
Esomeprazole	Nexium	Proton pump inhibitor	Gastroesophageal reflux
Fenofenadine	Allegra	Antihistamine	Allergies, asthma
Fluoxetine	Prozac	Antidepressant (SSRI)	Depression, obsessive/compulsive disorder
Fluconazole	Diflucan	Antifungal	Candidiasis, cryptococcal meningitis
Fluticasone	Flonase	Corticosteroid (inhaled)	Allergic rhinitis
Fluticasone propionate	Flovent	Corticosteroid (inhaled)	Asthma, COPD

GENERIC NAME	TRADE NAME	TYPE OF DRUG	COMMON USES
Furosemide	Lasix	Antihypertensive, diuretic	Edema, hypertension, congestive heart failure
Gabapentin	Neurontin	Anticonvulsant	Epilepsy
Gemfibrozil	Lopid	Antihyperlipidemic	Hypercholesterolemia, hyperlipidemia
Glipizide	Glucotrol XL	Antidiabetic	Noninsulin-dependent diabetes
Glyburide	Micronase	Antidiabetic	Noninsulin-dependent diabetes
Human insulin NPH	Humulin N	Blood glucose regulator	Diabetes mellitus
Hydrochlorothiazide	Dyazide	Antihypertensive	Hypertension, edema
Hydrocodone w/ acetaminophen	Vicodin	Analgesic (narcotic)	Moderate to severe pain
Ibuprofen	Advil, Motrin, Ibu	NSAID	Arthritis, mild to moderate pain, inflammation
Isosorbide mononitrate	Indur	Antianginal	Angina
Lansoprazole	Prevacid	Antigastric agent	Acid/peptic disorders, duodenal ulcer, erosive esophagitis
Latanoprost	Xalatan	Antiglaucoma agent	Glaucoma

GENERIC NAME	TRADE NAME	TYPE OF DRUG	COMMON USES
Levofloxacin	Levaquin	Antibiotic	Infections (e.g., skin, urinary tract, respiratory)
Levothyroxine	Levoxyl, Synthroid	Thyroid medication	Hypothyroidism, goiter
Lisinopril	Prinivil, Zestril	Antihypertensive	Hypertension, congestive heart failure
Loratadine	Claritin	Antihistamine	Allergies, asthma symptoms
Lorazepam	Ativan	Antianxiety	Anxiety, insomnia
Losartan	Cozaar	Antihypertensive	Hypertension
Medroxyprogesterone	Provera	Progestin	Amenorrhea, endometrial cancer, contraception
Metformin	Glucophage	Antidiabetic	Noninsulin-dependent diabetes
Metoprolol	Toprol-XL, Lopressor	Beta blocker	Angina, hypertension
Mometasone	Nasonex	Adrenocorticoid	Nasal allergy, nasal polyps
Montelukast	Singulair	Antiasthmatic	Mild to moderate asthma
Naproxen	Anaprox, Naproxyn	NSAID	Arthritis, mild to moderate pain, inflammation
Norgestimate /ethinyl estradiol	Ortho-Tri-Cyclen	Female sex hormone	Prevents pregnancy

GENERIC NAME	TRADE NAME	TYPE OF DRUG	COMMON USES
Olanzepine	Zyprexa	Antipsychotic	Psychosis, schizophrenia
Omeprazole	Prilosec	Antigastric agent	Acid/peptic disorders, duodenal ulcer, erosive esophagitis
Oxycodone	Oxycontin	Analgesic (narcotic)	Moderate to severe pain
Oxycodone w/ acetaminophen	Percocet	Analgesic (narcotic)	Moderate to severe pain
Pantoprazole	Protonix	Proton pump inhibitor	Gastroesophageal reflux
Paroxetine	Paxil	Antidepressant (SSRI)	Depression, anxiety, PTSD
Pencillin VK	Veetids	Antibiotic	Infections, especially upper respiratory
Pioglitazone	Actos	Antidiabetic	Noninsulin-dependent diabetes
Potassium chloride	K-Dur	Mineral supplement	Hypokalemia, muscle cramps
Pravastatin	Pravachol	Antihyperlipidemic	Hypercholesterolemia, hyperlipidemia
Prednisone	Deltasone	Corticosteroid	Many: adrenal insufficiency, inflammation, esophagitis
Promethazine	Phenergan	Antiemetic, antihistamine	Allergies, motion sickness, nausea/vomiting

GENERIC NAME	TRADE NAME	TYPE OF DRUG	COMMON USES
Propoxyphene w/ acetaminophen	Darvocet	Analgesic (narcotic)	Mild to moderate pain dermatitis, asthma, hemolytic anemia, others
Quinapril	Accupril	Antihypertensive	Hypertension, congestive heart failure
Rabeprazole	Aciphex	Proton pump inhibitor	Gastroesophageal reflux
Raloxifene	Evista	Anti-osteoporosis agent	Osteoporosis
Ramipril	Altace	Antihypertensive	Hypertension, congestive heart failure
Ranitidine	Zantac	Histamine H2 receptor	Duodenal and gastric ulcers, heartburn
Risperidone	Risperdal	Antipsychotic	Psychosis, schizophrenia
Rosiglitazone	Avandia	Antidiabetic	Noninsulin-dependent diabetes
Salmetrerol/ fluticasone	Advair	Bronchodilator	Bronchial asthma, COPD
Sertraline	Zoloft	Antidepressant (SSRI)	Depression, anxiety, panic disorder
Sildenafil citrate	Viagra	Anti-impotence agent	Erectile dysfunction
Simvastatin	Zocor	Antihyperlipidemic	Hypercholesterolemia, hyperlipidemia

GENERIC NAME	TRADE NAME	TYPE OF DRUG	COMMON USES
Sumatriptan	Imitrex	Antimigraine	Migraine
Tamsulosin	Flomax	Antihypertensive	Hypertension, congestive heart failure
Trazodone	Desyrel	Antidepressant	Depression
Triamcinolone	Azmacort	Corticosteroid (inhaled)	Asthma, COPD
Valsartan	Diovan	Antihypertensive	Hypertension, edema
Venlafaxine	Effexor	Antidepressant	Depression
Verapamil	Calan	Calcium channel blocker	Angina, hypertension, cardiac dysrhythmia
Warfarin	Coumadin	Anticoagulant	Prevention of blood clots
Zolpidem	Ambien	Sedative	Insomnia

ACE = angiotension-converting enzyme
CNS = central nervous system
COPD = chronic obstructive pulmonary disease
NPH = neutral protamine Hagedorn
NSAID = nonsteroidal anti-inflammatory drug
PTSD = post-traumatic stress disorder
SSRI = selective serotonin reuptake inhibitor

7

YOUR PRESCRIPTION: WHAT YOU SHOULD KNOW

Your doctor has given you a prescription and on your way to the pharmacy you look at the script and realize you can't decipher it: "T. 2 tabs ac tid," which means, "take 2 tablets before meals three times a day." Doctors and pharmacists have their own kind of Latin shorthand when it comes to prescriptions. As a consumer, you should know how to read your prescriptions or those of a loved one to help ensure proper dosing and use of these drugs. Below is a list of common abbreviations you may see on prescription labels, as well as some tips on how to be an informed consumer of prescription drugs.

How Often Should I Take It?

bid = twice a day
tid = three times a day
qid = four times a day
qh = every hour
qd = every day or once a day
q2h = every two hours
q3h = every three hours
q4h = every four hours
h.s. or qhs = at bedtime
q.o.d. = every other day
p.r.n. = as required
ut. dict. = use as directed

How Do I Take/Use This Drug?

po = by mouth
as = left ear
au = each ear
ad = right ear
os = left eye
od = right eye
ou = each eye
c or o = with food
s or ø = without food
ac = before meals
pc = after meals
tabs = tablets
caps = capsules

- Before you leave your health-care practitioner's office with your prescription, ask how to take the medication: how much, how often, whether to take it with or without food. Many medications cause negative reactions when taken with certain foods or alcohol.
- Make sure your health-care practitioner knows about any other medications (over-the-counter and/or prescription) you are taking, as well as any herbal remedies.
- Make sure your health-care practitioner has dated your prescription, as undated prescriptions cannot be filled.
- Note the number of refills on your prescription. Make sure you have enough refills to last you until your next doctor's appointment.
- Look for any warning labels or stickers on your prescription once you receive it. They should reinforce critical information you've been told, such as whether to take the medication with or without food, whether to avoid dairy products, or if you should avoid driving or operating machinery because the drug may make you drowsy.
- When your health-care practitioner gives you a prescription, ask if it is available in generic form. In many states, pharmacists can substitute a generic drug for a brand name, with the consumer's permission, if the health-care practitioner has not prohibited it on the prescription.

8

RECOMMENDED MEDICAL TESTS/SCREENINGS FOR ADULTS

The following recommendations for medical tests and screenings have been culled from various authoritative sources and should be viewed only as recommendations. Because every individual's health status is unique, you should always consult with your health-care professional to determine the test and screening options that best meet your needs.

Blood Pressure

A baseline blood pressure should be identified at age 18. Subsequently, blood pressure should be checked every two years; more frequently if the reading is borderline or high.

Breast Self-Examination

All women should examine their breasts monthly beginning at age 20. The recommended time for examination is three days after your period stops, or if you are no longer menstruating, do the examination on the first day of each month (or some other day each month that is easy to remember).

Chlamydia

All sexually active women ages 18–25 should be checked annually for Chlamydia. This can bc part of a yearly pelvic examination.

Cholesterol

The American College of Obstetricians and Gynecologists recommends having your cholesterol levels screened starting at age 45, then every three to five years if it is normal, yearly if it is not. The National Cholesterol Education Program and the US Preventive Services Task Force recommend having your cholesterol checked at five-year intervals if it is normal. If you are at high risk of coronary heart disease, your doctor may recommend you begin screening at a younger age.

Colon Cancer Screen

The American Cancer Society recommends that beginning at age 50, all men and women should undergo one of the following screenings.

- Yearly fecal occult blood test. If positive, a colonoscopy should be scheduled.
- Flexible sigmoidoscopy every 5 years.
- Preferred approach: yearly fecal occult blood test plus sigmoidoscopy every 5 years.
- Colonoscopy every 10 years.
- Double-contrast barium enema every 5 years.

Diabetes (Type II)

Regardless of your age, if you are at high risk of type II diabetes (obese, high blood pressure, family history of diabetes), you should be screened now. Otherwise, you should be screened every three years beginning at age 45.

Mammogram

The National Cancer Institute recommends women undergo a baseline mammogram at age 40, then have a mammogram every one to two years.

Osteoporosis Screening

The American College of Obstetricians and Gynecologists recommends that postmenopausal women 65 years and older undergo bone density testing every two years to screen for osteoporosis. Postmenopausal women who are younger than 65 and who have at least one risk factor for osteoporosis (family history of osteoporosis, history of fractures, low body weight, smoker) also be screened every two years.

Pap Smear

The US Preventive Services Task Force recommends that all women undergo a pap smear to screen for cervical cancer within three years of their first sexual encounter or at age 21, whichever comes first. Thereafter, women should have a pap smear at least every three years, more often if the doctor recommends it. (Note: A pap smear is typically done as part of an overall pelvic examination. Your doctor may recommend a yearly pelvic examination in the absence of a yearly pap smear.)

Physical Examination

- Every three years after age 30.
- Every two years after age 40.
- Every year after age 50.

Prostate Cancer Screen

The American Cancer Society, the National Comprehensive Cancer Network, and the American Urological Association recommend that all men undergo yearly PSA testing beginning at age 50. Men who have a family history of prostate cancer or who are black should begin screening at 40. However, some organizations, including the American College of Preventive Medicine, the National Cancer Institute, and the Centers for Disease Control and Prevention, do not endorse routine PSA testing.

Skin Cancer

A dermatologist or your general practitioner should examine your skin yearly beginning at age 18. You can perform your own self-examination at home every three months if you are at high risk for skin cancer.

Vaccinations

- Diphtheria-tetanus booster: every 10 years.
- Influenza: optional for people 50–65, then recommended yearly.
- Pneumococcal: yearly beginning at age 65.

Vision

The American Academy of Ophthalmology recommends the following vision screening routine if you are not encountering any vision problems.

- At least one examination between ages 20 and 39.
- Every two to four years between ages 40 and 64.
- Every one to two years beginning at age 65.

9

20 COMMON HOMEOPATHIC REMEDIES

The following is a list of some of the homeopathic remedies most commonly used in the United States. Please note that although some studies have been conducted on the effectiveness of these remedies, especially in Europe, and some have been shown to be effective, much of the evidence is anecdotal.

You should, however, always check with your health-care practitioner before taking a homeopathic remedy.

Aconite. Take during the early stages of a cold, fever, or inflammation; also for teething, sore throat, and dry cough.

Allium cepa. Useful as an acute cold and cough remedy, especially when symptoms include watery nasal discharge, sneezing, frontal headache, and watery eyes.

Apis. Use immediately after experiencing an insect bite, such as a bee or wasp sting or ant bite.

Arnica. Use immediately after experiencing physical trauma, to help reduce or prevent bruising or swelling.

Arsenicum. Take for indigestion or abdominal pain that may be accompanied by diarrhea or vomiting. Most effective for people who are fatigued, are very thirsty, and whose symptoms are worse at night.

Belladonna. Most effective for a throbbing or burning headache that begins suddenly.

Bryonia. Helpful for symptoms that come on slowly and worsen with slight movement, including constipation with hard stools, nausea and vomiting, chest pain and coughing, and very dry mucous membranes, any one of which is usually accompanied by headache.

Cantharis. Effective for treatment of cystitis, especially if you experience burning pain before, during, and after urination.

Chamomilla. Provides relief for teething children.

Euphrasia. Effective against fever with chills; headache accompanied by eye complaints; daytime cough; and excessive, watery eyes accompanied by a runny nose.

Gelsemium. Take when flu symptoms start, especially if they include chills.

Hypericum. Should be taken immediately after an injury to nerve-rich areas, such as a finger slammed in a door or a fall on the tailbone.

Ignatia. Helps those who are experiencing emotional pain, such as grief or sadness, and sleep difficulties.

Kali bichromicum. Effective against excessive mucus discharge from the throat and nose, which may be accompanied by headache.

Ledum. First-aid remedy for puncture wounds, black eyes, bites and stings, and chronic joint pain, such as occurs in arthritis.

Mercurius vivus. Useful for treatment of mouth ulcers.

Natrum muriaticum. Effective against headache associated with eyestrain, headache accompanied by chills and fever, and dry mucous membranes accompanied by cold sores on the lips.

Nux vomica. Take for relief of nausea, hangover, heartburn, and burping.

Rhus toxicodendron. Provides relief from inflammation related to arthritis, plus muscle aches and stiffness.

Ruta. Take for relief of injuries and strains that affect the tendons.

10

50 COMMON HERBAL REMEDIES

More than one-third of Americans use herbal supplements and related products. Here are 50 of the most common herbal remedies we are using. Before you take any herbal remedy, you should consult a knowledgeable health-care professional, as remedies may interact with any medications you are taking and/or be contraindicated for a medical condition you may have. Please note that although some studies have been conducted on the effectiveness of these remedies, especially in Europe, and many have been shown to be effective, some of the evidence is anecdotal.

Aloe (*Aloa barbadenis*). Topical treatment (gel) for burns and wounds.

Arnica (*Arnica montana*). Topical treatment (cream, ointment, gel) for sprains, bruises, and inflammation associated with insect bites and stings.

Astragalus (*Astragalus membranaceas*). Flu, common cold, minor infections. Available in capsules.

Bilberry (*Vaccinium myrtillus*). Vision problems, including cataracts, glaucoma, macular degeneration, poor night vision, retinopathy. Available in capsules and extracts.

Black cohosh (*Cimicifuga racemosa*). Menstrual and menopausal symptoms. Available in tincture and capsules.

Boswella (*Boswellia serata*). Pain and inflammation associated with arthritis. Available in capsules and as cream.

Butcher's broom (*Ruscus acluteatus*). Improves circulation; reduces swelling in legs and feet. Available as capsules and extract.

Cat's claw (*Uncaria tomentosa*). Enhances the immune system; relieves inflammation. Available in tincture and capsules.

Cayenne (*Capsicum annuum*). Enhances the immune system, relieves pain of arthritis. Available in cream and as capsules and tincture.

Celery (*Apium graveolens*). Reduces water retention, inflammation, gas, and high blood pressure. Available as an oil and seeds.

Chamomile (*Matricaria recutita*). Relieves nausea, indigestion, insomnia. Available in tincture and capsules.

Chaste tree berry (*Vitex agnus castus*). Relieves symptoms of premenstrual syndrome and menopause.

Cloves (*Caryophyllum aromaticus*). Effective in treatment of toothache and vomiting. Available as oil and the dried spice.

Dandelion (*Taraxacum officinale*). Reduces high blood pressure, water retention. Available as extract and capsules.

Devil's claw (*Harpagophytum procumbens*). Relieves pain, inflammation; reduces high cholesterol. Available in tincture and capsules.

Dong quai (*Angelica sinesis*). Relief of menstrual and menopausal symptoms. Available as tincture and capsules.

Echinacea (*Echinacea angustifolia, E. pallida, E. purpurea*). Relieves symptoms of cold and flu when taken at start of symptoms; also for urinary tract infections. Available in tincture and capsules.

Elderberry (*Sambucus canadensis*). Treatment of sore throat and fever. Available as dried herb and tincture.

Evening primrose (*Oenothera biennis*). Treatment of arteriosclerosis, asthma, eczema, endometriosis, high blood pressure, high cholesterol, premenstrual symptoms, rheumatoid arthritis. Available in capsules.

Eyebright (*Euphrasia officinalis*). Relief for inflamed or irritated eyes. Available as extract and capsules.

Fennel (*Foeniculum vulgare*). Relieves coughs, colds, gas, painful joints, and stomach cramps. Available as capsules and dried herb.

Fenugreek (*Trigonella foenum-graecum*). Treatment of high cholesterol, gastritis. Available as seed and capsules.

Feverfew (*Tanacetum parthenium*). Provides relief of migraine. Available as tincture and capsules.

Flaxseed (*Linun usitatissimum*). Relieves constipation, symptoms of irritable bowel syndrome. Available as seed and in capsules.

Garlic (*Allium sativum*). Effective against infections (bacterial, viral, and fungal), including athlete's foot, candidiasis, colds and flu; also treatment of high cholesterol, high blood pressure. Available as fresh garlic cloves, deodorized capsules, tablets.

Ginger (*Zingiber officinale*). Relieves symptoms of indigestion, motion sickness, nausea. Available as tincture, capsules, dried herb.

Ginkgo (*Ginkgo biloba*). Used to treat symptoms of aging, including memory problems, depression, confusion, ringing of the ears, macular degeneration, and poor circulation; also for migraines. Available in capsules.

Ginseng (*Panax quinquefolius, P. ginseng*). Relieves fatigue and improves concentration. Available in capsules.

Goldenrod (*Hydrastis canadensis*). Reduces inflammation of mucous membranes, helps fight infections. Available as tincture and capsules.

Grapeseed extract. Used as an antioxidant and to reduce inflammation. Available in capsules.

Green tea (*Camellia sinensis*). Reduces risk of heart disease and cancer. Available as dried tea, extract, capsules.

Gotu kola (*Centella asiatica*). May help improve memory and concentration; enhances circulation; reduces anxiety. Available in extract and capsules.

Hawthorn (*Crataegus monogyna, C. oxyacantha*). Treats symptoms related to heart problems, such as angina, arrhythmia, edema, high blood pressure, high cholesterol. Available as capsules.

Horse chestnut (*Aesculus hippocastanum*). Used to treat bruises, varicose veins, hemorrhoids. Available as extract and tablets.

Kava (*Piper methysticum*). Treats symptoms of depression, genitourinary infections, hyperactivity, insomnia. Available as tincture, extract, and capsules.

Lemon balm (*Melissa officinalis*). Helps reduce symptoms of stress, tension, insomnia. Available as dried herb.

Licorice (*Glycyrrhiza glabra*). Relieves cough, congestion, duodenal and stomach ulcers. Available as tincture and capsules.

Milk thistle (*Silybum marianum*). Used for liver conditions (e.g., hepatitis, cirrhosis); also helps protect against toxic effects of pollutants. Available as tincture and capsules.

Nettle (*Urtica dioica*). Relieves symptoms of hayfever. Available as extract and capsules.

Passion flower (*Passiflora incarnate*). Relieves symptoms of stress, headache. Available as extract.

Pau d'arco (*Tabebuia ipetiginosa*). Helps fight bacterial, viral, and fungal infections. Available as tincture and capsules.

Peppermint (*Mentha piperita*). Relieves symptoms of stomach pain and gas. Available as a tea.

Psyllium (*Plantago* spp.). Relieves constipation; reduces high cholesterol. Available as dried seeds, husks, and capsules.

Red clover (*Trifolium pretense*). Treats symptoms of menopause; may help prevent cancer. Available as extract and capsules.

St. John's wort (*Hypericum perforatum*). Treats burns, cuts; symptoms of depression, herpes. Available as tincture and capsules.

Saw palmetto (*Serenoa repens*). Used for prostate problems, hair loss. Available as extract.

Skullcap (*Scutellaria lateriflora*). Helps reduce symptoms of stress and anxiety; insomnia. Available as extract and capsules.

Turmeric (*Curcuma longal*). Treats athlete's foot, digestive disorders, heart disease, liver conditions; facilitates wound healing. Available as capsules.

Uva ursi (*Arctostaphylos uva-ursi*). Treats urinary tract infections. Available as capsules and tincture.

Valerian (*Valeriana officinalis*). Treatment of anxiety, stress, insomnia. Available as tincture and capsules.

11

SCHEDULE OF CHILDREN'S VACCINATIONS

The Centers for Disease Control and Prevention (CDC) recommends the following schedule of vaccinations for children as of April 1, 2004. For additional information about vaccinations, including precautions and contraindications, see *www.cdc.gov/nip/*.

Hepatitis B. First dose at birth or up to age 2 months if the mother is negative for hepatitis B.

Dose 2 at least four weeks after the first dose.
Dose 3 at least 16 weeks after the first dose and at least eight weeks after the second.
Dose 4 not before 24 weeks of age.

DTP (Diphtheria, tetanus, pertussis). Doses at 2, 4, and 6 months, then during the span of 15 to 18 months and 4 to 6 years.

Tetanus. At 11 to 12 years.

Haemophilus influenza type B. Doses at 2, 4, and 6 months, then during 12 to 15 months.

Polio. Doses at 2 and 4 months; then during 6 to 18 months and during 4 to 6 years.

MMR (Mumps, measles, rubella). Doses at 12 to 15 months, then at 4 to 6 years.

Varicella. At 12 to 18 months.

Pneumococcal. Doses at 2, 4, and 6 months; then during 12 to 15 months.

Influenza. Yearly doses starting at 6 months to 23 months in healthy children; starting at 6 months and continuing through childhood and adolescence for children with certain medical conditions, such as asthma, diabetes, sickle cell, HIV, among others.

12

STOCKING YOUR HOME MEDICINE CABINET

Your medicine cabinet—or home first aid kit—should contain the following basic ingredients to handle common medical emergencies that can occur around the house. A list of emergency numbers should always be next to each phone in the house.

- Adhesive bandages (various sizes)
- Adhesive tape
- Butterfly bandage
- Elastic bandages
- Triangular bandage
- Sterile gauze pads
- Sterile gauze rolls
- Sterile cotton balls
- Sterile eye patches
- Instant-activating cold packs
- Disposable sterile gloves
- Tissues
- Scissors with rounded tips
- Tweezers
- Safety pins
- Flashlight
- Medicine spoon
- Bulb syringe
- Aspirin, ibuprofen, acetaminophen (check expiration dates on bottles)
- Syrup of ipecac (this should be used only with the advice of a poison control center or doctor)
- Antibiotic cream

- Diphenhydramine (Benadryl)
- Calamine lotion
- Epinephrine kit (if anyone in your family is allergic to insect bites or to certain foods)

13

20 TOP HEALTH SPAS AROUND THE WORLD

Getting and staying healthy can be hard work, but it can be fun too. Health spas are a place where people can participate in activities that promote health, learn about and eat nutritious foods, and benefit from the services of complementary and/or alternative health-care providers. Like people, health spas come in all shapes and sizes and offer a broad variety of services. Here is an alphabetical list of some of the top spas from around the world.

Ananda Spa Resort
Tehri Garhwal, India
www.anandaspa.com

Brings together Western and Ayurvedic doctors, spa therapists, nutritionists. Offers Ananda yoga, spa treatments, Ayurvedic treatments, meditation, Swedish massage, aromatherapy, reflexology, hydrotherapy, exfoliations, body wraps, facials; personalized sessions for detox, destress, deep relaxation, weight loss; nutritional guidance; indoor pool, gym, squash court, golf.

Brenners Park Hotel & Spa
Baden-Baden, Germany
www.Brenners.com
+49(0)72-21-9000

Offers preventive medicine two-day check-up program, personal trainers. Facials, lymph drainage, body packs, reflexology, shiatsu, qi massage, lomi-lomi massage, Japanese steam bath; fitness room, indoor pool, water gymnastics, golf.

Cal-a-Vie Resort & Spa
San Diego, CA
www.cal-a-vie.com
866-772-4283

European spa philosophies; limited to 24 guests, staff-to-guest ratio: 4-to-1. Fitness classes, facials, seaweed wraps, hydrotherapy, Ayurvedic mud scrub, reflexology, magnet therapy, Vichy showers, hot stone massage, cooking classes, educational lectures.

Canyon Ranch
Two locations:
Lenox, MA
Tucson, AZ
www.canyonranch.com
800-742-9000

Staff-to-guest ratio: 3-to-1. Full-time doctors, nurses, and specialists in women's health, weight loss, cardiovascular health, and more. Health consultants in integrative medicine, herbal medicine, women's comprehensive health, and more. Massage, body wraps, reflexology, reiki, Thai massage, shiatsu, facials, Pilates, Ayurvedic yoga, personal training sessions, and more.

Chiv-Som International Health Resort ("Haven of Life")
Hua Hin, Thailand
www.hotelthailand.com/huahin/chivasom/

Nutritional analysis, health and fitness classes, personal trainers; reflexology, meditation, Thai and Swedish massage, body scrubs.

Double Eagle Resort & Spa
Juno Lake, CA
www.doubleeagleresort.com
760-648-7004

More than 50 specialty massages and skin care treatments. Medically directed team for lifetime wellness program, personal trainers. Fitness center, indoor pool; reflexology, hiking, horseback riding, yoga, Pilates, more.

Golden Door
PO Box 463077
Escondido, CA
www.goldendoor.com
800-424-0777

Staff-to-guest ratio: 4-to-1. Personal fitness guides, gourmet meals, gym, two pools, tennis, steam room, sauna, body scrubs, herbal wraps, tai chi, yoga, meditation, workshops, cooking classes.

The Greenbriar
White Sulphur Springs, WV
www.greenbriar.com
800-624-6070

Greenbriar Clinic features a medical clinic with doctors, nutritionists, clinical psychologists, two-day diagnostic and preventive medical examination. More than 50 offerings, including steam and sauna baths, massage, swimming, tennis, lawn games, body wraps, Vichy showers, facials, zephyr tubs, horseback riding, rafting, aerobic classes, more.

The Heartland
Gilman, IL
www.heartlandspa.com
800-545-4853

Nutritional program, yoga, tai chi, meditation, facials. Exercise room, indoor pool, tennis, biking, body sculpting, Pilates. Heartland Adventure (program that integrates health and fitness with personal challenge and development).

Hilton Head Health Institute
Hilton Head Island, SC
www.hhhealth.com
800-292-2440

Focus is on weight loss. Offers aromatherapy massage, therapeutic massage, reflexology, hot stone massage, body scrubs, water fitness, kickboxing, meditation, tai chi, yoga, Pilates, educational sessions.

Kohala Sports Club & Spa
Waikoloa, HI
www.kohalaspa.com
808-886-2828

European-style spa with Hawaiian comfort and hospitality. Massages, shiatsu, facials, scrubs, wraps, hydrotherapy, acupuncture, craniosacral, reiki, chi kung, biofeedback, hypnotherapy, stone massage, nutritional consultations.

Le Sport Spa
St. Lucia
www.lesport.com.lc
800-544-2883

Classes that change weekly; Ayurvedic treatments, nutritional consultations, personal trainers; meditation, yoga, golf, archery, fencing, tennis, scuba school, sailing, tai chi, more.

Mii Amo
Sedona, AZ
www.miiamo.com
888-749-2137

Personal meditation consultations, Ayurvedic consultations, special massages (Native American cleanse, watsu, stones, sports, lymphatic), shiatsu, reiki, crystal bath, qi gong, yoga, dance; past-life regression therapy, herbal detox wrap; weight room, aqua aerobics, cooking classes, lectures and workshops.

Miraval, Life in Balance
Catalina, AZ
www.miravalresort.com
800-232-3969

More than 100 treatment options, including massage, hydrotherapy, facials, Ayurvedic treatments, acupuncture, Trager, shiatsu, jin shin jyutsu, chi kung, yoga, horseback riding, tennis, golf, rock climbing, biking, hiking; Zen boot camp, meditation classes.

The Oaks at Ojai
Ojai, CA
www.oaksspa.com
800-753-6257

Nurse on duty; 1,000-calorie/day cuisine, fitness counseling and personal wellness training. Massage, watsu, facials, exercise classes, hiking, Pilates, qi gong, tai chi, yoga, Swedish massage, reflexology, acupuncture, reiki, aromatherapy.

Palazzo Arzaga
Brescia, Italy
www.palazzoarzaga.com
39(03)-068-0600

Medically supervised spa and fitness center, personalized spa programs, weight loss assistance. Antiaging facials, herbal wraps, therapeutic massage, body toning; sauna, gym, indoor mineral pool.

The Palms
Palm Springs, CA
www.palmsspa.com
800-753-7256

Nutritional counseling, aerobics, yoga, meditation, massages (Swedish, deep tissue, Panch Karma, Thai), facials, body wraps, body scrubs.

Rancho la Puerta
PO Box 463057
Escondido, CA
www.rancholapuerta.com
800-443-7565

North America's oldest spa. "Awaken the Spirit" program features guest experts in different areas in one-week programs. All natural cuisine; massage, hydrotherapy, seaweed wraps, herbal wraps, reflexology, facials, more. Tennis, 10 gyms, 3 pools, saunas.

Royal Parc
Evian-Les-Bains, France
www.royalparcevian.com
33(0)4-50-26-85-00

Has Better Living Institute that offers tailored programs. Massage, lymphatic drainage, cellulite massage, pressure therapy, Shirodhara, reflexology, fasciatherapy, yoga, reiki, Thai massage, shiatsu, Indian skin scrub; golf, swimming, sauna, gym, tennis, climbing, more.

Terme di Saturnia Resort and Spa
Tuscan Maremma, Italy
www.termedisaturnia.com/English/index.html

Known worldwide for its sulfurous water with healing powers. Offers programs in total beauty, antiaging revitalizing, aquarelax, thermal, program for men; steam bath, sauna, Turkish bath, hydrotherapy, horseback riding, golf, tennis, gym.

14

CHECK UP ON YOUR ALTERNATIVE PRACTITIONER

When choosing an alternative practitioner, you should investigate the provider's background and experience and, if applicable, licensing and certification. Many states have regulatory agencies or licensing boards for certain types of alternative practitioners. They may provide you with information about practitioners in your area. State, county, and city health departments are also possible sources for referrals to regulatory agencies or boards.

Another source of information is professional organizations for the type of practitioner you are seeking. Below is a list of such organizations for selected alternative medicine practices.

Acupuncture

NOTE: Acupuncturists are licensed by states, but in some states you can practice without a license. Depending on the licensing procedure in your state, you will need to contact either your state board of acupuncture, the board of medical examiners, or the department of health, licensing, or education. You may also contact the National Certification Commission for Acupuncture and Oriental Medicine (see "Traditional Chinese Medicine" on page 511).

AMERICAN ACADEMY OF MEDICAL ACUPUNCTURE
www.medicalacupuncture.org
4929 Wilshire Blvd., Suite 428
Los Angeles, CA 90010
323-937-5514

AMERICAN ASSOCIATION OF ORIENTAL MEDICINE
www.aaom.org
909 22nd St., PO Box 162340
Sacramento, CA 95816
916-451-6950

Alexander Technique

ALEXANDER TECHNIQUE INTERNATIONAL
www.ati-net.com
1692 Massachusetts Ave., 3rd Floor
Cambridge, MA 02138
888-668-8996

Aromatherapy

NATIONAL ASSOCIATION FOR HOLISTIC AROMATHERAPY
www.naha.org
4509 Interlake Ave. N., Suite 233
Seattle, WA 98103-6773
888-ASK-NAHA

Ayurvedic

NOTE: Ayurvedic practitioners are not licensed in the United States, and their practices are not regulated by any state or federal agency.

NATIONAL INSTITUTE OF AYURVEDIC MEDICINE
www.niam.com
584 Milltown Rd.
Brewster, NY 10509
845-278-8700

Biofeedback

ASSOCIATION FOR APPLIED PSYCHOPHYSIOLOGY AND BIOFEEDBACK
www.aapb.org
10200 W. 44th Ave., Suite 304
Wheat Ridge, CO 80033-2840
303-422-8436

Chiropractic

NOTE: Chiropractors are licensed in all 50 states.

AMERICAN CHIROPRACTIC ASSOCIATION
www.amerchiro.org
1701 Clarendon Blvd.
Arlington, VA 22209
800-986-4636

Herbalism

AMERICAN HERBALISTS GUILD
www.americanherbalistsguild.com
1931 Gaddis Rd.
Canton, GA 30115
770-751-6021

NOTE: Professional American Herbalists Guild members are identified by the term "herbalist AHG" after their name.

Homeopathy

NOTE: Three states license homeopaths—Arizona, Connecticut, and Nevada—and these individuals must have a medical or naturopathic degree to be a licensed homeopath.

NATIONAL AMERICAN SOCIETY OF HOMEOPATHS
www.homeopathy.org
1122 E. Pike St., Suite 1122
Seattle, WA 98122
206-720-7000

NATIONAL CENTER FOR HOMEOPATHY
www.homeopathic.org
801 N. Fairfax St., Suite 306
Alexandria, VA 22314
877-624-0613

Hypnotherapy

AMERICAN SOCIETY OF CLINICAL HYPNOSIS
www.asch.net
140 N. Bloomingdale Rd.
Bloomingdale, IL 60108-1017
630-980-4740

Massage Therapy

NOTE: The majority of states have licensing laws for massage therapists, and many cities have local regulations as well. These laws concern the legal educational requirements for massage therapists, which can range from state to state and city to city.

NATIONAL CERTIFICATION BOARD FOR THERAPEUTIC MASSAGE & BODYWORK
www.ncbtmb.com
8201 Greensboro Dr., Suite 300
McLean, VA 22102
800-296-0664

Naturopathy

NOTE: Currently naturopaths are regulated and licensed in 13 states: Alaska, Arizona, California, Connecticut, Hawaii, Kansas, Maine, Montana, New Hampshire, Oregon, Utah, Vermont, and Washington.

AMERICAN ASSOCIATION OF NATUROPATHIC PHYSICIANS
www.naturopathic.org
3201 New Mexico Ave., NW, Suite 350
Washington, DC 20016
866-538-2267

Nutritionists and Registered Dietitians

NOTE: All 50 states license registered dietitians, but few license nutritionists.

American Dietetic Association
www.eatright.org/Public/index.cfm
120 S. Riverside Plaza, Suite 2000
Chicago, IL 60606-6995
800-877-1600

Osteopathic

NOTE: All 50 states license osteopaths.

American Osteopathic Association
www.osteopathic.org
142 E. Ontario St.
Chicago, IL 60611
800-621-1773

Traditional Chinese Medicine

NOTE: The majority of states license traditional Chinese medicine practitioners.

National Certification Commission for Acupuncture and Oriental Medicine
(Oriental medicine, Asian bodywork, acupuncture, and Chinese herbology)
www.nccaom.org
11 Canal Center Plaza, Suite 300
Alexandria, VA 22314
703-548-9004

Tips on How to Choose an Alternative Practitioner

- Depending on the type of alternative therapy you are seeking, practitioners may need to fulfill certain practice requirements. Check out the practice requirements in your state, and choose a practitioner who meets them.
- Get referrals from a trusted physician or another alterna-

tive practitioner, a teaching hospital, medical college, or training facility. A therapeutic massage school, for example, should be able to recommend practitioners in your area. If possible, get referrals from several different sources.

- Ask family members, friends, and coworkers for recommendations.
- Get recommendations and/or referrals from professional associations.
- Ask how many years the professional has been practicing.
- If you have special medical issues (e.g., diabetes, osteoporosis, arthritis), make sure the practitioner you choose is familiar working with such conditions.

15

ADVANCE DIRECTIVES: WHAT EVERYONE SHOULD KNOW

Advance directives are documents, signed by competent individuals, that give specific directions to health-care providers about the kind of care and treatment they would like to receive in certain circumstances if they become incapacitated and unable to make medical decisions. Advance directives fall into two general categories: instructive and proxy. Instructive directives allow you to state your preferences regarding a particular treatment or types of treatment. Examples of instructive advance directives are living wills and any directive that focuses on specific treatments, such as a "no CPR" or "no tube feeding" directive.

A durable power of attorney for health care (DPAHC) is an example of a proxy advance directive. This type of directive allows you to name a patient advocate who will act for you and make sure your wishes are carried out should you become incapacitated. If you don't have an individual you trust will act on your behalf, then this type of directive is probably not for you.

Because laws about advance directives vary from state to state, you should become familiar with the laws in your state. You can ask your doctor, attorney, or state representative about the legal status and requirements of advance directives that affect you. You should also make it a yearly habit to review any advance directives you have to make sure you are still in agreement with what you have stated in them.

Do You Need an Advance Directive?

The short answer is, yes. Most advance directives are prepared by older people or those who are seriously ill. However, even people who are in good health should consider completing an advance directive. Given that an accident or critical illness can happen at any time, having an advance directive prepared can make it more likely that your wishes will be honored. Having an advance directive may also reduce uncertainty and stress experienced by family members who may be unsure about how you wish to be treated in case of a serious health situation.

Living Will

A living will contains your wishes concerning the type of treatment you want to receive—or not receive—in certain situations. It goes into effect when you are terminally ill (i.e., you have less than six months to live). A living will does not let you choose someone to make those decisions for you.

Living wills are often written using ambiguous language, such as "extraordinary means" and "unnaturally prolonging my life." Try to be as clear as possible when stating your wishes, as the more precise your instructions, the less likely there will be misinterpretation later. For example, name specific medical interventions you would or would not want to be used, such as blood transfusions, tube feeding, or CPR.

Durable Power of Attorney for Health Care

A durable power of attorney for health care (DPAHC) is a legal document that allows you to designate an individual (eighteen years or older) who will make health-care decisions for you should you be unable to do so yourself. This individual can be a family member, friend, or any other person you trust will carry out your wishes. A durable power document can be used to accept or refuse any medical treatment. It is best to have your wishes written down in the durable power so they are clear to everyone. Your advocate can only stop life-sustaining care if it is clearly stated in your durable power.

You are not required to have a durable power of attorney for health care. If you wish, you can tell your family, other trusted individuals, or your doctor what your wishes are. However, unless you have a durable power, none of these individuals have the legal authority to act on your behalf if you become incapacitated.

Do Not Resuscitate Order

A do-not-resuscitate (DNR) order is a request not to have cardiopulmonary resuscitation if your heart stops or if you stop breathing. You can verbally make your wishes known to your family and doctor, but it is best to have them in writing. Your DNR order should be part of your medical record. Doctors and hospitals in all states accept DNR orders.

How to Get an Advance Directive

If you want to prepare an advance directive, you can get sample forms from a hospital admitting office, your state medical society, your state representative or senator, or senior citizen groups, or you can use a computer software package for legal documents. Ask for a "Durable Power of Attorney for Health Care Form," a "Living Will Form," and/or a "Do Not Resuscitate Order Form." You can also simply write down your wishes on your own.

If you write your own directives or use a computer software package, you should make sure you follow your state laws. You may also want to have what you write reviewed by your doctor or an attorney. If you are drawing up a DPAHC, talk to the individual you have chosen to make sure he or she understands your wishes and is willing to be your advocate.

Once you are satisfied with the documents, have them notarized and give copies to your family and doctor. If you make any subsequent changes, you should have them notarized as well. By law, your family, physicians, and employees of your healthcare facility are not permitted to be witnesses for these documents.

If you are thinking about preparing an advance directive, consider the following questions:

- How do you feel about the use of surgery, resuscitation, drugs, tube feeding, and ventilators if you were to become terminally ill, if you were in a coma or unconscious and not likely to wake up, or if you were senile?

- Whom do you trust will make treatment decisions for you should you become incapacitated?
- What type of medical treatment would you want to receive if you had a severe stroke or other medical condition that left you dependent on others for your care?
- Which physical, mental, and social abilities do you feel are most important for you to enjoy living?

16

IMMUNIZATIONS FOR TRAVELING ABROAD

Before you hop that plane or freighter for lands far away, you may need to get an immunization or two. The immunizations presented here are standard. However, health conditions can change quickly in any country at any time. To make sure you have all the immunizations you need when traveling abroad, you should check with the Outbreak Notices, issued by the Centers for Disease Control and Prevention, at *www.cdc.gov/travel/blusheet.htm* or call 1-877-FYI-TRIP, for up-to-date information.

Standard Recommended Vaccinations

The Centers for Disease Control and Prevention recommends that all overseas travelers receive the following vaccinations 4 to 6 weeks before leaving on a trip.

- Hepatitis A or immune globulin.
- Hepatitis B: It is recommended you get this vaccine if you might be exposed to blood during your visit, are staying more than 6 months in the area, may receive medical treatment, or have sexual contact with the local population. Infants and 11- to 12-year-old children who did not receive the hepatitis series as infants should also be vaccinated before leaving on their trip.
- Rabies: Get the vaccine if you might be exposed to domestic or wild animals.
- Boosters for tetanus-diphtheria and measles and a single polio dose for adults.
- Typhoid: if you plan to visit developing countries in any of the following regions.

Africa (Central)

Includes Angola, Central African Republic, Cameroon, Chad, Democratic Republic of the Congo, Equatorial Guinea, Gabon, Sudan; Zambia.

- Yellow fever: A vaccination certificate may be required.
- See "Standard Recommended Vaccinations."
- Although there are no vaccinations for the following illnesses, be aware that there is a risk of contracting them in Central Africa: malaria (high risk year-round) and insect-borne diseases (dengue, filariasis, leishmaniasis, orchocerciasis).

Africa (East)

Includes Burundi, Comoros, Djibouti, Eritrea, Ethiopia, Kenya, Madagascar, Malawi, Mauritius, Mayotte, Mozambique, Reunion, Rwanda, Seychelles, Somalia, Tanzania, Uganda.

- Yellow fever: A vaccination certificate may be required for entry into some countries. Contact the CDC for up-to-date information. Yellow fever vaccination is recommended if you travel outside of urban areas. Also recommended is a meningococcal vaccine if you visit western Ethiopia from December through June.
- See "Standard Recommended Vaccinations."
- Although there are no vaccinations for the following illnesses, be aware that there is a risk of contracting them in East Africa: malaria (risk exists in all areas except in the cities of Addis Ababa, Ismara, and Nairobi, the islands of Reunion and Seychelles, and in highlands at altitudes greater than 2,500 meters) and insect-borne diseases (dengue, filariasis, leishmaniasis, orchocerciasis, trypanosomiasis, Rift Valley fever).

Africa (North)

Includes Algeria, Canary Islands, Egypt, Libyan Arab Jamahirya, Morocco, Tunisia.

- Yellow fever: A vaccination certificate may be required for entry into some North African countries if you are coming from tropical South America or sub-Saharan Africa.
- See "Standard Recommended Vaccinations."
- Although there are no vaccinations for the following illnesses, be aware that there is a risk of contracting them in North Africa: malaria (limited risk in parts of Algeria, El Faiyum area of Egypt, Libyan Arab Jamahirya, western Sahara, and Morocco) and insect-borne diseases (dengue, filariasis, leishmaniasis, orchocerciasis).

Africa (Southern)

Includes Botswana, Lesotho, Namibia, South Africa, St. Helena, Swaziland, Zimbabwe.

- Yellow fever: A vaccination certificate may be required in some countries if you are entering Southern Africa from tropical South America or sub-Saharan Africa.
- See "Standard Recommended Vaccinations."
- Although there are no vaccinations for the following illnesses, be aware that there is a risk of contracting them in Southern Africa: malaria (risk is year-found in northern Botswana, rural South Africa, all nonmountainous areas of Swaziland, and all of Zimbabwe except Harare and Bulawayo) and insect-borne diseases (dengue, filariasis, leishmaniasis, orchocerciasis, trypanosomiasis).

Africa (West)

Includes Benin, Burkina Faso, Cape Verde Islands, Cote d'Ivoire, Cambia, Ghana, Guinea, Guinea-Bissau, Liberia, Mali, Mauritania, Niger, Nigeria, Sao Tome and Principe, Senegal, Sierra Leone, Togo.

- Yellow fever: A vaccination certificate may be required; check with the CDC.

- Meningococcal: recommended when traveling to most West African countries from December through June.
- See "Standard Recommended Vaccinations."
- Although there are no vaccinations for the following illnesses, be aware that there is a risk of contracting them in West Africa: malaria (risk is high in all West African countries except for most of the Cape Verde Islands) and insect-borne diseases (dengue, filariasis, leishmaniasis, orchocerciasis, and trypanosomiasis).

Australia and South Pacific

Includes Australia, Christmas Island, Cook Island, Federated States of Micronesia, Fiji, French Polynesia (Tahiti), Guam, Kiribati, Marshall Islands, Nauru, New Caledonia, New Zealand, Niue, Northern Mariana Islands, Palau, Papua New Guinea, Pitcairn, Samoa, American Samoa, Solomon Islands, Tokelau, Tonga, Tuvalu, Vanuatu, Wake Island, Wallis, Utuna.

- All recommended vaccinations in the table, including hepatitis B for infants and children 11 to 12 years old.

The Caribbean

Includes Antigua, Barbuda, Bahamas, Barbados, Bermuda, Cayman Islands, Cuba, Dominica, Dominican Republic, Grenada, Guadeloupe, Haiti, Jamaica, Martinique, Montserrat, Netherlands Antilles, Puerto Rico, St. Lucia, St. Vincent and the Grenadines, St. Kitts-Nevis, Trinidad and Tobago, Turks and Caicos, Virgin Islands.

- Yellow fever: A vaccination certificate may be required for certain areas if you enter the Dominican Republic or Haiti from a sub-Saharan African or tropical South American country.
- See "Standard Recommended Vaccinations."
- Although there are no vaccinations for the following illnesses, be aware that there is a risk of contracting them in the Caribbean: malaria (risk is year-round in rural Dominican Republic and in Haiti) and insect-borne diseases (dengue, filariasis, leishmaniasis).

East Asia

Includes China, Hong Kong, Japan, Democratic People's Republic of Korea (North), Republic of Korea (South), Macao, Mongolia, Taiwan.

- Yellow fever: A vaccination certificate may be required to enter certain countries if you are coming from tropical South America or sub-Saharan Africa.
- See "Standard Recommended Vaccinations."
- Although there are no vaccinations for the following illnesses, be aware that there is a risk of contracting them in East Asia: malaria (in some parts of rural China) and insect-borne diseases (dengue, filariasis, Japanese encephalitis, leishmaniasis, plague).

Europe (Eastern)

Includes Albania, Armenia, Azerbaijan, Belarus, Bosnia and Herzegovina, Bulgaria, Croatia, Czech Republic, Estonia, Georgia, Hungary, Kazakhstan, Kyrgyzstan, Latvia, Lithuania, Moldova, Poland, Romania, Russia, Serbia and Montenegro, Slovak republic, Slovenia, Tajikistan, Turkmenistan, Ukraine, Uzbekistan.

- Yellow fever: A vaccination certificate may be required for entry into some countries if you are coming from tropical South America or sub-Saharan Africa.
- See "Standard Recommended Vaccinations."
- Although there are no vaccinations for the following diseases, be aware that there is a risk of contracting them in eastern Europe: malaria (risk only in the southern border areas of Azerbaijan and Tajikistan) and diphtheria (in Russia; check with the CDC before traveling there).

Europe (Western)

Includes Andorra, Austria, Azores, Belgium, Denmark, Faroe Islands, Finland, France, Germany, Gibraltar, Greece, Greenland, Iceland, Ireland, Italy, Liechtenstein, Luxembourg, Madeira, Malta, Monaco, Netherlands, Norway, Portugal, San Marino, Spain, Sweden, Switzerland, United Kingdom.

- Yellow fever: A vaccination certificate may be required for entry into certain countries if you come from tropical South America or sub-Saharan Africa.
- Vaccinations for hepatitis A, hepatitis B, and boosters for tetanus-diphtheria are recommended.
- Although there is no vaccination for tick-borne encephalitis, you are at risk of this disease if you spend time in wooded areas during the summer and/or consume nonpasteurized dairy products.

Indian Subcontinent

Includes Afghanistan, Bangladesh, Bhutan, India, Maldives, Nepal, Pakistan, Sri Lanka.

- Yellow fever: A vaccination certificate may be required to enter certain countries if you are coming from tropical South America or sub-Saharan Africa.
- See "Standard Recommended Vaccinations." Also recommended is a vaccination for Japanese encephalitis.
- Although there are no vaccinations for the following illnesses, be aware that there is a risk of contracting them in the Indian subcontinent: malaria (risk exists in some urban and many rural areas) and insect-borne diseases (dengue, filariasis, Japanese encephalitis, leishmaniasis, plague).

Mexico and Central America

Includes Belize, Costa Rica, El Salvador, Guatemala, Honduras, Mexico, Nicaragua, Panama.

- Yellow fever: A vaccination certificate may be required to enter certain countries if you are coming from tropical South America or sub-Saharan Africa.
- See "Standard Recommended Vaccinations."
- Although there are no vaccinations for the following illnesses, be aware that there is a risk of contracting them in Mexico and Central America: malaria (year-round risk in rural lowlands and some urban areas) and insect-borne diseases (dengue, filariasis, leishmaniasis, orchocenciasis, and Chagas disease).

Middle East

Includes Bahrain, Cyprus, Iran, Iraq, Israel, Jordan, Kuwait, Lebanon, Oman, Qatar, Saudi Arabia, Syrian Arab Republic, Turkey, United Arab Emirates, Yemen.

- Yellow fever: A vaccination certificate may be required for entry into some countries if you are coming from tropical South America or sub-Saharan Africa.
- See "Standard Recommended Vaccinations." Also recommended is a meningococcal vaccine if you are going to Mecca.
- Although there are no vaccinations for the following illnesses, be aware that there is a risk of contracting them in the Middle East: malaria (low risk in parts of Iran, Iraq, Oman, Saudi Arabia, Syrian Arab Republic, Turkey, United Arab Emirates, and Yemen) and insect-borne diseases (dengue, filariasis, leishmaniasis, orchocerciasis, plague).

South America—Temperate

Includes Argentina, Chile, Falkland Islands, Uruguay.

- Yellow fever: recommended if you travel outside urban areas in Argentina.
- See "Standard Recommended Vaccinations."
- Although there are no vaccinations for the following illnesses, be aware that there is a risk of contracting them in temperate South America: malaria (risk in Argentina in northern rural areas that border Paraguay and Bolivia) and insect-borne diseases (dengue, Chagas disease, leishmaniasis).

South America—Tropical

Includes Bolivia, Brazil, Colombia, Ecuador, French Guiana, Guyana, Paraguay, Peru, Suriname, and Venezuela.

- Yellow fever: A vaccination certificate may be required for certain countries.
- See "Standard Recommended Vaccinations."

- Although there are no vaccinations for the following illnesses, be aware that there is a risk of contracting them in tropical South America: malaria [contact the CDC to see if it is a problem in the area(s) to which you are traveling] and insect-borne diseases (dengue, filariasis, leishmaniasis, orchocerciasis, Chagas disease).

Southeast Asia

Includes Brunei Darussalam, Cambodia, Indonesia, Lao People's Democratic Republic (Laos), Malaysia, Myanmar (Burma), Philippines, Singapore, Thailand, Vietnam.

- Yellow fever: A vaccination certificate may be required for entry into some countries if you are coming from tropical South America or sub-Saharan Africa.
- See "Standard Recommended Vaccinations." Also recommended is a vaccination for Japanese encephalitis if you plan to visit rural areas for 4 weeks or longer or if there is a known outbreak of the disease in the area to which you plan to travel.
- Although there are no vaccinations for the following illnesses, be aware that there is a risk of contracting them in southeast Asia: malaria (year-round risk in some cities and in all rural areas except for Brunei Darussalam and Singapore) and insect-borne diseases (dengue, filariasis, Japanese encephalitis, plague).

ABOUT THE AUTHOR

EDWARD R. BURNS, M.D., brings an extraordinary breadth and depth of knowledge to the task of researching and writing this medical dictionary. He is a Professor of Medicine and Pathology at the Albert Einstein College of Medicine of Yeshiva University as well as Senior Associate Dean of the College. For many years he served as Director of Laboratories and Chief of the Division of Clinical Pathology at the same institution. Previously, Dr. Burns was Director of Clinical Hematology at the Hospital of the Albert Einstein College of Medicine. He has published extensively in prominent medical journals, authored a textbook of hematology, and reviews research published in scientific and medical journals, including the *American Journal of Clinical Pathology* and and *Archives of Internal Medicine.* He has received grants as principal investigator for various research projects and holds several patents in the field of diagnostic medicine. He is a member of Alpha Omega Alpha, the American Society of Hematology, and the New York Academy of Medicine. Dr. Burns lives in New York City with his wife, Chaya, a librarian specializing in children's literature, and is the proud father of four grown children.